Smart Innovation, Systems and Technologies

Volume 269

The Smart Innovation, Systems and Technologies book series encompasses the topics of knowledge, intelligence, innovation and sustainability. The aim of the series is to make available a platform for the publication of books on all aspects of single and multi-disciplinary research on these themes in order to make the latest results available in a readily-accessible form. Volumes on interdisciplinary research combining two or more of these areas is particularly sought.

The series covers systems and paradigms that employ knowledge and intelligence in a broad sense. Its scope is systems having embedded knowledge and intelligence, which may be applied to the solution of world problems in industry, the environment and the community. It also focusses on the knowledge-transfer methodologies and innovation strategies employed to make this happen effectively. The combination of intelligent systems tools and a broad range of applications introduces a need for a synergy of disciplines from science, technology, business and the humanities. The series will include conference proceedings, edited collections, monographs, handbooks, reference books, and other relevant types of book in areas of science and technology where smart systems and technologies can offer innovative solutions.

High quality content is an essential feature for all book proposals accepted for the series. It is expected that editors of all accepted volumes will ensure that contributions are subjected to an appropriate level of reviewing process and adhere to KES quality principles.

Indexed by SCOPUS, EI Compendex, INSPEC, WTI Frankfurt eG, zbMATH, Japanese Science and Technology Agency (JST), SCImago, DBLP.

All books published in the series are submitted for consideration in Web of Science.

More information about this series at https://link.springer.com/bookseries/8767

Joy Iong-Zong Chen · Haoxiang Wang ·
Ke-Lin Du · V. Suma
Editors

Machine Learning and Autonomous Systems

Proceedings of ICMLAS 2021

Springer

Editors
Joy Iong-Zong Chen
Department of Electrical Engineering
Dayeh University
Changhua, Taiwan

Haoxiang Wang
Go Perception Laboratory
Cornell University
Ithaca, NY, USA

Ke-Lin Du
Department of Electrical and Computer Engineering
Concordia University
Montreal, QC, Canada

V. Suma
Department of Information Science and Engineering
Dayananda Sagar College of Engineering
Bangalore, India

ISSN 2190-3018 ISSN 2190-3026 (electronic)
Smart Innovation, Systems and Technologies
ISBN 978-981-16-7995-7 ISBN 978-981-16-7996-4 (eBook)
https://doi.org/10.1007/978-981-16-7996-4

This Springer imprint is published by the registered company Springer Nature Singapore Pte Ltd.
The registered company address is: 152 Beach Road, #21-01/04 Gateway East, Singapore 189721, Singapore

This proceedings of ICMLAS 2021 is dedicated to organization staff, the members of the program committees and reviewers. Nevertheless, this conference is especially dedicated to the authors for contributing their research result to the conference.

Foreword

It is with great pleasure I write this Foreword to the Proceedings of the International Conference on Machine Learning and Autonomous Systems [ICMLAS 2021], which will be held on September 24–25, 2021 in Tamil Nadu, India.

This international conference encourages research students and young academics to connect with the more established academic community in a formal atmosphere to present, discuss, and share new and existing research works. Their contributions have helped to make the conference a successful event. The theme of this 1st international conference is "Intelligent Autonomous Systems", a topic that is gaining increasing attention in both academia and industries. Machine learning approaches play an essential role in acquiring information from observed data in order to accomplish adaptive sensing, modeling, planning, and control for autonomous systems. ICMLAS is an excellent venue for investigating the computing foundations for autonomous systems due to its proven track record of computing systems research in fields where performance and functional requirements necessitate the increasing integration of artificial intelligence and autonomous processes.

The success of ICMLAS is entirely dependent on the effort, ability, and energy of computational intelligence researchers who have authored and published articles on a wide range of issues. I hope readers will have a very satisfying research experience and take advantage of this opportunity to reconnect with researchers, industrialists and established new ones. Appreciation is also deserved to the program committee members and external reviewers who have devoted substantial time in analyzing and reviewing many papers, who have established and continue to set a high-quality standard for this conference.

The program includes invited talks, technical workshops, and discussions with eminent speakers on a wide range of scientific and social research topics. This comprehensive program allows all participants to meet and connect with one another. We wish you a productive and long-lasting experience at ICMLAS 2021. The conference will continue to thrive with your help and participation for a long time.

I wish all attendees of ICMLAS 2021 an enjoyable scientific gathering in India. I look forward to seeing all of you next year at the conference.

R. Rajesh
Conference Chair
Principal, Rohini College
of Engineering and Technology
Salem, India

Preface

ICMLAS 2021, the International Conference on Machine Learning and Autonomous Systems, is held at the Rohini College of Engineering & Technology, Tamil Nadu, India on 24–25, September 2021. ICMLAS is an interdisciplinary conference in the field of computing and information technology. This conference is a gathering of experts from all the fields of information technology and computer science that makes it a major occasion for computer science researchers to share their state-of-the-art research works, and also it creates a unique opportunity for academicians, researchers, and industrialists to present their research in an international forum.

This first year of ICMLAS has a new highlight. This conference includes attendees from different but related fields to create a unique opportunity through keynotes and technical sessions to exchange research ideas and information. The participants were almost from every part of the world with the background of academia/industry. The success of this conference relies on the high level of the received papers.

The proceedings of ICMLAS are a compilation of the high-quality accepted papers, which represent an increasing outcome of the conference. We are pleased to note that ICMLAS 2021 continues to adhere to Springer publishing standards in all aspects. This year, a total of 46 papers were accepted from 302 submissions. These papers were selected based on peer reviews by technical program committee members and other invited reviews following standard Springer publishing procedures. Further, the oral presentations were delivered over 2 days by organizing different parallel sessions along with a keynote session delivered by plenary speakers.

The proceedings represent the efforts of all contributors, reviewers, and conference organizers. We would like to thank all authors for submitting high-quality research works, all reviewers for their time and insightful remarks, and all members of the conference committees for volunteering their time generously throughout the last year. We acknowledge the financial and technical assistance provided by our sponsors, as well as the professional efforts Springer Conference Publishing Services in publishing the conference proceedings.

Changhua, Taiwan
Ithaca, USA
Montreal, Canada

Dr. Joy Iong-Zong Chen
Dr. Haoxiang Wang
Dr. Ke-Lin Du

Acknowledgements

We would like to express our appreciation and gratitude to all of the reviewers, who helped us to maintain the high quality in the manuscripts included in the proceedings published by Springer. We would also like to extend our thanks to the members of the organizing team for their hard work.

We thank our keynote speaker "Dr. R Kanthavel King Khalid University, Kingdom of Saudi Arabia" for sharing their expert research knowledge. The plenary talks delivered have covered the full range of the conference topics. The proceedings also recognize the efforts of all contributors, reviewers, and conference organizers.

We would like to express our appreciation to all contributors for submitting their highly accomplished work to all reviewers for their time and valuable comments and to all members of the conference committees for giving generously of their time over the past year. We thank our sponsors for financial and technical support, and we acknowledge Springer publications for providing timely guidance and assistance in publishing the conference proceedings.

Contents

About the Editors

Dr. Joy Iong-Zong Chen is currently a full professor of Department of Electrical Engineering Dayeh University at Changhua, Taiwan. Prior to joining the Dayeh University, he worked at the Control Data Company (Taiwan) as a technical manager since September 1985 to September 1996. His research interests include wireless communications, spread spectrum technical, OFDM systems, and wireless sensor networks. He has published a large number of SCI Journal papers in the issues addressed physical layer for wireless communication systems. Moreover, he also majors in developing some applications of the Internet of Thing (IoT) techniques, and Dr. Joy I.-Z. Chen owned some patents authorized by the Taiwan Intellectual Property Office (TIPO).

Haoxiang Wang is currently the director and lead executive faculty member of GoPerception Laboratory, NY, USA. His research interests include multimedia information processing, pattern recognition and machine learning, remote sensing image processing and data-driven business intelligence. He has co-authored over 60 journal and conference papers in these fields on journals such as Springer MTAP, Cluster Computing, SIVP; IEEE TII, Communications Magazine; ElsevierComputers & Electrical Engineering, Computers, Environment and Urban Systems, Optik, Sustainable Computing: Informatics and Systems, Journal of Computational Science, Pattern Recognition Letters, Information Sciences, Computers in Industry, Future Generation Computer Systems; Taylor&Francis International Journal of Computers and Applications and conference such as IEEE SMC, ICPR, ICTAI, ICICI, CCIS and ICACI. He is the guest editor for IEEE Transactions on Industrial Informatics, IEEE Consumer Electronics Magazine, Multimedia Tools and Applications, MDPI Sustainability, International Journal of Information and Computer Security, Journal of Medical Imaging and Health Informatics and Concurrency and Computation: Practice and Experience.

Dr. Ke-Lin Du is a research scientist at Center for Signal Processing and Communications, Department of Electrical and Computer Engineering, Concordia University, since 2001, where he became an affiliate associate professor in 2011. I investigated on signal processing, wireless communications and soft computing.

Dr. V. Suma has obtained her B.E. in Information Science and Technology, M.S. in Software Systems and her Ph.D. in Computer Science and Engineering. She has a vast experience of more than 17 years of teaching. She has published more than 183 international publications which include her research articles published in world class international journals such as ACM, ASQ, Crosstalk, IET Software and international journals from Inderscience publishers, from journals released in MIT, Darmout, USA. Her research results are published in NASA, UNI trier, Microsoft, CERN, IEEE, ACM portals, Springer and so on.

Chapter 1
Utilization of Sensors for Energy Generation Using Pressure or Vibrations

Sundeep Siddula, M. P. Chandra Mouli, M. Bhavya Sree, K. Harinadh, and P. Ramya Shree

Abstract With the progress of technology, the need for power is rapidly increasing, and rather than generating energy, which is currently inefficient. In order to preserve the generated energy from fossil fuels and use it for high loads, the development of various sensors for producing electrical energy for small loads has been adopted. These are referred to as piezoelectric material sensors. These sensors are used to transform the energy emitted by humans into usable electrical energy. As humans have a very high rate of energy output when working in day-to-day life, the exhibited energy is present in the form of pressure or stress, which is used to generate electrical energy, which is considered as a non- conventional energy generation source. The main aim of this research work is to utilize the exhibited energy to generate the electrical energy by using some sensors. This type of electrical energy is mostly utilized for streetlights, lamps, and small household appliances. The design and execution of footstep power production utilizing piezoelectric sensors is described in this article.

1.1 Introduction

With the advancement of trends in the technology, the electrical utilization has been increasing rapidly; this technological advancement is directly proportional to the utilization of electrical energy. The use of fossil fuels and other materials, which are traditional sources of power generation, has very high adverse effects on the environment with increased pollution and generation of harmful gases, etc., using conventional sources that require a large amount of land and equipment for power generation, which raises the cost of implementation. Renewable energy sources, such as solar, thermal, and hydro, have been adopted to provide the greatest quantity of energy at the lowest cost of implementation. However, while being economically friendly, they cannot be utilized for tiny appliances due to their ratings and consumption limitations. To use the generated electrical energy for small loads and appliances,

S. Siddula (✉) · M. P. C. Mouli · M. B. Sree · K. Harinadh · P. R. Shree
Vignana Bharathi Institute of Technology, Hyderabad, India

J. I.-Z. Chen et al. (eds.), *Machine Learning and Autonomous Systems*, Smart Innovation, Systems and Technologies 269, https://doi.org/10.1007/978-981-16-7996-4_1

piezo materials, which are small disc like structures, have been implemented. When these sensors are exerted by pressure, the voltage is generated and the energy will be utilized by the loads such as street lights, lamps, small drives, and low rated house hold appliances [1–13].

The word piezo was obtained from Greek word piezein, which means to press or to squeeze. The first demonstration was in the year 1880 by the curie brothers. They obtained the piezoelectricity by the combined knowledge of pyroelectricity along with the crystal structures to study the crystal behavior and demonstrated the same by using different crystals of quartz, topaz, cane sugar, and Rochelle salt, and most of the piezo sensors are of quartz type. When the sensor has subjected to pressure or stress the quartz has changed its dimensions and produces the electric charge. The most widely used piezo material was perovskite structured Pb, which is extensively used in various types of electromechanical sensors and actuators [14, 15]

1.2 Constructional Features

The components that are used for the foot step power generation are piezo sensors, microcontroller, rechargeable battery, LED, and LCD display. The connections are made according to the below Fig. 1.1.

1.2.1 Piezoelectric Sensors

Figure 1.2 shows the piezo sensors, which work on the principle of piezoelectric effect, i.e., when the pressure is exerted on the sensors, they produce an electric charge on the crystal surface like quartz. These sensors produce different rates of electric charges based on the position of pressure applied on the sensor. In order to obtain high amount of voltage, these sensors are connected in a series connection. With the increase in the number of series connection of sensors, the voltage gets increased.

When the sensor is excited to pressure, it produces an electrical output by rearranging the diploes. The rate of voltage generated is directly proportional to the pressure or stress that is being applied on the sensor.

The output voltage created by the sensor is determined by the location of the pressure applied to the sensor; if the pressure is applied to the sensor's midpoint, the maximum voltage is obtained and then used. Figure 1.3 depicts the voltages at various pressure points on the sensor.

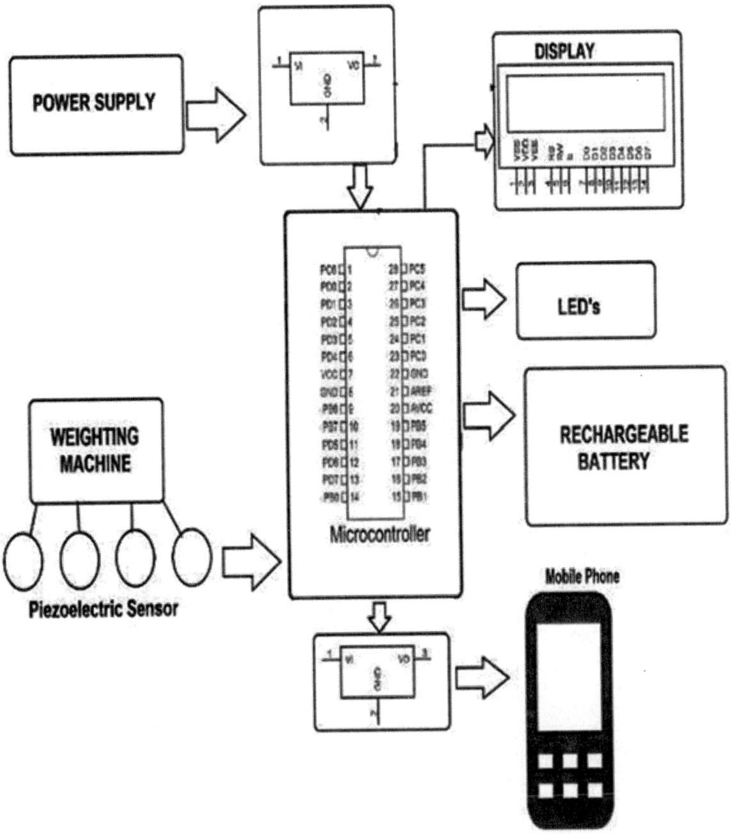

Fig. 1.1 Hardware circuit of implementation

1.2.2 *Microcontroller*

Figure 1.4 shows the microcontroller circuit device, which acts as the main CPU of the system, i.e., central processing unit, which has I/O ports and analog to digital converter pins, etc. This microcontroller is used to measure and display the no. of footsteps and the voltage generated when piezo sensor is squeezed. This also measures the battery voltage level by using analog to digital converter. Based on the SOC status of battery, the voltage generated by the piezo sensor will be used to charge the battery.

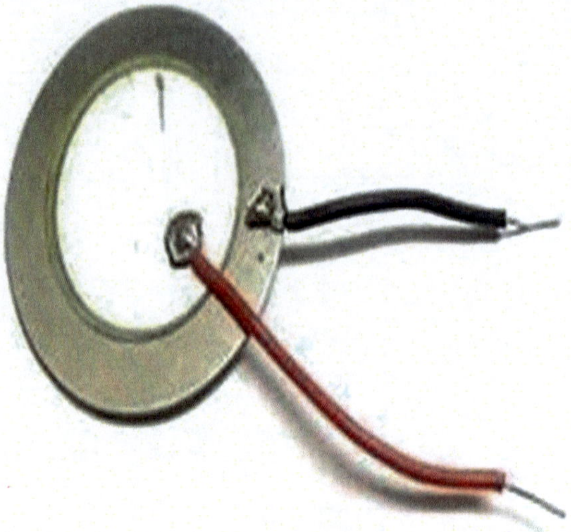

Fig. 1.2 Piezoelectric sensor

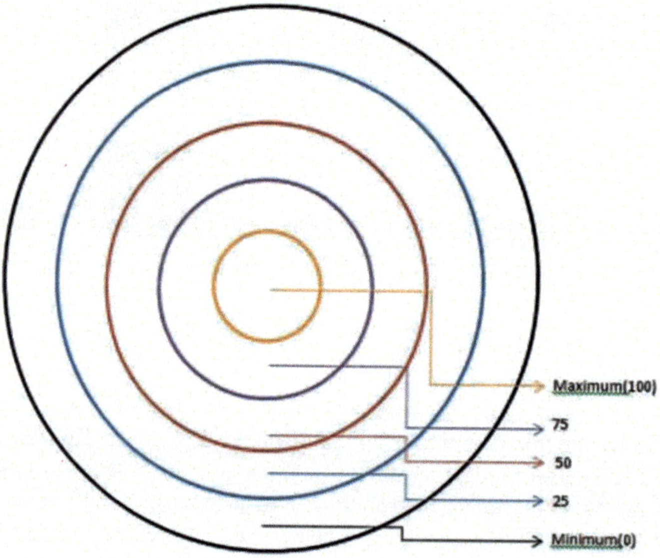

Fig. 1.3 Voltage pressure points on the Piezo sensor

Fig. 1.4 Microcontroller

1.2.3 Rechargeable Battery

Rechargeable battery is used as a storage or backup device, where the generated output power is stored and utilized, when there is an interruption in power supply or failure of the system.

1.2.4 LCD Display

Liquid Crystal Display [LCD] is a flat display, which uses light modulating properties of liquid crystals. This LCD is used to display the number of footsteps and the voltage that is obtained overall with piezo electric sensors.

1.2.5 LED

LED belongs to the family of diodes, where the light is being emitted. LEDs are placed on the piezo sensors, which are used to indicate that the voltage is generated by glowing the LED, when the pressure is being applied on the sensor and if the LED does not glow it indicates that the voltage is not generated.

1.3 Working

The footstep power generation works on the principle of piezoelectric effect principle. This can be explained as the ability of materials for generating electric charges in response to the force or stress applied on the sensor. The sensors are made of quartz material, which includes crystals that are neutral in nature. The atoms included inside them are not symmetrically distributed, but are similarly charged. When the pressure is being applied on the sensor, it gets compressed and when the pressure is removed, it gets decompressed. The mechanical vibration is detected by the crystals present in them during the compression and decompression processes, and the voltage is created by both net positive and negative charges that emerge on the opposite face of the crystal. The voltage is obtained by indicating the LED glow by mentioning that the voltage is generated. The obtained voltage is then fed to the rechargeable battery and loads through the microcontroller, which acts as a central processing unit, where the battery voltage level is measured by analog and digital converters present in the controller.

The SOC (State of charge) status level shows whether or not the battery should be charged. This SOC state is known as threshold voltage; if the voltage acquired is more than the threshold voltage, the load light illuminates; if the voltage obtained is less, the load lamp does not illuminate. As a result, electrical power is created and used. The voltage collected from the piezo sensor's footsteps is shown on the LED, displaying the number of footsteps and total voltage. The below Fig. 1.5 shows the hardware implementation of footstep power generation.

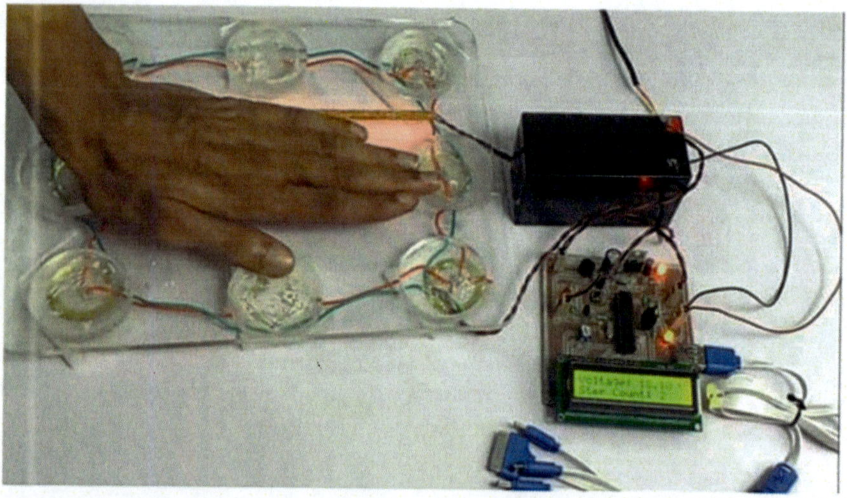

Fig. 1.5 Hardware implementation of footstep power generation

Table 1.1 Steps required to charge the battery up to 1 V

Applied weight (kg)	Steps require to charge 1 V
1.2	115
1.6	102
1.8	93
2.0	80

1.4 Observation Results

The Table 1.1 gives the information regarding steps require to charge 1V.

1.5 Conclusion

This paper concludes that as the demand for electrical energy increases, it is necessary to use the energy for small applications such as streetlights, LED bulbs, and so on, which preserves energy supplied by other sources that might be utilized for large appliances. As a result, piezo sensors are utilized to generate energy and to utilize it in a suitable perspective. Unlike other sources, the mechanical energy is abundant and reusable as it is produced by the humans every day in their working environment. Thus, it can be concluded that the efficient energy can be produced and utilized from the piezo sensors.

Acknowledgements Dr. Sundeep. Siddula, is an EEE faculty of Vignana Bharathi Institute of Technology (VBIT), Hyderabad, acknowledges DST for providing computational facilities under its FIST program at VBIT Hyderabad where the computational work has been carried out.

Appendix

Equipment	Type/rating	Quantity
Microcontroller	ATMega328	1 No
Piezoelectric sensors	PMN-PT	7 No
LCD display	16×2	1 No
Rechargeable battery	4 V, 1.5 Ah	3 No
LED's	–	As required

References

1. Patil, A., Jadhav, M., Joshi, S., Britto, E., Vasaikar, A.: Energy harvesting using piezoelectricity. In: 2015 IEEE International Conference on Energy Systems and Applications, Pune, India (2015)
2. Kamboj, A., Haque, A., Kumar, A., Sharma, V.K., Kumar, A.: Design of footstep power generator using piezoelectric sensors. In: 2017 International Conference on Innovations in Information, Embedded and Communication Systems (ICIIECS), Coimbatore, India (2017)
3. Design study of piezoelectric energy-harvesting devices for generation of higher electrical power using a coupled piezoelectric-circuit finite element method. IEEE Trans. Ultrasonic's Ferroelectr. Freq. Control **57**(2) (2010)
4. Basari, A.A., Awaji, S., Sakamoto, S., Hashimoto, S., Homma, B., Suto, K., Okada, H., Okuno, H., Kobayashi, K., Kumagai, S.: Evaluation onmechanical impact parameters in piezoelectric power generation. In: Proceedings of IEEE 10th Asian Control Conference (ASCC), 2015, pp. 1–6 (2015)
5. Meier, R., Kelly, N., Almog, O., Chiang, P.: Apiezoelectric energy-harvesting shoe system for podiatric sensing. In: Proceedings of IEEE 36th Annual International Conference of Engineering in Medicine and Biology Society (EMBC 2014), pp. 622–625 (2014)
6. Prasad, P.R., Bhanuja, A., Bhavani, L., Bhoomika, N., Srinivas, B.: Power generation through footsteps using piezoelectric sensors along with GPS tracking. In: 2019 4th International Conference on Recent Trends on Electronics, Information, Communication & Technology (RTEICT), Bangalore, India (2019)
7. Dalabeih, D., Haws, B., Muhtaseb, S.: Harvesting kinetic energy of footsteps on specially designed floor tiles. In: 2018 9th International Renewable Energy Congress (IREC), Hammamet, Tunisia (2018)
8. Smys, S., Basar, A. and Wang, H.: Artificial neural network based power management for smart street lighting systems. J. Artif. Intell. **2**(01), 42–52 (2020)
9. Bhalaji, N.: EL DAPP–an electricity meter tracking decentralized application. J. Electron. **2**(01), 49–71 (2020)
10. Zhang, T., Shi, X., Zhang, D., Xiao, J.: Socio-economic development and electricity access in developing economies: a long-run model averaging approach. Energy Policy **132**, 223–231 (2019)
11. Ang, C.K., Al-Talib, A.A., Tai, S.M., Lim, W.H.: Development of a footstep power generator in converting kinetic energy to electricity. In: 2018 International Conference on Renewable Energy and Environment Engineering (REEE 2018), Paris, France, vol. 80 (2019)
12. Anderson, J., Papachristodoulou, A.: A decomposition technique for nonlinear dynamical system analysis. IEEE Trans. Autom. Control **57**(6), 1516–1521 (2012)
13. Jin, Y., Sarker, S., Lee, K., Seo, H.W., Kim, D.M.: Piezoelectric materials for high-performance energy harvesting devices. In: 2016 Pan Pacific Microelectronics Symposium (Pan Pacific), Big Island, HI, USA (2016)
14. Triono, A.D. et al.: Utilization of pedestrian movement on the sidewalk as a source of electric power for lighting using piezoelectric censors. In: 2018 3rd IEEE International Conference on Intelligent Transportation Engineering (ICITE), Singapore (2018)
15. Panghate, S., Barhate, P., Chavan, H.: Advanced footstep power generation system using RFID for charging. Int. Res. J. Eng. Technol. **07**(02) (2020)

Chapter 2
Modeling and Optimization of Discrete Evolutionary Systems of Information Security Management in a Random Environment

V. A. Lakhno, D. Y. Kasatkin, O. V. Skliarenko, and Y. O. Kolodinska

Abstract A method for studying the stability of zero solutions of a system of nonlinear difference equations depending on the semi-Markov chain is proposed. The essence of the proposed method is to study the stability of moment equations that are deterministic for the above systems. The necessary optimality conditions for these systems are derived.

2.1 Introduction

Currently, information security remains one of the most dynamic areas of information technology development. Mainly, due to the growing number of attacks on information resources around the world. As the number of threats increases, it remains vital for companies to make sure their information systems are protected against these attacks.

Cybersecurity experts [1, 2] note that one of the most urgent tasks in the area of information security is threat assessment. Qualitative assessment of cyber threats is the basis for application of necessary tools and methods of information protection.

Many tools and methods for assessing information security (IS) and modeling possible attacks exist today, most prominent of which are given in [1–4]. Some of the tools and methods include: probability theory, fuzzy sets, game theory, graphs, states, Petri nets and random processes.

V. A. Lakhno (✉) · D. Y. Kasatkin
Department of Computer Systems and Networks, National University of Life and Environmental Sciences of Ukraine, Kyiv, Ukraine
e-mail: valss21@ukr.net

D. Y. Kasatkin
e-mail: d.kasatkin@nubip.edu.ua

O. V. Skliarenko · Y. O. Kolodinska
Department of Information Technology Cyber Security and Mathematical Sciences, Private Higher Education Institution European University, Kyiv, Ukraine
e-mail: sm@rty.pp.ua

J. I.-Z. Chen et al. (eds.), *Machine Learning and Autonomous Systems*, Smart Innovation, Systems and Technologies 269, https://doi.org/10.1007/978-981-16-7996-4_2

An application of the mathematical apparatus of Markov and semi-Markov random processes for threat assessment is still viable today. It has found wide application in theory and practice of information security. According to [2], Markov and semi-Markov processes can be used to assess the impact on information security of information and telecommunications systems from various types of attacks. This is especially true if an attack is a rare and independent event.

Based on this studying, the impact of attacks on an information system, the use of Markov and semi-Markov random processes is justified.

2.2 Review and Analysis of Previous Research

Most of the analyzed sources [3–6] identify coordinated hacker attacks on information systems (IS) of objects of informatization (hereinafter IOB) as the most significant threat.

Modern IOBs and their information security management systems (ISMS) can be characterized by a set of technical characteristics and architectures as queuing systems (QS). From all possible schemes of QS, we can consider particular examples: with losses, with expectation, with accumulation of final capacity, etc.

To ensure IS of IOB leading companies are implementing best practices ISMS. Such ISMS as a rule is built on the basis of basic requirements specified by ISO/IEC 27 k standards. It should be noted that the existing standards do not detail specific methodologies that allow the unambiguous formulation of design requirements for an ISMS for a specific organization such as an enterprise, government agency, etc. Instead they address various aspects of IS that need to be implemented to protect any business process, or in relation to the IS policy for the entire IOB [7–12]. In this case, it is desirable to have a formal model that describes its functional features in order to select priorities for providing IS IOB and deploying a functional ISMS.

The study of parameters of such models will provide an understanding of which particular aspects of an IS IOB should be prioritized. In turn, to understand the structure of a formalized model of an ISMS, it is necessary to consider which of the formal methods are more applicable in the study of ISMS. If an analogy can be found, it can be reasonably assumed that the formal design techniques regulated by ISO/IEC 27 k standards can be adapted to the task of creating an ISMS for an IOB as a whole.

Let's consider an ISMS for a hypothetical IOB. According to [11–15], ISMS is a "part of the organization's overall management system". Therefore, it is based on risk assessment for the IS. In fact, an ISMS is being designed, built, operated, monitored, reviewed, maintained and improved as new IOB-related tasks are surfacing. Such a definition gives grounds to speak of ISMS as an interpretation of a class of systems that are designed to repeatedly solve typical IS problems. Also, an analogy between ISMS and QMS is obvious.

Much of the mathematical models of the functioning of information security systems are built and described in terms of the general graphs theory. It is usually assumed that the system at a time can be one of the possible states and passes from this state to another under the influence of a random process, e.g. in case of attack. It is assumed that the laws of distribution of the state of the system in each of the possible states before the transition to another state are specified (or can be obtained as a result of statistical processing of the output data. Moreover, in many practical cases it can be formulated and solved the problem of finding the stationary distribution of probabilities of system states. However, theoretical and practical interest represents a decision of a more complex task. It is about finding the probability distribution system stays in its possible states at an arbitrary time after the commencement of a given initial state. The solution to this problem in the most general assumptions about the nature of random process that affects on a system is almost impossible. However, this decision can be obtained for an important separate case when this is a Markov process [16]. The problem is that technique that establishes a link between mathematical objects doesn't exist. However, this solution can be obtained for an important special case when the process is a Markov process. The problem is as follows. There is no technique that establishes a connection between two mathematical objects. The first object is the density distribution of the duration of the system in each of the states before the transition to another state. The second object is the desired functions that describe the dynamics of the probabilities of the system in its possible states during the attack. The problem of semi-Markov analysis is discussed in numerous publications. In [17], the problem of evaluating the efficiency of the system, the model of which is a queuing system with a heterogeneous input flow, is considered. In this case, the final distribution of state probabilities for the nested Markov chain is sought. In [18] a sample production system is studied using a semi-Markov models. The analysis concludes with the final calculation of the probability distribution system states.

In [19], the possibility of applying semi-Markov models for problems of analysis of computer networks, transport networks, objects of the Internet of Things is studied. The decision on the effectiveness is taken on the basis of the resulting probability distribution of the final states. The service system with non-Poisson input and non-exponential service is studied in [20] in order to obtain stationary performance characteristics. In [21], service system with semi-Markov input flow is studied. The analysis of the system ends with the calculation of the final probability distribution of states.

In [22], the analysis of the service system with arbitrary distribution of random service duration is carried out. To estimate the efficiency of the system, the obtained stationary distribution of state probabilities is used.

As a result of studying the known publications on the problem of analysis of semi-Markov systems, the following conclusion can be made. Known theoretical results of the study of semi-Markov systems are limited to the calculation of the final probability distribution of the states of the system. When solving some practical problems, this is enough. However, in many cases, for example, when solving problems of evaluating the efficiency of recoverable systems, it is essential to know the dynamics of the probability of the system being on a set of operational states. The same problem is important for multichannel critical service systems.

The stability or degree of security of an information security object can be assessed by measuring the accuracy of its operation before and after the occurrence of destabilizing factors, such as attacks. The smaller the difference between these two obtained values of accuracy, the higher the resistance of the system to attack. The following difficulties may arise when measuring the stability of an information security object. For example, for working web resources it is not always possible to observe the situations generated by destabilizing factors. In particular, it is difficult for information security specialists to predict real attacks on the system. It is even more difficult to use this fact to test the system, so usually, such states of the system are modeled using software simulations [18, 23]. For example, in [18, 22, 23], experiments are conducted on the basis of real data from open datasets with the addition of data generated by botnets, obtained by software modeling of attacks on the system. This allows you to determine the resistance of the system to different types of attacks and empirically measure the average cost of a successful attack. When assessing the stability of the system, it should be borne in mind that it will be different in relation to different destabilizing factors, in particular, the stability of the system can differ significantly for different types of information attacks.

Thus, in [23, 24, 25], the task of creating a stable system is divided into two tasks:

1. Ensuring the resilience of the system to information attacks.
2. Ensuring the stability of the system to internal problems of the system. For example, the problem of cold start, sparse data, etc.

In turn, the conclusions obtained from relevant research results will simplify the selection of the optimal version of information security management systems for a particular IOB.

The latter circumstance makes our research relevant.

The purpose of the study

Improvement of the models for information security management objects of informatization through the use of mathematical apparatus of Markov and semi-Markov processes, in particular, will enable the use of mathematical tools to calculate the probability of threats to information systems.

2.3 Methods and Models

The design, organization and application of ISMS and information security system (ISS) are in fact related to unknown events in the future and, therefore, always contain elements of uncertainty. There are other reasons for ambiguity as well, such as insufficient information for management decisions or socio-psychological factors. Therefore, for example, the design phase of an ISS or ISMS is naturally accompanied by considerable uncertainty. Its level can be reduced by using the most adequate models. Markov and semi-Markov processes can be used in the development of ISS and ISMS, as a universal tool for modeling the work of IOB of IS at the stages of identifying threats and channels of information leakage, as well as vulnerability and risk assessment. Thus, Markov and semi-Markov processes consist of a set of tools that:

(a) provide efficiency and quality of detection of potential channels of information leakage at IS objects, in processes and programs of IS, at information transfer on communication channels, due to incidental electromagnetic radiation and guidance (IEMRG), and also in the course of management of protection system;
(b) determine the assessment of vulnerabilities and risks for information on IS objects, in processes and programs of IS, when transmitting information through communication channels, at the expense of IEMRG, as well as in the process of the protection system managing.

According to the modern theory of evaluating the effectiveness of systems, in particular ISMS, the quality of the information security system is manifested only in the process of its intended use (target operation), so the most objective is the evaluation of the operational effectiveness. The set of indicators and criteria for assessing the effectiveness of ISS and ISMS should be based on the probability of the task being performed by the system (ensuring the required level of protection). The evaluation criteria are the concept of suitability and optimality. Suitability means compliance with all requirements for ISS and ISMS, and optimality—the achievement of one of the characteristics of extreme value under restrictions and conditions on other properties of the system. With the change of any parameter of the IOB IS, due to logical connections, one or more elements in the evaluation matrix change, which affects the generalized indicators, therefore, the general state of ISS and ISMS changes. Given the nature of these changes, we can assume that the functioning of the ISS is also a semi-Markov process, which allows us to describe changes in its state using a relatively simple mathematical model. Mathematical models of IS operation based on semi-Markov processes can be used to simulate attacks on IS, which will increase the effectiveness of threat management.

Our conclusion is that semi-Markov processes can be used in the development and description of the state of information security systems and ISMS. Models of information systems, ISMS and information security systems based on semi-Markov processes can be used to improve the accuracy of assessing the effectiveness of ISS, as well as in their development.

Differential and difference equations with random coefficients are, respectively, continuous and discrete evolutionary stochastic systems, which can be fully attributed to ISMS in a random environment and serve as mathematical models of random evolution, which have been intensively studied recently. In particular, in the area of information security, achieving high efficiency and methods of assessing the state of cybersecurity of enterprises is impossible without the use of stochastic, dynamic mathematical models. The process of modeling the security status of IOB, in general, is quite complex, but it is simplified in the case of using numerical methods for computers.

To describe the performance characteristics of operations, the results of which depend on a number of random factors (for example, the probability of a particular type of attack) use different probabilistic models (in particular, Markov and semi-Markov sequences), which take into account the accompanying probabilistic processes. Next, random Markov and semi-Markov chains that can be used in the design of ISMS for IOB will be considered as models.

A random process $x(t)$ in this case will be called Markov if for each moment of time the probability of any state of ISS or ISMS in the future depends only on the state of the system at the present moment and does not depend on how the system came to this state [12, 15].

In mathematical terms, a random process $x(t)$ is a Markovian process, if for any moments of time from the segment $t_1 < t_2 < ... < t_n$ from the segment $[0, T]$ the conditional distribution function of the "last" value $\xi(t_n)$ at fixed values $\xi(t_1), \xi(t_2), ..., \xi(t_{n-1})$ depends only on $x(t_{n-1})$, so at given values $x_i (i = 1, ..., n)$ this ratio is performed.

$$P\{\xi(t_n) < x_n|\xi(t_1) = \xi_1, ..., \xi(t_{n-1}) = \xi_{n-1}\} = P\{\xi(t_n) < x_n|\xi(t_{n-1}) = \xi_{n-1}\}$$

Markovian random processes are widely used in studies of queuing systems, examples of which can be information security management systems, intrusion detection systems and others. Semi-Markovian random processes are more general than Markov ones; they can be used to more accurately describe various phenomena, including the tasks that arise during the implementation of ISMS.

The random sequence of ISMS $\eta_n, n = \{0, 1, 2, ...\}$, with a finite number of states $Q_1, Q_2, ..., Q_q$ is called a semi-Markov chain, if the following conditions are met. Let $n = \{0, 1, 2, ...\}$, $0 = n_0 < n_1 < n_2 < ...$—sequence state changes moments η_n. Then the random sequence $\eta_{n_0}, \eta_{n_1}, ...$ forms a homogeneous Markov chain to describe the ISMS with transient probabilities.

$$p(k, s) = P\{\eta_{n_{j+1}} = Q_k|\eta_{n_j} = Q_s\},\ k, s = 1, 2, \ldots q,$$

and the probabilities of being the ISMS in states satisfy the conditions

$$P\{\eta_{n_{j+1}} - \eta_{n_j} = d|\eta_{n_0}, \eta_{n_1}, ...\eta_{n_k}, ...; n_{k+1} - n_k, j \neq k\} =$$
$$= P\{\eta_{n_{j+1}} - \eta_{n_j} = d|\eta_{n_j}; \eta_{n_{j+1}}\} = f_{\eta_{n_j}\eta_{n_{j+1}}}(d),\ (j, k, d = 0, 1, 2, ...),$$

where $f_{Q_i Q_j}(d) = f_{ij}$ are given numbers.

Thus, the semi-Markov process is characterized by the fact that after the jump at n_j the whole "prehistory" is forgotten and the change of the η_n depends only on the η_{n_j} value and difference $n - n_j$.

The paper considered a system of nonlinear difference equations, the right parts of which depend on the semi-Markov sequence, in the system takes the form

$$X_{n+1} = F(X_n, \eta_n),\ n = 0, 1, 2, \ldots \tag{2.1}$$

where η_n—just has being described a semi-Markov chain with given matrices of intensities of transition of ISMS states $Q(n) = \|q_{ks}(n)\|, (k, s = 1, 2, ..., q)$ and initial distribution η_0, a X_n—random m-measurable vector.

It is known that the intensities $q_{ks}(n)$ meet the following requirements:

$$q_{ks}(n) > 0,\ \sum_{n=0}^{\infty} q_s(n) = 1,\ \ if\ \ q_s(n) = \sum_{k=1}^{q} q_{ks}(n).$$

The equations for conditional and partial conditional distributions solutions of (1) by which the value gained for a second-order random point system solution, including:

$$\begin{aligned} M(n+1|X, s) &= \psi_s(n+1) X^{(s)}(n+1; X) X^{(s)*}(n+1; X) + \\ &+ \sum_{k=1}^{n} \sum_{j=1}^{q} M(n-k+1|X^{(s)}(k; X), j) q_{js}(k) \end{aligned} \tag{2.2}$$

where $M(n|X, s)$—conditional moments of the second-order system solution (2.1); $X^{(s)}(n; X)$—solution of $X_{n+1} = F_s(X_n)$, $F_s(X_n) = F(X_n, Q_s)$ with initial condition $X_0 = X$; $q_{js}(k)$—the probability of the first transition of the ISMS from the Q_s state to the Q_j state, provided that the process, for example, the consideration of a particular information security incident, during the time k was in the Q_s state. The function $\psi_s(n)$ determines the probability that the process η_n, passing to the state Q_s was in this state for time n without a jump.

Formula (2.2) allows you to investigate the stability of the zero solution of the system of Eq. (2.1) using numerical methods using recursions.

Various specific ISMS tasks (detection and analysis of information security risks, planning and practical implementation of processes to minimize the IS risk, monitoring of these processes, making the process of minimizing the risk of information necessary adjustments) have to deal with objects (including the contours of information security), whose work should be stable over time. Therefore, the study of sustainability is a very important direction in the area of information security. Let the system of Eq. (2.1) $F_i(0) = 0, i = 1, \ldots, q$, thus $X_n = 0$ for each $n = 0,\ 1,\ 2,\ \ldots$ is the equilibrium of the ISMS.

The equilibrium $X_n = 0$ of the system (2.1) call stable relative to the moments of the second order, if for each $\varepsilon > 0$ exists $\delta > 0$ such that for all solutions $X_n = X_n(\eta_n, X_0)$ with initial condition $|X| < \delta$ is true

$$E|X_n X_n^*| < \varepsilon \tag{2.3}$$

where $E(\cdot)$—expected value, sign * means transposition.

Theorem 2.1 Solution $X_n = 0$ of the system of Eq. (2.1) will be stable with respect to the moments of the second order, when for $\varepsilon > 0$ exists $\delta > 0$ such that from $|X| < \delta$ following: the solutions of the system (2.2) satisfy the inequality

$$|M(n|\ X,\ s)| < \varepsilon \tag{2.4}$$

for each $n = 0,\ 1,\ 2,\ \ldots, s = 1,\ 2,\ \ldots,\ q$.

Using numerical methods for computers using Theorem (2.1), we can investigate the stability of the solution $X_n = 0$ of the system of Eqs. (2.1) for various nonlinear functions $F_i(\cdot), i = 1,\ 2,\ \ldots,\ q$. To simplify the calculations, consider instead the system (2.1) linear difference equation.

$$x_{n+1} = a(\eta_n)x_n,\ n = 0, 1, 2, \ldots, \tag{2.5}$$

where η_n—semi-Markov chain with two states Q_1, Q_2.

Let $a(Q_1) = a_1, a(Q_2) = a_2, x_0 = 1$. We set the intensity $q_{ij}(k), i,\ j = 1,\ 2,\ k = 0,\ 1,\ 2,\ \ldots$.

The program for studying the stability of solutions of Eq. (2.5) at given values of a_1, a_2, x_0, accuracy e and intensities $q_{ij}(k)$ is written in C++ .

To use formula (2.2), you need to enter functions for calculation in the program $\psi_s(n), x^{(s)}(n+1; x)$ and $M(n|\ x,\ s)$. To find $\psi_s(n)$ using following formula

$$\psi_s(n) = 1 - \sum_{k=0}^{n-1}(q_{1s}(k) + q_{2s}(k)),\ n = 0, 1, 2, \ldots,\ s = 1, 2.$$

$x^{(s)}(n+1; x)$ having the problem statement, calculate as follows

$$x^{(1)}(n+1; x) = a_1^{n+1}x,\ x^{(2)}(n+1; x) = a_2^{n+1}x.$$

If $x = x_0 = 1$, then $x^{(1)}(n+1; x) = a_1^{n+1}, x^{(2)}(n+1; x) = a_2^{n+1}$.
Conditional moments calculated using recursion, given that

$$M(1|\ x,\ s) = F_s(x)F_s^*(x) = a_s^2 xx,\ s = 1, 2,$$

In particular when $x = x_0 = 1$ we have

$$M(1|\ 1,\ 1) = a_1^2,\ M(1|\ 1,\ 2) = a_2^2.$$

After calculating the conditional moments, we check the solution of the equation for stability by formula (2.4).

After numerous experiments, we obtain the following results:

(1) if $0 < a_1 < 1$ and $0 < a_2 < 1$, we have the stability of the solution of Eq. (2.5);

(2) if $a_1 > 1, a_2 > 1$, we have a solution instability;
(3) if one of the partial coefficients a_1, (but not both) is slightly greater than one, then for some values of transition probabilities and intensities have stability.

After processing the results of calculations and extending them to the case of the system of Eq. (2.1), an important conclusion can be made. Despite the fact that the solution $X_n = 0$ for some deterministic equations $X_{n+1} = F_i(X_n)$ is stable, and for others—unstable, but is "near" the stability limit, then for some functions $q_{ij}(n)$, that determine the semi-Markov process, the solution of the stochastic system (2.1) becomes stable.

A system of nonlinear discrete equations is also considered, the right-hand sides of which depend on the Markov and semi-Markov sequences simultaneously, i.e. the system of the form

$$X_{n+1} = F(X_n,\ \xi_n,\ \eta_n),\ n = 0, 1, 2, \dots \tag{2.6}$$

where ξ_n is a random Markov chain with a finite number of ISMS states $Q_1,\ Q_2,\ \dots,\ Q_q$, with a given initial distribution $p_k(\xi,\ 0)$ and transient probabilities $\Pi_{ks}(n), k, s = 1,\ 2,\ \dots,\ q$; η_n is a semi-Markov chain with finite number of states $k_1,\ k_2,\ \dots,\ k_p$, with given intensities $q_{ij}(n), i, j = 1,\ 2,\ \dots,\ p$ and initial distribution $p_k(\eta,\ 0)$. We believe that a random vector $X_0 = X_0(\omega)$ does not depend on the Markov and semi-Markov chains and its distribution is given. Sequences ξ_n, η_n are independent of each other.

The equations for conditional partial distributions of solutions of the system (2.6) are derived, by means of which the relations for the moments of the second order of the random solution of the system are obtained [2]. As a consequence, the equations for partial conditional distributions and moments of the second order of solutions of systems of nonlinear discrete equations with Markov coefficients are obtained. The stability of random solutions of the system of Eq. (2.6) is investigated by means of formulas for moments. The study also addressed the issues of practical aspects of ISMS control, which are related to the optimization of solutions of a nonlinear stochastic view control system.

$$X_{n+1} = F(X_n, \eta_n, U_n),\ n = 0, 1, 2, \dots \tag{2.7}$$

where η_n is a finite-valued semi-Markov process, $U_n - l$—dimensional control vector.

For the system (2.7) seeking optimal control form

$$U_n = S(X_n, \eta_n),\ n = 0, 1, 2, \dots \tag{2.8}$$

which minimizes the quality functional

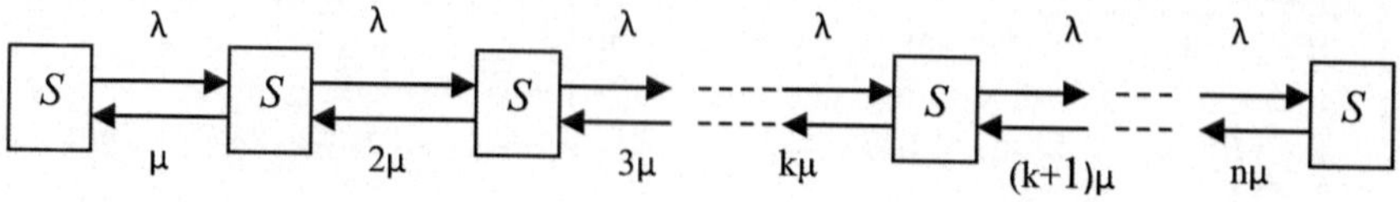

Fig. 2.1 Graph of ISMS states

$$I = \sum_{n=0}^{\infty} < W(X_n, \eta_n, U_n) > \tag{2.9}$$

Functional (2.9) will have a minimum if the stochastic Lyapunov functions along the solutions of the system (2.7)

$$V_s(x) = \sum_{n=0}^{\infty} < W(X_n, \eta_n, U_n)|X_0 = X,\ \eta_0 = Q_s >,\ s = 1, 2, \ldots, q \tag{2.10}$$

will have a minimum.

Necessary conditions for optimal control system (2.7) and its particular case when random process η_n is a Markovian process are derived. Construction of optimization models, such as the structure of ISMS enterprise or OBI and orders of the results allows best use of existing capacity to outline information security and achieve-high level of information security in terms of sustainability and dynamic attacking opposition. Suppose there are n channels that receive a stream of applications in the ISMS with intensity λ. The flow of services has an intensity of μ. The ISMS has the following states (number them according to the number of requests in the system): S_0, S_1, S_2, ...S_k, ..., S_n, where S_k is the state of the system when there are k requests, i.e. k service channels are occupied [3].

The graph of ISNS states is shown in Fig. 2.1.

The flow of orders sequentially translates the ISMS from any left state to the adjacent right with the same intensity λ. The intensity of the flow of services (ISMS performance), which translates the system from any right state to the neighboring left state, is constantly changing depending on the state. Indeed, if the ISMS is in state S_2 (two channels are busy), it can go to state S1 (one channel is busy) when it completes service or the first or second channel, i.e. the total intensity of their service flows will be 2μ. Similarly, the total service flow that transfers the ISMS from state S_3 (three channels occupied) to S_2 will have an intensity of 3μ, i.e any of the three channels can be released, etc.

From the point of view of the queuing theory, ISMS can be considered as a multiphase queuing system with queues. The structure of the model is shown in Fig. 2.2.

Computational experiment. As an example, consider a three-phase ISMS. The input stream in the system is a stream of requirements (orders), which successively undergo three stages of processing in the ISMS:

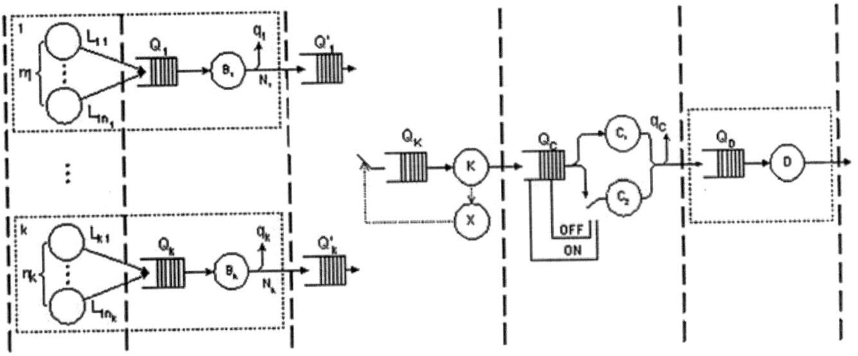

Fig. 2.2 The structure of the ISMS model

1. primary processing and control in the department;
2. transmission via communication channels to the main office;
3. processing in an automated system (e.g. SIEM).

Orders are entered from the workplaces of operators located in the branches. In the presence of errors, part of the applications after the initial control is eliminated (thinning of the incoming flow), and orders that have passed the initial control are received from the branch via data transmission channels to the mail server located in the central office.

The second stage of processing applications in ISMS is their logical control. It is also possible to weed out some of the applications due to their logical error.

Orders that have been processed on the mail server are processed by the automated system, where the third, final phase of application processing is implemented. After that, the orders are returned to the workplace of the operator from which they were sent. This completes the three-phase order processing cycle. We have a three-phase QMS as a result.

The first phase of service consists of k parallel systems of type $M|M|1$ (M means Markov process), the input of each of which receives the flow $L_j(j = 1, ..., k)$, which is the superposition $n_j(j = 1, \ldots, k)$ independent flows $L_{j_1}, \ldots, L_{j_n}, (j = 1, \ldots, k)$. The stream $L_{j_r}(j = 1, \ldots k, r = 1, \ldots, n_j)$ comes from the r-th source of applications of the j-th channel, independent of other sources. Applications are queued $Q_j(j = 1, \ldots, k)$, from which the device $B_j(j = 1, \ldots, k)$ is selected (Fig. 2.2). The service time on the device $B_j(j = 1, \ldots, k)$ is distributed exponentially in the General case with the parameter $m_j(j = 1, \ldots, k)$, which is confirmed by experimental data. The output stream of the device $B_j(j = 1, \ldots, k)$ is independently thinned with the probability $q_j(j = 1, \ldots, k)$. The flow obtained as a result of thinning, $N_j(j = 1, ..., k)$ enters the second phase of maintenance on the device C.

The input stream of the second service phase, implemented by the system $M|M|1$, is the superposition k of the independent output streams of the first service phase $N_j(j = 1, \ldots, k)$. From the queue Q_c, which includes applications, they are selected by the device C. The service time on the device C is distributed exponentially with the parameter m_c. The output stream of the device C is subjected to independent thinning with a probability q_c. The flow obtained as a result of independent thinning, N_c is the output stream of the second phase and enters the third phase of maintenance. The third phase of maintenance is implemented SUIB type $M|M|1$ (device D). Requests are received in the queue Q_D, from which are selected according to the discipline of service "the first request received is serviced first" by the device D. The service time on the device D is distributed exponentially with the parameter m.

After passing three phases of service, the application leaves the system.

Analysis of simulation results based on stochastic Petri nets allowed to identify places with insufficient bandwidth and determine the optimal system load. When analyzing communication protocols, Petri nets can be used to simulate a complex system that is under the influence of random factors.

Marking μ is the assignment of chips to the positions of the Petri net. A chip is a primitive concept of Petri nets, like positions and transitions. Chips are assigned (belong) to positions. The number and position of the chips when performing a Petri net may vary. Chips are used to determine the performance of the Petri net [2].

Marking μ of the Petri net $C = (P, T, I, O)$ is a function (P—position, T—transition, I—input, O—output), which reflects the set of positions P in the set of non-negative integers N. $\mu : P \to H$.

The marking μ can also be defined as an n–vector $\mu = (\mu_1, \mu_2, \ldots, \mu_n n)$, where $n = |P|$ and each $\mu_i \in N, i = 1, \ldots, n$. The vector μ determines for each position p_i of the Petri net the number of chips in this position. The number of chips in the position p_i is $\mu_i, i = 1, \ldots, n$. The relationship between the definitions of labeling as a function and as a vector is obviously established by the relation $\mu(p_i) = \mu_i$. Denoting it as a function is a little more general and therefore used much more often. Marked Petri net $M = (C, \mu)$ is a set of structure of Petri net $C = (P, T, I, O)$ and marking μ and can be written as $M = (P, T, I, O, \mu)$.

On the graph of the Petri net (Fig. 2.3), the chips are represented by a small dot in a circle representing the position of the Petri net [2].

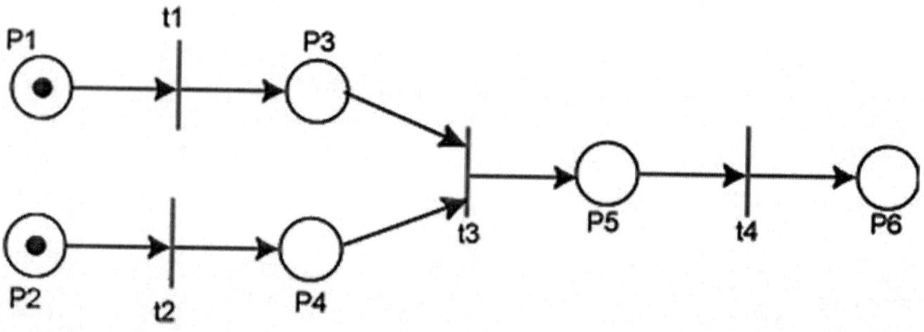

Fig. 2.3 Petri Net marking

To simulate the dynamics of the system states changing during the destabilizing factors are important to know not only the stationary probability distribution of the system, but the value of these probabilities at any given time.

2.4 Discussion of the Results of the Experiment

Comparative analysis of characteristics of an existing network performed using measurements and specifications of the system's design, which, in turn, are obtained by simulations, allows to identify shortcomings in the existing network's design and to draft reorganizations required to improve the most important functional characteristics of the ISMS. Stochastic network modeling performed has allowed us to identify and to draft optimal requirements necessary to reorganize the network equipment and communication channels.

2.5 Conclusions

The paper proposes a method for studying the stability of an information security management system of any object of informatization. The method is based on zero solutions of a system of nonlinear difference equations that depends on the semi-Markov chain. The essence of the proposed method is to study the stability of moment equations that are deterministic for the aforementioned systems for information security management. Necessary optimality conditions for these systems are derived.

References

1. Wu, D., Ren, A., Zhang, W., Fan, F., Liu, P., Fu, X., Terpenny, J.: Cybersecurity for digital manufacturing. J. Manuf. Syst. **48**, 3–12 (2018)
2. Hoffmann, R.: Markov models of cyber kill chains with iterations. In: 2019 International Conference on Military Communications and Information Systems (ICMCIS), pp. 1–6. IEEE (2019)
3. Yinka-Banjo, C., Ugot, O.A.: A review of generative adversarial networks and its application in cybersecurity. Artif. Intell. Rev. 1–16 (2019)
4. Zeng, R., Jiang, Y., Lin, C., Shen, X.: Dependability analysis of control center networks in smart grid using stochastic Petri nets. IEEE Trans. Parallel Distrib. Syst. **23**(9), 1721–1730 (2012)
5. Robidoux, R., Xu, H., Xing, L., Zhou, M.: Automated modeling of dynamic reliability block diagrams using colored Petri nets. IEEE Trans. Syst. Man Cybern.-Part A: Syst. Hum. **40**(2), 337–351 (2009)
6. Ficco, M.: Detecting IoT malware by Markov chain behavioral models. In: 2019 IEEE International Conference on Cloud Engineering (IC2E), pp. 229–234. IEEE (2019)

7. El Bouchti, A., Nahhal, T.: Cyber security modeling for SCADA systems using stochastic game nets approach. In: 2016 Fifth International Conference on Future Generation Communication Technologies (FGCT), pp. 42–47. IEEE (2016)
8. Xu, M., Hua, L.: Cybersecurity insurance: modeling and pricing. N. Am. Actuar. J. **23**(2), 220–249 (2019)
9. Hoffmann, R., Napiórkowski, J., Protasowicki, T., Stanik, J.: Risk based approach in scope of cybersecurity threats and requirements. Proc. Manufact. **44**, 655–662 (2020)
10. Abraham, S., Nair, S.: Exploitability analysis using predictive cybersecurity framework. In: 2015 IEEE 2nd International Conference on Cybernetics (CYBCONF), pp. 317–323. IEEE (2015)
11. Pokhrel, N.R., Tsokos, C.P.: Cybersecurity: a stochastic predictive model to determine overall network security risk using Markovian process. J. Inf. Secur. **8**(2), 91–105 (2017)
12. Wu, B., Maya, B.I.G., Limnios, N.: Using semi-Markov chains to solve semi-Markov processes. Methodol. Comput. Appl. Probab. 1–13 (2020)
13. Promyslov, V., Jharko, E., Semenkov, K.: Principles of physical and ınformation model ıntegration for cybersecurity provision to a nuclear power plant. In: 2019 Twelfth International Conference Management of large-scale system development (MLSD), pp. 1–3. IEEE (2019)
14. Sohal, A.S., Sandhu, R., Sood, S.K., Chang, V.: A cybersecurity framework to identify malicious edge device in fog computing and cloud-of-things environments. Comput. Secur. **74**, 340–354 (2018)
15. Abimbola, O.O., Odunola, A.B., Temitope, A.A., Aderounmu, G.A., Hamidja, K.B.: An improved stochastic model for cybersecurity risk assessment. Comput. Inf. Sci. **12**(4), 96–110 (2019)
16. Bulinskiy A.N., Shiryaev A.N., Teoriya sluchaynykh protsessov–M.: Fizmatgiz.–2005. – 364 s
17. Cao X.R.: Optimization of average rewards of time nonhomogeneous Markov chains. IEEE Trans. Autom. Control **60**(7), 1841–1856 (2015)
18. Dimitrakos, T.D., Kyriakidis, E.G.: A semi-Markov decision algorithm for the maintenance of a production system with buffer capacity and continuous repair times. Int. J. Product. Econ. **111**(2), 752–762 (2008)
19. Li, Q.L.: Nonlinear Markov processes in big networks. Special Matrices **4**(1), 202–217 (2016)
20. Li. Q.L., Lui J.C.S.: Block-structured supermarket models. Discret. Event Dyn. Syst. **26**(2), 147–182 (2016)
21. Okamura, H., Miyata, S., Dohi, T.: A Markov decision process approach to dynamic power management in a cluster system. IEEE Access **3**, 3039–3047 (2015)
22. Sanajian, N., Abouee-Mehrizi, H., Balcıoglu, B.: Scheduling policies in the M/G/1 make-to-stock queue. J. Oper. Res. Soc. **61**(1), 115–123 (2010). https://doi.org/10.1057/jors.2008.139
23. Ricci, F., Rokach, L., Shapira, B., Kantor, P.B. (Eds.): Recommender Systems Handbook. Springer, Boston, 842 p (2011). https://doi.org/10.1007/978-0-387-85820-3
24. Zhang, C., Liu, J., Qu, Y., Han, T., Ge, X., Zeng, A.: Enhancing the robustness of recommender systems against spammers. PLoS ONE **13**(11), e0206458 (2018). https://doi.org/10.1371/journal.pone.0206458
25. Kaur, P., Goel, S.: Shilling attack models in recommender system. In: International Conference on Inventive Computation Technologies (ICICT), Coimbatore, pp. 1–5 (2016). https://ieeexplore.ieee.org/document/7824865/

Chapter 3
Gender Identification Using Ensemble Linear Discriminant Analysis Algorithm Based on Facial Features

S. Jana, S. Thangam, and S. Selvaganesan

Abstract A wide range of applications rely on the capacity to extract information about people from images. Person identification for the surveillance or access control, gender and age estimation for constructing user models, and facial expression recognition are some of the most common applications that could provide relevant data for evaluating man–machine interfaces. In addition to the previously listed recognition tasks, early works were largely associated with psychological analysis, a method in which humans identify gender from faces studied. Our proposed system presents gender and age classification based on facial images using dimensionality reduction technique, linear discriminant analysis (LDA), along with subspace ensemble learning. Our experiments and results show that LDA and ensemble learning techniques can be used for estimating gender and age.

3.1 Introduction

Face is a significant biometric trait of humans since, unlike fingerprint authentication, it does not require the subject's participation. Automatically recognizing and analyzing face permits several applications in human–machine interactions and security functions. Useful information like quality, identity, age, gender, expression can be obtained from facial images. Gender classification and facial image analysis have been effectively employed in a variety of applications ranging from robotic–human

S. Jana (✉)
Electronics and Communication Engineering, Vel Tech Rangarajan Dr. Sagunthala R&D Institute of Science and Technology, Chennai, India
e-mail: drsjana@veltech.edu.in

S. Thangam
Computer Science and Engineering, Amrita School of Engineering, Bengaluru Campus, Amrita Vishwa Vidyapeetham, Bengaluru, India
e-mail: s_thangam@blr.amrita.edu

S. Selvaganesan
Information Technology, J J College of Engineering and Technology, Trichy, India

J. I.-Z. Chen et al. (eds.), *Machine Learning and Autonomous Systems*, Smart Innovation, Systems and Technologies 269, https://doi.org/10.1007/978-981-16-7996-4_3

interaction to biometrics. Gender classification is, in fact, a binary classification flaw in which one must determine whether a figure in a photograph is a boy or a girl. While humans can easily identify the gender of a person based on perception, it has not been feasible for a computer to do so. Gender classification from face pictures [1–5] has received abundant attention recently owing to its applications as well as rising demand for accuracy in computer program retrieval, demographic information assortment, human–computer interfaces, etc. Moreover, gender recognition when applied as a preprocessing step during face recognition results in a reduced number of faces to be verified making the process of face recognition faster. Similar to different pattern classification issues, two key steps for gender classification include feature extraction and pattern classification. From the perspective of feature extraction, there are four approaches. The first and the simplest of them is the exploitation of grayscale or color picture element vectors [1]. The second approach uses mathematical space transformation theory that has gifted some dimensional reduction techniques [6]. The limitation of this approach is their reduced efficiency with large variations in face orientation. The third method is to use texture information like wrinkle and complexion. The fourth method involves extracting features that can be applied in the classification of facial wrinkles and shapes. This is done by combining the facial feature detection with wavelet transform.

The method of establishing a subject's gender from body or face photographs is known as gender classification. Several approaches for gender recognition [7–9] have been proposed in the literature, including gait [10], iris [11], and hand form [12]. Recently, an intriguing method for gender classification from faces based on random views and taking occlusion into account was described [13]. In [14], the authors performed gender classification using Fisherface features, principal component analysis for dimensionality reduction, and subsequent classification as male and female using support vector machines. They obtained an accuracy of 88%. Apart from the common features such as iris and fingerprint, nose features were also used to identify the gender of a person [15]. Four features were extracted from nose, pre-processed, and used to classify using linear discriminant analysis. They demonstrated that, on an average, an accuracy of 77% could be obtained. Shakhnarovich et al. [16] in their system for real-time face detection used Adaboost algorithm for categorizing detected faces as Asian or no Asian as well as male or female.

On face photos obtained from real-world unconstrained video sequences, they demonstrated that their approach beats the SVM approach. Rai and Khanna, Khan et al. [17, 18] used a variety of neural network techniques to achieve gender classification goals.

In this study, we offer a new method that uses only a single sample to reliably distinguish human gender and classify gender photos from a database or in real-time works.

3.2 Related Works

In [10], Caifeng Shan, Shaogang Gong, and Peter W. McOwan et al. presented the application of computer vision in classifying human beings based on their gender in surveillance videos using gait as an attribute. Moreover, they claimed that they achieved an increased gender recognition rate of 97.2% in large datasets by fusing gait and facial features. They also stated that these two modalities could be fused at the characteristic level using a tool called Canonical Correlation Analysis (CCA) that is well suitable for combining two sets of measurements.

In [11] Vince Thomas, Nitesh V. Chawla, Kevin W. Bowyer, and Patrick J. Flynn proposed a machine learning system, which used the iris texture features to identify whether a subject is male or female. Because substantial research has already been conducted to segment and encode iris images, the scientists employed iris as a biometric. In addition, the existing methods to segment and encode iris images are conducive to creating a feature descriptor based on iris texture features. They developed a gender classification model using decision tree learning and achieved an accuracy of about 80%. They showed improved performance by eliminating biases. The machine learning framework that the authors developed involves feature selection, decision tree learning, and ensemble methods.

From arbitrary views and in the presence of occlusion, Matthew Toews and Tal Arbel et al. [13] provided a unique framework for recognizing, localizing, and categorizing faces in terms of visual features, such as sex or age. All of these objectives are incorporated into an appearance-based model for view-point invariance of an object based on scale-invariant properties. These characteristics are statistically quantified in terms of their frequency, appearance, geometry, and relationship with visually interesting attributes. Initially, an appearance model for the object category is learned, and then a Bayesian classifier is used to model the features that reflect the visual qualities. Unlike earlier techniques that assume data that is single viewpoint, upright, pre-aligned, and clipped from background distraction, the authors claim that the framework can be utilized in actual circumstances in the presence of perspective alterations and partial occlusion.

Based on the color FERET database, experimentation yields the first result for gender classification from arbitrary views, with an equivalent error rate of 16.3% [2]. Faces in crowded pictures from the CMU profile database are also shown to work reliably with the approach. When compared with the geometry-free bag-of-words model, their framework's geometrical information enhances categorization. When comparing Bayesian classification to support vector machines, it is clear that Bayesian classification outperforms them.

In [16], Gregory Shakhnarovich et al. presented an integrated system for face detection from an uncontrolled environment together with demographic analysis. The authors used Viola Jones face detection algorithm, the output of which is passed to a demographic classifier. The architecture of the demographic classifier is similar to that of face detector and produces error rates much lesser than known classifiers. The use of demographic information is to reduce the error rate in unconstrained and noisy sensing environments.

3.3 Proposed System

In this research, we offer a new method that uses only a single sample image per subject to properly classify human gender and age from a database or in real-time. The proposed algorithm uses LDA for the purpose of extraction and then uses hybrid Ensemble Learning for Gender and Age classification. The block diagram depicting the workflow in gender classification and age prediction is shown in Fig. 3.1. The details of the workflow are explained in the following section.

3.3.1 Workflow

3.3.1.1 Concept of Age and Gender Recognition

Face-based gender recognition is a method for determining whether a person is male or female using photographs of their faces. This is implemented using three processes, namely, training, testing, and matching excluding pre-processing of detected faces.

1. **Training**

   In this process, the human faces are detected from the captured images, and features of male subjects' and female subjects' faces are extracted separately. Using distinctions in shape, color, and intensity, feature extraction is used to

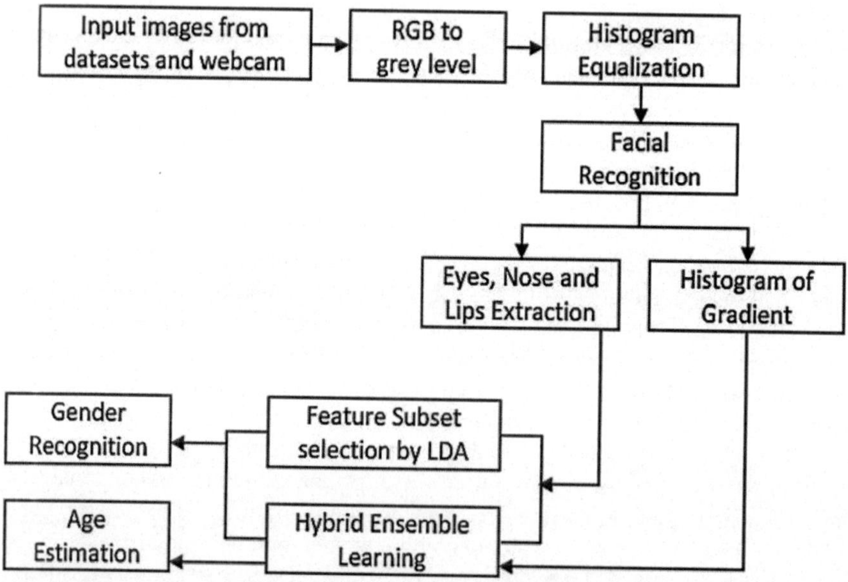

Fig. 3.1 Block diagram of gender classification and age estimation

provide useable information for recognizing objects or human faces. The steps in feature extraction include,

- Normalization,
- Mean calculation,
- Difference matrix,
- Co-variance matrix,
- Eigen face,
- Transfer function with LDA weight matrix.

2. **Testing**

   In this process, the face of the subject under test obtained after detection is pre-processed and features are extracted. It is similar to training process except that features of a single subject are considered.

3. **Matching**

   In this stage, test features are compared with the training features and correlation measure between the two is obtained to determine the matching between the two. If test image is matched with any one of the training data of male or female face images, then the corresponding gender is decided as the gender of the test subject. An estimate of the age of the subject under test is also obtained using the ensemble learning technique.

3.3.2 Preprocessing Techniques

The acquired images are transformed into monochrome images as a first step in pre-processing. A true color image is one in which each pixel is represented by three components, specifically the pixel scalar's red, blue, and green intensities. Intensity values are represented as a $M \times n \times 3$ array comprising uint8, uint16, single, or double values. To enhance the contrast of the image, Histogram equalization is performed. The normalized histogram for an image is obtained by dividing each pixel value by the maximum value, which are 255 in the case of an 8-bit image. Figure 3.2a shows the sample input color image. Figure 3.2b shows the grayscale image after conversion from color to grayscale. Figure 3.2c shows the histogram equalized image. Figure 3.2d shows the face segmented.

3.3.3 Feature Extraction

(a) **Locating the eyeballs and the corners of eyes**

   It is the position of the eyes that aids in the detection of a face. As a result, the property of valley points of brightness in eye-areas is employed to determine the

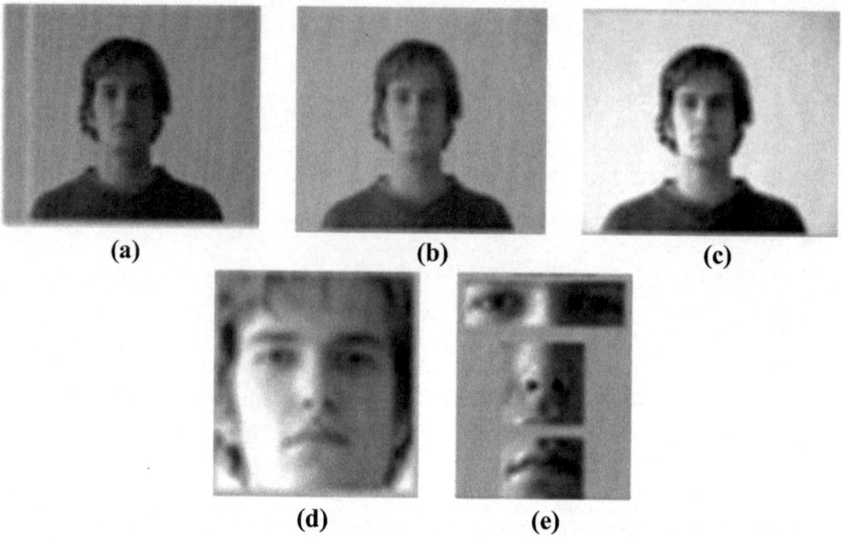

Fig. 3.2 Sample image after each pre-processing stage and face recognition **a** input color image **b** grayscale image **c** Histogram equalized image **d** output of face recognition module **e** eyes, nose, and lips extracted

location of eyes. To detect eyes, we use a combination of valley point searching, directed projection, and the symmetry of two eyeballs in our suggested system. The gradient picture is first projected on the top left and top right parts of the face, and the resulting normalized histogram is utilized to locate the eyes in the y direction using the valley point in the horizontal integral projection picture.

(b) **Identifying a nose area feature point**

Because the midpoint of two nostrils indicates a generally consistent position in the face, the feature for nose area is obtained. Keeping the two eyeballs as reference, the nose area is obtained by using integral projection of luminance. Initially, the integral projection curve in y direction is got by considering the strip region of two eyeballs width. The y coordinates of nostrils are then marked as the first valley point along the projection curve from the y coordinates of eyeballs.

(c) **Tracing the corners of the mouth**

In the process of face detection and recognition, the mouth is almost as important as the eyes. Facial expression changes cause changes in the shape and size of the mouth. Also, the presence of whiskers could interfere in the recognition of mouth area. However, the corners of mouth are less affected by changes in facial expression and hence can be used for feature points.

When a point has two dominant and different edge directions in its neighborhood, it is referred to as a corner. Due to the fact that a corner is invariant to lighting, translation, rotation, and other transformations, it is widely used as a feature in extracting the mouth region of human face. As a corner detector in images, the Harris Corner Detector, which is based on either autocorrelation of image intensity values or image gradient values, is frequently employed. The segments of the face, such as the eyes, nose, and lips, are shown in Fig. 3.2e.

3.3.4 Linear Discriminant Analysis (LDA)

Despite the fact that it is most typically used for pattern classification, linear discriminant analysis (LDA) can be used in feature selection as a useful metric to evaluate the separate ability of a feature subset. When applied to feature selection on High-Dimensional Small-Sized (HDSS) data with class-imbalance, LDA faces four obstacles, including singularity of scatter matrix, overfitting, overwhelming, and unreasonable computational complexity. We introduce an LDA-based feature selection method called Minority Class Emphasized Linear Discriminant Analysis (MCE-LDA) with a new regularization strategy to overcome the first three issues. Unlike traditional forms of regularization, which place a greater emphasis on the majority class, the proposed regularization places a greater emphasis on the minority class, with the goal of improving overall performance by reducing minority class overfitting and the overwhelming of the majority class to the minority class. To reduce computational costs, an incremental implementation of LDA-based feature selection has been introduced.

PCA is used to represent facial images without hair in a low-dimensional space in our method. Gender-related PCA features are subsequently selected using the Genetic Algorithm (GA). Eigen characteristics can be used to recreate the faces. Despite the lack of identify information in the reconstructed photos, they do reveal a lot about gender. This means that GAs can choose eigenvectors that primarily encode gender data.

When inter-class scatter is divided by intra-class scatter, LDA's purpose is to find a projection, x = M[y] (y is the input picture and M is the projection matrix), that yields the highest value. To prevent problems within singularities in the scatter matrix class), PCA is used to project the original space onto a smaller, intermediate space, and ultimately onto a final area.

HOG Feature Extraction.

In the recent past, among the various feature-based descriptors available, the most commonly used descriptor for object detection is Histogram of Oriented Gradients (HOG).

Gradient Computation.

In the process of gradient computation, the gradient values are initially obtained by applying the one-dimensional derivative mask in vertical and horizontal directions. The vertical mask is of 3×1 dimensions and the horizontal mask is of 1×3

dimensions as shown in Fig. 3.3. The vertical and horizontal masks are applied to detect the edges in the vertical and horizontal direction. A sample face image from ORL database is shown in Fig. 3.4a. The face image with the horizontal edges highlighted on application of horizontal gradient mask Dx to the sample image is shown in Fig. 3.4b. The face image with the vertical edges highlighted on application of vertical gradient mask Dx to the sample image is shown in Fig. 3.4c. The face image with both the horizontal and vertical edges highlighted on application of horizontal and vertical gradient mask to the sample image is shown in Fig. 3.4d.

$$D_x = \begin{bmatrix} -1 & 0 & 1 \end{bmatrix}$$

$$D_y = \begin{bmatrix} -1 \\ 0 \\ 1 \end{bmatrix}$$

Fig. 3.3 Vertical (D_y) and horizontal (D_x) mask for gradient computation

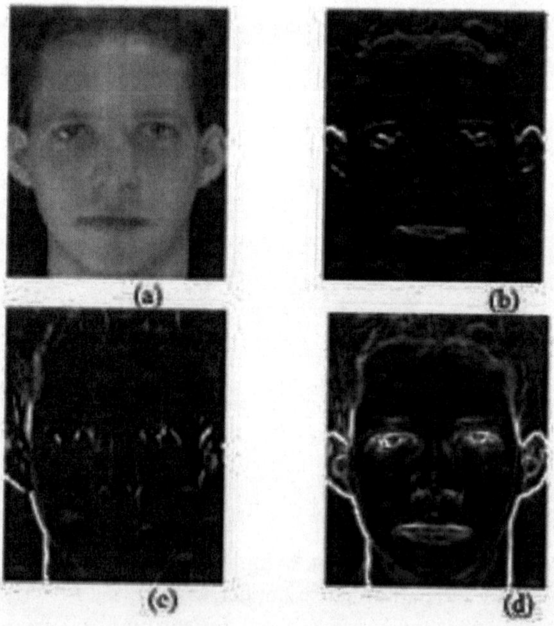

Fig. 3.4 **a** An example of ORL database face. **b** After Horizontal gradient operation. **c** After Vertical gradient operation. **d** With both horizontal and vertical gradients operation

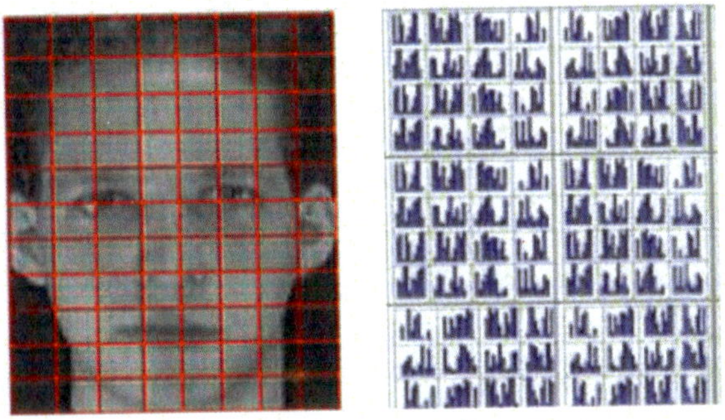

Fig. 3.5 **a** Division of face image into cells. **b** Histograms for each cell in the image

Orientation binning

The second stage in obtaining HOG features is orientation binning. The weight cast by each pixel within a cell, which can be either radial or rectangular in shape depending on the direction, is determined by the number of values collected during the gradient computation procedure. The number of channels varies depending on whether it's signed gradient or unsigned, it ranges from 0 to 3600 or 0 to 1800. Pixel plays a vital role in computing the gradient by contributing to the magnitude or a function of the magnitude. The performance of HOG features with a higher gradient magnitude yields more accurate results. Variations in gradient magnitude, such as the square of the gradient or the clipped form of the gradient, can also be useful. Figure 3.5a depicts a cell division image from the ORL database, whereas Fig. 3.5b depicts the histogram for each cell.

3.3.5 *Hybrid Ensemble Learning-Based Classification*

Ensemble learning is a method of solving a computational intelligence problem by intentionally generating and combining many models, such as classifiers or specialists. Ensemble learning's ultimate goal is to increase a model's performance in domains such as classification, prediction, functional approximation, and so on. Ensemble learning can also be used to assign a confidence level to the model's choice, Data fusion, incremental learning, non-stationary learning, and error correction are all techniques that can be used to select optimal (or nearly perfect) features. The main reason for employing ensemble-based classification in recent years is because there is no single classifier model that performs well among numerous models such as Multilayer Perceptron (MLP), Support Vector Machines (SVM), Decision trees, Naive Bayes classifier, and so on. In addition, the implementation of a classification method varies. Even when all other parameters are held constant, different initializations of MLPs may yield different decision limits. The usual method of selecting a classifier that gives minimum error on training data has proved to be wrong. Even in unsupervised learning method where the data is unseen, classification performance may be misleading in spite of cross-validation. Hence, using an ensemble of different models—rather than selecting one single model, and mixing their outputs by using some criteria for example, merely averaging them may reduce the possibility of choosing a poor classifier model.

3.4 Experimental Results

Tests were conducted on face images from IMDB-WIKI dataset [19, 20], which contains a large number of gender and age labeled face images for training. This dataset consists of the photos of around 100,000 of the most popular actors from where information regarding date of birth, name, gender were taken. Also, it contains around 62,328 profile images taken from the Wikipedia pages of the people along with the information like date of birth, name, gender. For each photo, information such as the time when the photo was taken, face score, and face location are also available. In case of more photos of a single person, second face score available along with the images is used to remove such images. The timestamp helps in determining the age of the person at the time of taking the photo from the difference computed between the date of birth and time stamp.

For the purpose of gender prediction, pre-trained models were used. The images in this database were taken from a standard laptop webcam at 640×480 pixel resolution. The input image, the intermediate images after grayscale conversion, image enhancement, facial parts such as eyes, nose, and lips detection, image showing the connectivity of the eyes, nose and lips, feature points forming the facial landmarks are shown in Fig. 3.6. For easier processing, the input image is transformed to a grayscale image. Figure 3.6a and b depicts the input and RGB to grayscale transformed images, respectively. For contrast enhancement, histogram equalization is applied to the gray image, and the result is displayed in Fig. 3.6c. Face is detected using viola-jones face detection algorithm. The detected face is shown in Fig. 3.6d. Figure 3.6e and f shows the facial point's connectivity and the facial landmarks, particularly the eyes, nose, and mouth, respectively. Figure 3.6g and h shows the vertical gradient and horizontal gradient of the image. Figure 3.6i shows the HOG image obtained for the processed input image. The HOG features are then used in the gender class identification and age estimation. With 80% of data used as training data and 20% of data used as testing data an accuracy of 98.1%. Experiments were also carried out using photos captured using webcam in real time and a slightly reduced accuracy of 94.9% was achieved. Literature survey [18] states that the accuracy obtained with Local Directional pattern features with SVM classification gave an accuracy of 95.05%. An accuracy of 93.5% has been reported in [18] using Fuzzy SVM for the classification of gender. A comparative study of classifiers namely Bayesian Classifier, NNs, and SVMs with LDA reveals the better performance of LDA over the remaining methods in gender classification and age estimation

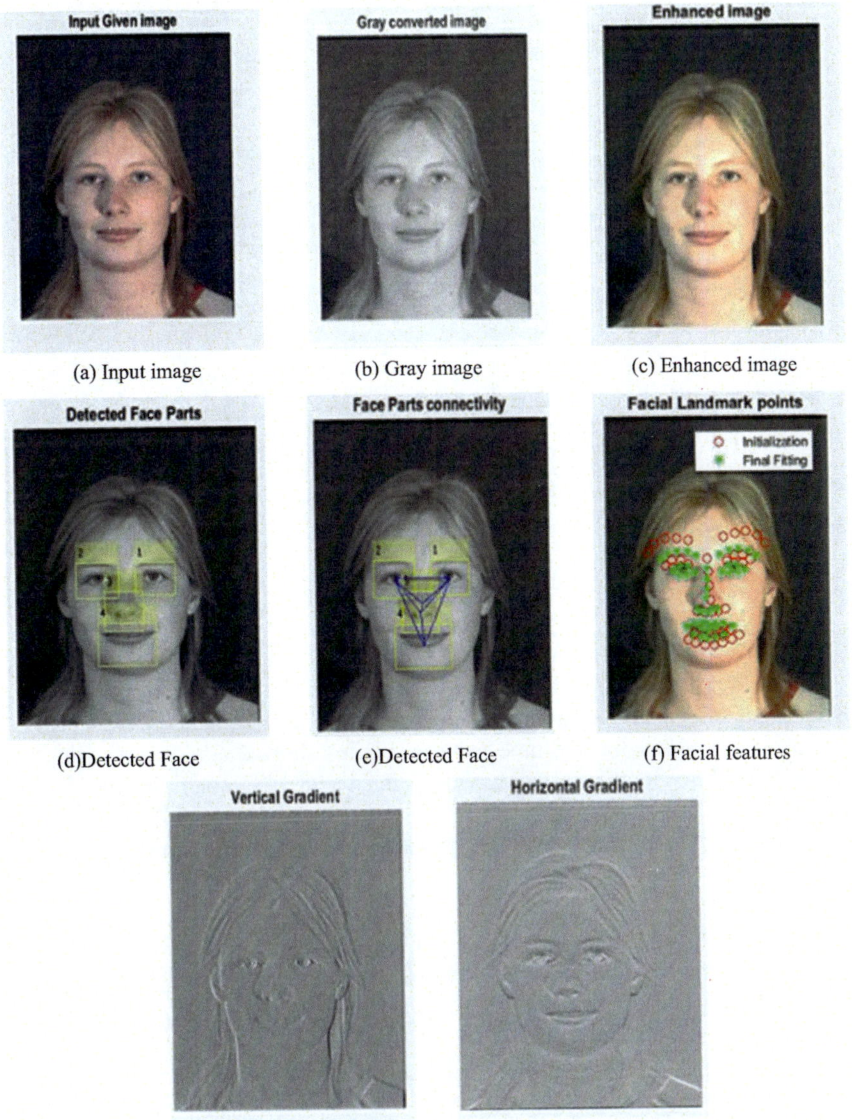

(a) Input image (b) Gray image (c) Enhanced image

(d)Detected Face (e)Detected Face (f) Facial features

(g)Vertical Gradient (h) Horizontal Gradient

Fig. 3.6 Image outputs of each process in gender class identification and age estimation

3.5 Conclusion

We used a hybrid ensemble learning-LDA to classify gender based on facial traits and segments in this work. To categorize a person's gender from a recognized facial image, the face, eyes, nose, and mouth are first recognized using the Viola Jones technique, and then significant pixels in the image are extracted using LDA. The hybrid

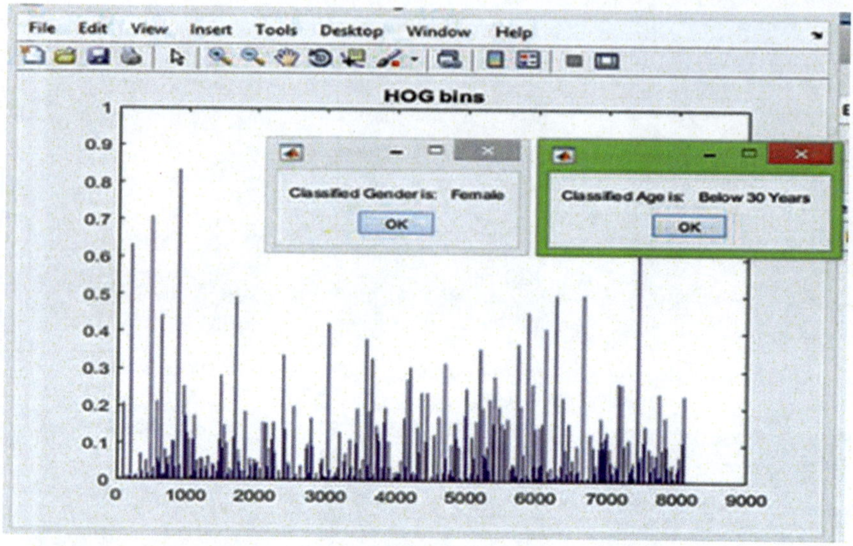

i)HOG bins

Fig. 3.6 (continued)

ensemble learning method is applied to the retrieved features. With photos from camera and datasets, the system is evaluated with various inputs and different poses. Experimental results from real-time webcam photos showed a 94.9% recognition rate, whereas offline images from datasets revealed a 98.1% recognition rate.

References

1. Moghaddam, B., Yang, M.: Learning gender with support faces. IEEE Trans. Pattern Anal. Mach. Intell. **24**(5), 707–711 (2002)
2. B. Golomb, D. Lawrence, and T. Sejnowski, "SEXNET: A Neural Network Identifies Sex From Human Faces.," NIPS, 1990.
3. Mitsumoto, S.T.: Hiroshi, "Male/female identification from 8 x 6 very low resolution face images by neural network,." Pattern Recogn. **29**(2), 331–335 (1996)
4. Yang, M., Moghaddam, B.: Support vector machines for visual gender classification. In: Proceedings of 15th International Conference on Pattern Recognition, pp. 1115–1118 (2000)
5. Makinen, E., Raisamo, R.: Evaluation of gender classification methods with automatically detected and aligned faces. IEEE Trans. Pattern Anal. Mach. Intell. **30**(3), 541–547 (2008)
6. Khan, A., Baig, A.R.: Multi-objective feature subset selection using non-dominated sorting Genetic Algorithm. J. Appl. Res. Technol. **13**(I), 145–159 (2015)
7. Aji, S., Jayanthi, T., Dr. Kaimal, M.R.: Gender identification in face images using KPCA. In: 2009 World Congress on Nature & Biologically Inspired Computing, NaBIC 2009, Coimbatore (2009)
8. Chacko, V.R., Kumar, M.A., Soman, K.P.: Experimental study of gender and language variety identification in social media. Adv. Intell. Syst. Comput. **750**, 489–498 (2019)

9. Vinayakumar, R., Kumar, S.S., Premjith, B., Poornachandran, P., Padannayil, S.K.: Deep stance and gender detection in Tweets on Catalan Independence@Ibereval 2017. In: IberEval 2017 Evaluation of Human Language Technologies for Iberian Languages Workshop 2017, Murcia, Spain (2017)
10. Shan, C., Gong, S., McOwan, P.W.: Learning gender from human gaits and faces. In: IEEE Conference on Advanced Video and Signal Based Surveillance, pp. 505–510 (2007)
11. Thomas, V., Chawla, N.V., Bowyer, K.W., Flynn, P.J.: Learning to predict gender from iris images. In: IEEE International Conference on Biometrics: Theory, Applications, and Systems, pp. 1–5 (2007)
12. Amayeh, G., Bebis, G., Nicolescu, M.: Gender classification from hand shape. In: IEEE Computer vision and Pattern Recognition Workshop, CVPR, pp. 23–28 (2008)
13. Toews, M., Arbel, T.: Detection, localization and sex classification of faces from arbitrary viewpoints and under occlusion. IEEE Trans. PAMI **31**(9) (2009)
14. Fatkhannudin, M.N., Prahara, A.: Gender classification using fisherface and support vector machine on face image. Signal Image Process. Lett. **1**(1), 32–40 (2019)
15. Wu, J.H., Huang, P.S., Jian, Y.J., Fang, J.T.: Gender classification by using human nose features. Biomed. Eng. Appl. Basis Commun. **28**(05) (2016)
16. Shakhnarovich, G., Viola, P.A., Moghaddam, B.: A unified learning framework for real time face detection and classification. In: Proceedings of IEEE International Conference on Automatic Face and Gesture Recognition, pp. 14–21 (2002)
17. Rai, P., Khanna, P.: Gender classification techniques: a review. Adv. Comput. Sci. Eng. Appl. **166**, 51–59 (2012)
18. Khan, S.A., Nazir, M., Akram, S., Riaz, N.: Gender classification using image processing techniques: a survey. In: IEEE 14th International Multitopic Conference (INMIC), pp. 25–30 (2011)
19. Rothe, R., Timofte, R., Van Gool, L.: Deep expectation of apparent age from a single image. In: IEEE International Conference on Computer Vision Workshops (2015)
20. Rothe, R., Timofte, R., Van Gool, L.: Deep expectation of real and apparent age from a single image without facial landmarks. Int. J. Comput. Vision **26**, 144–157 (2018)

Chapter 4
Software Stack for Autonomous Vehicle: Motion planning

Sonali Shinde and Sunil B. Mane

Abstract An autonomous vehicle can operate itself without any human interaction through the capacity to sense its surroundings. It requires a navigation plan to travel avoiding collision and following all the traffic rules at the same time. Navigation issue is a computational issue to discover the succession of legitimate designs that move the object from source to destination. In the robotics field, this term is known as motion planning or path planning. There exist well-recognized methodologies for this problem; however, by applying some helpful heuristics, a better version of driving API can be designed. This study develops a complete software architecture needed for autonomous vehicles. It gives brief insights about available techniques for each module involved in motion planning and possible optimizations to achieve better results. The global planner uses a road network stored in open drive format to find the most eligible global path for ego vehicle. Whereas reactive local planner uses surrounding information to plan better paths avoiding static and dynamic obstacles. At last, the results are introduced in examination with existing methods to show improved measurements accomplished by this framework.

4.1 Introduction

4.1.1 Definition

An autonomous vehicle is a well-equipped vehicle for sensing its surroundings and driving without human inclusion. A human driver isn't required to be present in the vehicle to take control of driving under any circumstances. An autonomous vehicle can accomplish all the driving tasks that a human driver does.

S. Shinde (✉) · S. B. Mane
Department of Computer Engineering and Information Technology,
College of Engineering, Pune, Maharashtra, India
e-mail: sunilbmane.comp@coep.ac.in

J. I.-Z. Chen et al. (eds.), *Machine Learning and Autonomous Systems*, Smart Innovation, Systems and Technologies 269, https://doi.org/10.1007/978-981-16-7996-4_4

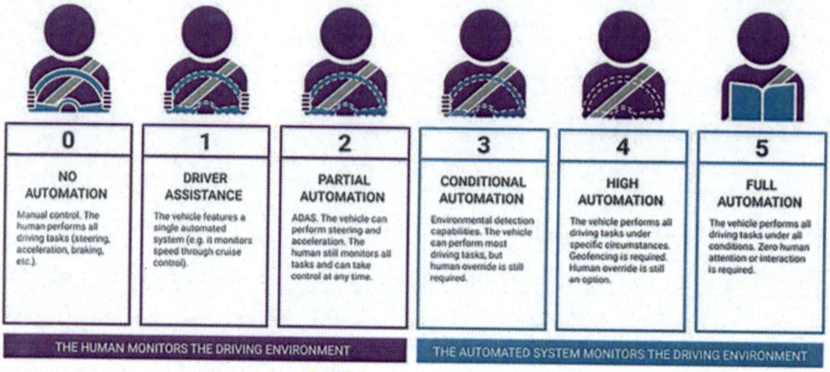

Fig. 4.1 Levels of driving automation [15]

4.1.2 Motivation

Intelligent vehicles advancement was restricted before the 90s on the grounds that diminished interests in the field [1]. Over the most recent couple of many years, both industry and the scholarly community have invested efforts into creating advancements for self-governing driving. The Society of Automotive Engineers (SAE) has characterized 6 levels of driving computerization. They range from Level 0 (completely manual) to Level 5 (completely self-ruling) [15] (Fig. 4.1).

4.1.3 Benefits of Autonomous Cars

The situations for quality life improvements of human beings and convenience are boundless. The older and the truly incapacitated would have freedom and easy access to transportation. In an transportation report [3], specialists have stated three trends—vehicle automation, vehicle electrification, and ride sharing. If these trends are incorporated simultaneously, they will help achieve maximum utilization of autonomous vehicles.

4.1.4 Necessity of Simulation

On the primary look, autonomous vehicles appear to be only a straightforward continuation of the advancement of help frameworks which help the driver to follow the desired track, holding the separation to different vehicles and staying away from mishaps, with the vision of avoiding 80 percent of accidents, all things considered, in light of the fact that they are principally brought about by human mistakes. Be that as it may, there is a gigantic test concerning the necessities of framework execution and unwavering quality for this progression. Schöner [2] A realistic simulation environment is a basic apparatus for building up a self-driving vehicle, since it permits us to guarantee that our vehicle will work securely before we even advance foot in it. We can test our vehicle in circumstances that would be hazardous for us to test on actual streets.

4.2 Related Work

There have been many techniques and methodologies developed/studied for different tasks involved in autonomous vehicle software development. Some of them are discussed briefly in the following section.

Researchers from Toyota research institute, Computer vision center, Barcelona, and Intel labs [4] have introduced CARLA. CARLA is an open-source simulator for autonomous driving research. CARLA has been created starting from the earliest stage to help advancement, preparing, and approval of self-governing metropolitan driving frameworks. Notwithstanding publicly available code and conventions, CARLA gives rich advanced resources (metropolitan designs, building structures, vehicles) which were produced considering this reason and can be utilized unreservedly. The simulation stage underpins adaptable particular of sensor suites and natural traveling conditions. This assists with examining the presentation of three ways to deal with autonomous driving: an exemplary particular pipeline, a start to finish model prepared through impersonation learning, and a start to finish model prepared by means of reinforcement learning.

Perez et al. [6] introduced a survey of motion planning strategies. A portrayal of the strategy utilized by field expert groups. They briefly talk about the comparative study among motion planning strategies. It covers important topics such as collision detection and avoidance techniques. It greatly helps to comprehend all the challenges to be addressed in all these motion planning strategies. Tianyu and Dolan [5] introduced a movement organizer for ego vehicle on-street driving, particularly on highways. It talks briefly about state lattice. An engaged hunt is acted in the recently distinguished locale wherein the optimal trajectory is likely to exist. This paper mainly gives insight into a computationally proficient planner to manage dynamic conditions conventionally. It uses Dynamic Programming method to investigate in local space and locate paths that can lead to desirable moves. At that point, an engaged direction search is led utilizing the "generate-and-test" approach, and the best path selection is done dependent on the basis of score of each path; computed considering various path planning parameters. Another crude for motion planning of autonomous cars is presented by Piazzi and Bianco [7]. It is a totally parameterized quintic spline, signified as rpspline, that permits injecting interpolated points satisfying the second-degree geometry overall. Issues, for example, minimality, regularity, evenness, and Jexibility of these G2-splines are tended to in the piece. This research is greatly associated to plan smooth paths for self-driving cars.

Vivek et al. [8] present near examination between LQR (Linear Quadratic Regulator), Stanley, and MPC (Model Predictive Controller). These are controllers for path following application acting under level 4 driving autonomy. Then controllers are tested under a bunch of weather conditions to measure the controller accuracy. The initial results are carried out in MATLAB and later actual testing is performed on IPG car to observe control behaviors. Stanley is more of a instinctive guiding controller and the later ones are developed further. All control activities are determined by upgrading the vehicle models. A comparative study is presented by observing deviation from the recommended path and measuring state errors by applying mentioned controllers under various kinematic variations. This research provides a definite thought regarding controllers with respect to their utilization, favorable circumstances, and impediments in this application.

4.3 Problem Statement

Identify primary components of autonomous vehicle software architecture. Develop self-driving API to navigate ego vehicle to follow recommended path in the simulator. Perform planning tasks in autonomous driving, including global and local planning. Finally integrate all these modules, to build a full autonomous vehicle driving API and to navigate from given source to destination adhering to all the traffic rules and protecting vehicle consistently.

4.4 Proposed System

4.4.1 Simulator

This paper suggests to use the CARLA simulator and to implement this system. It features profoundly detailed virtual universes with streets, buildings, weather, and vehicle and pedestrian actors. The entire simulation can be controlled with an external client who can be used to send commands to the vehicle. Best of all, CARLA democratizes autonomous driving R&D [4].

1. Version : CARLA 0.9.10.
2. Client API: Python 3.7.
3. Spawn player to act as ego vehicle: Tesla Model13 car.
4. Source location: Spawn location of ego vehicle.
5. Destination location: Select random spawn point.
6. Town selection: Town03(US), Town01(Europe).

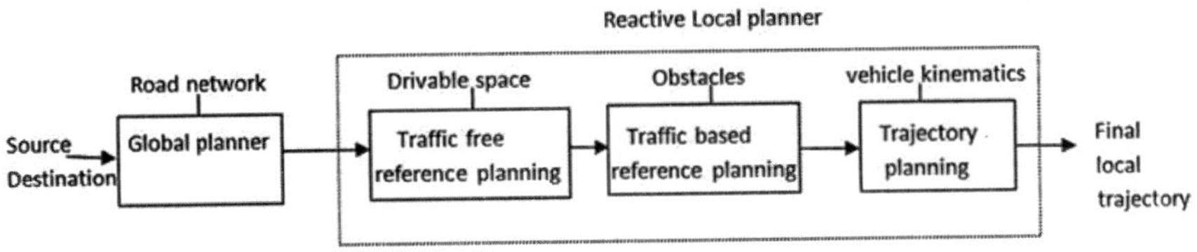

Fig. 4.2 Path planning architecture

4.4.2 *Path Planning*

Path planning deals with searching a plausible path and contemplating the geometry of the vehicle and its environmental factors, the kinematic constraints, and others that may influence the feasible paths (Fig. 4.2).

Broadly the path planning is divided into the following:

1. Global planning.
2. Local planning.

1. Global Planning Global planning is the most significant planning level. It aims at finding the shortest and optimum path from source to destination considering the road network for the underlying world. Global planning takes into account the basic constraints of the environment such as road map, path length, turning angles, and lane changes. The global path is formulated even before the vehicle steps into the actual driving task.

Global planning process:

1. Gather topology information from CARLA town, keeping way-points at 2unit distance (unit set as per open drive format).
2. Create road map using topology.
3. Convert road map into network diagraph.
4. Perform graph base A* search to find the shortest route among source and destination.

Heuristic function with admissibility property: distance heuristics.
Weight: accumulated arc length of path and number of signals encountered.

2. Local Planning Local planning generates dynamic strategies based on the real-time surrounding inputs coming from the dynamic environment. Both phases are significant subjects of exploration with the local planning having more open-ended inquiries. Local paths are getting refreshed continuously while executing the previous plan.

The system uses the legacy version of the state lattice planner as a base for local planner.

Heuristics to apply over legacy version-

1. Use road information to determine drivable space.
2. Optimize paths only to the necessary length.
3. Collision detection for the entire path.

Though every phase in local planning is highly interconnected, this paper subdivides local planner into the following tasks:

1. Traffic-free reference planning (Gathering road information).
2. Traffic-based reference planning (Behavior planning).
3. Trajectory planning (Velocity planning).

2.1 Traffic-Free Reference Planning It includes identifying all feasible paths between current vehicle position and local destination; considering all the intermediate state space for all way-points encountered in between. These recognitions are done based on the road network, by calculating drivable space.

1. Calculate Look Ahead Distance (LAD)
   LAD = look ahead time * open loop speed of ego vehicle.
   Look ahead time is a programmer's choice based on how far the ego vehicle needs to look.
2. Determine Look Ahead Point (LAP) such that
   arc length between(ego vehicle position, LAP) $\approx$ LAD.
3. Now, generate local paths using state lattice planning; considering LAP as reference way-point.
   Every intermediate point is represented as [x,y,z] tuple.
4. For each path point, store details about the associated signal, stop sign, and speed constraints (for later use)

2.2 Traffic-Based Reference Planning Local planner needs to select the best path among all available paths. A maneuver is a property associated with vehicle movement, incorporating the speed and location of the vehicle. Instances of maneuvers include taking turns, moving straight, moving to another lane, decelerating, following a lead vehicle, and overtaking. The best path needs to be chosen out of all available local paths, by considering static and dynamic obstacle collision checking and adhering to the global path.

1. Eliminate local paths leading to the static collision.
2. Select the best path out of static collision-free paths.
3. Perform dynamic collision checking based on the velocity profile.

Static collision checking Occupancy grid map plays a very crucial role in collision checking. An occupancy grid is a grid map plotted surrounding ego vehicle at center position [9]. This paper uses the 2D version of it. Every grid cell demonstrates if the obstacle is present or not. Conventionally Bayesian updates of occupancy grid beliefs are used to denote grid status. It rather stores obstacle information in circular form in each affecting grid. Occupancy grid needs to be refreshed at every time step to assure correct time to collision is taken into account and collision states are updated.

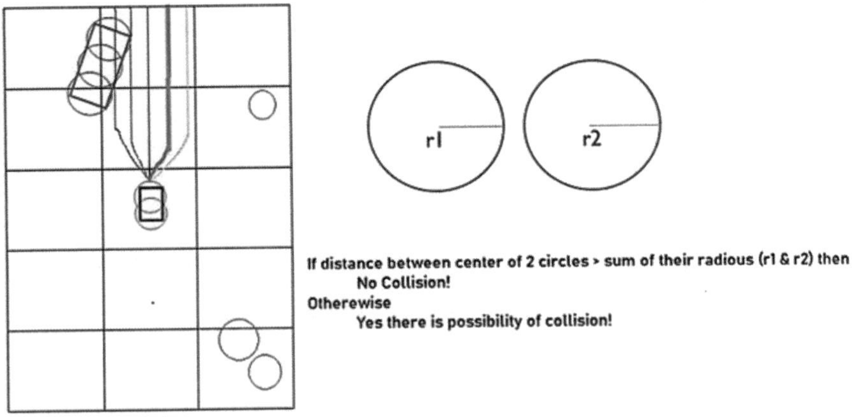

Fig. 4.3 Occupancy grid, circle to circle collision checking

Circle to circle collision checks will be performed for each point in each path to determine if the path is collision-free or not [11]. Steps to prepare occupancy grid map for current frame:

1. Prepare 30 × 60 empty grid around ego vehicle.
   Ego vehicle position = [0,0]
   Cell dimension = 5 × 5 unit.
2. Represent each static obstacle into packing circles.
   Tuple representing each circle: [x, y, r], where x, y denotes local co-ordinates (with reference to ego vehicle) of obstacle and r represents the radius.
3. Locate grid cell for each circle using 2D grid indexing. If any circle is affecting multiple grid cells, circle information will be replicated in each affecting grid cell; marking that circle as OCCUPIED.

The dark green path shown in the figure is selected as the best path. Because initial 3 paths (red) are eliminated due to collision possibilities.

This paper suggests the usage of multi-threading technique for parallel collision checking of multiple paths (Fig. 4.3).

Dynamic collision checking This API looks for dynamic obstacles in vicinity of 30 m. Velocity time graphs are generated for each dynamic obstacle using motion equations

$$s = ut + 0.5at^2 \tag{4.1}$$

given u = initial speed of moving obstacle, a = acceleration rate, and t = time

At each point in the time graph, consider the possibility of all feasible paths. Based on outcomes, the closest point affecting the dynamic vehicle is determined as the collision point. By applying a circular collision checking strategy, a new approximate collision point is determined. Time to collision is calculated using relative velocities of dynamic obstacle and ego vehicle. Finally, this data is sent to velocity planner for smooth trajectory planning [9].

2.3 Trajectory Planning Trajectory generation or velocity profile generation of ego vehicle. Velocity planner gets start and goal velocity and respective navigation state configurations. Similar to injecting interpolated way-points, it interpolates those velocities over all intermediate vehicle states, so the kinematic constraints of ego vehicle are satisfied and navigation is smooth.

Four important velocities under consideration are as follows:

1. Reference velocity provided by traffic-based reference planner (Vr).
2. The velocity of first dynamic obstacle affecting ego vehicle (Vd).
3. Velocity required to maintain vehicle stability (Vs).
4. Velocity imposed by CARLA way-point (speed limit constraints on road) (Vk).

Velocity profile for ego vehicle must satisfy the following constraint:

$$Vi \leq min(Vr, Vd, Vs, Vk) \tag{4.2}$$

for every ith way-point in final trajectory [9].

4.4.3 Controller

A robust control strategy is a key requirement of autonomous vehicle driving API. Control code consists of two controllers, lateral and longitudinal controllers, respectively. The lateral controller is needed to control the steering of the ego vehicle. And the longitudinal controller is needed to control the speed/velocity of the ego vehicle. Stanley controller is one of the most popular lateral path tracking controller. It was used by the Stanford racing team to win the second Darpa Grand Challenge event. Stanley controller relies on the front axle as a point of reference. It corrects for a large cross-track error and reaches to the goal state. Finally, the vehicle safely tracks the path in the final stages of the simulation.

Paper suggests applying PID control to the longitudinal vehicle model. It is given in the continuous-time domain. The velocity error is fed into the high-level controller, and the intended acceleration of the vehicle is output. To implement such a controller in software, we discretize it by altering the integral to a summation over time steps of a defined length. If neither the reference acceleration nor the estimated vehicle acceleration is available, the derivative term can be approximated with the finite difference over a set time step. The low-level controller creates the throttle and breaking signals in response to the high-level controller's computed acceleration [8, 9].

4.4.4 Code Conversion

Programming stack is at first implicit Python—for fast cycle—and afterward changed over to C++ for speed and direct equipment access. C++ programming language is regularly utilized for autonomous vehicle PC programming. It doesn't consume a lot of memory asset and contains next to no pointless code to make it run gradually. This is helpful when managing code that must have the option to rehash in a quick and decided manner. C++ is some of the time challenging, yet it performs very well on Linux, Macintosh, and Windows frameworks.

4.5 Limitations of Proposed System

This paper is using town01 and town03 maps from CARLA which are addressing Europe and the US Nation, respectively. To work with Indian traffic control rules, a suitable road network needs to be prepared using an open DRIVE format. Moreover, these towns provided by the simulator are imaginary. To test this implementation in real world, the respective road network needs to be loaded in advance.

During behavior planning, if a stop signal is encountered, the behavior planner considers this point as look ahead point for that particular iteration. It restricts the vision of ego vehicle.

This paper follows the cautious behavior of the driving agent. Be that as it may, there are better hybrid renditions conceivable dependent on different practices like aggressive or normal conduct (Fig. 4.4).

4.6 Simulation Results and Analysis

Simulation results are presented in below diagram.

(a) Global Path Generation Given source and destination global path (colored in white) is generated by the global planner.
(b) Local Path Generation Possible alternative local paths are generated in drivable space.
(c) Avoiding Static Obstacles The ego vehicle follows the best collision-free path and hence static collision is avoided.
(d) Following Lead Vehicle Ego vehicle is following lead vehicle considering the relative speed and distance.

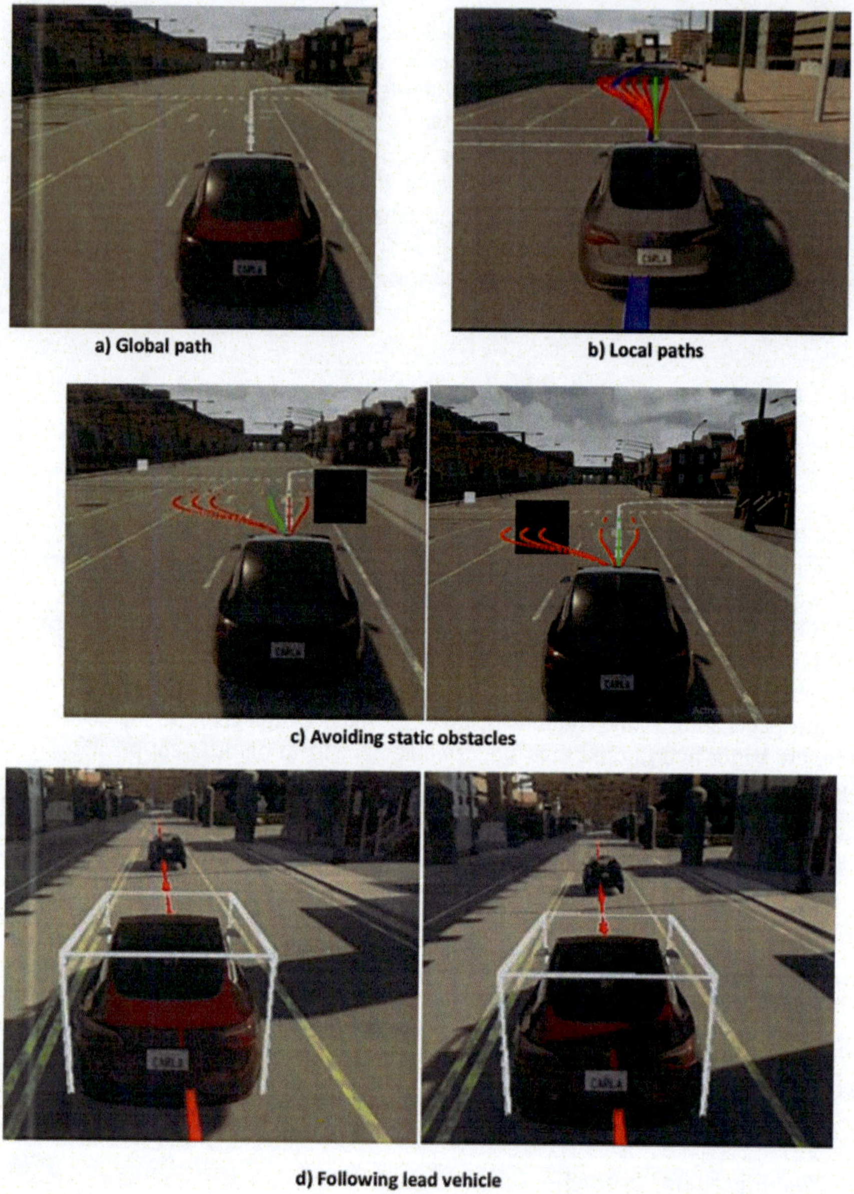

Fig. 4.4 Simulation results

4.6.1 Analysis

It's hard to perform a direct comparison on computation, for heuristic-based approaches. Comparison based on modifications suggested by this study is presented as follows:

1. Based on the Dijkstra algorithm. The A* graph search with network diagraph of the road map is heuristic reducing computation time [6].
2. Optimizing only the required length of path saves considerable computation cost.
3. When probabilistic occupancy grid is used with static obstacles, the intersection between ego vehicle and basic component falling in an occupied grid cell is evaluated. If a collision is confirmed, then a weight is associated with the element, according to its associated quantity of occupancy [16].
   If we look into the case where the static obstacle is a car and it affects 4 grids. The probability-based approach will have to look into finer details to determine what part of the vehicle may collide with the ego vehicle to choose collision-free path. But the circular obstacle approach presented in this paper will store actual centers of circle and [x,y] co-ordinates, we can determine collision status with minimal and very basic calculations.

4.7 Conclusion and Future Work

A realistic simulation environment is a basic apparatus for building up a self-driving vehicle, since it permits us to guarantee that our vehicle will work securely before we even advance foot in it. It enables testing of the vehicle in circumstances that would be hazardous for us to test on actual streets.

The proposed software stack serves as a driving API for autonomous vehicle in CARLA. It perceives all inputs from the environment and then processes all this perception knowledge, generates a feasible path, and sends instructions to the car's actuators, which control acceleration, braking, and steering. Obstacle avoidance techniques, predictive response system, predefined rules, and object identification inputs help the API to navigate vehicle safely. This paper has introduced simulations for level 2 automation. This will assist with robotizing the supply chain in the present day world. A well-tested vehicle in a simulation environment can catch defects earlier. Significant reasons for mishaps, including inebriated or diverted driving, won't be factors with self-driving vehicles. So self-driving vehicles can lessen mishaps altogether.

As a future work, the high-level language code needs to be transformed into lower level language code desired by hardware under consideration. Functional API needs to be updated to adapt real-time services, like mapping city road network in open drive format, so vehicle can be tested for that city. This API is highly modular and supports progressive vehicle testing under various scenarios.

References

1. Shladover, S., Desoer, C., Hedrick, J., Tomizuka, M., Walrand, J., Zhang, W., McMahon, D., Peng, H., Sheikholeslam, S., McKeown, N.: Automated vehicle control developments in the PATH program. IEEE Trans. Veh. Technol. **40**, 114–130 (1991)
2. Schöner, H.P.: Simulation in development and testing of autonomous vehicles. In: Bargende, M., Reuss, H.C., Wiedemann, J., (eds.), 18. Internationales Stuttgarter Symposium. Proceedings. Springer Vieweg, Wiesbaden (2018). https://doi.org/10.1007/978-3-658-21194-3_82
3. Three Revolutions in Urban Transportation. In: Transportation Matters, ITDP (2017). https://www.itdp.org/2017/05/03/3rs-in-urban-transport/
4. Dosovitskiy, A., Ros, G., Codevilla, F., López, A.M., Koltun, V.: CARLA: An Open Urban Driving Simulator (2017). arXiv:abs/1711.03938
5. Gu. T., Dolan, J.M.: On-road motion planning for autonomous vehicles. In: Su, C.Y., Rakheja, S., Liu, H., (eds.), Intelligent Robotics and Applications. ICIRA 2012. Lecture Notes in Computer Science, vol. 7508. Springer, Berlin (2012). https://doi.org/10.1007/978-3-642-33503-7_57
6. González, D., Pérez, J., Milanés, V., Nashashibi, F.: A review of motion planning techniques for automated vehicles. IEEE Trans. Intell. Transp. Syst. **17**(4), 1135–1145 (2015). https://doi.org/10.1109/TITS.2015.2498841
7. Piazzi, A., Guarino Lo Bianco, C.: Quintic G/sup 2/-splines for trajectory planning of autonomous vehicles. In: Proceedings of the IEEE Intelligent Vehicles Symposium 2000 (Cat. No.00TH8511), 2000, pp. 198–203 (2000). https://doi.org/10.1109/IVS.2000.898341
8. Vivek, K., Milankumar, A.S., Gumtapure, V.: A comparative study of Stanley, LQR and MPC controllers for path tracking application (ADAS/AD). In: IEEE International Conference on Intelligent Systems and Green Technology (ICISGT), vol. 2019, pp. 67–674 (2019). https://doi.org/10.1109/ICISGT44072.2019.00030
9. Waslander, S., Kelly, J.: Self-driving Cars Specialization. Offered by University of Toronto. https://www.coursera.org/specializations/self-driving-cars
10. González-Sieira, A., Mucientes, M., Bugarín, A.: A state lattice approach for motion planning under control and sensor uncertainty. In: Armada, M., Sanfeliu, A., Ferre, M. (eds.), ROBOT2013: First Iberian Robotics Conference. Advances in Intelligent Systems and Computing, vol. 253. Springer, Cham (2014). https://doi.org/10.1007/978-3-319-03653-3_19
11. González, D., Pérez, J., Lattarulo, R., Milanés, V., Nashashibi, F.: Continuous curvature planning with obstacle avoidance capabilities in urban scenarios. In: 17th International IEEE Conference on Intelligent Transportation Systems (ITSC), pp. 1430–1435 (2014). https://doi.org/10.1109/ITSC.2014.6957887
12. Dolgov, D., Thrun, S., Montemerlo, M., Diebel, J.: Path planning for autonomous driving in unknown environments. In: Khatib, O., Kumar, V., Pappas, G.J. (eds.), Experimental Robotics. Springer Tracts in Advanced Robotics, vol. 54. Springer, Berlin (2009). https://doi.org/10.1007/978-3-642-00196-3_8
13. Ekinci, M., Thomas, B., Gibbs, F.: Knowledge-based navigation for autonomous road vehicles. Turk. J. Electr. Eng. Comput. Sci. **8**, 1–29 (2000)
14. Kuwata, Y., Karaman, S., Teo, J., Frazzoli, E., How, J., Fiore, G.A.: Real-time motion planning with applications to autonomous urban driving. IEEE Trans. Control Syst. Technol. **17**, 1105–1118 (2009)
15. The 6 Levels of Vehicle Autonomy Explained. Article published by synopsys.com https://www.synopsys.com/automotive/autonomous-driving-levels.html
16. Rummelhard, L., Nègre, A., Perrollaz, M., Laugier, C.: Probabilistic grid-based collision risk prediction for driving application. In: ISER, Marrakech/Essaouira, Morocco. ffhal-01011808f (2014)
17. Jacob, I.J., Ebby Darney, P.: Artificial bee colony optimization algorithm for enhancing routing in wireless networks. J. Artif. Intell. **3**(01), 62–71 (2021)
18. Smys, S., Basar, A., Wang, H.: Artificial neural network based power management for smart street lighting systems. J. Artif. Intell. **2**(01), 42–52 (2020)

Chapter 5
Transformers for Speaker Recognition

Kayan K. Katrak, Kanishk Singh, Aayush Shah, Rohit Menon, and V. R. Badri Prasad

Abstract Audio, when seen in retrospect, though extensively researched upon, hasn't been fruitful enough in this era of informatization. Though technological advances in this field have proven to be effective in many applications, we see a tone that is outmoded. One important application of speaker recognition technology is in biometrics where a person's voice is used to identify them uniquely using different characteristics like the behavior and physiological nature of the speaker's audio and its implicit waveform. So far, we have seen commendable performances at the identification of a person, but they come with constraints in dialect with language modeling and lack of critical resources. We try to remove this constraint and make it truly universal. We use the essential property of attention from transformer technology and a novel combination of denoising techniques and data augmentation on a self-made dataset which contains speech samples of the Indian accent to achieve this.

5.1 Introduction

Audio, the main sensory information we receive, helps us perceive our surroundings. Almost every person or event around us has a different and unique sound. Audio can be broken down into 3 main attributes which help us distinguish between them. These attributes are Amplitude—loudness of the audio, frequency—pitch of the audio and timbre—the quality of the audio also known as the identity of the sound [1]. As humans we are subconsciously trained to recognize and distinguish different speakers and sound events. However, giving audio recognition to a machine can be very challenging. Factors like background noise or transducer noise can significantly impact the performance of a model. Dialect related features like various amplitudes varying over time for different speakers is a possible constraint in making any speaker

K. K. Katrak (✉) · K. Singh · A. Shah · R. Menon · V. R. Badri Prasad
PES University, Bangalore, India

V. R. Badri Prasad
e-mail: badriprasad@pes.edu

J. I.-Z. Chen et al. (eds.), *Machine Learning and Autonomous Systems*, Smart Innovation, Systems and Technologies 269, https://doi.org/10.1007/978-981-16-7996-4_5

recognition system [2, 3] truly universal. One of the many crucial challenges to overcome is the limited number of samples of the same user that can be trained in the algorithm to classify in real time.

The above challenges have been addressed in this paper. The pre-processing of the audio files is crucial in determining the efficiency of the model. The spectral gating noise reduction technique helps us clean the data and remove the noise, transducer as well as background, from the audio samples [4]. To help scale our dataset, without having the need to collect multiple samples, is achieved using data augmentation techniques like time stretching and pitch shifting to create new and different training samples and broaden the scope of our trainable samples [5, 6]. An important feature that needs to be extracted, when studying audio, is the Mel-Frequency Cepstral Coefficients (MFCCs) [7, 8]. After the samples have been pre-processed, we begin extracting the features like MFCCs from the audio. The decision of using previously experimented models to train our pre-processed training samples, to determine the efficiency of our methods, was crucial to making a fool proof pre-processing embedding. On surveying models, we found that the Long Short Term Memory (LSTM) model is more effective than Convolution Neural Network (CNN) models when taking input as an audio stream because of the state-based nature of sequential data and since LSTM has memory retention capabilities, it shows better classification accuracies [7, 9]. Speaker audio has Spectro-temporal characteristics which are essential for its classification and can be better identified using LSTMs. We further utilize the attention mechanism of transformers. Transformers can extract information in terms of context from speech and represent them in terms of embeddings. These embeddings are then manipulated by the transformer to clearly differentiate between speakers and give the entire prediction subsystem better prediction power.

Through this paper, we propose the following workflow (Fig. 5.1) that was simultaneously implemented while experimenting with the transformer model. This model along with the novel combinations of data pre-processing and augmentation techniques achieved an accuracy of 94.6% with the Indian accented self-made dataset. Our research was aimed at finding a more efficient and computationally cost-effective method to help identify a speaker using minimal features of his voice. This was achieved by using a combination of pre-processing techniques, namely, data augmentation and the spectral noise gating algorithm. All these methods helped to enhance the accuracy of the model and gave a better understanding of what was set out to be achieved. The traditional use of vanilla Recurrent Neural Networks (RNN) has become obsolete in its idea of propagation and combination of signals in the final hidden state, which was in fact the main power of an RNN. The idea of using transformers for this made much more of an impact on our understanding of the problem.

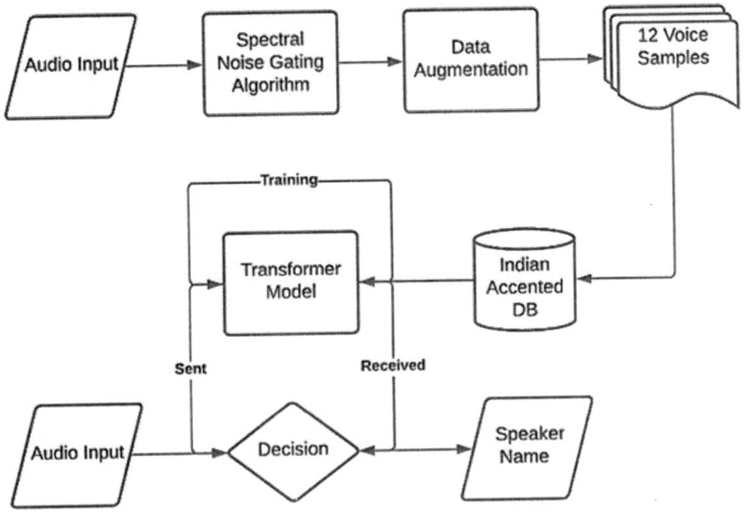

Fig. 5.1 Proposed workflow

5.2 Related Work

This section is divided into different modules which help describe the core idea of our model starting with Pre-processing and data augmentation, different deep learning models and then showing the work on transformers.

5.2.1 Pre-processing and Data Augmentation

This section will shed light on different models studied to clean our audio samples and addresses the challenge of data scarcity.

Pohjalainen et al. [4] gave us an overview of the techniques that are available at our dispense to help pre-process the audio signal better so that it helps our machine learning systems to better process the signal, in this case, audio. It highlights the different techniques for audio noise removal methods in the cepstral domains and a comparison of these with the standard techniques. The proposed methods in this paper are found to outperform the generic noise removal methods. In the simulated environment, Adaption when combined with temporal smoothing outruns all the other approaches.

Kiapuchinski et al. [10] impart a good understanding on how to remove noise from clips whose recording is taken in a natural environment using the spectral noise gating technique. The technique used, removes noise in real time. Some important features, statistical in nature, are computed along with the recording. This method was used to remove the additional work of cleaning the audio post recording to classify bird

Fig. 5.2 Spectral gating algorithm [4]

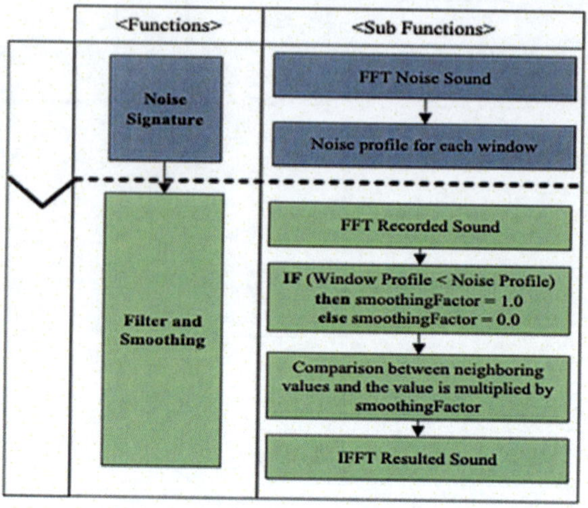

species from the recordings of bird songs. This method is crucial in the classification as classification is highly impacted with the type of audio that is taken as input to the model. While laboratory tests, which are in a simulated environment work extremely well with the machine learning model, 50–60% of the audio is affected by the surrounding sound of the wind and other sounds which we classify as noise. These are natural and cannot be tamed when it comes to audio input and must be handled carefully to help improve recognition accuracy (Fig. 5.2).

In most of the methods, a manual and a more tedious process is rendered to eliminate noise by using hardware like noise filters, etc. to make the audio ready to be processed. These are some challenges when it comes to real-time classification with the application of automatic methods.

As a mere drawback, the processing time of the complete embedded environment which includes the noise reduction, feature extraction, and the embedded classifier was found to exceed normal timelines in a real-time system and is also mentioned as a possible future work option in this domain.

Salamon et al. [6] attempt to use audio data augmentation to solve the impediment of lack of data and explore the effects of various augmentations on the outcomes of a CNN architecture for environmental sound classification. The dataset used by them is the UrbanSound8K dataset. Four audio data augmentation techniques were applied that resulted in 5 augmentation sets:

Time Stretching (TS): to increase or reduce the speed of the audio sample while the pitch remains the same.
Pitch Shifting (PS1): to elevate or reduce the pitch of the sample.
Pitch Shifting (PS2): additional shifting of the sample by larger values since the initial shifting proved to be useful.
Background Noise (BG): the sample is amalgamated with another recording featuring noises in the background containing several types of acoustic scenes.

Dynamic Range Compression (DRC): the dynamic range of the sample is compressed using four arguments.

The above-listed augmentations were applied to the UrbanSound8K dataset which consists of environmental noises which are grouped in 10 different classes. The effects of the augmentations on each different class were observed. The deltas in the accuracies of the classification model due to the various augmentation techniques are applied as visualized in Fig. 5.3.

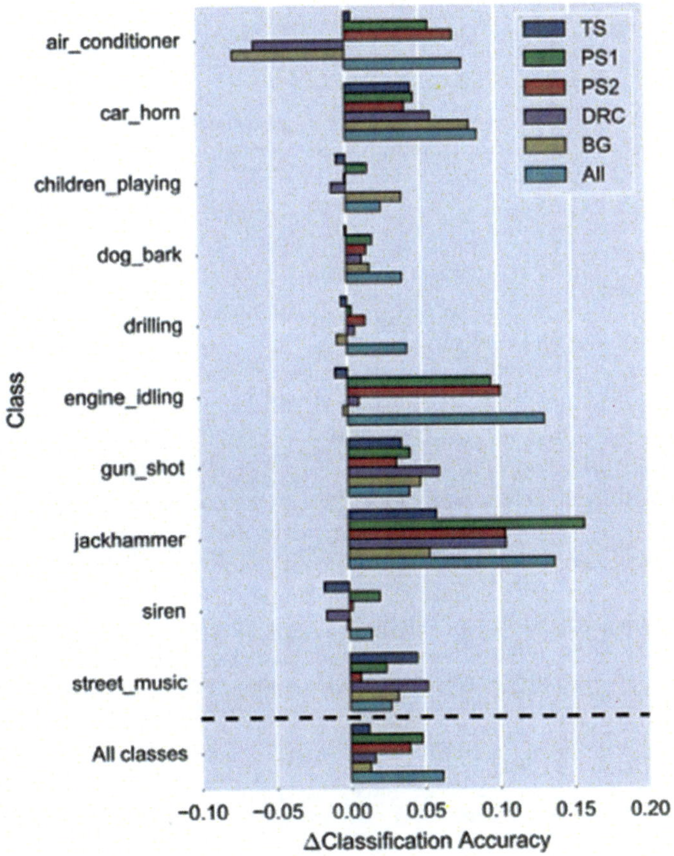

Fig. 5.3 Results of data augmentation [6]

5.2.2 Models

This section advocates the use of Deep Learning techniques when dealing with audio and what are the steps followed to do so, from audio processing to feature extraction to training.

Purwins et al. [7] broaden our scope into domains of arts like music and sounds of the environment that are used in it for their commonality and possible cross-fertilization between areas. It highlights general methods and key references in this domain. This paper reviews representations like audio waveforms and the Mel Spectrogram used along with CNN's and LSTM variants and additional audio specific neural network models. In many ways, the renaissance of deep learning was triggered in 2012 through breakthroughs in Image and Audio classification giving rise to research into CNNs, LSTMs, and deep feedforward networks. Even though signal processing approaches solved many problems in recognition in past decades [11], deep learning, in the recent years, has shown to outperform legacy models of audio classification and signal processing. Initially, being widely used in image classification, deep learning has now found its way into the area of sound classification too. As a result, methods like Gaussian MMs, Hidden MM's have been beaten in their nature of working by methods of Deep Learning, provided there exists a sufficient supply of reasonable and relevant data.

Lezhenin et al. [9] try to classify environmental sounds by using deep layered networks; primarily the LSTM neural network and weighing its performance against the standard CNN. Environmental sounds by their nature are difficult to characterize by employing a neural network. However, the presence of strong Spectro-temporal patterns makes classification possible. LSTM's take advantage of this feature for classification and provide results superior to other neural networks.

LSTM models work on retaining memory of previous layers so that the impact of these layers will not be forgotten while calculating the weights of current layers (Vanishing Gradient Problem). This is achieved by using a memory cell which considers the weights of all other previous layers. Total number of layers is 3 (2 LSTM + 1 dense) with a SoftMax activation function. The Loss function employed was Categorical Loss entropy and the optimizer used was Adam. Training for 10 distinct classes was done for 20 epochs using fivefold cross validation.

The crux of our research is to determine the speaker of the audio input as opposed to the general type of audio sampled, in which case a CNN using data augmentation taking CRP images with Mel-spectrograms can show state-of-the-art accuracies.

5.2.3 Transformers

In this section, we will see how transformers are impacting our understanding of this subject.

Devlin et al. [12] enlist details about how bidirectional models can carry more semantic and contextual information than traditional methods for language modeling and subsequent tasks. The limitation of many state-of-the-art language models is that word vectors are represented one dimensionally which makes it a hindrance for the type of pre-training we can do with respect to the models. This will snowball into further sub-optimal results because fine-tuning is a crucial part for detection heads or any other task, to be accurate in terms of predictions.

Bidirectional Encoder Representations from Transformers or BERT improves the pre-training process by bypassing the unidirectionality constraint of regular language models [13]. It employs a masked language model pre-training phase that masks certain words within the training language vocabulary and the loss function hinges on predictions made for these masked words. The BERT system works essentially on two processes: pre-training and fine-tuning. During the pre-training stage, the model is subjected to large amounts of unlabeled data that is used to perform several pre-training tasks. Once pre-training is done, the weights and hyperparameter information are stored for the next stage, i.e., fine-tuning. This uses labeled data to fine-tune the said weights and hyperparameters for improving overall accuracy by performing several downstream tasks (Fig. 5.4).

The downstream tasks employ self-attention, more specifically Bidirectional Self-attention. Self-attention refers to the machine comprehension process where words of a sentence are embedded with relationship vectors to the words before it and these vectors/features are trained as sentence progression continues. Bidirectional implies that word vectors are embedded not only for words before the current word, but also those after it as explained in Fig. 5.5.

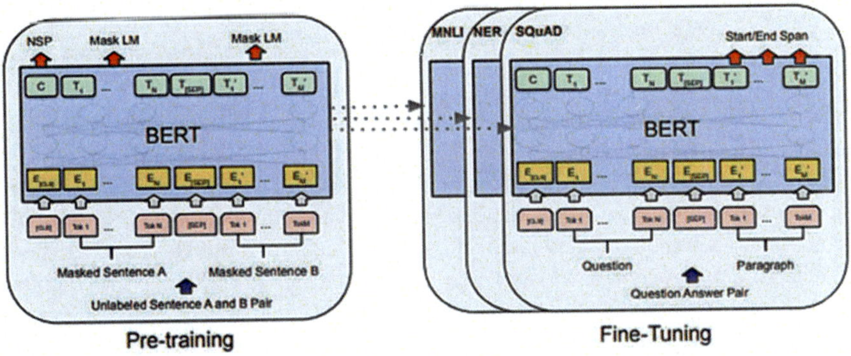

Fig. 5.4 BERT Architecture [12]

Fig. 5.5 Attention in a sentence [13]

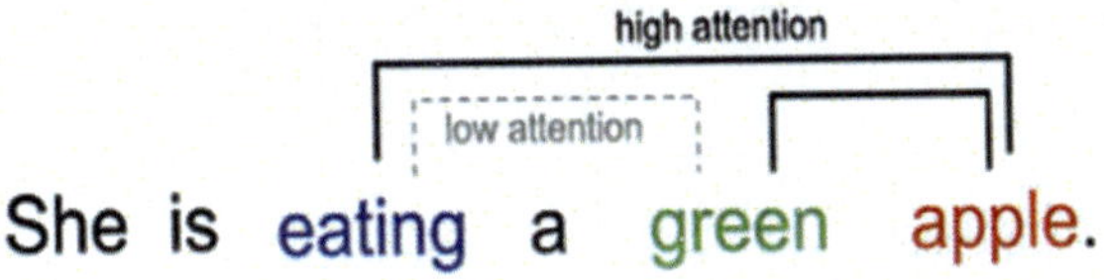

5.3 Methodology

5.3.1 *Dataset*

Our model works with a limited dataset which keeps building as and when a new speaker is added into the system using data augmentation. This means that we have a constantly growing dataset, which over time will improve the model's accuracy. We first tested our models with an already existing dataset of speakers. We used the LibriSpeech dataset, which comprised of 251 different speakers, all with the US English accent. Each speaker recording is around 8–30 min of clean, read speech which gave us a lot of data to train and test on. The audio samples are jumbled, and the corpus is split into 70% for training, 15% for validation, and the rest for testing.

Our self-made dataset mainly comprises Indian accented English. On recording each sample, they are passed through a spectral gating algorithm for noise reduction followed by 4 data augmentation techniques, i.e., Time Shifting, Time Stretching, Pitch Shifting, and Noise Addition, each consisting of different factors, resulting in 12 different copies of our input sample, the differences between which, are not noticeable to the naked ear but helps increase our dataset and improve training.

This tackles our problem of data scarcity and not having to ask each user to record minutes' worth of audio to improve the model. Every time a user works on the model, he automatically populates the dataset hence improving the model in each run.

5.3.2 *Pre-processing and Augmentation*

As our survey states [4, 10, 14], this section in our research is very crucial to the final outcome and in improving the accuracy of our model. A spectral noise gate is now commonly used in audio subsystems to remove background noise with considerable precision [15]. A Short Time Fourier Transform is applied on the audio clip [16] with a window length of 2048 and hop length of 512. Once applied, statistics like the mean and std. dev of the noisy audio clip are calculated (in frequency). A threshold is determined based on the loudness that our model can use to differentiate between noise and important signal. We then used a mask that can be applied to the original audio clip where it attenuates the noise (low frequency) which is determined by the threshold (1.5 std. dev. of the mean noisy audio clip) we have given to our model. This mask, smoothed with a filter over the time and frequency domains, is then applied to the parts which cross the threshold. We recover the clean audio clip by then applying an inverse STFT. The same has been depicted in Fig. 5.6.

Regarding populating the dataset, data augmentation techniques were implemented. These take a sample audio and modify its characteristics by a small factor. The techniques we used were Time Shifting, Time Stretching, Pitch Shifting, and Noise Addition. This creates 12 different copies of our input sample, each differing

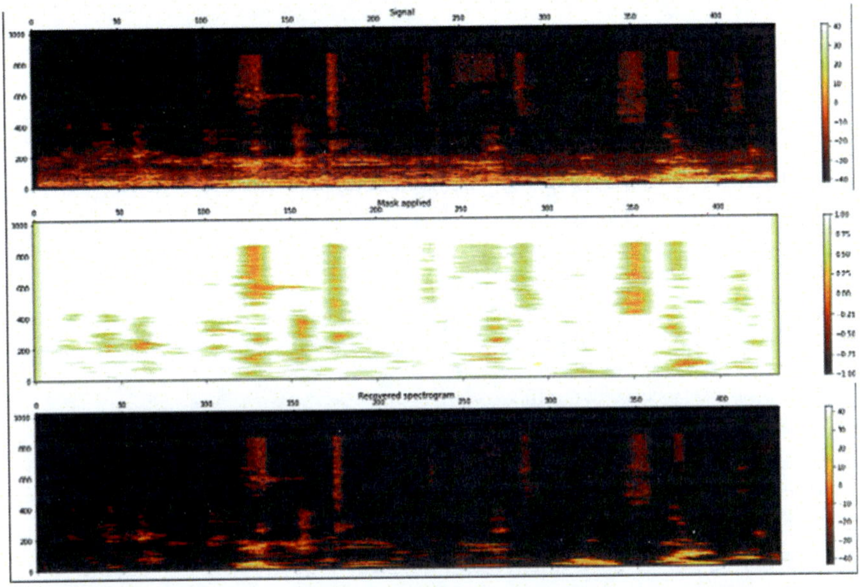

Fig. 5.6 (Top) Audio input represented with a spectrogram. (Middle) Mask applied to noisy audio clip. (Bottom) Recovered Spectrogram after removal of noise

by small factors not noticeable to the naked ear but helps increase our dataset and improve training.

5.3.3 Model

A basic vanilla RNN model was built and trained on the LibriSpeech dataset. After witnessing the model's success, we began to train it with our own recordings, which is a corpus of Indian accented audio samples. We then create a transfer learning setup where the initial weights are shifted and customized during the second stage of learning where the user input is given for classification. Furthermore, the second iteration of the model is embedded with a transformer. The transformer extracts attention from speech from the user input. Attention refers to context stored dynamically for every user input, what was said before and after a said audio frame, and how it was said. This creates a discriminatory metric for the prediction subsystem and is hence able to utilize feature embeddings and attention vectors of those embeddings to make accurate predictions of the user class.

Vanilla RNN Model. We began the implementation with training and testing the LibriSpeech dataset on a vanilla RNN model with 3 ReLU activation layers, followed by dropout layers after every activation layer and finally a SoftMax layer to predict the output. The loss function used is a categorical cross entropy function with an Adam

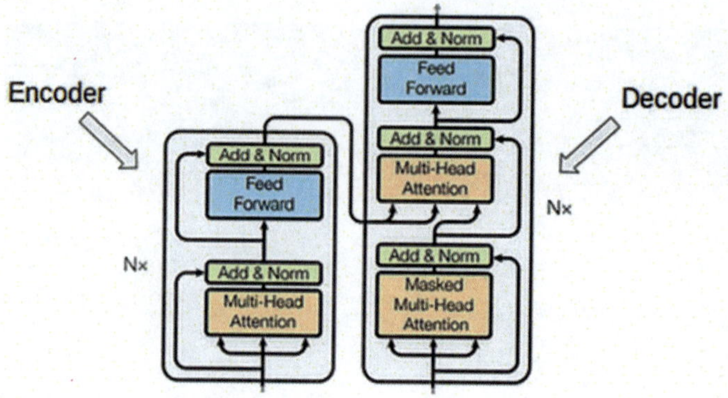

Fig. 5.7 Encoder and decoder in transformer [20]

optimizer. We also added the EarlyStopping callback function to avoid overfitting the model.

Transformer Model. Transformers help transform one sequence into another with the help of an Encoder and a Decoder, but this differs from the previously described RNN model [17, 18]. It uses attention mechanism without any RNN which has been proved to improve the performance [19]. However, its usage has only been tested in the field of text summarization, translation, and other Natural Language tasks [13].

In an Attention-based memory mechanism, at every encoder step, a hidden state which carries information about the context is carried and transformed. These hidden states are summed up with respect to some weights and are used as context in the form of a vector. Finally, the decoder then generates a target sequence from this vector. Each Encoder consists of a multi-headed attention block and a feedforward block and the Decoder is similar in architecture except for the addition of the masked multi-headed attention block (Fig. 5.7).

The high-level view above shows the architecture which consists of several such blocks stacked together, the number of which, is often a hyperparameter. It is important here, to understand the concept of self-attention, which is crucial at the decoder stage. Self-attention allows the model to look at the other words in the input sequence to get a better understanding of a certain word in the sequence. A transformer's main technical detail to help it understand better is its 'Attention'. This attention mechanism analyses the input sequence and decides at every step which other parts of the sequence are truly important [20].

In our model we have used the attention technique: Scaled dot-product attention with Multi-headed attention, where we have several self-attention layers within the transformer which run in parallel (Fig. 5.8).

Scaled Dot-Product Attention input consists of queries and keys of dimension d_k, and values of dimension d_v. The Query, Key, and Value are packed into matrices Q, K and V respectively [18]. Scaled dot-product attention is thus

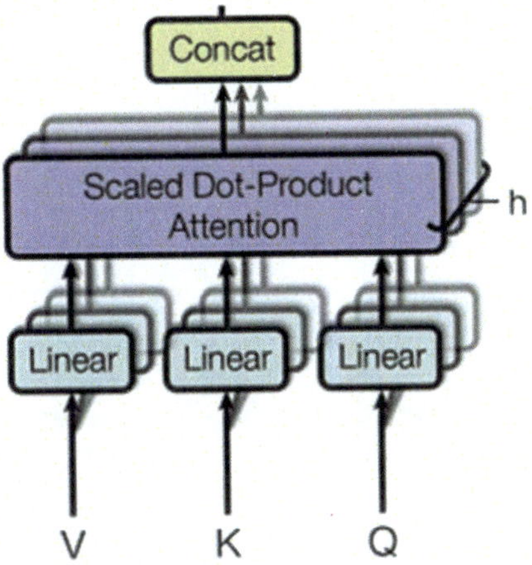

Fig. 5.8 Multi-headed attention with scaled dot-product attention vectors [17]

$$\text{Attention}(Q, K, V) = \text{Softmax}\left[\frac{QK^TV}{\sqrt{d_k}}\right] \tag{5.1}$$

Multi-headed attention learns these parameters independently and linearly, to increase computational efficiency. Multi-headed attention allows the model to jointly attend to information from different representation subspaces at different positions [18]. With a single attention head, averaging inhibits this.

$$\text{MultiHead}(Q, K, V) = \text{Concat}(\text{head}_1, \ldots, \text{head}_h)W^0 \tag{5.2}$$

where

$$\text{head}_i = \text{Attention}(QW_i^Q, KW_i^K, VW_i^V) \tag{5.3}$$

$$W_i^Q, W_i^K, W_i^V \in \mathbb{R}^{d_{\text{model}} \times d_k} | W^0 \in \mathbb{R}^{hd_v \times d_{\text{model}}} \tag{5.4}$$

The input is sent to the Input Embedding and Positional Encoding layers. This produces a representation in terms of word vectors and therein captures position. After the Key, Query, and Values are sent to the encoder the representation is additionally embedded with the attention score.

The Position Encoding is computed independently of the input sequence [13]. These are fixed values that depend only on the max length of the sequence. For instance, the first item is a constant code that indicates the first position, the second item is a constant code that indicates the second position, and so on. These constants are computed using the formula below, where *pos* is the position of the word in the sequence, d_{model} is the length of the encoding vector (same as the embedding vector) and i is the index value into this vector.

$$PE_{(pos,2i)} = \sin\left(\frac{pos}{1000^{\frac{2i}{d_{\text{model}}}}}\right) \tag{5.5}$$

$$PE_{(pos,2i+1)} = \cos\left(\frac{pos}{1000^{\frac{2i}{d_{\text{model}}}}}\right) \tag{5.6}$$

At the decoder stack, the target sentence is fed to the Output Embedding and Positional Encoding. The resulting target is then normalized and fed to the Query parameter in the Encoder-Decoder Attention of the primary Decoder.

In the transformer model as well, we use a SoftMax activation layer in the final output layer, Categorical Cross entropy as the loss function and an Adam Optimizer. We also add in the EarlyStopping callback function to avoid overfitting of the model.

5.4 Experimentation Results and Discussion

While both the models show commendable performances, as shown in Table 5.1, on running the Vanilla RNN model with the LibriSpeech dataset, an accuracy of 98.2% was observed. When implemented with our own dataset and combination of pre-processing techniques, the model accuracy was observed to be 97.3% with the vanilla RNN. The transformer model with the LibriSpeech dataset gave us an accuracy of 84.33%. On further experimentation with our self-made Indian accented dataset, an accuracy of 94.6% was achieved.

While the number of classes has increased from a minimum of 4 to a maximum of 251, it was found that the accuracy saw a decline from 94.6 to 84.33%. It was observed that the tradeoff between the exponential increase in the number of classes versus the minimum drop in accuracy gives an assurance that the model can be scalable and achieve its desired outcome simultaneously.

On comparing our model with the techniques and features used in the audio field today, we see that our model has a promising accuracy as compared to the rest. Reducing the dependencies on many features of the user's speech, we have made a foolproof model with a novel combination of pre-processing techniques.

Table 5.2 lists the accuracies of present models and the features they are trained on, to classify audio samples, along with the addition of our own model accuracies, to show a distinction between our proposed method and the present norm.

Table 5.1 Accuracy trade-offs between models and datasets

Model/dataset	LibriSpeech dataset (%)	Our dataset (%)
Vanilla RNN	98.2	97.3
Transformer	84.33	94.6

Table 5.2 Classification accuracy on audio datasets

Classifier	Features	Accuracy (%)
SVM	Mel-bands and MFCC	70
CNN	log mel-spectogram	73
CNN + Aug	log mel-spectogram	79
CNN (GoogleNet)	Mel-spectogram, MFCC, CRP images	93
RNN	Mel-spectogram	98.2
Proposed transformer model	Mel-spectogram	94.6

5.5 Conclusion

The novel combination proposed in this paper makes it a seminal work in the field of audio with the Indian accent. We observed an accuracy of 94.6% which shows promise and we believe that combinations of better noise reduction algorithms, better pre-processing techniques, and some other deep learning models combined with our model can help improve on the accuracies achieved by this research. This research was aimed at building a good model with considerable accuracy and we believe that further experimentations and forensics, will help make this infallible and, help the field of Audio in its endeavors to improve indefinitely.

References

1. https://towardsdatascience.com/sound-event-classification-using-machine-learning-8768092beafc
2. Mamyrbayev, O., Mekebayev, N., Turdalyuly, M., Oshanova, N., Medeni, T.I., Yessentay, A.: Voice identification using classification algorithms, intelligent system and computing. IntechOpen, August 21, 2019. https://doi.org/10.5772/intechopen.88239
3. Yu, Y.: Research on speech recognition technology and its application. Int. Conf. Comput. Sci. Electron. Eng. **2012**, 306–309 (2012). https://doi.org/10.1109/ICCSEE.2012.359
4. Pohjalainen, J., Ringeval, F.F., Zhang, Z., Schuller, B.: Spectral and cepstral audio noise reduction techniques in speech emotion recognition. In: Proceedings of the 24th ACM international conference on Multimedia (MM 2016), pp. 670–674. Association for Computing Machinery, New York, NY, USA. https://doi.org/10.1145/2964284.2967306
5. Nanni, L., Maguolo, G., Paci, M.: Data augmentation approaches for improving animal audio classification from the ecological informatics journal. Ecol. Inform. December 2019. https://doi.org/10.1016/j.ecoinf.2020.101084
6. Salamon, J., Bello, J.P.: Deep convolutional neural networks and data augmentation for environmental sound classification. IEEE Signal Process. Lett. **24**(3), 279–283 (2017). https://doi.org/10.1109/LSP.2017.2657381
7. Purwins, H., Li, B., Virtanen, T., Schlüter, J., Chang, S., Sainath, T.: Deep learning for audio signal processing. IEEE J. Sel. Top. Signal Process. **13**(2), 206–219 (2019). https://doi.org/10.1109/JSTSP.2019.2908700

8. Muda, L., Begam, M., Elamvazuthi, I.: Voice recognition algorithms using mel frequency cepstral coefficient (MFCC) and dynamic time warping (DTW) techniques. J. Comput. **2**(3), March 2010. ArXiv Prepr. arXiv:1003.4083
9. Lezhenin, I., Bogach, N., Pyshkin, E.: Urban sound classification using long short-term memory neural network. In: Conference on Computer Science and Information Systems, pp. 57–60 (2019). https://doi.org/10.15439/2019f185
10. Kiapuchinski, D.M., Lima, C.R.E., Kaestner, C.A.A.: Spectral noise gate technique applied to birdsong preprocessing on embedded unit. IEEE Int. Symp. Multimedia **2012**, 24–27 (2012). https://doi.org/10.1109/ISM.2012.12
11. Manoharan, S., Ponraj, N.: Analysis of complex non-linear environment exploration in speech recognition by hybrid learning technique. J. Innov. Image Process. (JIIP) **2**(4), 202–209 (2020). https://doi.org/10.36548/jiip.2020.4.005
12. Devlin, J., Chang, M.-W. et al.: BERT: pre-training of deep bidirectional transformers for language understanding. In: Gill, C. (eds.) Computation and Language, Cornell University Automatic Speaker Recognition using Transfer Learning. https://doi.org/10.18653/v1/n19-1423
13. https://towardsdatascience.com/bert-explained-state-of-the-art-language-model-for-nlp-f8b21a9b6270
14. Berdibaeva, G.K., Bodin, O.N., Kozlov, V.V., Nefed'ev, D.I., Ozhikenov, K.A., Pizhonkov, Y.A.: Pre-processing voice signals for voice recognition systems. In: 2017 18th International Conference of Young Specialists on Micro/Nanotechnologies and Electron Devices (EDM), pp. 242–245 (2017). https://doi.org/10.1109/EDM.2017.7981748
15. https://timsainburg.com/noise-reduction-python.html#noise-reduction-python
16. https://towardsdatascience.com/extract-features-of-music-75a3f9bc265d
17. https://medium.com/inside-machine-learning/what-is-a-transformer-d07dd1fbec04
18. https://en.wikipedia.org/wiki/Transformer_(machine_learning_model)#:~:text=The%20Transformer%20is%20a%20deep,as%20translation%20and%20text%20summarization
19. https://towardsdatascience.com/music-genre-classification-transformers-vs-recurrent-neural-networks-631751a71c58
20. Vaswani, A., Shazeer, N., Parmar, N., Uszkoreit, J., Jones, L., Gomez, A.N., Kaiser, L., Polosukhin, I.: Attention is all you need. In: NIPS 2017. ArXiv Prepr. arXiv:1706.03762

Chapter 6
Real-Time Face Mask Detection Using MobileNetV2 Classifier

A. Vijaya Lakshmi, K. Praveen Kumar Goud, M. Saikiran Kumar, and V. Thirupathi

Abstract Since the flare-up of the COVID-19 pandemic, generous advancement has been made in the field of PC vision. The coronavirus COVID-19 pandemic is producing a worldwide health crisis, according to the World Health Organization, and the most effective safety approach is to wear a face mask in communal locations (WHO). Several methods and strategies have been used to develop face detection models. The suggested method in this research employs deep learning, TensorFlow, Keras, and OpenCV to detect face masks. Since it is somewhat asset proficient to introduce, this model might be utilized for wellbeing contemplations. The SSDMNV2 approach uses Single-Shot Multi-box Detector as a face detector and MobilenetV2 architecture as a framework for the classifier. MobileNets are built on a simplified design that builds low-weight deep neural networks using depth-wise separable convolutions. Using a face mask detection dataset, a real-time face mask identification from a live stream using OpenCV is accomplished. Our objective is to use computer vision and deep learning to determine whether or not the individual in the picture/video transfer is wearing a face veil.

6.1 Introduction

The COVID-19 pandemic has had a long-term influence in a variety of locations across the world. COVID-19 transmission is slowed by wearing face masks, according to scientists. Every medical expert, healthcare organization, medical practitioner, and researcher is looking for effective vaccines and treatments to combat this terrible disease, but no breakthrough has been recorded yet [1]. Water droplets from the infected person disseminate into the atmosphere and infect those nearby [2]. The virus spreads through intimate contact, as well as in congested environments. Donning a mask during this epidemic is an essential preventative precaution [3], and it is especially important in times when maintaining social distance is difficult. COVID-19 is observed to spread mainly among persons close to one another. As a

A. Vijaya Lakshmi (✉) · K. Praveen Kumar Goud · M. Saikiran Kumar · V. Thirupathi
Vardhaman College of Engineering, Hyderabad, India

J. I.-Z. Chen et al. (eds.), *Machine Learning and Autonomous Systems*, Smart Innovation, Systems and Technologies 269, https://doi.org/10.1007/978-981-16-7996-4_6

result, the CDC advised that everyone over the age of two wear a mask in public places. As a result, limiting the probability of transferring this fatal virus can significantly minimize the virus's spread and illness severity. Surveillance of big groups of individuals, on the other hand, is getting increasingly challenging. Anyone who is not wearing a face mask is detected as part of the surveillance procedure.

Face mask recognition has proven to be a tough task in the realm of image processing. With the advancements of CNN [4] and deep learning [5] extremely it is possible to obtain great accuracy in picture categorization and item identification. Technology developed by Hanvon, veiled face acknowledgment has a sensitivity of about 85% [6]. The sensitivity was achieved using synthetic sample data, which is not the situation in our work, which includes combined live and synthetic photos.

In this paper, we provide a computer vision and AI-based mask face identification technique. In this article, a face mask detection model called SSDMNV2 and the MobileNetV2 architecture is utilized to classify images is proposed [7]. SSDMNV2 can tell the difference between pictures with masks on face recognition and photographs without masks on face recognition. To slow the transmission of the corona, the developed framework might be used in combination with video surveillance to detect those who aren't using face masks. The detection of face masks is a difficult job for the current suggested face detector models. This is because mask-wearing faces have a wide range of accommodations, blockage levels, and mask varieties. There were several causes for the current face mask identification model's low performance relative to regular ones, the first of which was a shortage of acceptable data resources. Second, wearing a mask on one's face causes a certain noise, which exacerbates the detecting process. Even so, a huge dataset is required to create an efficient face mask identification model.

The remainder of this paper is coordinated as follows. The accompanying segment, Sect. 6.2, will go through comparable work suggested and done in the realm of recognizing face masks. Section 6.3 describes the dataset, methodology, and technology utilized to create this Face mask detection model. Section 6.4 presents the results of the experiments, and Sect. 6.5 presents the conclusion.

6.2 Related Work

Gray-scale face photographs have already been the focus of several academics and analysts [8]. Some depended on AdaBoost, a strong classifier for training, while others were fully based on pattern recognition frameworks and required no prior knowledge of the face model [9]. Basically, Viola-jones didn't operate in low-light situations. As an outcome, analysts began investigating for another model that could discriminate between faces and veils on the face. Various face mask detection collections have been created in the past to obtain a better understanding of how these systems function. Recent datasets built by collecting online pictures include WiderFace [10] and IJB-A [11].

Face detector methods take cues from the user's input and then utilize a variety of deep learning methods to CNN-based classification [12]. More CNN-based 3D models evolved as technology progressed [13]. The SSDMNV2 face mask detection model was created with OpenCV and TensorFlow deep neural network modules, and the MobileNetV2 classifier has been used for image classification.

6.3 Proposed Methodology

The initial step in determining if someone has worn a mask is to train the algorithm using a relevant dataset. In Sect. 6.3.1, we spoke about the dataset in more detail. The MobilenetV2 [13, 14] classifier is used as a pre-trained model to detect regardless of whether the individual is wearing a veil. After the classifier has been trained, the SSDMNV2 model is utilized to recognize faces continuously. Figure 6.1, a sequence diagram, shows the technique adopted in this work.

6.3.1 Dataset Used

A mixture of open-source datasets and photos were utilized to train the model in a specific method. The dataset used in this work basically contains 1376 photos partitioned into two classifications: cover wearing pictures (686) and non-veil wearing photographs (690).

6.3.2 Pre-processing and Data Augmentation

The database from the secret facial recognition and application had high-quality audio, and the images of the database had several duplicates. They were sorted, and all

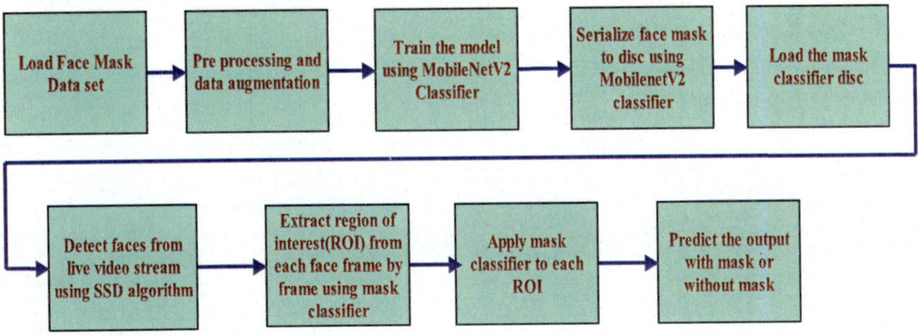

Fig. 6.1 The proposed methodology flow diagram

duplicates were manually extracted. Cleaning, finding, and correcting database errors eliminates the negative effects of any predictive model. This segment explains how data is processed before it is used for data training. To arrange the list in dictionary order, first develop an alphanumerical function. After that, the list is categorized into numerical order, and the images are then processed into scale. The list is then translated to a NumPy array so that counting can be done quickly. OpenCV was used to enhance the bulk of the photos. To transform the raw input photos into clean versions, the pre-processing technique is used.

Following that, the technique of image augmentation is employed to minimize errors. Dataset used in this work basically contains 11,042, and this contains two types of photos: mask-wearing images (5521) and non-mask-wearing images (5521).

6.3.3 Categorisation of Images Using MobileNetV2

The classification challenge was solved using the MobileNetV2. To prevent the loss of previously learnt features, the foundation layers are then frozen. Then additional coachable layers are stacked, which are educated upon obtained given field of study to find the characteristics that distinguish a mask-wearing face from a mask-free face. The weights are then stored when the model has been fine-tuned. Using pre-trained models saves time and money by allowing us to make use of weights that are already biased without sacrificing previously learned features.

MobileNetV2 expands on MobileNetV1's concepts by employing depth-wise separable convolution as an efficient building element. MobilenetV2, on the other hand, adds two additional aspects to the architecture.

1. Between the layers, there are bottlenecks in the form of linear bottlenecks.
2. Bottleneck-to-bottleneck linkages made quickly.

Figure 6.2 depicts the basic structure of MobileNetV2. The layers used by MobileNetV2 are as follows.

- Convolutional Layer

The convolution layer utilizes a sliding window method to help in the extraction of features from pictures. It facilitates feature map creation. The output Y is obtained by convolutioning two functional matrices of input X and kernel H. Figure 6.3 shows the convolution operation.

$$Y(T) = (X * H)(x) = \int_{-\infty}^{+\infty} X(T) \times H(T - x) dT$$

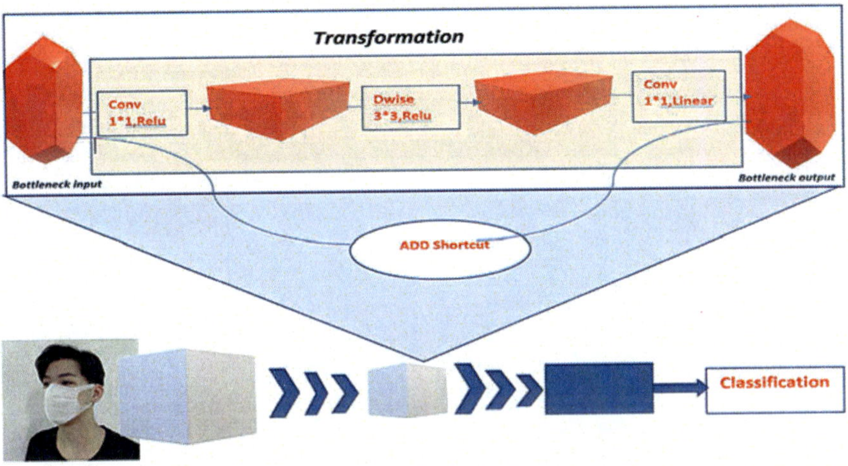

Fig. 6.2 The basic structure of MobilenetV2

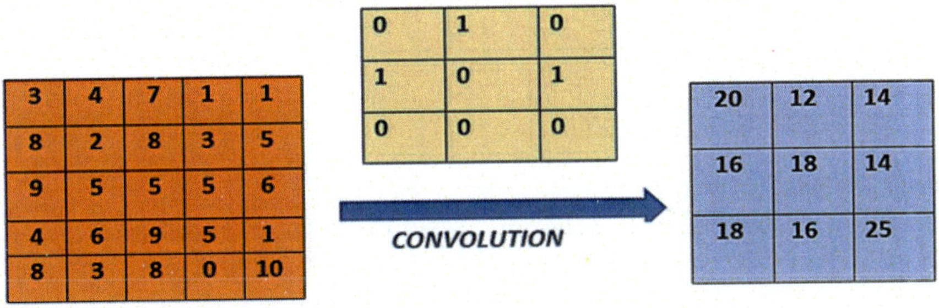

Fig. 6.3 Convolution operation

- Pooling Layer

The use of pooling techniques can speed up calculations by reducing the dimensionality of the input. There are several types of pooling procedures that may be used, some of which are described here.

i. Max pooling: Max pooling can be obtained by implementing a max filter on non-overlapping subregions of the original representation, as shown in Fig. 6.4.
ii. Average Pooling: In this Instead of a maximum value, the average for each block is calculated. The average pooling procedure is depicted in Fig. 6.5.

- Dropout Layer

By removing random biassed neurons from the model, helps to prevent overfitting that can occur during training. These neurons can be found in both hidden and visible levels. The dropout ratio may be modified to alter the chances of a neuron being dropped.

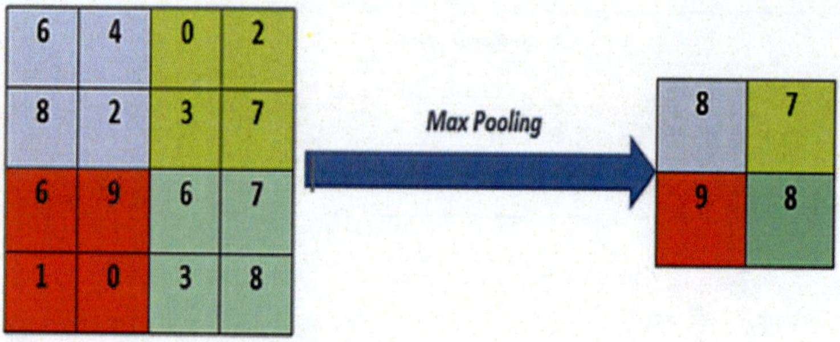

Fig. 6.4 Max pooling operation

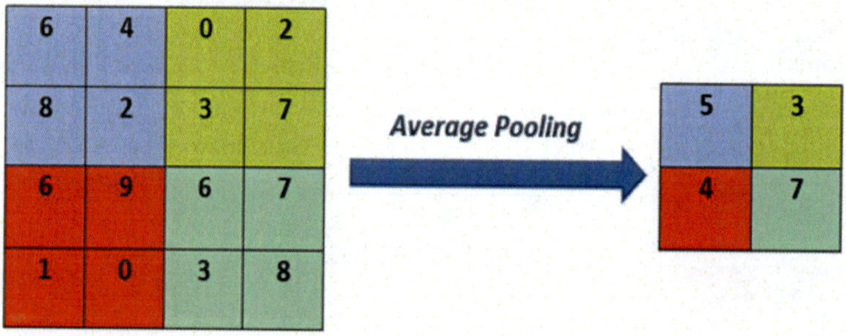

Fig. 6.5 Average pooling operation

- Non-linear Layer

The convolutional layers are usually followed by these layers. Rectified Linear Units are among the most often utilized non-linear functions (ReLU).

- Fully Connected Layer

In multi-class or binary classification, these layers assist in categorizing the provided images.

6.3.4 Face Detection Using a Single-Shot Multi-box Detector

The procedure for determining whether or not the individual on the video is wearing a face mask. The YOLO is practically identical to the Single-Shot Multi-box Detector in that it uses Multi-box to detect numerous items in a single shot. It has a substantially quicker object detecting system with reasonable accuracy. Two types of models are made available by OpenCV.

i. Caffe implementation
ii. Tensor flow implementation

We employed the Caffe model to implement the SSDMNV2 model to recognize faces. The face mask classifier receives these results as input. Face detection can be done in real time without consuming a lot of resources while using this method.

6.3.5 Algorithms that Describe the Entire Pipeline

Using the two algorithms shown below, the suggested SSDMNV2 technique has been effectively presented. Before being trained on the complete dataset, the photos were preprocessed. Algorithm 1, shown in Fig. 6.6, took images and their pixel values as input, which was then scaled and normalized. To boost accuracy, the photos were then submitted to a data augmentation process. It was then subjected to the MobilenetV2 model classifier. The model learned in the previous section was then used in both categories of still images and real-time webcam in Algorithm 2, as shown in Fig. 6.7.

6.4 Experimental Results

All of the tests were performed on a laptop with an i7CPU, 16 GB of RAM. In this study, Python 3.7 was chosen for the creation and implementation of several experimental trials. The predictions for two video streams are shown in Figs. 6.8 and 6.9. These are the video stream predictions generated by the SSDMNV2 model utilizing the MobileNetV2 classifier. The wearing of a mask is portrayed by the square green box on the top, with an accuracy score; in contrast to the fallacious method, the red square box represents wearing no mask.

6.5 Conclusion

An innovative face mask detector is important in public healthcare as technology improves and new trends arise. MobileNet serves as the backbone of the architecture, which may be utilized for both high and low computation scenarios. A face mask acquisition model known as SSDMNV2 was used in the training and creation of picture databases separated into groups of masked and non-masked persons. In this model, the OpenCV neural communication approach achieved good results. One of the distinguishing elements of the recommended solution is using the MobilenetV2 image separator to correctly separate pictures.

Fig. 6.6 Algorithm 1

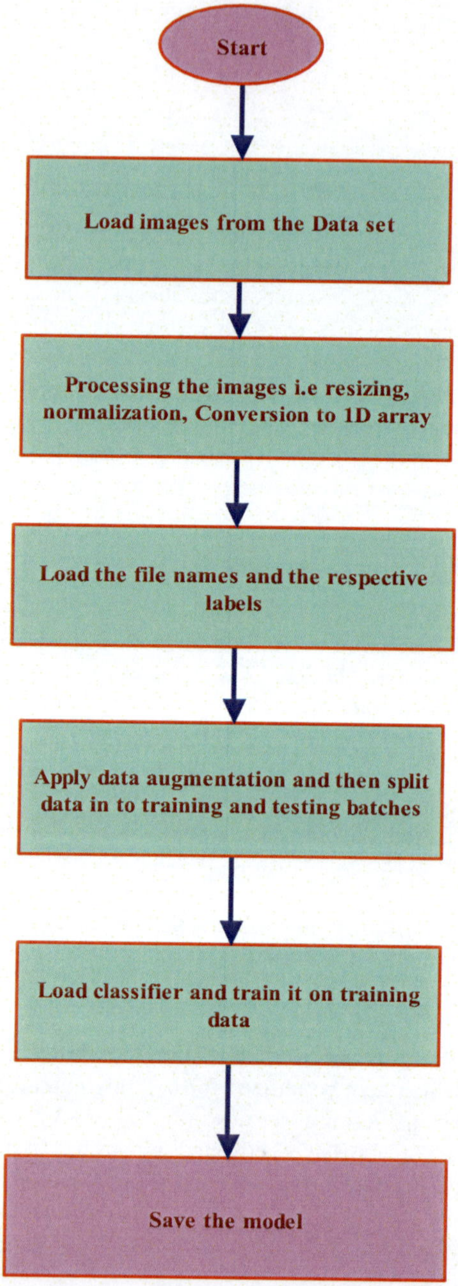

Fig. 6.7 Algorithm 2

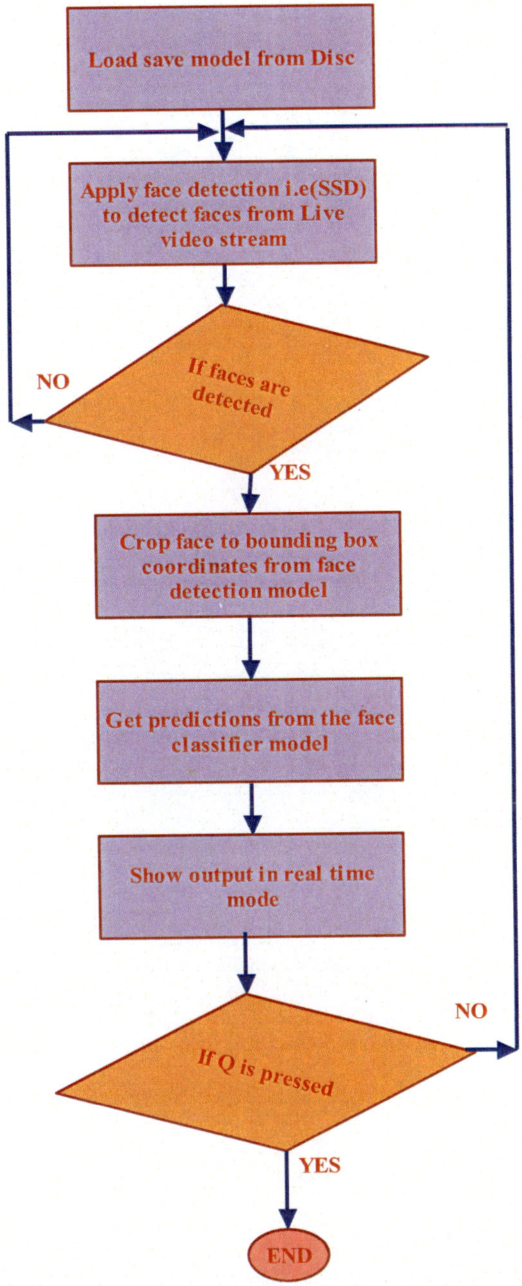

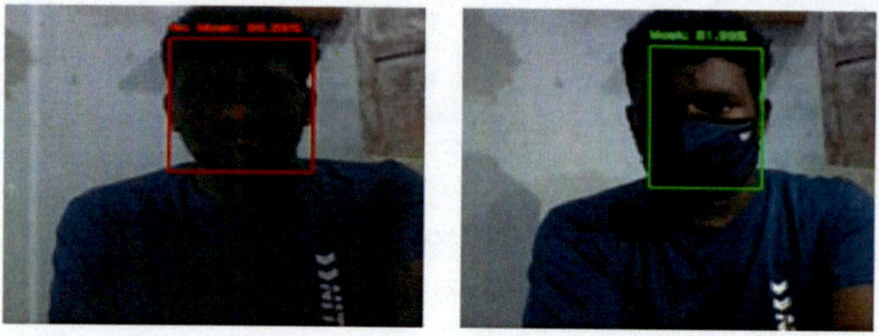

Fig. 6.8 Single face detection on a live video stream with and without a mask

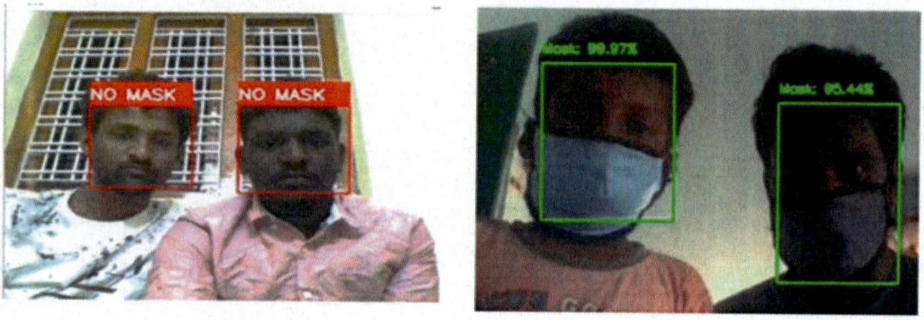

Fig. 6.9 Multiple faces detection on a live video stream with and without a mask

References

1. Megahed, N.A., Ghoneim, E.M.: Antivirus-built environment: lessons learned from Covid-19 pandemic. Sustain. Cities Soc. **61**, Article 102350 (2020)
2. Kumar, P., Hama, S., Omidvarborna, H., Sharma, A., Sahani, J., Abhijith, K. V., ...Tiwari, A.: Temporary reduction in fine particulate matter due to anthropogenic emissions switch-off during COVID-19 lockdown in Indian cities. Sustain. Cities Soc. **62**, Article 102382 (2020)
3. Rahmani, A.M., Mirmahaleh, S.Y.H.: Coronavirus disease (COVID-19) prevention and treatment methods and effective parameters: a systematic literature review. Sustain. Cities Soc. Article 102568 (2020)
4. Lawrence, S., Giles, C.L., Tsoi, A.C., Back, A.D.: Face recognition: a convolutional neural-network approach. IEEE Trans. Neural Netw. **8**(1), 98–113 (1997)
5. Ahmed, I., Ahmad, M., Rodrigues, J.J., Jeon, G., Din, S.: A deep learning-based social distance monitoring framework for COVID-19. Sustain. Cities Soc. Article 102571 (2020)
6. Li, C., Wang, R., Li, J., Fei, L.: Face detection based on YOLOv3. Recent Trends in Intelligent Computing, Communication and Devices, pp. 277–284. Springer, Singapore (2020)
7. Nguyen, H.: Fast object detection framework based on mobilenetv2 architecture and enhanced feature pyramid. J. Theor. Appl. Inform. Technol. **8**(5) (2020)
8. Ojala, T., Pietikainen, M., Maenpaa, T.: Multiresolution gray-scale and rotation invariant texture classification with local binary patterns. IEEE Trans. Pattern Anal. Mach. Intell. **24**(7), 971–987 (2002)

9. Kim, T.H., Park, D.C., Woo, D.M., Jeong, T., Min, S.Y.: Multi-class classifier-based adaboost algorithm. In: International Conference on Intelligent Science and Intelligent Data Engineering, October, pp. 122–127. Springer, Berlin, Heidelberg (2011)
10. Yang, S., Luo, P., Loy, C.C., Tang, X.: Wider face: a face detection benchmark. In: Proceedings of the IEEE Conference on Computer Vision and Pattern Recognition, pp. 5525–5533 (2016)
11. Klare, B.F., Klein, B., Taborsky, E., Blanton, A., Cheney, J., Allen, K., ... Jain, A.K.: Pushing the frontiers of unconstrained face detection and recognition: LARPA Janus Benchmark A. In: Proceedings of the IEEE Conference on Computer Vision and Pattern Recognition, pp. 1931–1939 (2015)
12. Ren, S., He, K., Girshick, R., Sun, J.: Faster R-CNN: towards real-time object detection with region proposal networks. Adv. Neural Inform. Process. Syst. 91–99 (2015)
13. Li, H., Lin, Z., Shen, X., Brandt, J., Hua, G.: A convolutional neural network cascade for face detection. In: Proceedings of the IEEE Conference on Computer Vision and Pattern Recognition, 5325–5334 (2015)
14. Sandler, M., Howard, A., Zhu, M., Zhmoginov, A., Chen, L.C.: Mobilenetv2: inverted residuals and linear bottlenecks. In: Proceedings of the IEEE Conference on Computer Vision and Pattern Recognition, pp. 4510–4520 (2018)

Chapter 7
Spoken Language Identification for Native Indian Languages Using Deep Learning Techniques

Rushikesh Kulkarni, Aditi Joshi, Milind Kamble, and Shaila Apte

Abstract In this paper, we present a Spoken Language Identification System (LID) for native Indian languages. LID task aims to determine the spoken language in a speech utterance of an individual. In this system, 'Resemblyzer' is used for feature extraction, which derives a high-level representation of a voice in the form of a summary vector of 256 values. Experimentation is done on 'IndicTTS Database', developed by IIT Madras which comprises 13 languages, and 'Open-source Multi-speaker Speech Corpora' database developed by the European Language Resources Association which consists of 7 languages. The work consists of training Deep Neural Networks (DNN), Recurrent Neural Networks with Long Short-Term Memory (RNN–LSTM) and Gaussian Mixture Model (GMM) on each database for 1.5 and 5 s feature lengths and comparing their performances with each other and in each scenario.

7.1 Introduction

Spoken Language Identification (LID) is an upcoming research field of audio signal processing. It aims to determine the spoken language in a speech utterance of an individual. From our experimental studies, it was observed that having a LID block as a precursor to a speaker authentication system greatly improves the performance of

R. Kulkarni
Department of Electronics Engineering, Vishwakarma Institute of Technology, Pune, India
e-mail: rushikesh.kulkarni17@vit.edu

A. Joshi (✉) · M. Kamble
Department of Electronics and Telecommunication Engineering, Vishwakarma Institute of Technology, Pune, India
e-mail: aditi.joshi17@vit.edu

M. Kamble
e-mail: milind.kamble@vit.edu

S. Apte
Anubhooti Solutions, Pune, India

J. I.-Z. Chen et al. (eds.), *Machine Learning and Autonomous Systems*, Smart Innovation, Systems and Technologies 269, https://doi.org/10.1007/978-981-16-7996-4_7

the system. Such LID systems also find applications in various real-world problems such as automatic speech to text, call centres, and accent detection systems.

A LID system while functioning similarly to the human brain for language detection, there are some key differences that can be observed in the two.

Das and Roy [20] from the moment a baby is born, its starts learning the languages being spoken in its vicinity. This is an implicit training of phonemes, syllables, and sentence structure. Based on this training that occurs throughout the childhood, a child can not only easily recognize the language he/she was trained for, but also can guess languages from the same family of languages. For example, a person speaking Marathi can recognize languages like Hindi, Sanskrit, Gujarati, etc. (Indic Languages). On the other hand, it is very difficult for that person to recognize languages such as Tamil, Kannada, and Telugu. This is because these languages are from a different family of languages (Dravidian Languages).

Machines can be trained to classify all languages efficiently if it can understand each of these languages. But training such a machine is not only financially expensive but also requires lots of time and human resources.

A more viable LID system can be developed in either of the two ways—implicit and explicit LID systems. Implicit LID systems do not require segmented and labelled speech data. Such systems rely only on the true label of the language along with raw speech as input for classification. When a system is developed which requires segmented and labelled speech corpus it is known as explicit LID.

We have focused on native Indian languages for the LID task. We have developed a system where we have used a novel approach for feature extraction. Later we have performed a comparative study to analyse different techniques. Emotions, accent, recording device, speed all these factors impact a lot in identifying the language.

This paper is divided into 5 sections. First section focuses on the introduction, second section provides a comprehensive summary about the research done in this field. Later, section three shows the methodology followed. Fourth section demonstrates the results and finally, in the fifth section analysis is done based on results found.

7.2 Literature Review

A lot of work has been done so far in the Automatic Spoken Language Identification domain. With the progress of Deep Learning in the last decade, the use of various Deep Learning techniques has increased exponentially.

In [1], Deep Neural Networks (DNN) are used for Automatic Language Identification (LID) from short-term acoustic features. Mel-Frequency Cepstral Coefficients (MFCC) have been used to derive features of speech signals from four local Indian languages—Kannada, Hindi, Tamil, and Telugu in [2]. For Classification, Support Vector Machines and Decision Tree Classifiers have been used which gave accuracies of 76% and 73% respectively on 5 h of training data. In [3], Convolutional

Neural Networks (CNN) are used for Language Identification. The languages that can be classified are—German, Spanish, and English. Filter Banks are used, which extract features from the representations of the signal in the frequency domain. The main contribution of [4] is in experimenting with different acoustic features to identify the best feature set that can enable a classifier to discriminate Indian Spoken Language. 18 Feature sets are examined are 13 MFCC Feature vectors, 13 MFCC + 13 $\triangle$MFCC, 13 MFCC + 13 $\triangle$ MFCC + 13 $\triangle$2 MFCC each with window size 20, 100, and 200 ms, with and without a moving window. Sisodia et al. [5] assessed Ensemble Learning models to classify the following spoken languages—German, Dutch, English, French, and Portuguese. Ensemble learner models are designed using bagging, Adaboosting, random forests, gradient boosting, and extra trees. The authors have presented a comparative study between Deep Neural Networks (DNN), Convolutional Neural Networks (CNN) for Spoken Language Identification (LID) is presented where Support Vector Machines (SVM) are treated as baseline [6]. The performances of the fusions of the mentioned methods are also discussed. To evaluate the performance of the system, NIST 2015 i-vector Machine Learning Challenge task is used where the aim is recognition of 50 in-set languages. In [7], the problem of Language Identification is solved in the image domain instead of the traditional audio domain. For the same, a hybrid Convolutional Recurrent Neural Network is used that operates on spectrogram images of the provided audio snippets. Draghici et al. [8] tried to solve the problem of Language Identification is solved using Mel-spectrogram images as input features. A performance comparison is done for this strategy between Convolutional Neural Network (CNN) and Convolutional Recurrent Neural Network (CRNN). This work is based on Paper no. 7 and differs in a modified training strategy to ensure equal class distribution and efficient memory usage. Ganapathy et al. [9] made the use of bottleneck features from a Convolutional Neural Network for LID task is discussed. The Bottleneck features are used in tandem with the conventional acoustic features and performance is evaluated. Through experiments, it was found that average relative improvements of up to 25% are obtained on the system with bottleneck features compared to the system without them. In [10], the authors present an open-source, end-to-end, LSTM-RNN system running on limited computational resources (a single GPU) that outperforms a more contemporary reference i-vector system by up to 26% increase in performance when both are tested on a subset of the NIST Language Recognition Evaluation with 8 target languages and a feature length of 3 s.

Researchers have worked on different techniques to identify the spoken language. Systems which are independent of speech are developed for LID tasks [11], systems using manual transcription are also developed. Different classifiers such as KNN, Random Forest, SVM, RNN, Naïve Bayes are used. Further, authors have worked on different feature vectors to understand the language. Mel-Frequency Cepstral Coefficients (MFCC) with delta and delta-delta coefficients are extracted from audio signals to retrieve relevant information [12]. Here, it is observed that performance of the system is better with RNN-LSTM compared to DNN. In addition, a novel LID system was proposed based on the architecture of TDNN followed by LSTM-RNN long with MFCC features [13]. For the analysis purpose the dataset used is—(1) NIST LREO7 (14 languages of 5 language cluster: Arabic, English, Slavic, Iberian, and Chinese (2) YouTube (17 Arabic dialects). Research has been done to compare various techniques and understand the pros and cons of every technique [14]. Speaker Identification is a task which is usually done with the use of speakers first language (L1). Although, researchers have worked on developing a system which will identify the language using the speaker's second language [15]. Here, authors have initially created Listen, Attain and Identify (LAI) system. It consists of Bidirectional Gated Recurrent Unit. This takes log-Mel Filter bank features as input. Later, this system is modified to and a combination of Convolutional Neural Networks, Gated Recurrent units, and a fully connected DNN layer is used to create CGDNN architecture. It was observed that time required to train CGDNN model was 6 times lesser than LIA model. Further, research is done using Dynamic Hidden Markov network (DHMnet). The DHMnet is a never-ending learning system and provides high resolution model of the speech space [16]. Here, experimentations are carried out for three languages, English, Japanese and Chinese. In addition to this, system was designed using features—MFCC, Delta MFCC, and Double Delta MFCC [4] and ANN classifier was used for dataset—All India Radio (9 languages). Here, 5 cross-validation approach is used to evaluate the performance of LID system. Different approaches such as use of deep bottleneck network (DBN) are used in the DNN architecture for doing the task of LID. Here, an internal bottleneck acts as a feature extractor. Two layers—DBN-TopLayer and DBN-MidLayer are used to extract features. MFCCs and its derivatives and senone posteriors are used as the frame-level features. A new Hellinger kernel-based similarity measure between utterances is also proposed [17]. GMM models can be a good classifier for the task of LID [18]. i-vector paradigms are used for understanding the audio [6].

In most of the research performed in this domain, either MFCCs and its derivatives, filter banks or i-vectors have been used as features. While all these features have proved to be accurate in much of the literature, it is time consuming to extract these features. Hence, while trying to develop a real time system, these features are unsuitable. Resemblyzer, on the other hand, provides a 256-value feature vector for a given audio sample within a few milliseconds and thus, can be utilized in a real time system. In the following sections, we present our work wherein, audio files from 2 different databases are used by the Resemblyzer to generate a feature vector. This feature vector is then used for training DNN, RNN–LSTM, and GMM models. Finally, a test database is used to gauge the performance of the trained models and in turn, the performance of the Resemblyzer feature vector as a feature.

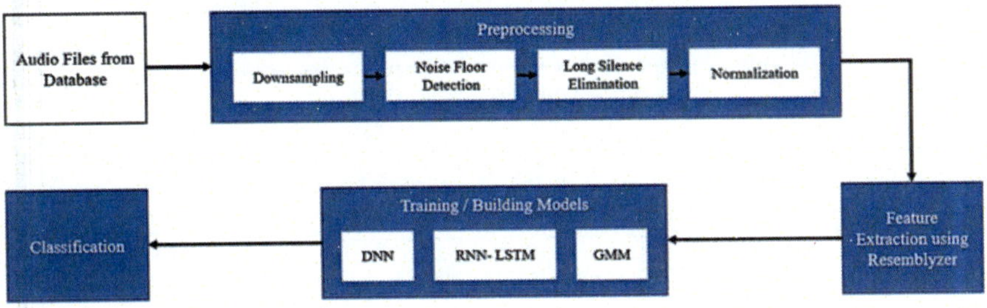

Fig. 7.1 Workflow

7.3 Methodology

7.3.1 Workflow

The process followed in this work has been shown in Fig. 7.1. The audio files are taken from the databases. These audio files are then preprocessed to ensure uniformity is maintained across all the files. The preprocessed files are then passed to the Resemblyzer which creates the 256-value feature vector of the input audio. This feature vector is used for training the three different models—DNN, RNN–LSTM, and GMM. These trained models are finally used for classification/prediction. For classification, separate set of audio files are used from the same database which were not used for testing.

The details regarding each of the process steps are elaborated in the following sections.

7.3.2 Database

In our project, we have worked on two databases—'IndicTTS' by IIT Madras and 'Open-source Multi-speaker Speech Corpora' by European Language Resource Association. Both the databases contain speech files of uneven lengths which vary between 1 and 15 s. All speech files have been preprocesses and later truncated to attain uniform file lengths. During pre-processing, speech files <1.5 s have been discarded. Post truncation, all speech files in both the databases have a length of either 1.5 or 5 s.

7.3.2.1 IndicTTS Database

IndicTTS [19] is a database created by IIT Madras. It is a special corpus of Indian languages covering 13 major languages of India, viz., Assamese, Bengali, Bodo,

Table 7.1 IndicTTS database language wise distribution

Sr. no	Language	Male (in h)	Female (in h)
1	Assamese	12.95	14.5
2	Bengali	10.05	5.01
3	Bodo	0	4
4	Gujarati	10.92	10.33
5	Hindi	5.16	5.18
6	Kannada	3.4	3.95
7	Malayalam	9.7	8.19
8	Manipuri	10.61	10.14
9	Marathi	0	4.8
10	Odia	4.47	4.27
11	Rajasthani	9.82	10.24
12	Tamil	10.3	10.03
13	Telugu	4	23.9

Gujarati, Hindi, Kannada, Malayalam, Manipuri, Marathi, Odia Rajasthani, Tamil, Telugu. It comprises 10,000+ spoken sentences/utterances each of mono and English recorded by both Male and Female native speakers. The speech files in this database are available in .wav format with a sampling rate of 48 kHz along with the corresponding text as transcription. This dataset boasts of large amounts of training data which is beneficial for training Neural Networks. It suffers from the fact that for each language, audios are recorded by only 1 female and 1 male speaker.

The language wise data present in the IndicTTS database is as shown in Table 7.1.

7.3.2.2 Open-Source Multi-speaker Speech Corpora Database

The European Language Resources Association created the 'Open-source Multi-speaker Speech Corpora for Building Gujarati, Kannada, Malayalam, Marathi, Tamil, and Telugu Speech Synthesis Systems' (OMSC) [12]. There are free high-quality multi-speaker speech corpora in this database for Bengali, Gujarati, Kannada, Malayalam, Marathi, Tamil, and Telugu, which are seven of India's twenty-two official languages spoken by 374 million native speakers. Except for Marathi, which is a female-only database, the majority of the corpora contain at least 2000 recorded lines from female and male native speakers of the language. But the important feature of this database is that it consists of recordings from multiple speakers and hence, is a better measure for Language Identification. Similar to the IndicTTS database, the speech files in this database are available in .wav format with a sampling rate of 48 kHz.

The distribution of languages in the OMSC database for feature lengths = 1.5 and 5 s is seen in the following pie charts. As previously mentioned, due to uneven

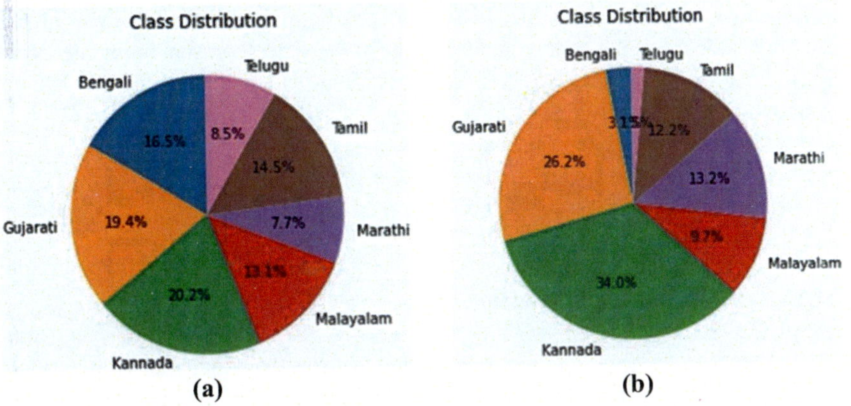

Fig. 7.2 OMSC database language wise distribution for Feature Length: = **a** 1.5 s and **b** 5 s

lengths of the audio files, some audio files have to be truncated to attain uniform audio length which leads to reduction in data and different distributions which can be seen in the two pie charts (Fig. 7.2).

7.3.3 Pre-processing and Feature Extraction

7.3.3.1 Pre-processing

After procurement of the databases, all the audio files were first preprocessed. During pre-processing, we first perform downsampling. All the audio files have been down-sampled to 16 kHz as their sampling frequency. Later, Noise Floor Detection is performed to eliminate any white noise present in the audios. We have also performed Long Silence Elimination to remove silences from the audio. Due to Noise Floor Detection and Long Silence Elimination, the length of the audio gets reduced. Thus, a 1.5 s audio which should have been passed for feature extraction has to be discarded. Only audios which have a length of 1.5 s or above, after pre-processing can be passed for feature extraction. Finally, all audios are normalized. Normalization is done to maintain the same audio level across all audio files. All such normalized audios are passed to the Resemblyzer for feature extraction.

7.3.3.2 Feature Extraction

After pre-processing, we used Resemblyzer [21, 22] to derive a 256-value feature vector from the audios. Feature vectors for audio lengths 1.5 and 5 s with a window of half the audio length were extracted. The feature vectors were stored as rows in a

.csv file with each of the 256 values in a separate column. A unique name was given to each row specifying the gender and language of the speech present in the file from which the feature vector was extracted. The .csv file was then loaded onto the python console which was then used for training the models.

7.3.4 Training

7.3.4.1 Gaussian Mixture Model

A combined .csv file containing all the extracted features from males and females for each language was supplied to create a GMM model for that language. This model was then saved using relevant python libraries and then was loaded for calculating the predictions. During predictions, each test file was passed through each language's GMM model. Here, the per-sample average log-likelihood was calculated with each GMM model. After comparing all the scores, the language model with highest value of the average log-likelihood was deemed to be the language of the test file. Training was performed for both 5 and 1.5 s file length.

7.3.4.2 Deep Neural Networks

We used Deep Neural Networks (DNN) as one of our training techniques. For systems wherein the available data is large, DNNs are a good solution for prediction tasks. Considering the abundant databases that we were using, DNNs seemed to be a good solution.

The model architecture of the DNN is as shown in Fig. 7.3.

The input layer consists of 256 neurons, in accordance with the input vector having 256 values. We have used 4 fully connected layers with decreasing number of neurons in each subsequent layer. The input layers and hidden layers have 'Rectified Linear Neuron (ReLu)' as their activation function. Since this task is a problem of multiple classes, we used 'Categorical Cross-Entropy' as the loss function and 'softmax' as the activation function in the output layer. The entire model was trained with a batch size = 32 and epochs = 30.

7.3.4.3 Recurrent Neural Network-Long Short-Term Memory

After analysing the data for DNN and GMM, we also performed analysis for LID task using RNN–LSTM. Since we are working with a time series data, problems like vanishing gradient that is usually encountered in other networks can be solved using RNN–LSTM. We trained the RNN on both the feature lengths of 1.5 and 5 s. The RNN–LSTM architecture is as shown in Fig. 7.4.

```
_________________________________________________________________
Layer (type)                 Output Shape              Param #
=================================================================
dense_7 (Dense)              (None, 256)               65792
_________________________________________________________________
dense_8 (Dense)              (None, 256)               65792
_________________________________________________________________
dense_9 (Dense)              (None, 128)               32896
_________________________________________________________________
dense_10 (Dense)             (None, 64)                8256
_________________________________________________________________
dense_11 (Dense)             (None, 32)                2080
_________________________________________________________________
dense_12 (Dense)             (None, 13)                429
=================================================================
Total params: 175,245
Trainable params: 175,245
Non-trainable params: 0
```

Fig. 7.3 DNN architecture

```
_________________________________________________________________
Layer (type)                 Output Shape              Param #
=================================================================
lstm (LSTM)                  (None, 1, 256)            525312
_________________________________________________________________
lstm_1 (LSTM)                (None, 1, 256)            525312
_________________________________________________________________
lstm_2 (LSTM)                (None, 1, 128)            197120
_________________________________________________________________
dropout (Dropout)            (None, 1, 128)            0
_________________________________________________________________
time_distributed (TimeDistri (None, 1, 64)             8256
_________________________________________________________________
time_distributed_1 (TimeDist (None, 1, 32)             2080
_________________________________________________________________
time_distributed_2 (TimeDist (None, 1, 16)             528
_________________________________________________________________
flatten (Flatten)            (None, 16)                0
_________________________________________________________________
dense_16 (Dense)             (None, 13)                221
=================================================================
Total params: 1,258,829
Trainable params: 1,258,829
Non-trainable params: 0
```

Fig. 7.4 RNN–LSTM architecture

The model consists of 3 LSTM layers, a dropout layer with 50% probability, 3 Time-Distributed layers along with one flattening layer to be fed to the output layer as a one-dimensional vector. Time-Distributed layers have 'ReLu' as their activation function while 'softmax' is used in the output layer. Like DNN, 'Categorical Cross-Entropy' is the loss function used. The entire model was trained with a batch size = 32 and epochs = 30.

7.4 Results and Discussion

As discussed in the methodology, a total of 4 DNN, 4 RNN, and 4 GMM models were trained across all the scenarios. For testing each model, a set of samples from that same database with the same feature length are used to ensure none of the factors such as accent, recorder model are uniform. Moreover, all these testing samples were kept separate from the training database and were not used while training. Their results are documented in this section. In this section, we shall be assessing these results based on some metrics. The metrics used in this section are based on the number of T_P = True Positive, F_P = False Positive, T_N = True Negative, and F_N = False Negative. They can be derived as follows:

a. Accuracy $= (T_P + T_N)/(T_P + T_N + F_P + F_N)$
b. Precision (P) $= T_P/(T_P + F_P)$
c. Recall (R) $= T_P/(T_P + F_N)$.
d. F1-score $= 2\,[(P \times R)/(P + R)]$

The results obtained from the 12 models are compared as follows. A confusion matrix is shown which shows the percentage of true labels and the predicted labels per language.

7.4.1 *IndicTTS Database*

7.4.1.1 Feature Length = 1.5 s

a. DNN (Accuracy = 98.34%)—Out of the 13 languages, 12 are classified to almost 100% accuracy except for 'Bengali', which has around 70% accuracy. Bengali is mostly confused with Telugu. The overall accuracy of this model is 98%.
b. RNN (Accuracy = 98.43%)—LSTM—Like DNN, Bengali is the most misclassified language with 70% accuracy. The overall accuracy is 98%.
c. GMM (Accuracy = 95.48%)—GMM is the worst performer in the 1.5 s feature length category for the INDIC database. It has an accuracy of 95% which means that it has misclassified around 600 audio samples (Fig. 7.5).

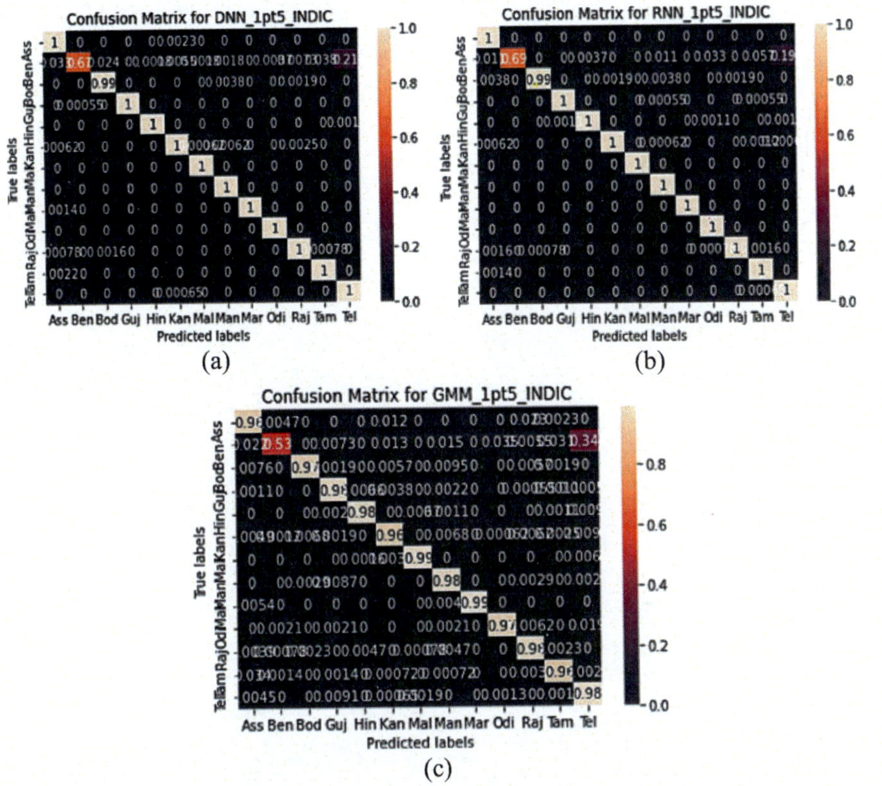

Fig. 7.5 Confusion matrix for IndicTTS, Feature Length = 1.5 s for **a** DNN, **b** RNN–LSTM, and **c** GMM

A comparison of the 3 models for IndicTTS Database for feature length of 1.5 s based on Precision, Recall, and F1-score is shown in Table 7.2.

7.4.1.2 Feature Length = 5 s

a. DNN (Accuracy = 97.78%)—Prevalent from the confusion matrix and the table present to the right, when considering the DNN with 5 s feature lengths, Bengali continues to be misclassified, but now Malayalam becomes the most misclassified language with only 44% accuracy. Malayalam is mostly confused with Gujarati which is weird considering they belong to different families of languages.
b. RNN–LSTM (Accuracy = 97.87%)—The performance of Bengali language increases drastically with the use of RNN–LSTM, but Malayalam continues to fare worse, now misclassifying across multiple languages. The remaining 12

Table 7.2 Analysis for IndicTTS, Feature Length = 1.5 s

Label	Precision			Recall			F1-score		
	DNN	RNN–LSTM	GMM	DNN	RNN–LSTM	GMM	DNN	RNN–LSTM	GMM
Assamese	0.95	0.97	0.82	1	1	0.96	0.97	0.98	0.88
Bengali	1	1	0.97	0.67	0.69	0.53	0.8	0.82	0.69
Bodo	0.97	1	0.97	0.99	0.99	0.97	0.98	0.99	0.97
Gujarati	1	1	0.98	1	1	0.98	1	1	0.98
Hindi	1	1	0.98	1	1	0.98	1	1	0.98
Kannada	1	1	0.98	1	1	0.96	1	1	0.97
Malayalam	1	1	0.98	1	1	0.99	1	1	0.99
Manipuri	0.99	0.97	0.89	1	1	0.98	0.99	0.99	0.94
Marathi	1	1	1	1	1	0.99	1	1	1
Odia	1	0.96	0.96	1	1	0.97	1	0.98	0.96
Rajasthani	0.99	1	0.97	1	1	0.98	0.99	1	0.98
Tamil	0.98	0.97	0.98	1	1	0.96	0.99	0.99	0.97
Telugu	0.93	0.94	0.86	1	1	0.98	0.96	0.97	0.92

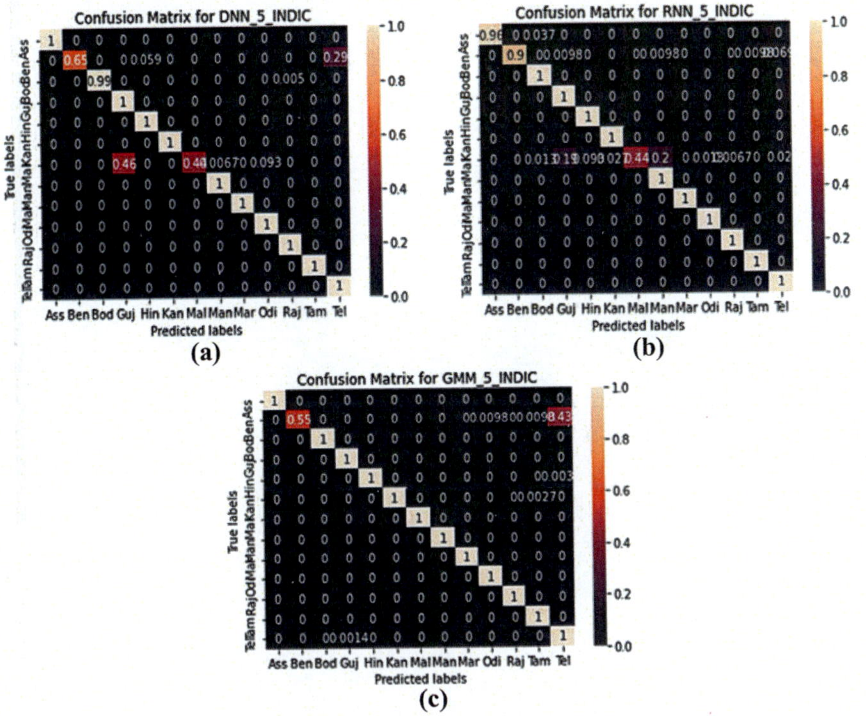

Fig. 7.6 Confusion matrix for IndicTTS, Feature Length = 5 s for **a** DNN, **b** RNN–LSTM, and **c** GMM

languages all have scores around 99%. Overall accuracy is better than DNN at 98%.

c. GMM (Accuracy = 98.87%)—Surprisingly, GMM is the best model in the INDIC, 5 s category with an overall accuracy of 99%. This might be because due to lesser data being available, the deep learning models are not able to train efficiently. Moreover, since there is only one speaker per language, the GMM model does a better job at generalizing the language despite lesser data (Fig. 7.6 and Table 7.3).

7.4.2 OMSC Database

7.4.2.1 Feature Length = 1.5 s

a. DNN (Accuracy = 98%)—Despite having multiple speakers across all the languages, the DNN does a good job of proper classification with an overall accuracy of 98%. The worst classified language is Telugu which still has a 94% classification accuracy.

Table 7.3 Analysis for IndicTTS, Feature Length = 5 s

Label	Precision			Recall			F1-score		
	DNN	RNN–LSTM	GMM	DNN	RNN–LSTM	GMM	DNN	RNN–LSTM	GMM
Assamese	1	1	1	1	0.96	1	1	0.98	1
Bengali	1	1	1	0.65	0.9	0.55	0.79	0.95	0.71
Bodo	1	0.99	1	0.99	1	1	1	0.99	1
Gujarati	0.93	0.97	1	1	1	0.95	0.96	0.98	1
Hindi	0.98	0.96	1	1	1	1	0.99	0.98	1
Kannada	1	0.99	1	1	1	1	1	1	1
Malayalam	1	1	1	0.44	0.44	1	0.61	0.61	1
Manipuri	0.5	0.03	1	1	1	1	0.67	0.06	1
Marathi	1	1	1	1	1	1	1	1	1
Odia	0.82	0.97	0.99	1	1	1	0.9	99	0.99
Rajasthani	1	1	1	1	1	1	1	1	1
Tamil	1	1	1	1	1	1	1	1	1
Telugu	0.96	0.99	0.94	1	1	1	0.98	0.99	0.97

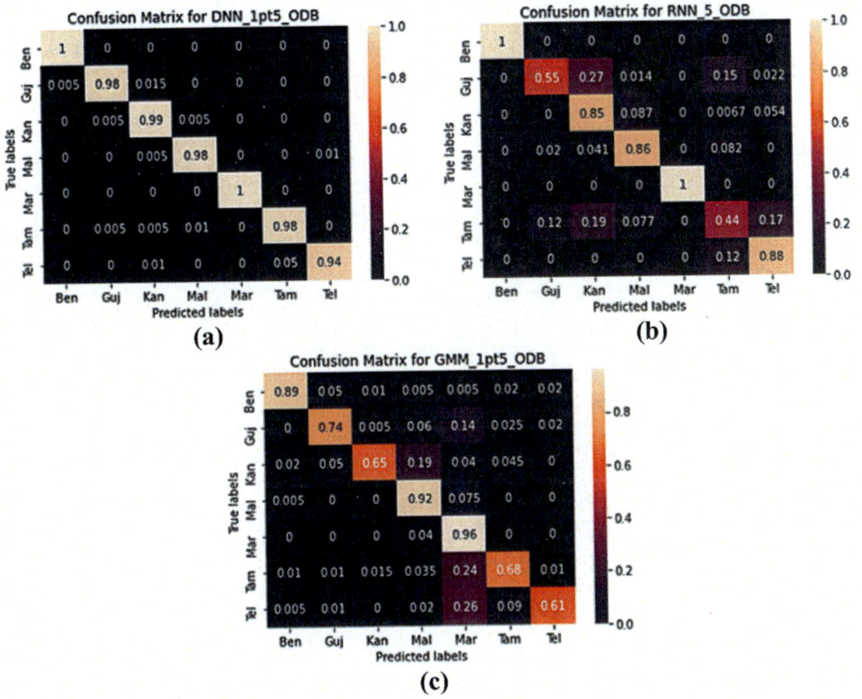

Fig. 7.7 Confusion matrix for OMSC, Feature Length = 1.5 s for **a** DNN, **b** RNN–LSTM, and **c** GMM

b. RNN-LSTM (Accuracy = 97.2%)—Performance of RNN–LSTM is almost similar to DNN with DNN only slightly performing better by doing a better job of classifying Bengali than RNN–LSTM.
c. GMM (Accuracy = 75.86%)—The GMM fares are inferior in this category with an accuracy of only 76%. This might be due to the fact that there are multiple speakers in the database and the Deep Learning models do a much better job at generalizing the features of the language than GMM (Fig. 7.7 and Table 7.4).

7.4.2.2 Feature Length = 5 s

a. DNN (Accuracy—74%)—The presence of lack of data is apparent from the low accuracy of the model. With very less data available for training, DNN, which heavily relies on data is not able to generalize the features of the languages efficiently and causes misclassification almost across all languages. DNN is not the solution when data available is low.
b. RNN–LSTM (Accuracy—70.92%)—RNN is also not a good solution in low data scenarios. Despite being less data sensitive than a DNN, RNN-LSTM is not

Table 7.4 Analysis for OMSC, Feature Length = 1.5 s

Label	Precision			Recall			F1-score		
	DNN	RNN–LSTM	GMM	DNN	RNN–LSTM	GMM	DNN	RNN–LSTM	GMM
Bengali	0.99	0.99	0.96	1	1	0.89	1	1	0.92
Gujarati	0.99	0.96	0.86	0.98	0.95	0.74	0.98	0.96	0.8
Kannada	0.97	0.96	0.96	0.99	0.96	0.65	0.98	0.96	0.78
Malayalam	0.98	0.97	0.74	0.98	0.99	0.92	0.98	0.98	0.82
Marathi	1	0.96	0.24	1	1	0.96	1	1	0.38
Tamil	0.95	0.97	0.79	0.98	0.96	0.68	0.97	0.97	0.73
Telugu	0.99	0.98	0.92	0.94	0.95	0.61	0.96	0.96	0.74

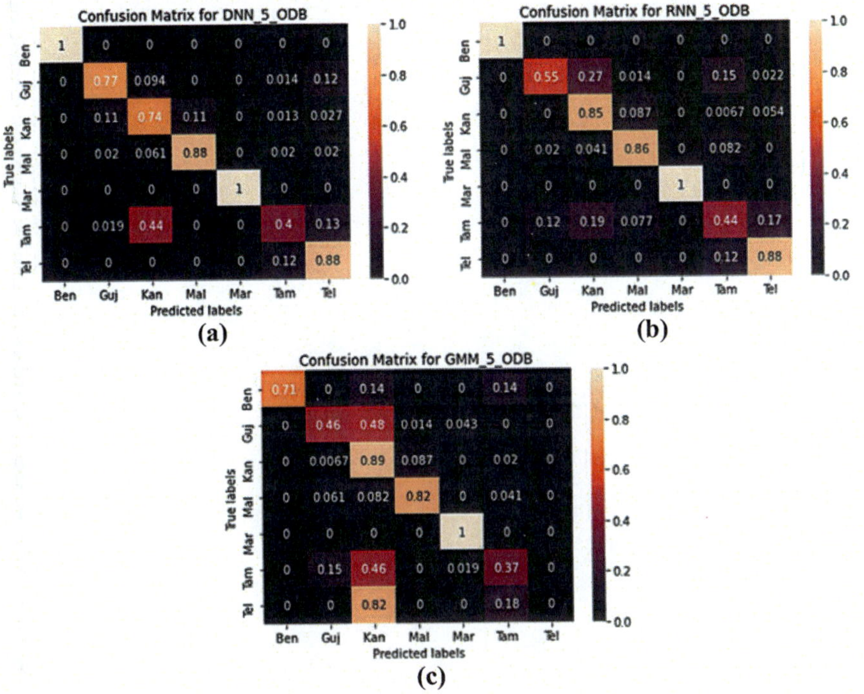

Fig. 7.8 Confusion matrix for OMSC, Feature Length = 5 s for **a** DNN, **b** RNN–LSTM, and **c** GMM

a good solution with an overall accuracy of 71% compared to the DNN having an accuracy of 74.

c. GMM (Accuracy—63.77%)—GMM performs the worst in this category. Lowest amount of data coupled with multiple speakers proves to be a difficult scenario for using GMM as can be seen by the paltry accuracy of 64% (Fig. 7.8 and Table 7.5).

It should be noted that out of all the scenarios, OMSC Database with feature length = 5 s has the least amount of data available. For comparison, the testing phase for this scenario involved classification of 424 test segments. On the other hand, the testing phase for IndicTTS Database with feature length = 1.5 s involved classification of 12,237 test segments.

7.4.3 *Discussion*

The novel approach in this project is the usage of 'Resemblyzer' for feature extraction over the more conventional features used. Resemblyzer gives a 256-value vector

Table 7.5 Analysis for OMSC, Feature Length = 5 s

Label	Precision			Recall			F1-score		
	DNN	RNN–LSTM	GMM	DNN	RNN–LSTM	GMM	DNN	RNN–LSTM	GMM
Bengali	1	1	1	1	1	0.71	1	1	0.83
Gujarati	0.86	0.92	0.84	0.77	0.55	0.46	0.81	0.68	0.6
Kannada	0.74	0.72	0.55	0.74	0.85	0.89	0.74	0.78	0.68
Malayalam	0.72	0.69	0.73	0.74	0.86	0.82	0.79	0.76	0.77
Marathi	1	1	0.59	1	1	1	1	1	0.74
Tamil	0.75	0.45	0.75	0.4	0.44	0.4	0.53	0.45	0.53
Telugu	0.34	0.43	0	0.88	0.88	0	0.49	0.58	0

which can be readily used for input to the training models. The high accuracies observed in the database for various models is a sign that this can indeed be used as a good feature. But one of the major drawbacks is that Resemblyzer works effectively in a subject dependent LID system where the language will be recognized correctly if there has been a mention of that language by a particular speaker in the model's training history. The 'Resemblyzer' feature vector performs poorly in a subject independent LID model for the models trained and tested under the scope of this work.

In the previous section, we looked at the confusion matrices for all the 12 models encompassing various scenarios. We have also assessed the results based on various metrics such as precision, recall, f1-score, and accuracy. In this section, we shall compare the accuracies of these models primarily on the basis of their databases to further understand the behaviour of these models.

7.4.3.1 IndicTTS Database

As discussed in the previous sections, the main characteristic of the Indic Databases is that it has a very large amount of data for a large number of languages. But one of the major drawbacks of this database is that for each language, the audio files are recorded by only one male and one female speaker. Thus, due to the lack of diversity of speakers, its results may be somewhat biassed towards the speaker along with the language spoken. The performances of DNN, RNN, and GMM for both feature lengths are shown in the bar graph in Fig. 7.9.

The DNN and RNN models make more sense of the data of a short 1.5 s length. The abundance of data present here aids the deep learning models. On the contrary, the GMM model is neither able to take advantage of the single speaker factor nor can generalize the language classes and hence, its performance is inferior to DNN and RNN.

On the contrary, the increase in feature length to 5 s comes with decrease in the amount of data. Hence, the deep learning models fail to perform. On the other hand,

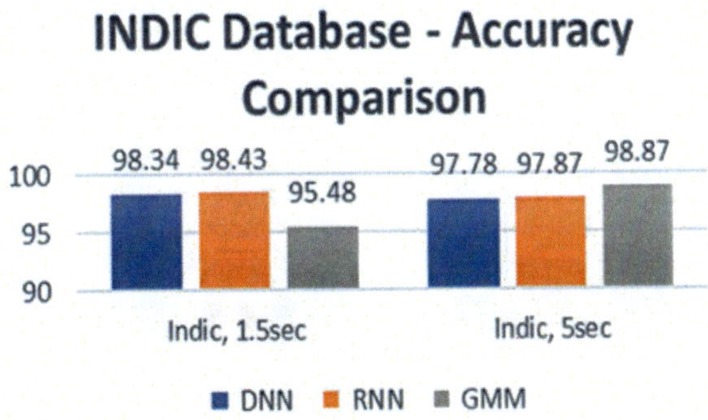

Fig. 7.9 IndicTTS database—accuracy comparison

the longer audio length is beneficial for the GMM as it performs much better than its counterparts here.

7.4.3.2 OMSC Database

The OMSC database is a much better test for the LID task thanks to the multiple speaker recordings present. Here, the performance of each model can be scrutinized more appropriately. The performances of DNN, RNN, and GMM for both feature lengths are shown in the bar graph in Fig. 7.10.

With the feature length at 1.5 s and abundant data available, the performance of DNN and RNN is much better than GMM. It can be stated that DNN and RNN have been able to overcome the multiple speaker barrier and accurately identify the features of the language to classify them with a high accuracy.

The lowest amount of data was present for the 5 s category and hence, both DNN and RNN–LSTM have low accuracies. GMM is not able to improve its performance with longer feature lengths and suffers from a low accuracy of 64%.

The accuracies of all the models can be seen together in the following bar chart in Fig. 7.11.

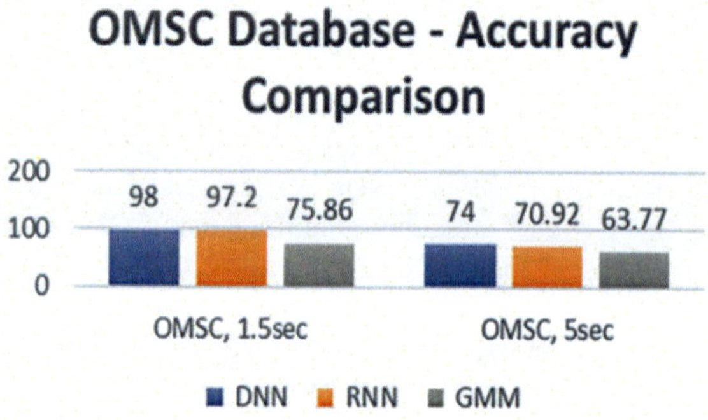

Fig. 7.10 OMSC database—accuracy comparison

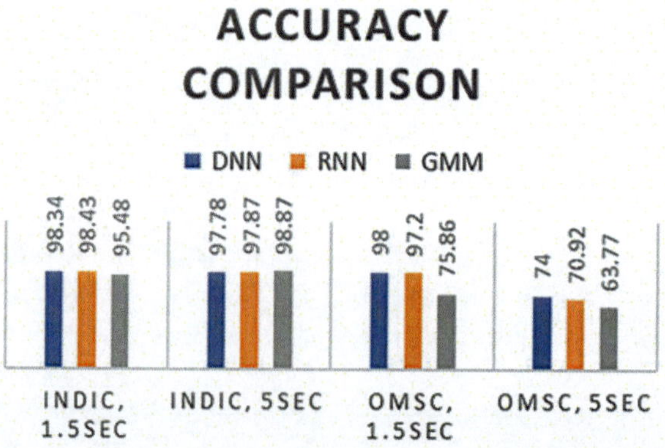

Fig. 7.11 Accuracy comparison—all models

7.4.4 *Future Scope*

Through this project, we have been able to successfully build a LID system using the 'Resemblyzer' feature vector as the novel feature.

The presence of a large amount of data is crucial for building a LID system with a high accuracy. So far, we have been able to develop a speaker dependent system wherein if the speaker's recordings are present in the system, the chances of correctly classifying the language are very high. This type of system can be employed in organizations or banks wherein only the members of that organization or holders of bank accounts in a particular bank utilize the LID system for their needs. A dynamic system can be developed where any new member is made to speak certain lines in a certain language. These audio files will be added to the training models so that the system is able to correctly classify the language that the new member is conversing in.

Since we observed a low accuracy for speakers outside the database, this area remains open for research to develop a speaker independent system. The models can be further tuned, new variations and architectures may be added for the same. The utility of 'Resemblyzer' in such a system is also an area which requires further research.

Under the scope of the current work, we have been able to implement a speaker dependent LID system using deep learning techniques and validated its performance through testing across various scenarios.

7.5 Conclusion

LID system has been successfully implemented using the Resemblyzer feature vector. Various AI techniques have been used to validate the results. While comparing the system for INDIC database, best results have been procured for the feature length of 1.5 s with DNN accuracy of 98.34% and RNN accuracy of 98.44%. Similarly, for the OMSC database, best results have been procured for the feature length of 1.5 s for DNN—98%. DNN has been consistent in its performance provided it is supplied with abundant training data.

References

1. Lopez-Moreno, Gonzalez-Dominguez, J., Plchot, O., Martinez, D., Gonzalez-Rodriguez, J., Moreno, P.: Automatic language identification using deep neural networks. In: 2014 IEEE International Conference on Acoustics, Speech and Signal Processing (ICASSP), Florence, Italy, pp. 5337–5341 (2014). https://doi.org/10.1109/ICASSP.2014.6854622
2. Venkatesan, H., Venkatasubramanian, T.V., Sangeetha, J.: Automatic language identification using machine learning techniques. In: 2018 3rd International Conference on Communication

and Electronics Systems (ICCES), Coimbatore, India, pp. 583–588 (2018). https://doi.org/10.1109/CESYS.2018.8724070
3. Mukherjee, S., Shivam, N., Gangwal, A., Khaitan, L., Das, A.J.: Spoken language recognition using CNN. In: 2019 International Conference on Information Technology (ICIT), Bhubaneswar, India, 2019, pp. 37–41 (2019). https://doi.org/10.1109/ICIT48102.2019.00013
4. Aarti, B., Kopparapu, S.K.: Spoken Indian language classification using artificial neural network—an experimental study. In: 2017 4th International Conference on Signal Processing and Integrated Networks (SPIN), Noida, pp. 424–430 (2017). https://doi.org/10.1109/SPIN.2017.8049987
5. Sisodia, S., Nikhil, S., Kiran, G.S., Sathvik, P.: Ensemble learners for identification of spoken languages using mel frequency cepstral coefficients. In: 2nd International Conference on Data, Engineering and Applications (IDEA), Bhopal, India, pp. 1–5 (2020). https://doi.org/10.1109/IDEA49133.2020.9170720
6. Heracleous, P., Takai, K., Yasuda, K., Mohammad, Y., Yoneyama, A.: Comparative study on spoken language identification based on deep learning. In: 2018 26th European Signal Processing Conference (EUSIPCO), Rome, Italy, pp. 2265–2269 (2018). https://doi.org/10.23919/EUSIPCO.2018.8553347
7. Bartz, C., Herold, T., Yang, H., Meinel, C.: Language identification using deep convolutional recurrent neural networks. In: Liu, D., Xie, S., Li, Y., Zhao, D., El-Alfy, ES (eds.) Neural Information Processing. ICONIP 2017. Lecture Notes in Computer Science, vol. 10639. Springer, Cham (2017). https://doi.org/10.1007/978-3-319-70136-3_93
8. Draghici, A., Abeßer, J., Lukashevich, H.: A study on spoken language identification using deep neural networks (2020). https://doi.org/10.1145/3411109.3411123
9. Ganapathy, S., Han, K., Thomas, S., Omar, M., Van Segbroeck, M., Narayanan, S.: Robust language identification using convolutional neural network features. In: Proceedings of the Annual Conference of the International Speech Communication Association. INTERSPEECH (2014)
10. Zazo R, Lozano-Diez A, Gonzalez-Dominguez J, Toledano DT, Gonzalez-Rodriguez J (2016) Language identification in short utterances using long short-term memory (LSTM) recurrent neural networks. PLoS One **11**(1), e0146917. https://doi.org/10.1371/journal.pone.0146917
11. Saikia, R., Singh, S.R., Sarmah, P.: Effect of language independent transcribers on spoken language identification for different Indian languages. In: 2017 International Conference on Asian Language Processing (IALP), Singapore, pp. 214–217 (2017). https://doi.org/10.1109/IALP.2017.8300582
12. Kozhirbayev, Z., Yessenbayev, Z., Karabalayeva, M.: Kazakh and Russian languages identification using long short-term memory recurrent neural networks. In: 2017 IEEE 11th International Conference on Application of Information and Communication Technologies (AICT), Moscow, Russia, pp. 1–5 (2017). https://doi.org/10.1109/ICAICT.2017.8687095
13. Sun, L.: Spoken language identification with deep temporal neural network and multi-levels discriminative cues. In: 2020 IEEE 3rd International Conference on Information Communication and Signal Processing (ICICSP), Shanghai, China, pp. 153–157 (2020). https://doi.org/10.1109/ICICSP50920.2020.9232093
14. Morgan, D.P., Riek, L., Mistretta, W.J., Scofield, C.L., Grouin, P., Hull, F.: Experiments in language identification with neural networks. In: [Proceedings 1992] IJCNN International Joint Conference on Neural Networks, Baltimore, MD, USA, vol. 2, pp. 320–325 (1992). https://doi.org/10.1109/IJCNN.1992.226968
15. Ubale, R., Qian, Y., Evanini, K.: Exploring end-to-end attention-based neural networks for native language identification. In: 2018 IEEE Spoken Language Technology Workshop (SLT), Athens, Greece, pp. 84–91 (2018). https://doi.org/10.1109/SLT.2018.8639689
16. Markov, K., Nakamura, S.: Language identification with dynamic hidden Markov network. In: 2008 IEEE International Conference on Acoustics, Speech and Signal Processing, Las Vegas, NV, USA, pp. 4233–4236 (2008). https://doi.org/10.1109/ICASSP.2008.4518589
17. Song, Y., Cui, R., Hong, X., Mcloughlin, I., Shi, J., Dai, L.: Improved language identification using deep bottleneck network. In: 2015 IEEE International Conference on Acoustics, Speech

and Signal Processing (ICASSP), South Brisbane, QLD, Australia, pp. 4200–4204 (2015). https://doi.org/10.1109/ICASSP.2015.7178762.
18. Muralikrishna, H., Sapra, P., Jain, A., Dinesh, D.A.: Spoken language identification using bidirectional LSTM based LID sequential senones. In: 2019 IEEE Automatic Speech Recognition and Understanding Workshop (ASRU), Singapore, pp. 320–326 (2019). https://doi.org/10.1109/ASRU46091.2019.9003947
19. https://www.iitm.ac.in/donlab/tts/database.php
20. Das, H., Roy, P.: A deep dive into deep learning techniques for solving spoken language identification problems in speech signal processing (2018). https://doi.org/10.1016/B978-0-12-818130-0.00005-2
21. https://github.com/resemble-ai/Resemblyzer
22. Wan, L., Wang, Q., Papir, A., Moreno, I.: Generalized end-to-end loss for speaker verification, pp. 4879–4883 (2018). https://doi.org/10.1109/ICASSP.2018.8462665

Chapter 8
Smart Attendance System Using Machine Learning Algorithms

M. Nitin Chowdary, V. Sujana, K. Satvika, K. Lakshmi Srinivas, and P. S. Suhasini

Abstract In face recognition, the difficult task is to detect the face from the existing database, and the problems may arise due to bad image quality, bad lighting, or the person not facing the camera, etc. The face of the same person will change for different expressions. If the person's age changes, the size, and color of the person may also change, which makes face recognition more difficult. In this paper, the proposed system android application is used for face recognition for attendance system. The proposed system can be used in colleges and schools for attendance systems. The human face will be recognized from the image uploaded to the server of that particular class; this can be done by using an algorithm. Here we are using the HAAR cascade classifier to identify the face by removing noise and crop the face in the image. Then, at that point the HOG (Histogram Of Gradients) calculation is utilized to extricate the features from the detected face. In the last step, we use SVM (Support Vector Machine) classifier to identify the person from our database. When the person is found in the image, then the attendance will be automatically allocated in the excel sheet of that particular class.

8.1 Introduction

In this day and age, time is perhaps the most valuable thing. Through the various phases of human advancements, a man consistently attempted to search for approaches to make his work simpler. The man was perpetually discontent, so the idea of automation appeared. One key example is the replacement of manual attendance system with automated attendance system. The participation checking framework has made crafted by the instructors considerably easier. Face identification and acknowledgment are fundamental in numerous applications, one of which is the Attendance Management System. We humans can recognize faces that are very familiar to us. Although we may fail or confuse to recognize people sometimes. Sometimes we remember that we have seen the person but fail to recollect their name.

M. N. Chowdary (✉) · V. Sujana · K. Satvika · K. L. Srinivas · P. S. Suhasini (✉)
ECE Department, V R Siddhartha Engineering College, Vijayawada, India
e-mail: sarahsuhasini@vrsiddhartha.ac.in

J. I.-Z. Chen et al. (eds.), *Machine Learning and Autonomous Systems*, Smart Innovation, Systems and Technologies 269, https://doi.org/10.1007/978-981-16-7996-4_8

In this model, we are going to recognize faces with the help of a computer. Face recognition is a system that is used to confirm or identify the identity of an individual. Due to the widespread usage of smartphones and cameras, face recognition became more successful in the smart society. The picture of the individual may be captured live using a webcam, or it can be an image that is recorded through a video [9].

It was believed that Woodrow Wilson Bledsoe was the father of face recognition. He used a tablet named RAND to record the coordinates of the face. This tablet is equipped with a stylus that releases electromagnetic waves. He used that system to manually record the coordinates, then by making use of the database, he tried to recognize the faces, but it cannot produce accurate results. Later many people proposed their systems in the field of face recognition. Even Facebook also uses face recognition that helps people detect the faces in the images uploaded by Facebook users. Different kinds of techniques are used to implement a facial recognition system like the HOG algorithm, R-CNN (region-based convolution neural networks). Holistic methods take the whole face as input, and face recognition is done by converting them into vectors.

In this paper, we are using the HOG algorithm for face recognition. It is carried out in two steps, i.e., face detection and face recognition. Face detection can be done by removing unnecessary noise from an image. For this purpose, we are using the HAAR cascade classifier. Classifier means a model or an algorithm that classifies the given input to different categories based on attributes. There are different types of classifiers like naïve Bayes, perceptron, Decision trees, logistic regression, etc. The use of a classifier depends on the problem to be solved. Likewise, we are using the HAAR cascade classifier for face detection. Once the face is detected, we extract the features from the face using HOG (Histogram of Oriented Gradients). Here gradient is nothing but an arrow-like structure used to identify the basic structure of the face, including eyes, lips, nose, etc. Then we are using SVM (Support Vector Machine) to recognize the face and produce the recognition label (here the label is the name of identified person).

8.2 Literature Survey

Face recognition is grabbing more attention in the present world of technology. If we take a daily life example, it is being used for unlocking phones. However, face recognition and detection can be carried out in different ways. The authors in [1] proposed a face recognition with image enlargement technique. It uses Principal Component Analysis (PCA) algorithm [9] for face recognition. For image enlargement, three well-known techniques, namely, Nearest Neighbor, Bilinear and Bucolic are investigated. First, an input image is down sampled to six different resolutions. The down-sampled image is then enlarged to its original size using above mentioned three image enlargement techniques. Another method for a face recognition system is by using DCT (Discrete Cosine Transform) coefficients-based face descriptor

[2]. In this work, the face descriptor's dimensional size decreases using Predictive Linear Discriminate Analysis (PDLA), and the KNN (K-Nearest Neighbors) is implemented for verification. Accuracy can be significantly improved by using sparse representation and feature fusion [3]. Firstly, the training samples and test samples are pre-processed by Gray image conversion, histogram equalization, scaling, smooth filtering. Secondly, the LBP, Gabor, and HOG features of face images are extracted and finally, the RSC classification test is carried on the partial samples. Similarly, the authors in [6] proposed a face acknowledgment framework dependent on Stacked Convolutional Auto encoded and Sparse Representation (SCAE) is utilized to separate the profound highlights as the highlights extricated by conventional techniques are straightforward and rudimentary. In [4], A face recognition algorithm based on Prewitt and CNN (Convolution Neural Network) is proposed. It used L2 regularization, and Dropout methods which is used to prevent overfitting. Another system for face recognition is via the Lorentzian metric [5]. In this system, a new Lorentz Face Recognition (LFR) method based on the unique properties of Lorentz space was developed. An automatic attendance management system that helps in replacing the manual method, i.e., the attendance without human interference was proposed in [7]. However, the model failed in capturing all the student's faces in a classroom as the camera is fixed at a specific place. Another application of the face recognition system in attendance systems is proposed in [8], where a hand-held device is used to mark the attendance without the intervention of lecture time. A sensor is used to mark the presence, where a student needs to place their finger over it. It is a portable device and can be easily carried. An outline of various face acknowledgment strategies has been penned down in [10]. It features the methods that give great effectiveness for brightening changes also, for extraordinary natural conditions.

8.3 Proposed Architecture

A folder in which code and images are to be kept is created. A folder with student images will be created inside that folder. Some pictures of different students are to be clicked (one person per picture) and rename the picture to student name (e.g., nitin.jpeg or sujana.jpeg) and store all of them in the image folder which is created earlier. It is necessary to ensure that each image will have only ONE face in it, they can't be group pictures and they are all in JPEG format with filenames ending in.jpeg and they need to be trained in order to extract the required features (Fig. 8.1).

In the face recognition-based system, there might be a chance that camera does not capture the image properly and in some cases when a camera is fixed in the class, it might miss some of the students while capturing. In order to overcome these drawbacks, phone cameras which in minimum will have 16 megapixels can be used so that the clarity of image or video will be high, hence it increases the accuracy. Instead of taking attendance manually lecturer can just take a photograph of class with their phone and send it to local host which is created by using mobile application. Image will be sent automatically into image enhancement mode and is

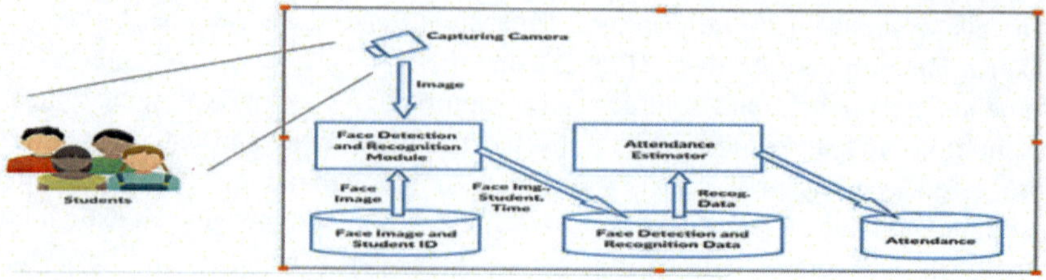

Fig. 8.1 Flow process for group image (During Recognition)

detected, recognized and is compared with the stored and trained database and then finally the attendance is updated in the local server. End of the day any student can check the status of their attendance and it becomes easier to generate monthly reports and eases the work of lecturer.

8.4 Motivation Behind the Work

In today's world, security is one of the significant issues. Many technologies are being developed to enhance security issues. One of the technology is face recognition. Why it has more significance means it does not need any wanted action from the user in identification. Using this technology, many real-world cases are being solved like criminal identification, etc. By using face recognition technology, we can also ease the mankind work like marking attendance for every class.

Now we are going to discuss the different techniques of face recognition and the limitations of these techniques. One of the techniques used to implement face recognition is principal component analysis (PCA). The principal component analysis uses eigenvectors to remove noise and then by using eigenfaces generated, it will magnify the image to recognize. It does not extract any facial features. The proper centered face is necessary, which means recognizing the face in only the front view. It also fails to recognize the faces in different orientations and lightning conditions. In this proposed work, we are overcoming this limitation.

Linear discriminant analysis is also used for face recognition. In this technique, projections of images are defined by fisher faces and this is known as fisher space. When we want to recognize a face it is fed into the fisher space and then uses KNN (k-nearest neighbors) algorithm for identification. It works better under lightning conditions but accuracy is a limitation.

Face recognition is also done by using artificial neural networks. A function named as radial basis function artificial neural network is integrated with a non-negative function matrix factorization. For the simplification of this process computation has been sped up. It is used for recognizing face images with partial distortion and occlusion. The main disadvantage of this approach uses more samples which makes it more complex. The results are also not accurate.

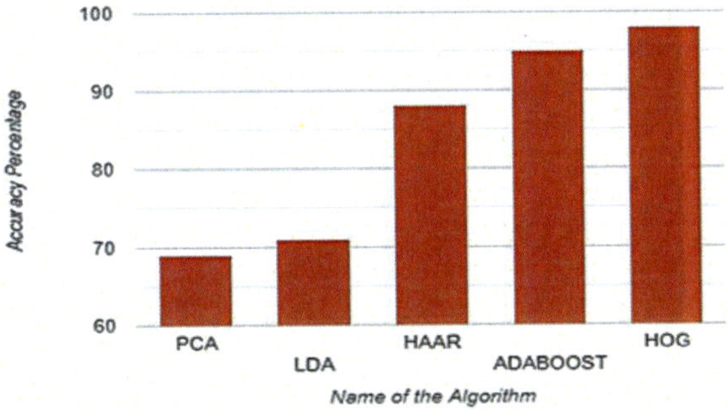

Fig. 8.2 Pixel representation of an image

In this work, we are using the HOG algorithm for face recognition. It involves two stages face detection and face recognition. Face detection can be done by removing unnecessary or noise from an image. For this purpose, we are using the HAAR cascade classifier. Once the face is detected, we extract the features from the face using HOG (Histogram of Oriented Gradients). Then we are using SVM (Support vector Model) to recognize the face and produce the recognition label (here, the label is the name of identified person). The proposed system has the following extra features like Face recognition with high accuracy, introducing an attendance system to record turnout time, recognizing the face in different lighting conditions, and different orientations (Fig. 8.2).

8.5 Design and Technical Description

The block diagram of the model is as follows.

From the block diagram (Fig. 8.3), we can see the flow of the process. First, we provide an image then the HAAR cascade classifier will detect the face by removing noise. Later HOG extractor is used to extract the features and an SVM model is used to recognize the face and provide the name of the person as a recognized label. OpenCV has more than 2500 libraries that consist of different machine learning algorithms. Different features of the OpenCV library capture video and images. Processing images means filtering and transforming, reading and writing images, detecting faces, objects, eyes, cars, etc. in the image, analyzing video in slow motion,

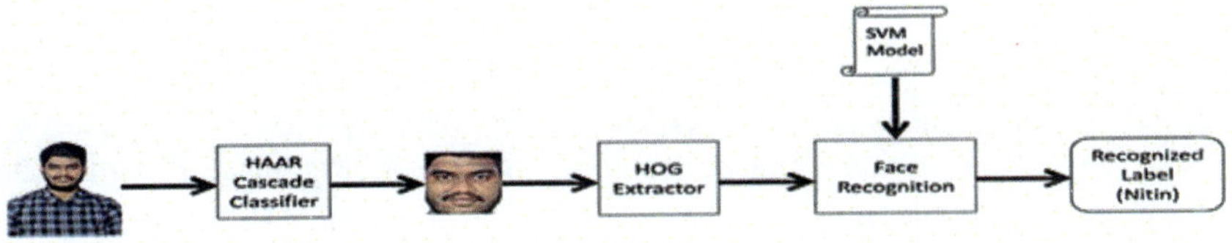

Fig. 8.3 Marking Gradients

Fig. 8.4 HOG representation

etc. Some of the main libraries are highgui, objdetect, features2d, calib3d, video I/O, video, etc. The required modules are cmake 2. Dlib 3. Face recognition 4. NumPy 5. OpenCV python 6. Pip.

8.6 Methodology

8.6.1 *Input*

The first and the front most advance in any of the model is to give the input. There are many kinds of input like the input may be in form of text, numbers, images, etc. depending on the type of the problem. As our problem is face recognition we are taking the images of the people as input to our problem (Fig. 8.4).

8.6.2 *Face Detection*

In order to recognize the face first we have to detect where the face is located in the image. Because we always do not make sure that the provided image consists of only the face of the person. There may be some other objects in the image. Moreover, many of the images will contain the entire image of the person. But we do not need the entire image of the person to recognize his/her face. So we need to remove the unnecessary things in the image. This step is known as removing noise or noise removal. Noise is nothing but the corrupted pixels in the image that may lead to inaccurate results. Noise can be created during the time of photo capture or during the time of the transmission. Here we are going to use HAAR cascade classifier for the purpose of face detection.

8.6.3 HAAR Cascade Classifier

Paul et al. proposed "Rapid object detection using boosted cascade of simple features" and discussed the HAAR cascade classifier in the year 2001. Cascade function is utilized to prepare a great deal of positive and negative pictures. Positive pictures are the pictures that contain the pictures which we need our classifier to distinguish. Negative pictures are the Images of all the other things, which do exclude the article we need to identify. For noise removal, HAAR features are used. Different HAAR features are edge features, line features, four-rectangle features (Figs. 8.5, 8.6 and 8.7).

By applying the above features to input image, the result will be as follows (Fig. 8.8).

The flow of process in HAAR cascade for input image is as shown in Fig. 8.9.

The output of the above step will be as shown in Fig. 8.10.

The flow of process in HAAR cascade for group image is as shown in Fig. 8.11.

The next step in the process is to extract the features of the face using HOG extractor.

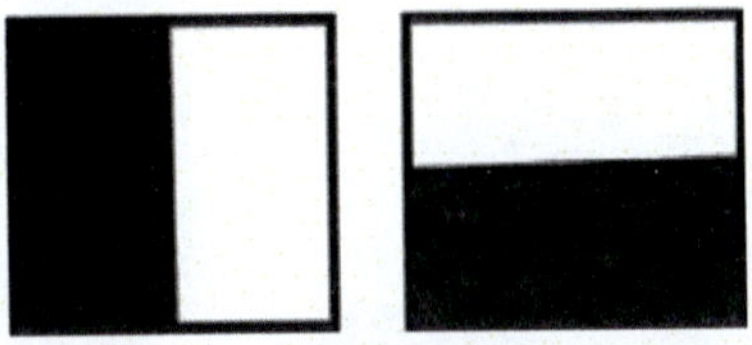

Fig. 8.5 Edge features

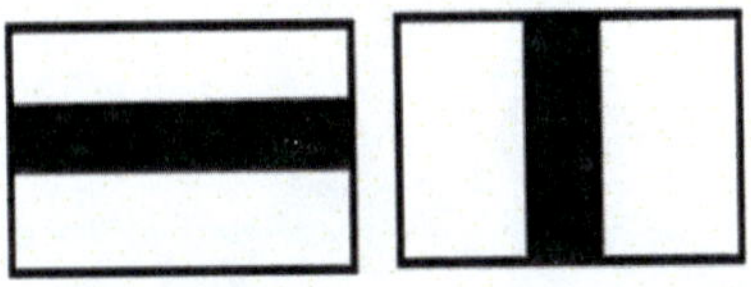

Fig. 8.6 Line features

Fig. 8.7 Four-rectangle features

Fig. 8.8 HAAR features applied to input image

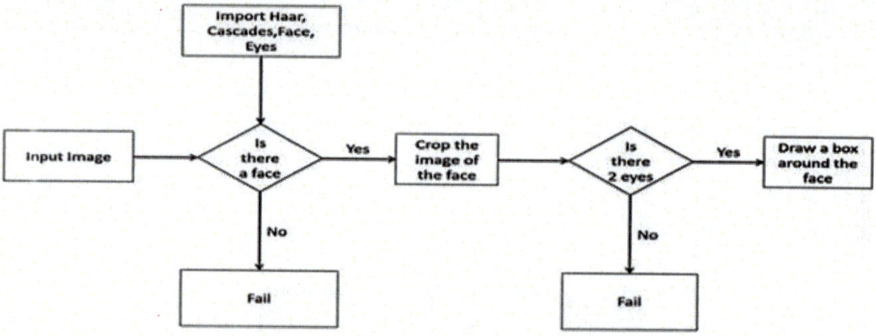

Fig. 8.9 Flow process for input image (During Training)

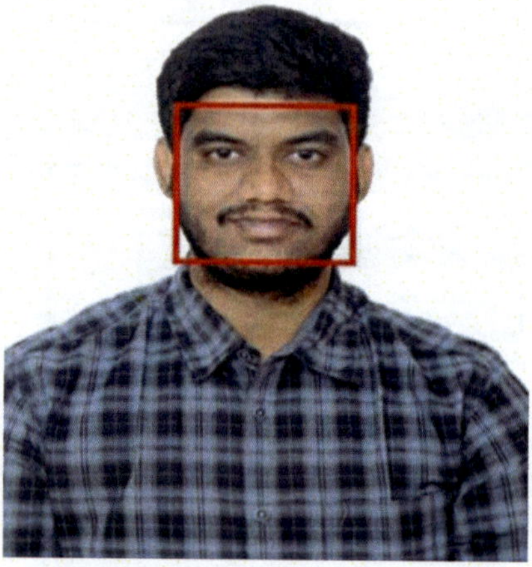

Fig. 8.10 Cropped face

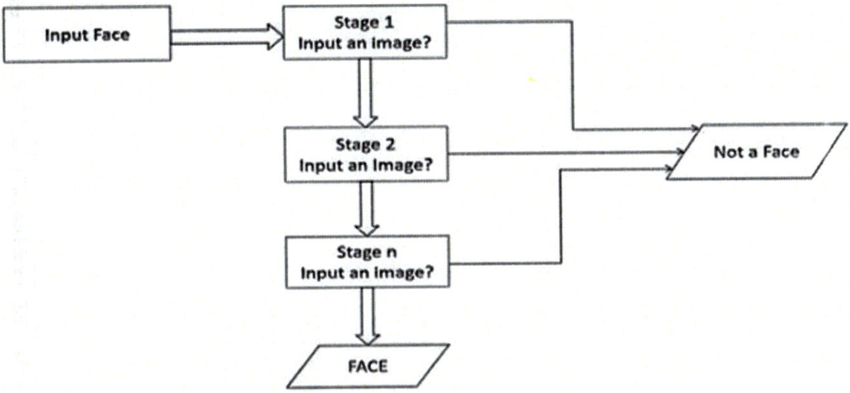

Fig. 8.11 Flow process for group image (During Recognition)

8.6.4 HOG (Histogram of Gradients Algorithm)

Now we will take the detected face and extract features from the face using a histogram of gradients. For that, first we need to find the gradients in the image. We all know that a cluster of pixels forms an image. When we move toward right one pixel at a time, we might find a modification in pixel value from black to white, which means that is changed from lower to higher pixel number. This change is known as gradient. Changing from lighter tone to darker tone is a negative gradient and vice versa. As the name suggests horizontal gradient represents left to right and vertical gradient represents top to down.

The end result gives the basic structure of a face in a simple way (Fig. 8.12).

The image position might not be the same in all cases. Sometimes only half of the person's face may be visible, or it may be tilted. But it is important to recognize the face in such scenarios also. To overcome this, we wrap every picture in such a way that the lips and eyes are always in sample place in the image. We identify 68 particular points known as landmarks on a face. For example, bridge of the nose, the inner edge of each eyebrow, top of the chin, etc (Fig. 8.13).

Finally, in this step the features of the face that means measurements of the face will be generated (Figs. 8.14 and 8.15).

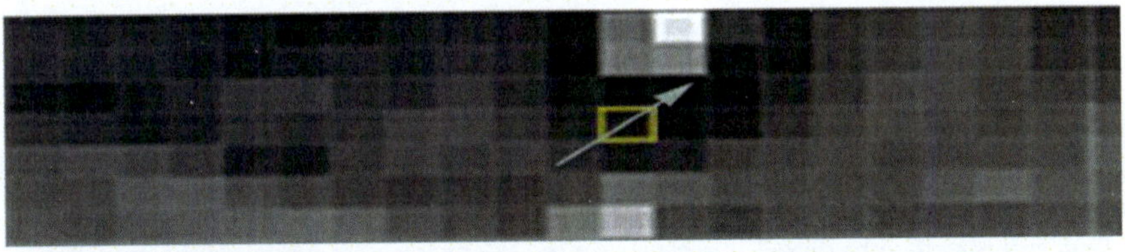

Fig. 8.12 Pixel representation of an image

Fig. 8.13 Marking gradients

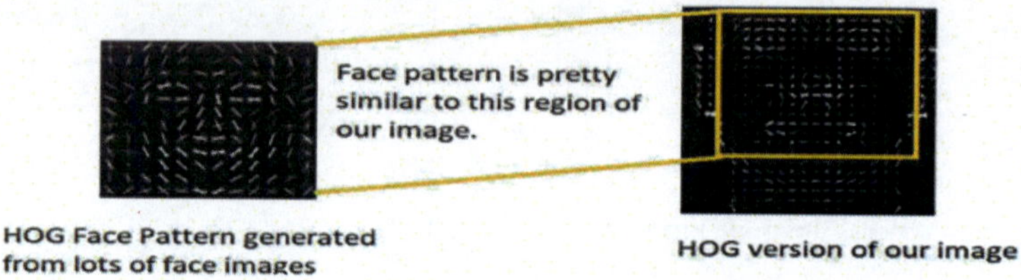

Fig. 8.14 HOG representation

Fig. 8.15 Facial landmarks estimation

128 Measurements will be Generated for the face shown in Fig. 8.16.

8.6.5 *SVM (Support Vector Machine) Classifier*

The final step is actually quite simple. The only task is to find the person with facial measurements close to those of our test image. Here SVM classifier does two tasks.

Fig. 8.16 Generating measurements of above image

Firstly, it does some extremely intricate data transformations. Secondly, it classifies your data based on the outputs or labels you have defined. This SVM classifier takes milliseconds to run and it displays the person's name as the result. The main moto of Support Vector Machines is to solve traditional two-class pattern recognition problem. SVM is being adapted to face recognition, by altering the interpretation of classifier's output and planning a representation of facial images which are consistent with a two-class problem. Traditional SVM returns the class of the object, i.e., binary value SVM algorithm gets trained by altering the issue in different space. This algorithm distinctly captures the differences between two facial images. The withdrawal from view-based approaches encrypts every facial image as a discrete face view. Talking about the different space, the two compelling classes are the differences between images of two different individuals and differences between images of same individual. These are the two classes which would be the input to SVM algorithm. SVM algorithm creates a decision surface classifying the two classes. In face recognition, we re-explicate the decision surface to create similar metric between two facial images, which assists us to build face recognition algorithms.

8.7 Results

In this proposed system, we developed an android app. This app consists of features like taking an image of peoples and uploading the images to the server. Whenever the administrator wants class attendance, they will run the Python code. The algorithm HAAR Cascade classifier and HOG Extractor will recognize the faces in the image and update the attendance in the particular class excel sheet.

- The faculty must install the Android app that we have developed for the attendance System.
- Start the WAMP server. It helps us in providing a flexible web development environment and helps us in creating web applications. Alongside, PhpMyAdmin allows you to manage your databases easily.

Once the WAMP server is started, open the command prompt and use the command "ipconfig" and note down the IPv4 Address, which is the IP Address of the server and it is to be entered in the android application that is created for this model to establish the connection (Figs. 8.17 and 8.18).

Once the connection is established, it redirects to the home screen of the android application (Fig. 8.19).

Click on "TAKE SNAP" and after the snap is taken, either click on "SECTION1" or "SECTION2" depending on the section to which you want to upload the image (Fig. 8.20).

Fig. 8.17 UwAmp

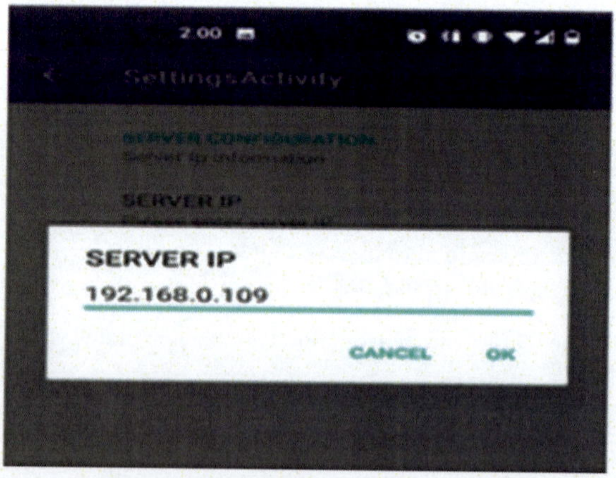

Fig. 8.18 Connecting to server

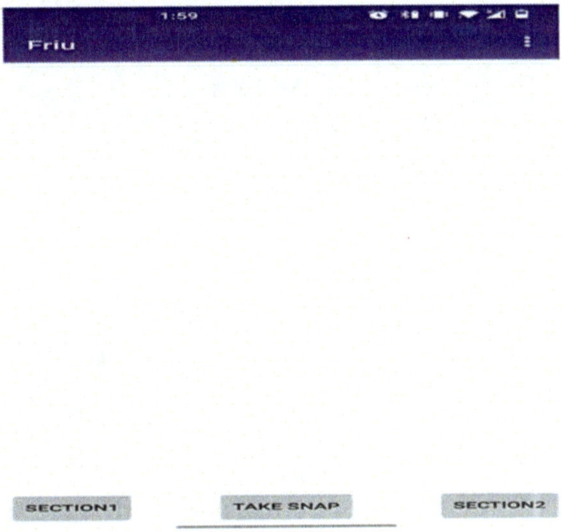

Fig. 8.19 Application home screen

Fig. 8.20 Files uploaded to server

8.7.1 Libraries

Libraries that are needed to be installed in IDE for performing some of the key operations include face recognition, listdir, etc. These libraries will simplify the need of training and detection operations needed to be performed in the proposed system.

8.7.2 Database

A particular folder is to be created within the path, which stores the images of students (one student face per each image) (Fig. 8.21).

Fig. 8.21 Database folder

Fig. 8.22 Input image

8.7.3 *Training*

List of images will be taken as input and corresponding encoding operation is carried out. However, as mentioned earlier in the methodology encoding information will be generated. Encoding information including the sequences and corresponding name of the face in that particular image are stored in two separate text files. Finally after the execution of the code, i.e., after performing the process as mentioned in the methodology, two text files containing the encoding sequences and the corresponding names will be created in the project folder (Fig. 8.22).

8.7.4 *Detection*

The leading step in the proposed system is to detect the faces in a group image and identify the faces by using the corresponding encoding sequences and names, obtained as a result of training stage.

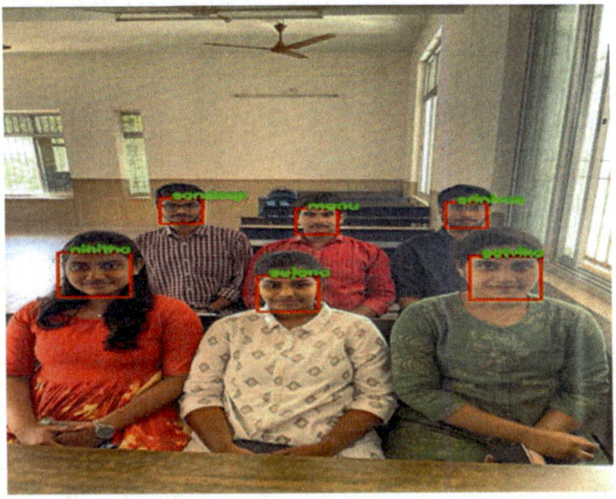

Fig. 8.23 Output

8.7.4.1 Input

Input Image is uploaded to the server through the mobile application developed. The name of the image will be a timestamp, using which specific time will be identified on a particular day so that attendance will be marked for that particular period on a particular day. Figure 8.19 shows the image before being applied as an input to the detection code.

8.7.4.2 Output

After the execution of the entire code for detection that is designed for the proposed system, it will result in a new image in the project folder which highlights the student's face identified with a rectangular box along with their name (Figs. 8.23 and 8.24).

The students presented in the input image will get the attendance marked in an excel sheet as shown in Fig. 8.20a.

8.8 Conclusion and Future Scope

Finally, we have developed an android application and combined it with a better methodology for face recognition that provides high accuracy and identifies a person in different angles also in different lighting conditions. This model can be widely implemented in colleges and schools, which makes the lecturer's work easier and quicker. In the future also it may be extended further. It can be used in hospitals. It can be used for different purposes like preventing fraud activities in ATMs. Face recognition can be used in identifying duplicate voters, verifying driving license,

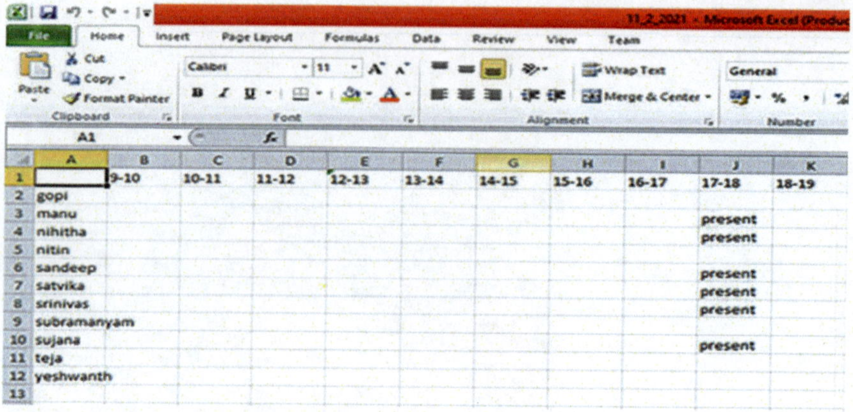

Fig. 8.24 Attendance marked in an Excel sheet

and identifying terrorists at the airport, railway stations, and malls. The world is advancing in technology; face recognition can be used to identify or find a person on a social media platform. There are many applications involving face recognition in areas like subways and other transportation outlets, which use our face as a credit card to pay for transportation charges this would save time and reduce traffic.

References

1. Halidu, M. K., Bagheri-Zadeh, P., Sheikh-Akbari, A., & Behringer, R. (2019, June). Pca in the context of face recognition with the image enlargement techniques. In *2019 8th Mediterranean Conference on Embedded Computing (MECO)* (pp. 1–5). IEEE.
2. Wijaya, I. G. P. S., Husodo, A. Y., & Arimbawa, I. W. A. (2016, October). Real time face recognition using DCT coefficients based face descriptor. In *2016 International Conference on Informatics and Computing (ICIC)* (pp. 142–147). IEEE.
3. Jiang, C., Wang, M., Tang, X., & Mao, R. (2019, November). Face recognition method based on sparse representation and feature fusion. In *2019 Chinese Automation Congress (CAC)* (pp. 396–400). IEEE.
4. Xiao, J., & Weiwei, G. (2019, October). Face recognition algorithm based on Prewitt and convolutional neural network. In *2019 3rd International Conference on Electronic Information Technology and Computer Engineering (EITCE)* (pp. 2007–2009). IEEE.
5. Kerımbekov, Y., & Bılge, H. Ş. (2017, May). Face recognition via Lorentzian metric. In *2017 25th Signal Processing and Communications Applications Conference (SIU)* (pp. 1–4). IEEE.
6. Chang, L., Yang, J., Li, S., Xu, H., Liu, K., & Huang, C. (2018, November). Face recognition based on stacked convolutional autoencoder and sparse representation. In *2018 IEEE 23rd International Conference on Digital Signal Processing (DSP)* (pp. 1–4). IEEE.
7. Nandhini, R., Duraimurugan, N., & Chokkalingam, S. P. (2019). Face recognition based attendance system. *International Journal of Engineering and Advanced Technology (IJEAT), 8*(3S).
8. Mohamed, B. K., & Raghu, C. V. (2012, December). Fingerprint attendance system for classroom needs. In *2012 Annual IEEE India Conference (INDICON)* (pp. 433–438). IEEE.

9. Khan, M., Chakraborty, S., Astya, R., & Khepra, S. (2019, October). Face Detection and Recognition Using OpenCV. In *2019 International Conference on Computing, Communication, and Intelligent Systems (ICCCIS)* (pp. 116–119). IEEE.
10. Patel, R., Yagnik, S.B.: A literature survey on face recognition techniques. International Journal of Computer Trends and Technology (IJCTT) **5**(4), 189–194 (2013)
11. Sujay, S. N., Reddy, H. M., & Ravi, J. (2017, December). Face recognition using extended LBP features and multilevel SVM classifier. In *2017 International Conference on Electrical, Electronics, Communication, Computer, and Optimization Techniques (ICEECCOT)* (pp. 1–4). IEEE.
12. Damale, R. C., & Pathak, B. V. (2018, June). Face recognition based attendance system using machine learning algorithms. In *2018 Second International Conference on Intelligent Computing and Control Systems (ICICCS)* (pp. 414–419). IEEE.
13. Sushama, M., & Rajinikanth, E. (2018, April). Face recognition using DRLBP and SIFT feature extraction. In *2018 International conference on communication and signal processing (ICCSP)* (pp. 994–999). IEEE.
14. Bhele, S.G., Mankar, V.H.: A review paper on face recognition techniques. International Journal of Advanced Research in Computer Engineering & Technology (IJARCET) **1**(8), 339–346 (2012)
15. Singh, A., Sharma, S., Singh, S.: Android Application Development using Android Studio and PHP Framework. International Journal of Computer Applications **975**(8887), 5 (2016)

Chapter 9
Novel Approach to Phishing Detection Using ML and Visual Similarity

Preet Sanghavi, Achyuth Kunchapu, Apeksha Kulkarni, Devansh Solani, and A. Anson

Abstract Phishing involves attempts to trick the user by extracting crucial important information that should otherwise not be leaked. Data like bank account numbers, social media accounts, company revenue reports, online transactions secrets are some of the examples that an attacker hopes to extract by exploiting the vulnerability of the user. In today's day and age, it is very important to warn the populace against such attacks and provide relevant security awareness. Through our research, we aim to provide a brief study and novel insights into how machine learning and image visualization can be used to detect phishing webpages using a two-part implementation technique. The first part involves an analysis and overview of different machine learning classifiers across different datasets and features to find the most relevant technique to detect phishing websites efficiently. The second part introduces a novel approach of how discrete Fourier transform and image comparison can be used to compare suspicious webpages with real ones. This study aims at providing a new direction for researchers to understand how machine learning and automation can be inculcated in phishing detection to avoid any data breach.

P. Sanghavi (✉)
Dwarkadas J. Sanghvi College Of Engineering, Mumbai, India

A. Kunchapu
Vellore Institute of Technology, Vellore, India

A. Kulkarni
Vivekanand Education Society's Institute of Technology, Mumbai, India

D. Solani
Nagindas Khandwala College, Mumbai, India

A. Anson
JSPM's RSCOE, Pune, India

J. I.-Z. Chen et al. (eds.), *Machine Learning and Autonomous Systems*, Smart Innovation, Systems and Technologies 269, https://doi.org/10.1007/978-981-16-7996-4_9

9.1 Introduction

Attackers try to deceit people by trying to gain access to private information. Phishers have realized that the best way to get private data is through webpages that are highly similar to their legitimate counterparts. In this paper, we will try to overcome this problem using two approaches. The first tackles phishing using machine learning whereas the second part uses visual similarity for phishing detection.

9.1.1 Machine Learning Algorithms in Phishing Detection

Machine learning is a branch of computer science that studies learning algorithms based on examples. Classification is a kind of task that necessitates the use of ML algorithms to learn how to label problem domain instances with a class label. Classifying a website as 'spam' or 'not spam' is a simple example. In this study, we have considered four ML algorithms for our approach to detect phishing websites. They are Logistic Regression, MLP Classifier, DTree Classifier, and Sequential Classifier.

- Logistic Regression: Given particular variables, LR is employed to predict the output with two different values, zero or one, agree or disagree, and so on. The logistic model accepts real-valued inputs and predicts the likelihood of them falling into the default class (class 0). We can use the output as a forecast for the default class (class 0) if the probability is greater than 0.5, otherwise, the prediction is for the opposite class (class 1). It's commonly used to visualize data and highlight the connection between one binary that is reliant and a notional number of independent features.
- MLP Classifier: The perceptron is the first step in building a neural network. In simple words, the perceptron takes inputs, multiplies them by some weights, and then sends them through an activation function (such as logistic, relu, tanh, or identity) to get an output. A multi-layer perceptron model is constructed by combining the layers of these perceptrons to form neural networks. The input, hidden, and output layers are the three layers of a neural network. The data are received directly by the input layer, while the appropriate output is created by the output layer. Hidden layers are the layers in between that perform the intermediary calculation. Starting with the input and ending with the output. Except for the input nodes, each node has an activation function that is utilized to get the output of the node using inputs from other nodes that are weighted.
- Decision Tree Algorithm: In the realm of machine learning, DTree is one of the most extensively used algorithms. The decision tree algorithm is simple to comprehend and implement. A parameter is represented by an inner node, a decision command is represented by a branch, and in a decision tree, the leaf node symbolizes the conclusion. The root node is the topmost node in the tree. It splits the data into groups according to the parameter value. It splits the tree recursively, which is called recursive partitioning. When contrasted to the use of a neural network, it

takes less time to train. The temporal complexity of the decision tree is determined by the dataset's size and the number of records and parameters. DTrees are capable of managing large amounts of dimensional data with the utmost precision.

- Sequential Classifier: Neural networks are a rough approximation of the human brain's learning process. Neurons are the building blocks of an Artificial Neural Network, and layers are created by them. Each epoch, the weights connected with the neurons and accountable for the overall predictions are modified. Each Neural Network has a cost function that gets smaller as the learning progresses. The best weights for which the cost function gives the best results are then used. In Keras, the simplest technique to build a model is sequential. It allows you to layer-by-layer construct a model. Each layer has weights that match the weights of the one above it.

9.1.2 Image Comparison in Phishing Detection

Fast Fourier Transform: There are several transformations possible like discrete cosine transform, discrete Fourier transform as seen in [15]. In our study, we make the use of an optimization of Discrete Fourier Transform. There are many fast and efficient optimizations to this DFT algorithm. One of such examples is the fast Fourier transform that has a faster and efficient way of evaluating DFT. FFT reduces the time complexity from $\mathcal{O}(n^2)$ to $\mathcal{O}(n \log n)$.

In FFT, while compression takes place in the replaced domain, coefficients that have small as well as bigger frequency values are preserved for the next step. In the next step, the high-frequency values are quantized meaning that they are gotten rid of from the converted images. This data manipulation technique contributes to the concept of image condensation and compaction.

Image Comparison Metrics: Image comparison is another approach for visually classifying images as representatives of other images. In this study, the quality of the photograph is an important factor. In phishing detection, labeling images with their ground truth values is a difficult task; therefore, images are assessed by using reference metrics like mean squared error. In this paper, we have used the following two similarity metrics for image similarity after passing our images through Fourier transformations.

Mean-squared error: The mean-squared error between two images can be given by formula (9.1).

$$MSE = \frac{1}{mn} \sum_{i=0}^{m-1} \sum_{j=0}^{n-1} [I(i, j) - K(i, j)]^2 \quad (9.1)$$

This metric is generally used as a evaluator of photograph characteristic metric. For two images, if the value of MSE is close to zero, it means that the two images are similar to each other. However, one common problem while employing MSE is that it has heavily relied upon the intensity of pixel values.

Moreover, to involve in more deeper evaluations of different images, another metric called as the SSIM or structured similarity index method can be employed as noted in [14]. This metric can help us study structure and similarity between features of two different images for efficient image classification as legitimate or not. SSIM: The SSIM value between two images (x and y) can be given by the formula (9.2).

$$\mathrm{SSIM}(x, y) = \frac{(2\mu_x\mu_y + c_1)(2\sigma_{xy} + c_2)}{(\mu_x^2 + \mu_y^2 + c_1)(\sigma_x^2 + \sigma_y^2 + c_2)} \tag{9.2}$$

Structural similarity index method is a model to perceive senses that uses image destruction as a matter of difference in terms of the body of the image. Factors such as luminance, masking of contrast are also inculcated that helps direct the importance of SSIM along heavily connected pixels. These pixels can be used to extract crucial information from the data. Luminance factor helps understand parts of an image that are not clearly visible along the edges whereas contrast masking is used for places where the texture of data is not clearly defined. This is the second metric used in our study for calculating the similarity of legitimate and suspicious webpages.

9.2 Literature Review

9.2.1 Machine Learning in Phishing Detection

In this paper, authors [1] explained the methods to catch phishing sites using ML algorithms. They have compared three machine learning techniques, DTree, Random Forest algorithm (RF), and Support Vector Machine for accuracy and false positive rate and false-negative rate by splitting the training and testing dataset into three different ratios. Authors [2] proposed a classification model to classify the website phishing. In this paper, four classification algorithms, namely, Naive Bayes, Regression Tree, K-nearest neighbor, and SVM are utilized to compare confusion matrix, success rate percent, and error rate percent, additionally, compared the page rank of benign websites and phishing websites Authors [3] explained an approach to find phishing behavior using page rank, primary domain, subdomain, path domain, and Alexa rank. The training was done using 9661 phishing websites, and the testing was done on a dataset that was collected from Phishtank. It contains 1000 phishing sites and 1000 legitimate sites. In this paper, authors [4] proposed classification models, logistic regression, Decision tree, Random Forest, ada-booster, KNN, Neural Network, SVM, Gradient boosting, and XGBoost. Different metrics train time, test time, accuracy, recall, and F1 score are compared for these different classifiers. Machine learning algorithms are trained and tested on a dataset from Phishtank, which has about 11,000 sample websites. The authors [5] of this research examine the characteristics of phishing websites and propose two types of features for web phishing detection: original feature and interaction feature. Then, to detect phishing websites,

DBN was introduced, and the detection model and algorithm for DBN were discussed. Authors [6] presented an approach to identify the phishing website using KNN, DTree, and Random forest algorithms and finally compared the accuracy of all three algorithms used and concluded that they are getting better accuracy with Random forest algorithm. In this paper, authors [7] use classification models Pruned Decision tree, SVM, Naive Bayes, and Neural Network. Various metrics for these different classifiers, TPR, FPR, and accuracy scores are compared. The dataset used for training and test the machine learning algorithms comes from the University of California, Irvine ML Repository. In the previous studies, most of the research done by different authors focus on the algorithm and its accuracy. But in our research, we have identified how important the choice of features in the dataset is and their impact on the corresponding algorithm to get better accuracy and how accuracy changes based on the number of features used for three different datasets.

9.2.2 Visual Similarity-Based Phishing Detection

As seen in [16], Henri et al have summarized the working and applications of Fourier transformations on images alongside a brief overview of different Fourier techniques for hasting the time complexity thereby creating a platform for fast Fourier transformation. A demonstration of how fast Fourier transform has helped significantly in greatly decreasing the magnitude of mathematical calculations helps us turn our attention to this methodology and apply it in phishing detection based on visual similarity. Moreover, as seen in [17], James W. Cooley and his team helped shed light on the applications of fast Fourier transform, a wide array of different functions of this method are also realized like signal processing, spectral analysis, or solutions to differential equations. Finally, we used SSIM for checking structural similarity because as seen in [18], structural similarity index is highly advanced photo quality metric. The results are highly accurate and advanced to the point where human observation fails to detect differences.

9.3 Proposed Pipeline and Implementation

9.3.1 Pipeline and Overview

Our research proposes a novel twofold phishing detection technique. In the first part, if our pipeline involves phishing detection using machine classification, each of our three datasets discussed in the next subsection will be passed through a pipeline. In the pipeline, the data are cleaned and features are extracted after which our five classification models are trained and tested. In the second part, legitimate and fake webpage images are passed through Fourier transformation after which the RMS

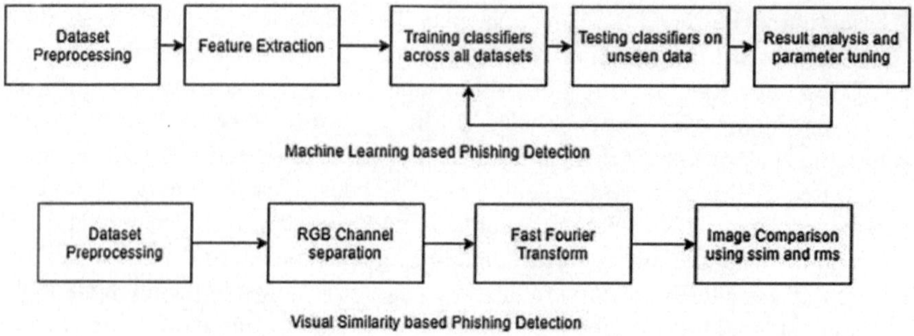

Fig. 9.1 Twofold proposed pipeline

Table 9.1 Count of features in various datasets

Dataset	Count of features
Dataset 1	12
Dataset 2	28
Dataset 3	27

and SSIM values of the images are calculated. Both the folds of our study can be summarized, as shown in Fig. 9.1.

9.3.2 *Dataset Description*

Machine Learning Datasets: The three datasets used in this Machine Learning Techniques were downloaded from different sources (OpenML.org and Mendeley Data). The data include nine features derived from the URL. The properties give data such as the popup window, URL anchor, age of the domain, Number of dots, length of URL, IP, and Internet traffic, etc. Every feature value contains binary or ternary categorical values. The binary value indicates that the value allocated to a feature is determined by whether or not that feature exists inside the URL. The value ascribed to ternary features is determined by the presence of the feature in a specified ratio. In all three datasets, the output variables consist of two binary values in which '0' refers to Legitimate and '1' refers to Phishing. Tables 9.2, 9.3, 9.4 list the feature description in three datasets that we employed in this study. Table 9.1 refers to the number of the features available in the various datasets

Image Comparison Data: For the image visualization part of our implementation, we have collated data from different sources like OpenPhish and PhishTank. Fake websites images have been collected from the aforementioned websites whereas their visual counterparts have been scraped from the internet.

Table 9.2 Description of dataset 1 features

Feature	Meaning
ldl_url	IDL of URL
dld_url	DLD of URL
urlLen	URL length
domainlength	Domain length
pathurlRatio	Path to URL ratio
ArgUrlRatio	Argument to URL ratio
domainUrlRatio	Domain URL ratio
URL_sensitiveWord	Sensitive words in URL
NumberRate_URL	Number rate in URL
URL_DigitCount	Number of digits in URL
URL_Letter_Count	Number of letters in URL
SymbolCount_URL	Number of symbols in URL
URL_Type_obf_Type	Result

9.3.3 Dataset Preprocessing

Machine Learning Datasets: Before using machine learning techniques, the vectors should be normalized to guarantee that each feature has an equal impact on the classification process. We find the maximum and minimum values for each feature, and for each value of this feature.

$$NormalisedValue = (CurrentValue - MinValue)/(MaxValue - MinValue) \tag{9.3}$$

After normalization, all feature values fall into the 0 to 1 range, and each feature contributes equally to the classification outcome.

Image Comparison Data: Here, all the images are resized to the same dimensions to avoid any errors in channel classification for FFT as well as for image comparison to calculate RMS and SSIM values.

9.3.4 Training and Testing for Machine Learning Classification

All three datasets after data preprocessing are split in the ratio of 80:20 for training and testing the machine learning algorithms. Also, the number of features is reduced one by one and trained the models again to check the model performance on less number of features and also on important features. All three dataset features are

Table 9.3 Description of dataset 2 features

Feature	Meaning
length_url	URL length
length_hostname	Hostname Length
ip	IP Address
nb_dots	Number of dots
nb_hyphens	Number of hyphen
nb_at	Number of at symbols
nb_qm	Number of question mark symbols
nb_and	Number of 'AND'
nb_or	Number of 'OR'
nb_eq	Number of equal to symbols
nb_underscore	Number of underscore symbols
nb_percent	Number of percent symbols
nb_slash	Number of slash symbols
nb_star	Number of star symbols
nb_colon	Number of colon symbols
nb_comma	'Number of ampersand symbols'
nb_semicolumn	Number of semi colon symbols
nb_dollar	Number of dollar symbols
nb_space	Number of spaces
nb_www	Number of 'www'
nb_com	Number of 'com'
nb_dslash	Number of double slash symbols
http_in_path	Http in path
shortening_service	URL Shortening
domain_registration_length	Domain registration length
domain_age	Domain age
web_traffic	Web traffic of URL
page_rank	Pagerank of URL
status	Result

numeric values. So, it is easy to train the data on Logistic Regression, MLP Classifier, Decision Tree, and Sequential Classifier.

To find how important the feature is we have considered a random number of features to train the algorithm. When an important feature is removed from training, it gives less accuracy.

Logistic Regression For Implementing Logistic Regression we have used sklearn library *LogisticRegression* class. Several optional parameters define the behavior of the model and method in the LogisticRegression class. For our implementation, we

Table 9.4 Description of dataset 3 features

Feature	Meaning
NumDots	Number of dots
SubdomainLevel	Subdomain level
PathLevel	Path Level
UrlLength	Length of URL
NumDash	Number of dash symbols
NumDashInHostname	Num of dash symbols in host name
AtSymbol	Presence of '@' symbols
TildeSymbol	Presence of ' ' symbols
NumUnderscore	Number of '_' symbols
NumPercent	Number of percent symbols
NumQueryComponents	Number of query components
NumAmpersand	Number of ampersand symbols
NumHash	Number of hash symbols
NumNumericChars	Number of numeric characters
NoHttps	Number of HTTPS
RandomString	Random string
IpAddress	IP address
DomainInSubdomains	Domain in subdomain presence
DomainInPaths	Presence of domain in paths
HttpsInHostname	Presence of HTTPS in host name
HostnameLength	Length of hostname
PathLength	Length of path
QueryLength	Length of query
NumSensitiveWords	Number of nonsensitive words
EmbeddedBrandName	Brand name presence
PctExtHyperlinks	Presence of external hyperlinks
PctExtResourceUrls	Presence of external resource URL

have selected maximum iterations for solvers to converge as 2000 and the remaining parameters with default values. Logistic Regression training is quick and gave promising results. When Logistic Classification was performed on dataset 1 which has important features compared to the other two datasets gave promising results.

MLP Classifier Unlike previous classification methods such as Logistic and Decision Tree Classifier, MLP Classifier does classification using an underlying Neural Network. For Implementing MLP Classifiers, we used sklearn library class—*MLPClassifier*. The main parameters for MLP are hidden layer sizes, max iterations, activation function, solver and random state. For this implementation, we have used default values while training the three datasets. The MLP Classifier will take more time than logistic regression to classify the phishing websites and benign websites as it includes hidden layers. As predicted, when the number of features is reduced, the accuracy of the MLP Classifier gets reduced.

Decision Tree A decision tree is a flowchart-like structure in which each of the internal nodes represents a feature test output, Each leaf node represents a class label (decision made after computing all features), and branches represent feature combinations that lead to those class labels. The categorization rules are represented by the pathways from root to leaf. *DecisionTreeClassifier* was created using scikit-learn. The Classifier class will have inputs like cost function, splitter, and others. The Gini cost function is used to determine how pure the nodes are, with node purity referring to the amount of mixed training data provided to each node. In this Implementation for Decision tree classifier, default inputs were considered.

Sequential Classifier Sequential is the model type we'll be employing. In Keras, the simplest technique to build a model is sequential. It allows you to layer-by-layer construct a model. Dense is a common layer type that works well in most situations. All nodes in the previous layer connect to the nodes in the current layer in a dense layer.

The Input Shape will depend on the number of features we are using on the model, as features reduce the Input shape will also reduce accordingly. The output layer is the final layer. It only has one node. The learning rate is controlled by the optimizer. As an optimizer, we'll use the name 'adam.' For the most part, Adam is a nice optimizer to employ. Throughout the training, the Adam optimizer adjusts the learning rate, and loss functions as binary_crossentropy. As predicted, the time consumed by sequential classifiers is more than others.

9.3.5 FFT and Image Comparison

In this part of our pipeline, we present a novel method for distinguishing between a potentially phishing website with the original website. Our main aim is to determine whether the webpages are similar to an extent such that the potentially fake website can be considered as a phishing website. This is true because one of the crucial observations made during our research is that based solely on visual comparison a website can be found out as a phishing or legitimate website. Thus, in order to get a higher confidence in the prediction provided by the machine learning models in the previous section, the webpage evaluator can compare the two webpages based on their RMS and SSIM values to determine how similar the webpages are. If the visual similarity value is high, evaluator will be alerted whereas, in contrast, the pages will be regarded as dissimilar. Steps for image-based webpage comparison are done as follows:

- Step 1: Give input of suspicious URL and legitimate URL webpage images.
- Step 2: Pass the images through the Fourier transformation for image compression. In order to compare both the images, we need to perform Fourier transforms on

the image so that fewer comparisons are involved and the speed and efficiency of the system can be improved. In order to compute FFT, we have to compute each of the three channels of the image separately. Each channel of the original image has been transformed and the transformed channels are then merged back to get the original RGB image.

- Step 3: In this step, the RMS and SSIM values of the images are calculated.
- Step 4: Compare SSIM and RMS values to the threshold value for phishing to detect phishing using Algorithm 1.

Algorithm 1: Algorithm for detecting phishing webpages based on image similarity

$x \leftarrow ssimValue$
$y \leftarrow rmsValue$
$phishingFlag \leftarrow False$
if $x \geq 0.6$ **then**
  $phishinglFag \leftarrow True$
else
  if $x \geq 0.4 \% y \leq 1000$ **then**
    $phishingFlag \leftarrow True$
  else
    $phishingFlag \leftarrow False$
  end if
end if

9.4 Results and Accuracy

9.4.1 Machine Learning Classification

The findings of various machine learning algorithms are presented in this section. In 80:20 ratio, the three datasets are separated into training and testing sets. Each classifier is trained using a training set, and the performance of the classifiers is evaluated using a testing set. The accuracy score of classifiers was calculated to assess their performance. The results of our preliminary research are Phishing URLs and domains have distinct properties from regular URLs and domains. When compared to other URLs and domain names on the Internet, phishing URLs and domain names are much shorter. Many of the phishing URLs included the brand name of the company being impersonated. We used Python to create four classifiers: Logistic Regression, Decision Tree, MLP Classifier, and Sequential Classifier. Phishing URLs were detected using these classifiers. The accuracy of all these classifiers is calculated and compared by reducing the number of features in each dataset.

Table 9.5 Accuracy comparison between classifiers using dataset 1

Number of features	Logistic	MLP classifier	DTree	Sequential classifier
7	88.02	90.85	92.42	92.48
8	87.99	91.99	92.45	92.19
9	88.9	92.4	92.48	93.16
10	88.87	91.2	92.68	93.1
11	89.19	92.25	92.02	93.36
12	91.47	95.4	95.77	96.2

Table 9.6 Accuracy comparison between classifiers using dataset 2

Number of features	Logistic	MLP classifier	DTree	Sequential classifier
23	81.06	82.24	88.15	86.06
24	81.23	83.36	89.05	85.32
25	81.97	85.37	91.94	86.41
26	82.89	88.2	93.42	88.68
27	83.02	88.46	94.99	88.85
28	85.98	92.55	96.12	93.68

Table 9.7 Accuracy comparison between classifiers using dataset 3

Number of features	Logistic	MLP classifier	DTree	Sequential classifier
22	84.65	90.3	86.05	89.75
23	84.7	90.35	85.85	89.45
24	85.35	90.75	86.55	90.2
25	85.2	90	85.85	90.4
26	85.39	94.75	87.75	94.85
27	85.55	94.65	87.65	94.85

Table 9.5 shows that when all the features from the dataset are used, Logistic Regression, MLP Classifier, Decision Tree, and Sequential classifiers provide high accuracy. The process was continued to apply on other datasets as well (Table 9.6).

After comparing results with Dataset 1, the classifiers are trained on more features, the second dataset contains more features than the first dataset and as expected the algorithms gave better accuracy upon increasing the number of features (Table 9.7).

From all the three comparisons, the first dataset gives better accuracy results in Logistic Regression, MLP Classifier, Decision Tree, and Sequential classifier as shown in Table 9.5. This tells us two key points while training the machine learning algorithms. One point to notice is that the accuracy can be increased by increasing

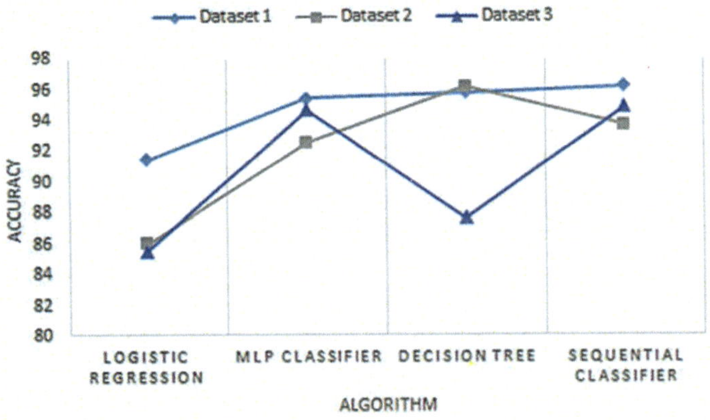

Fig. 9.2 Results graph for different datasets across classifiers

the number of features, and the second is by using the important features like NumberRate_URL denoting the strength of a page's link. Figure 9.2 shows the classifier's overall performance with three datasets with complete number of features.

9.4.2 Image Comparison

The results for visual similarity can be seen in Table 9.8 where the company indicates the brand from which the website images were scraped for the corresponding fake webpages. Company is followed by the real and the fake webpage used for our study

Table 9.8 SSIM and MSE value comparison between real and fake webpages

Company	REAL	FAKE	MSE	SSIM
amazon.png	login_german_real.png	login_german_fake.png	1069.09	0.72
amazon2.png	login_english_real.png	login_english_fake.png	699.87	0.82
apple.png	userid_real.png	userid_fake.png	8913.18	0.6
dhl.png	login_real.jpg	login_fake.jpg	13930.73	0.7
dropbox.png	home_real.png	home_fake.png	5221.85	0.76
instagram.png	login_real.png	login_fake.png	939.24	0.84
linkedin.png	signin_real.png	signin_fake.png	30675.28	0.51
linkedin2.png	register_real.jpg	register_fake.jpg	14465.36	0.69
microsoft.png	signin_real.png	signin_fake.png	1212.12	0.96
trade me.png	home_real.png	home_fake.png	9091.03	0.6
yahoo.png	home_real.jpg	home_fake.jpg	17759.5	0.5
yahoo2.png	home_real2.png	home_fake2.png	10028.07	0.5

alongside their corresponding RMS and SSIM values. Our results are based on the findings of RGB images and not only grayscale images. This is done mainly by segregating each of the three red, blue, and green channels. Fast Fourier Transform is applied on each channel after which multichannel image comparison has been implemented. Based on our findings, the SSIM value seems to be a really good estimator as to how the images can be compared based on their structural similarity. Table 9.8 shows the variation of SSIM and RMS values for different legitimate and bogus webpages.

9.5 Conclusion and Future Scope

Based on our research, any machine learning algorithm combined with a classifier can detect web phishing. We, currently, have classifiers that detect phishing websites with high accuracy, but our research shows that classifiers work better when more features are used as training data, and when the most significant features are utilized as training data. The challenges in this area will be that spammers will continue to improve the URLs and design of phishing websites so that they resemble legitimate websites. As a result, it is necessary to improve existing features and introduce new ones for phishing detection.

Moreover, for image similarity, we have concluded our study with a threshold SSIM value for image similarity. This value can be used as a predefined threshold that can be learned further even more on a larger amount of dataset. The future scope of this research can look more into how different parameters can be tuned based on the dataset and features it involves. Additionally, a large corpus can be created in order to learn the SSIM threshold value in greater depth. Our work involves the usage of only a few samples because our research involved scraping for images ourselves for real as well as fake ones, thereby limiting our study to a particular extent; however, if a large dataset is involved, a machine learning model can be trained to accurately understand how the threshold value varies and its optimized value.

References

1. Mahajan, R., Siddavatam, I.: Phishing website detection using machine learning algorithms. Int. J. Comput. Appl. **181**, 45–47 (2018). https://doi.org/10.5120/ijca2018918026
2. James, J., Sandhya, L., Thomas, C.: Detection of phishing URLs using machine learning techniques, 304–309 (2013). https://doi.org/10.1109/ICCC.2013.6731669
3. Nguyen, L. To, B., Nguyen, H., Nguyen, M.: Detecting phishing web sites: a heuristic URL-based approach, 597–602 (2013). https://doi.org/10.1109/ATC.2013.6698185
4. Shahrivari, V., Darabi, M., Izadi, M.: Phishing detection using machine learning techniques (2020). arXiv abs/2009.11116
5. Yi, P., Guan, Y., Zou, F., Yao, Y., Wang, W., Zhu, T.: Web phishing detection using a deep learning framework. Wirel. Commun. Mob. Comput. **2018**, 1–9 (2018). https://doi.org/10.1155/2018/4678746

6. Mehanović, D., Kevrić, J.: Phishing website detection using machine learning classifiers optimized by feature selection. Traitement du Signal **37**(4), 563–569 (2020). https://doi.org/10.18280/ts.370403
7. Kulkarni, A., Brown, L.L., III.: Phishing websites detection using machine learning. Int. J. Adv. Comput. Sci. Appl. (IJACSA) **10**(7) (2019)
8. Zhang, N., Yuan, Y.: Phishing detection using neural network - CS229 lecture notes, 2012 - cs229.stanford.edu (2012)
9. Selvakumari, M., et al.: Phishing website detection using machine learning and deep learning techniques. J. Phys.: Conf. Ser. **1916**, 012169 (2021)
10. Akila, D.: Phishing websites detection using machine learning **8**, 111–114 (2019). https://doi.org/10.35940/ijrte.B1018.0982S1119
11. Shikalgar, S., Sawarkar, S.D., Narwane, S.: Detection of URL based phishing attacks using machine learning. Int. J. Eng. Res. Tech. (IJERT) **08**(11) (2019)
12. Wikipedia, https://en.wikipedia.org/wiki/Logistic_regression
13. PluralSight, https://www.pluralsight.com/guides/classification-keras
14. Sara, U., Akter, M., Uddin, M.: Image quality assessment through FSIM, SSIM, MSE and PSNR-A comparative study. J. Comput. Commun. **7**, 8–18 (2019). https://doi.org/10.4236/jcc.2019.73002
15. Shekokar, N., Shah, C., Mahajan, M., Rachh, S.: An ideal approach for detection and prevention of phishing attacks. Proc. Comput. Sci. **49**, 82–91 (2015). https://doi.org/10.1016/j.procs.2015.04.230
16. Nussbaumer, H.J.: The fast fourier transform. In: Fast Fourier Transform and Convolution Algorithms. Springer Series in Information Sciences, vol. 2. Springer, Berlin (1981). https://doi.org/10.1007/978-3-662-00551-4_4
17. Cooley, J.W., Lewis, P.A.W., Welch, P.D.: The fast fourier transform and its applications. IEEE Trans. Educ. **12**(1), 27–34 (1969). https://doi.org/10.1109/TE.1969.4320436
18. Flynn, J.R., Ward, S., Abich, J., Poole, D.: Image quality assessment using the SSIM and the just noticeable difference paradigm. In: Harris, D. (Ed.), Engineering psychology and cognitive ergonomics. Understanding human cognition. EPCE 2013. Lecture Notes in Computer Science, vol. 8019. Springer, Berlin (2013). https://doi.org/10.1007/978-3-642-39360-0_3
19. Sungheetha, A., Sharma, R.: 3D image processing using machine learning based input processing for man-machine interaction. J. Innovat. Image Process. **3**(1), 1–6 (2021). https://doi.org/10.36548/jiip.2021.1.001

Chapter 10
Text Summarization of Articles Using LSTM and Attention-Based LSTM

Harsh Kumar, Gaurav Kumar, Shaivye Singh, and Sourav Paul

Abstract This research work focuses on developing a model for performing text summarization of articles, depending on the novelty of the information that is arranged in the final summary, where summarization is roughly categorized into two types: extractive and abstractive summarization. In today's fast-paced society, most people disregard if they are deemed to be rather extensive due to a lack of time. Therefore, having a brief overview of this article would not only encourage readers to read it, but if the topic is of interest, it would but also allow them to go deeper into it. As a result, this research work intends to bridge this research gap with the proposed model. Initially, this research project developed a reference RNN model and tested its accuracy. It was discovered that the summary loses a lot of real terms and reduces and confuses the overall meaning, therefore it was decided to build an LSTM model for the same and test it with three distinct datasets. The first experiment is carried out with a small dataset, whereas the second experiment is carried out with a comparatively bigger dataset by applying LSTM to it. We decided to carry it forward and improve the model by increasing the size of our dataset and choosing a large dataset with an attention layer to the proposed LSTM model by comparing the accuracy of all the models with the existing model, and it will be evaluated by using ROUGE metrics and proving the proposed solution to the world. Furthermore, the future research directions will be provided in Future Scope regarding where and how the proposed model can achieve more efficiency in the future.

H. Kumar (✉) · G. Kumar · S. Singh · S. Paul
Department of Computer Science and Engineering, Sikkim Manipal Institute of Technology, Sikkim Manipal University, Gangtok, Sikkim, India
e-mail: harsh_201800494@smit.smu.edu.in

S. Paul
e-mail: sourav.p@smit.smu.edu.in

J. I.-Z. Chen et al. (eds.), *Machine Learning and Autonomous Systems*, Smart Innovation, Systems and Technologies 269, https://doi.org/10.1007/978-981-16-7996-4_10

10.1 Introduction

Nowadays, the digital articles are providing a lot of potential insights to the readers. However, when individuals encounter longer paragraphs, they often avoid reading them, believing it to be a waste of time. To overcome this challenge, NLP (Natural Language Processing) technology has been introduced. It is a subset of Artificial Intelligence (AI) that deals with how humans communicate with computers by using natural language. The primary aim of NLP makes a lot of things easier to understand by reading, deciphering, understanding, and most importantly, it does not reduce the value of human language, which is a primary reason for using NLP. To deduce meaning from human languages, they rely on machine learning. NLP is present in almost everything we interact with today. It finds its application in Google Translate. It also finds its use in various personal assistant applications existing today such as Google's Assistant—OK Google, Alexa from Amazon, Siri from Apple, and Cortana from Microsoft. Consequently, Microsoft Word also uses NLP to check grammatical accuracy of texts. NLTK (Natural Language Toolkit) is one of the most popular tools for developing NLP models. It is supported by an active community as well as a variety of language processing tutorials, example datasets, and resources. It also provides various tasks like stemming, classification, tokenization, etc. Various libraries, tools, and services such as GenSim, natural language toolkit, beautiful soup, regular expression, Numpy, Pandas, TensorFlow, and Keras are used quite often to derive all the methodologies and provide the basic functionalities in NLP.

10.1.1 Motivation

With the advancement and development of technology across the world, a lot of information and guides are passed along with those technologies, which necessitate research for continued technological advancement in industries, and something that can summarize those big and large data into a shorter version of it could be very useful in the industry and popular as well. As a result, Text Summarization models must be developed. Seeing today's period of combating issues like COVID-19, or other COVID pandemic iterates like as SARS, or any other terrible illness, or any other significant news or problem that is happening across or near us that need our immediate attention. There are tons of information passed down through internet, newspapers, articles. Seeing that a major issue we tried to pursue something related to the same as our research work that implements some novel ideas to develop and compare the accuracies between it and its pre-existing models present in the market today.

The proposed research is carried out in three different stages with three different experiments working for the same cause, which will be followed up later in this paper.

10.1.2 What is Text Summarization?

Text summarization with time has become an important and timely tool for interpreting large text information and assisting in many ways in today's fast-growing information age. Moreover, it is very difficult for human beings to manually summarize those large documents of text. There is an abundance of text material available on the internet [1, 2]. However, generally, the Internet helps us with more information than what is needed and is asked of it and of which most of them is found to be trivial [3]. Hence, a twofold problem is seen to be encountered, which is: looking for relevant papers among a vast number of documents accessible and absorbing a big amount of relevant data. The ultimate objective of automated text summarizing is to take a given source text and transform it into a shorter version while maintaining its information richness and overall meaning [4] In concise terms, Text Summarization is condensing of the source text into a smaller and compact version of it preserving its overall meaning and useful information having minimum loss in the overall information content. Manually summarizing those lengthy texts is quite tough for us.

Extractive text summarization and abstractive text summarization are the two major categories of text summarization approaches.

10.1.2.1 Extractive Text Summarization

Only the most essential sentences or phrases from the original reference material are recognized and extracted. This method entails extracting all of the important terms from the reference material and then combining them to create a summary. Extractive summarization extracts or duplicates the particular portions of the original text based on scores derived by using statistical or linguistic criteria [5]. This method involves no changing to the reference text.

10.1.2.2 Abstractive Text Summarization

The process of abstractive summarization involves rewriting the source text to generate new phrases or sentences, which may not be present in the original text. This is considered as a challenging assignment for a computer since it may not be present in the original text. Abstractive text summarization necessitates advanced machine learning algorithms and substantial natural language processing (NLP) since it necessitates a full understanding of the material to generate the summary [5]. As a result, this technique is more difficult than what we observed in extractive summarization because Abstractive Summarization involves the utilization of real-word data and semantic classification analysis. On the other hand, this model creates whole new sentences, and hence it is superior to the former. The summary supplied here is more intelligible and efficient because it is similar to human-generated summary.

10.2 Background and Related Work

This section discusses about the existing text summarization models. Most of the research on extractive text summarization has been done, but in recent years, abstractive methods are preferred to generate new sentences. Many models have been made on abstractive summarization like the one that has several RNN. First RNN encodes a sequence of symbols into its fixed-length vector representation, and the other RNN decodes the representation into another sequence of the symbol [6]. To overcome this limitation, Mike Schuster and Kuldip K. Paliwal proposed Bidirectional RNN. The Bidirectional RNN can be trained without the limitation of using input information just up to a preset future frame. This is accomplished by training it in a positive and negative direction of time simultaneously [7, 8]. The primary capsules in the capsule network that are the inputs of capsules in the first layer of capsules can consist of arbitrary dimensions. They may be obtained from a recurrent network's hidden state or a convolution layer. The LSTM and CNN outputs are outperformed in a substantial manner by both capsule network and SVM algorithms [9].

Other models like Abstractive text summarization uses LSTM and CNN-based deep learning proposed by Shengli Song, Huang Haitao, Ruan Tongxiao, which can construct new sentences by exploring more fine-grained fragments than sentences, namely, semantic phrases and different from existing abstraction-based approaches.

On reviewing past experiments, we found out that accuracy was problem and so we decided to build our Text summarizer using attention-based LSTM, which provides more concise and accurate results. In Text to speech synthesis (TTS), RNN is used to gather speech signals simultaneously, but it is replaced by a transformer multihead attention mechanism to improve accuracy. The attention fails in autoregressive error accumulation and suffers from slow inference between the frames. Theoretically, appending continuous features of acoustics to the training network with the help of bidirectional LSTM for speech synthesis is easier [10].

10.2.1 Approach

10.2.1.1 RNN

The result of the preceding step's output will be taken and supplied as an input to the next step in a Recurrent Neural Network (RNN). The most important and the prime feature of Recurrent Neural Network is the Hidden State, which remembers the information about a sequence.

By giving all layers, the same weights and biases, RNN transforms independent activations into dependent activations. This reduces the complexity of increasing parameters by giving input as an output to the next layer.

For calculating its current state, at each (t), the h(t) of the RNN (where t = time step and h(t) = hidden state) is updated by:

$$\mathbf{h}_{(t)} = f\left(\mathbf{h}_{(t-1)}, x_t\right).$$

where f is a function called non-linear activation function. f can be as simple as an element-wise logistic sigmoid function and can be equally complex as a long short-term memory (LSTM) [6].

The $\mathbf{h}_t$ calculated above becomes $\mathbf{h}_{t-1}$ for next time step.

After all the time steps are completed, calculate the output by:

$$\mathrm{Y}_f = \mathrm{W}_{yh} h_t.$$

where w_{yh} is the output layer's weight.

An error is created when this output is compared to the real output. After that, the mistake is backpropagated to the network, the weights are adjusted, and the RNN is trained.

RNNs maintain a vector of activations for each time step, which makes the RNN extremely deep [11]. Due to this, the structure of RNN becomes complex causing it to face problems like vanishing gradient and exploding. The problem of vanishing gradient and exploding is solved using LSTM as it continues adjusting weight until it gets the optimal set of weights, and the scaling factor is set to 1.

10.2.1.2 LSTM

LSTM stands for Long Short-Term Memory networks. They were introduced by Hochreiter and Schmidhuber in 1997 [12]. The long-term dependence problem that RNNs have generated is specifically avoided with LSTMs. It is nearly second nature to them to remember things for long periods of time [13–16].

1. Calculate the current hidden state by first taking the element-wise hyperbolic tangent of the current internal cell state vector and then performing element-wise multiplication with the output gate.

10.2.1.3 LSTM with Attention Layer

The decoder generally makes predictions after looking and fetching outputs from the encoder in a simple Encoder–Decoder architecture. But attention-based architecture attends and attenuates every hidden state from each of the encoder nodes at every time step, and then finally makes predictions after deciding which one is more informative.

Purpose and Benefits of adding Attention mechanism to LSTM:

1. Adding attention to LSTM model solves the problem of fixed-length internal representation and provides more contextual information to decoder, i.e., it pays attention to specific input vectors based on the attention weights.

2. Local attention is described as concentrating just on a tiny portion of the source positions for each target word. Given the current state of the decoder, local attention selects the best-aligned position in the input sequence. The aligned position can be chosen monotonically or predictively. Also, local attention is less expensive than global attention.

10.2.1.4 Text Summarization in Python

```
#Set up the Decoder
decoder_inputs = Input(shape=(None,))
dec_emb_layer = Embedding(y_voc_size, latent_dim,trainable=True)
dec_emb = dec_emb_layer(decoder_inputs)

#LSTM using encoder_states as initial state
decoder_lstm = LSTM(latent_dim, return_sequences=True, return_state=True)
decoder_outputs,decoder_fwd_state, decoder_back_state = decoder_lstm(dec_emb,initial_state=[state_s, state_t])

-------------------
#Set up the Attention Layer
attn_layer = AttentionLayer(name='attention_layer')
attn_out, attn_states = attn_layer([encoder_outputs, decoder_outputs])

#Concat attention output and decoder LSTM output
decoder_concat_input = Concatenate(axis=-1, name='concat_layer')([decoder_outputs, attn_out])

#Dense layer
decoder_dense = TimeDistributed(Dense(y_voc_size, activation='softmax'))
decoder_outputs = decoder_dense(decoder_concat_input)

# Define the model
model = Model([encoder_inputs, decoder_inputs], decoder_outputs)
model.summary()
```

10.3 Proposed Work

10.3.1 Methodology

10.3.1.1 Data Preparation or Collection

This research work has considered three different datasets and has trained them using LSTM and attention-based LSTM. The first dataset consists of 1300 articles, second dataset consists of 80,000 articles, and the major dataset that we took, i.e., Article Food Review dataset from Kaggle consists of 100,000 data. These articles are then extracted and saved as a CSV file.

10.3.1.2 Data Pre-Processing/Cleaning

The next phase is data pre-processing, which follows data collection. Data pre-processing aids in the reduction of unwanted data and noise.

Stop words like a, an, the, etc., are removed as they do not add value to the meaning of a sentence.

```
stop_words = set(stopwords.words('english'))
```

We use contraction mapping dictionary to map the contracted words with their intended meanings. It helps in better understanding and improved accuracy of the model. Data cleaning is implemented along with the removal of stop words to make data ready for our model.

```
contraction_map= {"ain't": "is not", "aren't": "are not","can't": "cannot", "couldn't": "could not"}
```

Finally, tokenizing the sentences.

```
#prepare a tokenizer for reviews on training data
x_tokenizer = Tokenizer()
x_tokenizer.fit_on_texts(list(x_tr))

#convert text sequences into integer sequences
x_tr    =   x_tokenizer.texts_to_sequences(x_tr)
x_val   =   x_tokenizer.texts_to_sequences(x_val)

#padding zero upto maximum length
x_tr    =   pad_sequences(x_tr,  maxlen=max_len_text, padding='post')
x_val   =   pad_sequences(x_val, maxlen=max_len_text, padding='post')

x_voc_size   =   len(x_tokenizer.word_index) +1
```

```
[ ] #preparing a tokenizer for summary on training data
y_tokenizer = Tokenizer()
y_tokenizer.fit_on_texts(list(y_tr))

#convert summary sequences into integer sequences
y_tr    =   y_tokenizer.texts_to_sequences(y_tr)
y_val   =   y_tokenizer.texts_to_sequences(y_val)

#padding zero upto maximum length
y_tr    =   pad_sequences(y_tr, maxlen=max_len_summary, padding='post')
y_val   =   pad_sequences(y_val, maxlen=max_len_summary, padding='post')

y_voc_size  =   len(y_tokenizer.word_index) +1
```

10.3.1.3 Data Training

The hyper-parameters such as batch size, loss function, and the number of epochs are used to train the model. Our input data are tokenized, and we have limited the input size to 80 and output size to 10 in order to ensure a consistent training input. In our model, three layers of LSTM are added along with encoder and decoder.

10.3.1.4 Data Evaluation

In this step, training and testing loss graphs are plotted to determine accuracy and loss. We have compared the accuracy between our experiments using ROUGE metrics R_1, R_2, and $R_{L.}$

10.4 Result Evaluation

10.4.1 Model Training and Experiment

10.4.1.1 Experiment 1

In this experiment, we took COVID-19 news dataset from Kaggle [17], which contained almost 1300 data, which was less, but we just wanted to check if our model is working or not and to our surprise this model did work, but results were not satisfying because of its small size which we were prepared for. We took batch size of 512 and epochs of 50 and started training the dataset later testing the same. We used LSTM to train in our first dataset.

Here, since the dataset was small thus, training took very less time, and the result was not that efficient. But still, it managed to get a sparse_categorical_accuracy of 66.56%.

Finally, when it came to summarize the history of loss and accuracy, we came up with:

Figure 10.1a clearly shows the graph that has a very less learning rate at testing, whereas Figure 10.1b shows the gap between the testing and training accuracy, which is not that much, i.e., the gap is less, and validation accuracy is a little overfitting here. Also, the testing curve goes above the training curve showing the inefficiency of our final output.

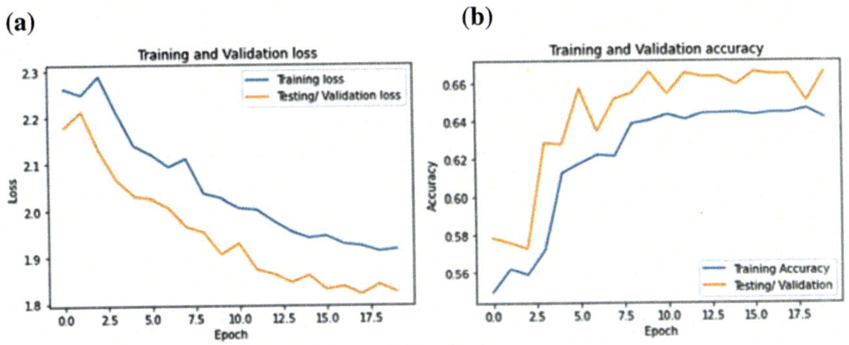

Fig. 10.1 **a.** Training versus Validation Loss Curve, **b.** Training versus Validation Accuracy Curve

10.4.1.2 Experiment 2

In this experiment, seeing the poor performance of our previous dataset, we increased its size by merging it with another article review dataset [18]. We took COVID-19 news and other article review dataset from Kaggle, which contained almost 80,000 data, but to check the efficiency, we first factored it using data frame to almost 10,000 data, which was bigger than the earlier experiment, since we wanted to check for our efficiency and to our decision, we did get improved results than before, and the data were able to be trained and learned. We proceeded with the same LSTM approach. It did take more time to train, since the data were now more than what they were before. We took batch size of 512 and epochs of 50 and started training the dataset later testing the same. Again, we followed the same approach, but it took nearly 3 h to get the model trained and ready.

Here, since the dataset was bigger than the previous thus, training took more time. This time the accuracy was less than before but it is learning. We can say so by seeing and noting the decreasing pattern of loss and increasing pattern of sparse categorical accuracy. The epochs did not complete because of the large data and GPU shortage in cloud but still managed to give more sparse accuracy.

Finally, when it came to summarize the history of loss and accuracy, we came up with:

Figure 10.2a clearly shows the graph that has a high learning rate at testing, whereas, Figure 10.2b shows the gap between the testing and training accuracy that increases in the beginning and later cease to increase and is not very much, i.e., the gap between training and testing is less, which is great, and validation accuracy is a little overfitting here. But the output was better but not that great.

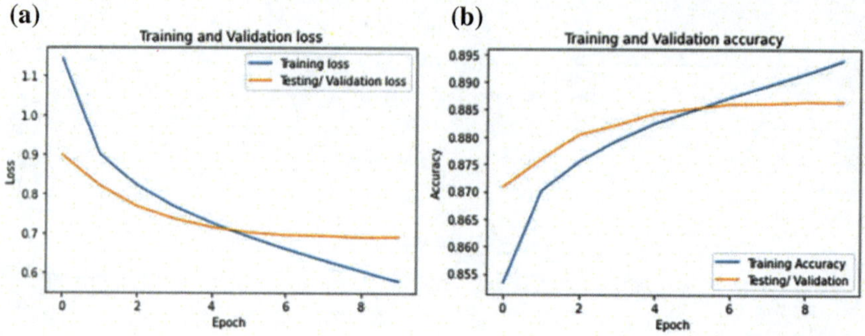

Fig. 10.2 **a.** Training versus Validation Loss Curve, **b.** Training versus Validation Accuracy Curve

10.4.1.3 Experiment 3

In this experiment, seeing the performance of our previous dataset, we further again increased its size taking Food Article Review dataset from Kaggle [19]. We took this dataset that has 100,000 data, but to check the efficiency, we first factored it using data frame, which was bigger than the earlier experiment, since we wanted to check for our efficiency and to our decision, we did get improved results than before, and the data were able to be trained and learned. We proceeded with the same LSTM approach. It did a lot more time than before to train, preferably 6 h since the data were now far more than what they were before. We took batch size of 128 and epochs of 50 and started training the dataset later testing the same. Again, we followed the same approach, but this time used another layer with LSTM to get improved summary. We used attention layer and finally concatenated the output using concat layer and decoded the output, and hence finally, we summarized it.

Here, since the dataset was far bigger than the previous thus, training took more time. This time the accuracy was lesser than before but it is learning since due to more time the GPU stopped in between but the results were more effective than the rest. We can say so by seeing and noting the decreasing pattern of loss and increasing pattern of sparse categorical accuracy.

Finally, when it came to summarize the history of loss and accuracy, we came up with:

Figure 10.3a clearly shows the graph that has a higher learning rate at testing, whereas Fig. 10.3b shows the gap between the testing and training accuracy that increases in the beginning and later cease to increase and is not very much, i.e., the gap between training and testing is really less, which is great, and also validation accuracy is a little overfitting here but higher than the previous experiments, which is again, great. Finally, the output was also better than the other experiments.

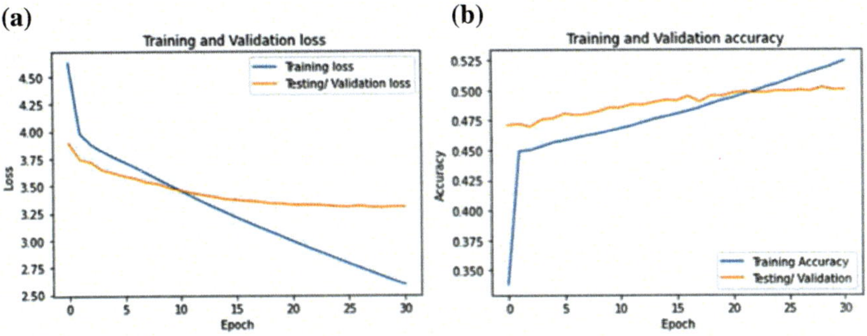

Fig. 10.3 **a.** Training versus Validation Loss Curve, **b.** Training versus Validation Accuracy Curve

10.4.1.4 Discussion

We can conclude that after using the attention layer, the results were improved far better than the rest. Thus, attention mechanism adds remarkable value to the model and to our solution, which is great. In fact, further experiments can also be done over this, which is shared by us in Future Scope section later.

Furthermore, the evaluation will clear more doubts and makes the picture clear using ROUGE Metrics below [20, 14].

10.4.2 Quantitative Analysis Using ROUGE Metrics

ROUGE is nothing but a set of metrics that is used to rate and estimate machine translations and automatic summarization of texts. It compares the total number of overlapped words against the words of reference and system summaries [20].

ROUGE (Recall) = X/t_1 where, X is the no, overlapping words, t_1 is total words, in the reference summary
ROUGE (Precision) = X/t_2 where, X is the no, overlapping words, t_2 is total words, in the reference summary

Generally, Rouge-N and Rouge-L are used to compare the granularity of texts. ROUGE-N is the unit that measures n-gram overlap.

1. The overlap of unigrams between the system summary and the reference summary is referred to as ROUGE-1.
2. The overlap of bigrams between the system and reference summaries is referred to as ROUGE-2.

ROUGE-L is the unit that measures the longest matching sequence of words using LCS.

Keywords: Recall-Oriented Understudy for Gisting Evaluation (ROUGE), n-grams (unigram, bigram, trigram, and higher order n-gram overlaps).

Model	Rouge 1	Rouge 2	Rouge L
Base_Paper	0.2052	0.0548	–
LSTM (Covid-19)	0.6376	0.0854	0.6376
LSTM	0.1309	0.0245	0.1298
Attention_LSTM	0.3652	0.0141	0.3649

Our best-performing models, according to the data above, are those with LSTM as encoder and decoder cells, with ROUGE-1 scores of 0.3652, ROUGE-2 scores of 0.0141, and ROUGE-L scores of 0.3649.

10.5 Conclusion and Future Work

This research work has replaced the RNN model with LSTM and then LSTM with attention layer in this research project, and this research work has attempted to construct a model that can summarize articles, news, or suitably long paragraphs based on the user's requirements. Furthermore, the proposed model can also give a concise title for an article if the output summary words are reduced. The proposed research work has successfully explored RNN with LSTM using Seq2Seq model, which is an encoder–decoder model. The proposed LSTM model contains multiple layers, which solve the vanishing gradient issue over using RNN by providing memory cells and by retaining information for a longer period. We have compared the accuracy between three different datasets of different sizes and have concluded finally by evaluating the experiments with ROUGE metrics by calculating mean R_1, R_2, and R_L scores of the proposed models.

It is a hot topic and has insane work applications in the industry, and it requires more scope for further improvement by using a huge dataset and increasing the number of sub-LSTM layers (which, for our case, was three), and then later, attention layer is used to focus on certain parts of the sentence and making valuable predictions for the same and finally concatenating the output to generate more effective and concise summary. Also, ROUGE metrics can be used to evaluate the scores and contrast between ours to check the validity. The attention layer was, thus, proved to be better than the rest of the layers and models in our research.

References

1. https://www.analyticsvidhya.com/blog/2019/06/comprehensive-guide-text-summarization-using-deep-learning-python/
2. https://towardsdatascience.com/a-quick-introduction-to-text-summarization-in-machine-learning-3d27ccf18a9f
3. Abu-Jbara, A., Radev, D.: Coherent citation-based summarization of scientific papers. In proceedings of the 49th Annual HLT, HLT'11, pp. 500–509. Association for Computational Linguistics (2011)
4. Gupta, V., Lehal, G.S.: A survey of text summarization extractive techniques. Emerg. J. Technol. Web Intell. **2**(3): 258–268 (2010). https://doi.org/10.4304/jetwi.2.3.258-268
5. Suleiman, D., Awajan, A.: Deep learning based abstractive text summarization: approaches, datasets, evaluation measures, and challenges. Math. Probl. Eng. **2020**. https://doi.org/10.1155/2020/9365340
6. Cho, K., Merriënboer, van B., Gulcehre, C., Bahdanau, D., Bougares, F., Schwenk, H., Bengio, Y.: Learning Phrase Representations using RNN Encoder–Decoder for Statistical Machine Translation, June 2014. https://doi.org/10.3115/v1/D14-1179
7. Schuster, M., Paliwal, K.K.: Bidirectional Recurrent Neural Networks, IEEE., (Volume: 45, Issue: 11, Pgs. 2673–2681. https://doi.org/10.1109/78.650093
8. Feigenblat, G., Roitman, H., Boni, O., Konopnicki, D.: Unsupervised query-focused multi-document summarization using the cross entropy method. In: Proceedings of the 40th International ACM SIGIR, pp. 961–964 (2017)
9. Manoharan, J.S.: Capsule network algorithm for performance optimization of text classification. J. Soft Comput. Paradigm (JSCP) **3**(1), 1–9
10. Adam, E.E., Babikir,: Deep learning based NLP techniques in text to speech synthesis for communication recognition. J. Soft Comput. Paradigm (JSCP) **2**(04), 209–215 (2020)
11. Jozefowicz, R., Zaremba, W., Sutskever, I.: An empirical exploration of recurrent network architecture Publication:ICML'15: Proceedings of the 32nd International Conference on International Conference on Machine Learning Volume 37, July 2015, pp. 2342–2350
12. https://colab.github.io/posts/2015-08-Understanding-LSTMs/
13. https://becominghuman.ai/a-simple-introduction-to-natural-language-processing-ea66a1747b32
14. https://towardsdatascience.com/the-ultimate-performance-metric-in-nlp-111df6c64460
15. https://www.geeksforgeeks.org/long-short-term-memory-networks-explanation/
16. https://medium.com/swlh/a-simple-overview-of-rnn-lstm-and-attention-mechanism-9e844763d07b
17. https://www.kaggle.com/saadaljebrreen/corona-various-news-covid19
18. https://www.kaggle.com/sunnysai12345/news-summary
19. https://www.kaggle.com/snap/amazon-fine-food-reviews
20. https://www.freecodecamp.org/news/what-is-rouge-and-how-it-works-for-evaluation-of-summaries-e059fb8ac840/

Chapter 11
Hand Landmark-Based Sign Language Recognition Using Deep Learning

Jerry John and Bismin V. Sherif

Abstract As human being's communication is a very crucial part of our life, it is essential for sharing thoughts and ideas for our survival. But unfortunately for the speaking and hearing-impaired minority, it is difficult or impossible to communicate through speech. But most of the common people have no knowledge about sign language and its interpretations. This is one of the major problems faced by these kinds of people during their communications. Getting an interpreter is not easy every time. To solve this issue, a model is developed using neural networks for finger-spelling based on the various hand gestures. In this user-independent model, CNN-based models are trained using a set of image and hand skeleton datasets. The skeleton dataset is created using the hand image dataset and is done to improve the accuracy of the model. By using this model, a speech-impaired person can easily communicate with a person who has no knowledge of it, the model translating the hand gestures to sentences in English. The main issue with the current system is that there are some groups of alphabets, whose sign language symbols look alike. This makes the image-based classification model a bit difficult to correctly classify and the accuracy of the prediction can also be low. This issue can be solved using a hand skeleton classifier. As in the hand skeleton image, the position and structure of each finger can be identified more accurately than from an image. This makes the neural network model to learn better and can classify with higher accuracy of about 98.6.

11.1 Introduction

Sign languages are languages that use visual means of communication through hand signals, hand gestures, facial expressions, and body language to convey the message. These languages help us to talk to the people who cannot communicate easily. Sign

J. John (✉) · B. V. Sherif
APJ Abdul Kalam Technological University, Thiruvananthapuram, Kerala, India
e-mail: mtech20ai07@mgits.ac.in

Muthoot Institute Of Technology and Science, Kochi, India

J. I.-Z. Chen et al. (eds.), *Machine Learning and Autonomous Systems*, Smart Innovation, Systems and Technologies 269, https://doi.org/10.1007/978-981-16-7996-4_11

languages are often thought as regards to the deaf community. But, there are many situations where people with certain disabilities use them like:

(1) Autism: people with autism have a hard time to communicate with others. So to some extent, they can use sign language to communicate. This can help them with speech development, interactions and to learn new words. Thus, sign language helps them to communicate.
(2) Apraxia of speech : it is a motor speech disorder, which affects the messages from the mouth from the brain. This makes the person difficult to communicate with others. So they are benefitted but the sign language.
(3) Cerebral palsy : this affects the way a person moves their mouth, jaw and head making them difficult to speak. Thus, they can use sign language to communicate.
(4) Down syndrome : delay in development and can affect the learning ability of the person. Sign language is used to improve their communication skills.
(5) Speech impairment :- people suffering from any sort of speech impairment can use sign language to communicate.

The use of sign language doesn't end here there are many conditions where people lose their speaking and hearing abilities through some diseases, trauma etc. So, there are many people who are in need of these sign languages.

But even after studying the sign language, it may be difficult for them to talk to normal people as there are less than 1

People start to learn sign language by learning the alphabet A–Z by using hands. This use of hand for each letter for written alphabets is called as finger-spelling. In American Sign Language (ASL), even though it uses the same alphabets as the English language they are not the same, ASL has its own unique structure and unique grammar.

As there are many types of sign languages, it is so difficult for them to communicate with people with the same condition from a different country and to the normal people to are unaware of the language, which makes it difficult for them to live independently. If they need to go somewhere even to a hospital, restaurant, or even a coffee shop, they need somebody else's help to make others understand what they are saying, which makes it so much harder for them to lead a normal life. So we normally use the help of a 'translator' to help in this condition. But it cannot be used everywhere as this translator needs to travel everywhere with them or it takes time for the translator to come. So to solve all these problems, we can more easily use a translator app, which helps us to understand what the person is speaking in sign language. As this is an app, it can be used anywhere anytime, which gives more value to our time and money. It can also be used by people having different sign languages to understand each other. By this innovative digital technology, we can identify their gestures and the app converts them to sign language into text.

11.2 Literature Survey

Research on sign language recognition has been started since 2000. It is mainly based on two approaches, i.e. sensor-based and vision-based approaches. The sensor-based approach is very costly and complicated compared to vision-based approach. In the vision-based approach, features are obtained from the input images or videos. In the image-based research by Rangel Daryl et al. [9], we can see works like Two Hand Indian Sign language (THISL) classification. They used 9,100 image datasets and acquired an overall accuracy of 91.72% on various machine learning models.

In most of the sign language research work based on video datasets in which it can be seen has classified moving gestures, i.e. the signs that are represented with a movement especially by V Gomathi et al. [5]. As we are looking into the image dataset, so are concentrating more on image-based works. Shadman Shahriar et al. [6] propose a system where the input image is processed by a skin colour model. Its background is removed and noises are filtered via a mask filtering system. Then the required hand portion is extracted and trained using a Convolutional Neural Network. Research work done by Shahin Alom et al. [7] which mainly concentrating on recognizing digits in the sign language using a combination of convolutional neural networks and support vector machine. Along with that works on local sign languages like Bengali sign language [8] by Mehedi Hasan et al., Arabic sign language etc. can also be seen.

Sang-Ki Ko et al. [2] proposed a system in which the key points are extracted, then it is standardized by mean and standard deviation of the key points. This is used as the input to recurrent neural network. Research work proposed by Razieh Rastgo et al. [1] used CNN-based models to estimate the key points, and key points are connected using a midpoint algorithm. CNN and LSTM are used to model dynamics of hand gestures. The best hand pose estimation and hand landmark detection can be done on the research work by Fan Zhang Valentin Bazarevsky et al. [4] in which a mediapipe framework is developed for detecting those features.

11.3 Problem Identified in the Existing Solution

The problems that came across during the literature survey are mentioned below.

1. As there are thousands of languages in the world, the sign language community also has 150–300 different types of sign language around the globe and can only find a dataset for a few of these sign languages. Proper dataset is one of the important factors in building a machine learning or neural network model. Building a strong dataset manually is a very time-consuming and hectic job, and it causes many inconsistencies in the dataset also, for example the size of the image or the frame rate of a video can vary. A good program that can create a consistent dataset automatically with less human intervention can improve this problem.

2. In most of the finger-spelling (or dactylology is the representation of the letters of a writing system, and sometimes numeral systems, using only the hands) works, they are using image dataset as the input for creating the model. For most cases, this model can work perfectly, but there are some symbols (letters), which look almost similar. The accuracy of correct prediction of such symbols (letters) is very less. The symbols for letters K, R, U, and V can be easily confused by the model and the accuracy can be very low.

11.3.1 Solution Suggested

For the proposed solution, Indian Sign Language Dataset is being used, which contains 1200 images of each English alphabet. A separate program is developed so that it can automatically create a new dataset, using this it's creating a dataset for Indian sign language words, which can be recognized using images. In the proposed system, it is able to convert the signs shown in a camera into sentences using finger-spelling techniques. For building hand pose recognition models, image dataset and hand key point dataset are used. Using these two datasets, separate models are created. A program is developed to identify 21 hand landmarks using a media pipe. Each point will have x, y, z coordinates. The program will convert this reading to a csv file. After that csv file is used to train the model (Figs. 11.1, 11.2).

Coloured images have a lot of features that are not required for this project. Feeding coloured images will take a lot of resources and time for training models and a lot of computation power will be required for processing colour images, so that the colour image is first converted into gray scale image. For the project, the main features to be considered are the outlines of the hand, finger position, etc. Due to that reason, we will apply thresholding to the gray scale image. (Thresholding is a type of image segmentation, where the pixel of an image is changed to make the image easier to analyse. In thresholding, an image is converted from colour or grayscale into a binary image, i.e. one that is simply black and white. Most frequently, thresholding is used as a way to select areas of interest in an image, while ignoring the parts we are not concerned with.) For achieving this state, we are using Gaussian Filtering (it is used to blur images and remove noise and detail.)

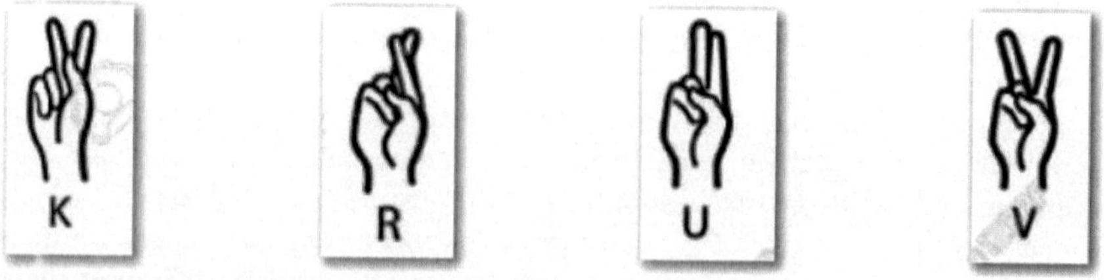

Fig. 11.1 Similar symbols in sign language

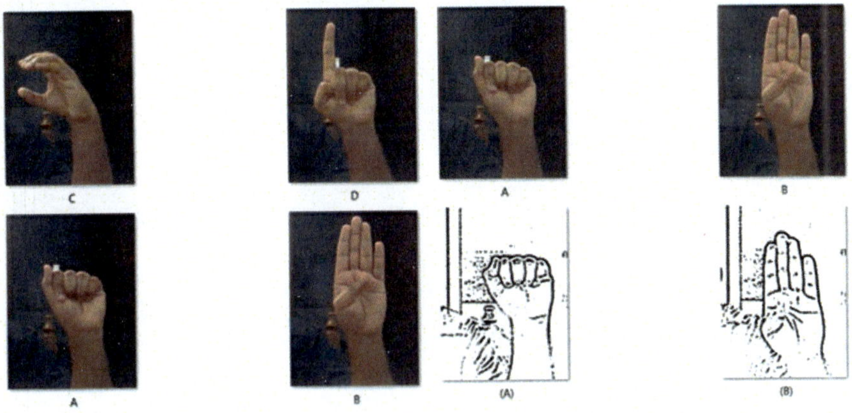

Fig. 11.2 Raw image dataset and image after applying Gaussian filtering

For image classification models, different methods like CNN, YOLO will be implemented to check the performance and accuracy. After getting the predictions from both models, different methods are experimented to check which method will provide the best accuracy. Some of them are: 1. Giving each model results in equal preference. 2. Giving the image model more preference over those signs with high classification accuracy and for the low accuracy signs using the hand landmark model. 3. Separately creating a model for a group of letters with which the model is getting confused with. So that when the main model classifies the symbol as any of this group, it will pass that input image to a special model. This method can be tested in both cases.

11.4 Methodology

The work can be classified into four parts

1. Data collection
2. Data pre-processing
3. Model creation
4. Gesture classification.

11.4.1 Data Collection

The Indian Sign Language Dataset is used for this work. The dataset is collected from Kaggle datasets, where all the sign language images are present from A to Z. Each letter is having almost 1200 images of 200*200*3 pixels in RGB format. The second method for data collection is by creating a data set. For this purpose, a python program is built with OpenCV, which basically consists of a window showing the live feed from the webcam. Each alphabet is assigned to the corresponding letters on the keyboard. To create a new dataset that particular symbol has to be shown on the webcam and need to press which letter is that on the keyboard. As the number of times the keys are pressed, that many number of images dataset is created. A separate folder is created in the system automatically by the program to save this image dataset created.

11.4.2 Data Pre Processing

Two base models are built for this purpose. In the first model (that is a raw image model) by giving the RGB image as the input. This image is then converted into gray scale image as there is no need for colour parameters for this model. A Gaussian filter is applied to this image to remove all the unnecessary parts and to concentrate on the edges and fingers which is required in this work. This image is then sent to CNN for building this model (Figs. 11.3, 11.4 and 11.5).

For the second model (i.e. hand skeleton-based model), the RGB image is converted into a hand skeleton detected image. This is done using the media pipe framework. The media pipe framework will detect 21 hand landmarks and draw a connection line between them, so that a skeleton-based structure is formed. This image is then used as the input for the CNN model.

A separate model is built for a group of letters, which are often misclassified. This group includes [D, R, U], [T, K, D, I], [S, M, N]. For each group, two separate models are created using the image model and hand selection model. For these models, images of those particular letters are only given as the input (Fig. 11.6).

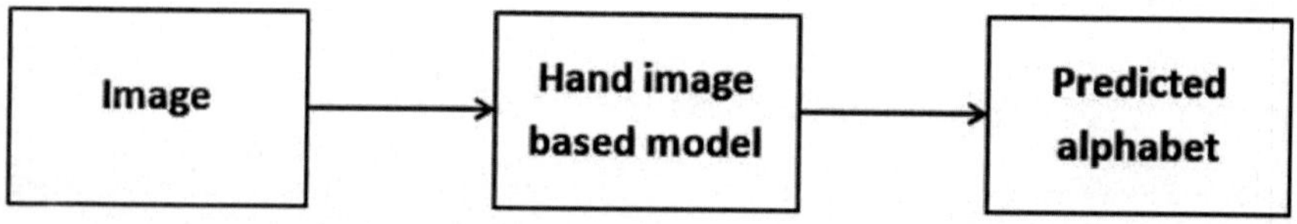

Fig. 11.3 Hand image-based model

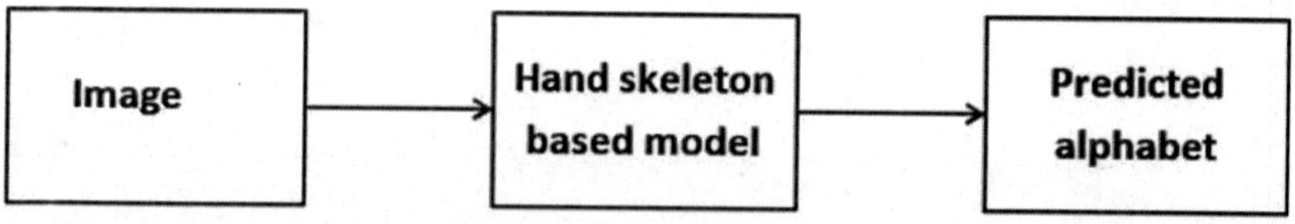

Fig. 11.4 Hand skeleton image-based model

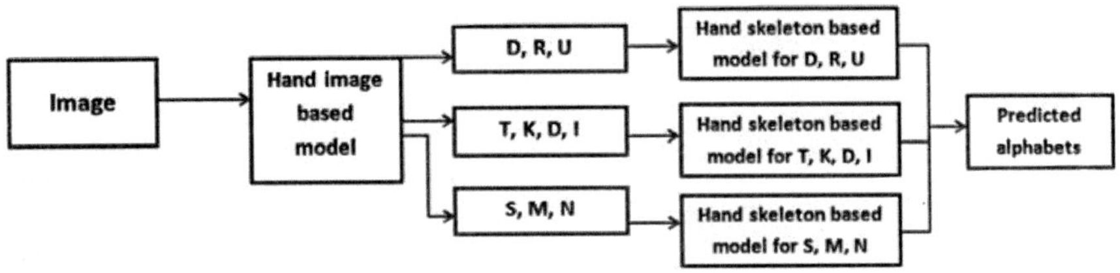

Fig. 11.5 Hand image-based mode with hand skeleton submodels

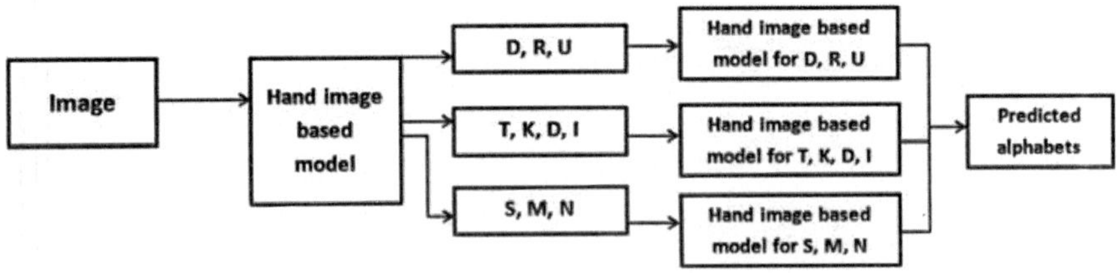

Fig. 11.6 Hand image-based mode with hand image based submodels

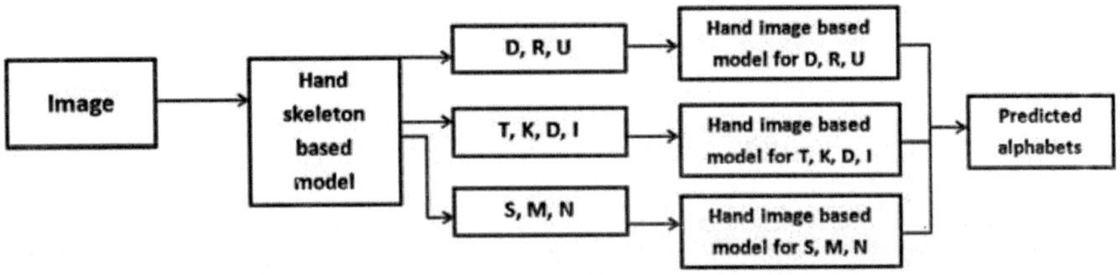

Fig. 11.7 Hand skeleton-based mode with hand image based submodels

11.4.3 Model Creation

We are using a CNN-based model. The first layer used is a convolution layer, which takes the image of size 128 * 128, 32 filters are used in this layer. This will generate a 126*126 pixel output from that layer. Then apply a pooling layer, which will downsample using max pooling, which results in a 63*63 pixel image. Next another convolution layer is applied as same as the first convolutional layer with 32 filters and get an output image with 60 * 60 pixels. Then as max pooling layer is applied and which will reduce the image into 30 * 30 pixels. After which the image is flattened and is given as the input to the densely connected layer. In the output layer, it is having 26 neurons as it is having 26 classes to classify (Fig. 11.7).

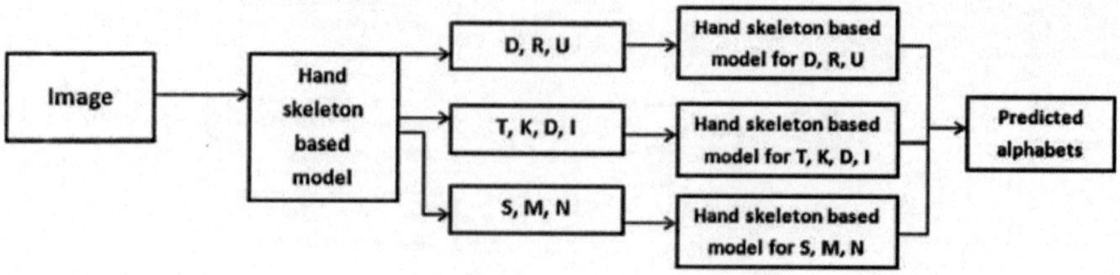

Fig. 11.8 Hand skeleton-based mode with hand image-based submodels

11.4.4 Gesture Classification

Gesture classification is the final step, where the use of the above build models is to predict the gesture. An OpenCV-based python program is developed to get the live feed from the webcam. Any gestures can be shown to the webcam, the image is captured, and this image is used as the input for CNN models. The output from the model (i.e. English alphabet corresponding to the gesture shown) is displayed on the screen (Fig. 11.8).

11.5 Results and Observation

It is observed that the hand-based image model got an accuracy of 97.2%, and the hand skeleton-based model got an accuracy of 97.9%. It can be observed that there is an increase in the accuracy as the dataset is converted to a hand skeleton dataset. It is also observed that the main models with submodels can perform better with better accuracy, like the hand image model as the main model and hand skeleton as the submodel get an accuracy of 98.2%, when the hand image as the submodel gets an accuracy of 97.8%. But the best performance can be seen in the model where the main model and submodels with the hand skeleton dataset are being used and have an accuracy of 98.6% (Fig. 11.9).

11.6 Applications

This model can be used in areas where hand pose detection is needed, for example it can be used to translate sign language alphabets using finger-spelling and words whose sign can be identified by an image. It can also be used to identify different mudras in classical dance, gesture-controlled robots and automated homes, game control, human–computer interaction etc. In the future, we will be able to control devices using gestures and poses, for such applications, accurately identification of the poses and landmarks are very important (Table 11.1).

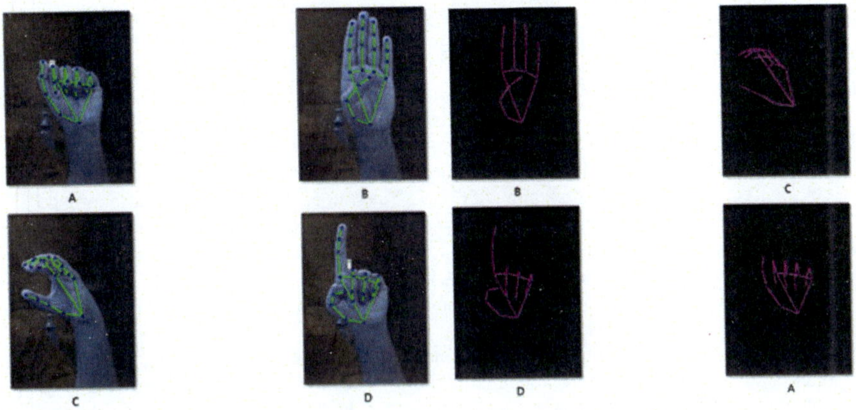

Fig. 11.9 Detecting hand landmarks and hand skeleton-based on hand landmarks

Table 11.1 Performance parameters

Models	Accuracy
Hand-based image model (Model 1)	97.2
Hand skeleton-based model (Model 2)	97.9
Hand image model with hand skeleton submodels (Model 3)	98.2
Hand image model with hand image submodels (Model 4)	97.8
Hand skeleton-based model with hand image submodel (Model 5)	97.8
Hand skeleton-based model with hand skeleton submodel (Model 6)	98.16

11.7 Conclusion

In this research work, a functional real-time vision-based Indian sign language recognition for speech impaired people has been developed for Indian Sign Language alphabets. We are trying to achieve better accuracy using image and hand skeleton datasets, and this dataset is created using a separate program on the sign language image data set. As the hand landmark points only concentrate on the 23 landmark points, issues faced can easily be avoided for image dataset such as complex background, light conditions, etc. An accuracy of 98.6% is obtained using the hand skeleton-based models. For future work, face and body landmarks can also be included for identifying facial expressions for each gesture.

References

1. Rastgo, R., Kiani, K., Escalera, S.: Hand sign language recognition using multi-view hand skeleton. Expert Syst. Appl. **150** (2020)
2. Ko, S.-K., Son, J.G., Jung, H.: Sign language recognition with recurrent neural network using human key point detection. In: Conference on Research in Adaptive Systems (2018)
3. Ge, L., Ren, Z., Li, Y., Xue, Z., Wang, Y., Cai, J., Yuan, J.: 3D hand shape and pose estimation from a single RGB image. In: Proceedings of the IEEE/CVF Conference on Computer Vision and Pattern Recognition (2019)
4. Zhang, F., Bazarevsky, V., Vakunov, A., Tkachenka, A., Sung, G., Chang, C.-L., Grundmann, M.: MediaPie hands: on-device real-time hand tracking. In: Computer Vision for Augmented and Virtual Reality, Seattle (2020)
5. Muthu Mariappan, H., Gomathi, V.: Real-time recognition of Indian sign language. In: 2019 International Conference on Computational Intelligence in Data Science (ICCIDS) (2019)
6. Shahriar, S., Siddiquee, A., Islam, T., Ghosh, A., Chakraborty, R., Khan, A.I., Shahnaz, C., Fattah, S.A.: Real-time american sign language recognition using skin segmentation and image category classification with convolutional neural network and deep learning. In: TENCON 2018 - 2018 IEEE Region 10 Conference (2018)
7. Alom, S., Hasan, J., Wahid, F.: Digit recognition in sign language based on convolutional neural network and support vector machine. In: 2019 International Conference on Sustainable Technologies for Industry 4.0 (STI) (2019)
8. Hasan, M., Srizon, A.Y., Hasan, A.M.: Classification of Bengali sign language characters by applying a novel deep convolutional neural network. In: 2020 IEEE Region 10 Symposium (TENSYMP) (2020)
9. Daroya, R., Peralta, D., Naval, P.: Alphabet sign language image classification using deep learning. In: TENCON 2018 - 2018 IEEE Region 10 Conference (2018)
10. Chaikaew, A., Somkuan, K., Yuyen, T.: Thai sign language recognition: an application of deep neural network. In: 2021 Joint International Conference on Digital Arts, Media and Technology with ECTI Northern Section Conference on Electrical, Electronics, Computer and Telecommunication Engineering (2021)
11. Hein, Z., Htoo, T.P., Aye, B., Htet, S.M., Ye, K.Z.: Leap motion based myanmar sign language recognition using machine learning. In: 2021 IEEE Conference of Russian Young Researchers in Electrical and Electronic Engineering (ElConRus)
12. Samal, B.R., Panda, M.: Integrative review on vision-based dynamic Indian sign language recognition systems. In: 2021 1st Odisha International Conference on Electrical Power Engineering, Communication and Computing Technology (ODICON) (2021)
13. Raval, J.J., Gajjar, R.: Real-time sign language recognition using computer vision. In: 2021 3rd International Conference on Signal Processing and Communication (ICPSC)
14. Sonawane, P., Shah, K., Patel, P., Shah, S., Shah, J.: Speech to Indian sign language (ISL) translation system. In: 2021 International Conference on Computing, Communication, and Intelligent Systems (ICCCIS) (2021)
15. Malakan, Z.M., Albaqami, H.A.: Classify, detect and tell: real-time American sign language. In: 2021 National Computing Colleges Conference (NCCC)
16. Grover, Y., Aggarwal, R., Sharma, D., Gupta, P.K.: Sign language translation systems for hearing/speech impaired people: a review. In: 2021 International Conference on Innovative Practices in Technology and Management (ICIPTM)
17. Karuppusamy, P.: Building detection using two-layered novel convolutional neural networks. J. Soft Comput. Paradigm (JSCP) **3**(01), 29–37 (2021)
18. Manoharan, S., Ponraj, N.: Analysis of complex non-linear environment exploration in speech recognition by hybrid learning technique. J. Innovat. Image Process. (JIIP) **2**(04), 202–209 (2020)
19. Sungheetha, A., Sharma, R.: 3D image processing using machine learning based input processing for man-machine interaction. J. Innovat. Image Process. (JIIP) **3**(01), 1–6 (2021)

20. Mugunthan, S.R., Vijayakumar, T.: Design of improved version of sigmoidal function with biases for classification task in ELM domain. J. Soft Comput. Paradigm (JSCP) **3**(02), 70–82 (2021)

Chapter 12
Tamil Language Handwritten Document Digitization and Analysis of the Impact of Data Augmentation Using Generative Adversarial Networks (GANs) on the Accuracy of CNN Model

Venkatesh Murugesh, Aditya Parthasarathy, Gokul P. Gopinath, and Anindita Khade

Abstract Recognition of handwritten documents is an essential part of today's world for various reasons such as analysis, data extraction, etc. Recognition of handwritten in Indian regional languages, for, e.g., Tamil character is very difficult, due to variations in size, style, and orientation angle. Limited availability of handwritten character dataset makes it difficult to achieve high accuracy for all the characters present in the language. In this paper, we propose a deep learning approach to recognize and classify handwritten characters of the Tamil language. The proposed system can be divided into five phases: expansion of the dataset and data augmentation with the help of Generative Adversarial Networks (GAN), preprocessing of the input image to reduce the noise, segmentation of the characters, feature extraction, and classification of the character done using Convolutional Neural Networks (CNN). We have achieved an accuracy of 94.03% and with data augmentation using GAN, an accuracy of 97% was achieved. The proposed system, along with the recognition of characters, also converts the image of the handwritten document into an editable word document or a PDF file.

V. Murugesh (✉) · A. Parthasarathy · G. P. Gopinath · A. Khade
Department of Computer Engineering, SIES Graduate School of Technology, Navi Mumbai, India
e-mail: venkatesh.murugesh17@siesgst.ac.in

A. Parthasarathy
e-mail: aditya.parthasarathy17@siesgst.ac.in

G. P. Gopinath
e-mail: gokul.gopinath17@siesgst.ac.in

A. Khade
e-mail: anindita.khade@siesgst.ac.in

J. I.-Z. Chen et al. (eds.), *Machine Learning and Autonomous Systems*, Smart Innovation, Systems and Technologies 269, https://doi.org/10.1007/978-981-16-7996-4_12

12.1 Introduction

Digital data has become an integral part of today's world. Though terabytes over terabytes of digital data are being generated every minute, there is a lot of offline data, present in the handwritten form that is not yet digitized, especially the documents written in Indian regional languages. Despite the advent of digital media and technology, handwritten documents continue and will continue to exist. There is ample data and knowledge that can be extracted from these texts and used for applications such as digitizing government documents, digitizing and preserving ancient texts, automated address readers at post offices, automated cheque readers, etc. Thus, handwritten text recognition is one of the challenging tasks of computer vision with considerable practical interest. Though thorough research has been done in this field for English and other Latin-based languages, not much progress has been made towards handwritten characters in Indian languages. Tamil is one of the oldest languages in the world with several million speakers in the southern Indian state of Tamil Nadu and Sri Lanka [1]. Recognition of handwritten Tamil characters is a difficult task, due to variations in size, style, and orientation angle of the written characters. Moreover, there are several pairs of similar-looking characters. Unlike English and other Latin-based languages, Tamil has many characters; 247 to be precise. The breakup of the characters can be done as follows: there are 12 vowels, 18 consonants, and 1 unique character, along with 216 compound characters. Due to such a diversity in the number of classes, and the other aforementioned issues, the classification of Tamil characters becomes a Herculean task.

12.2 Related Work

In [1–3], the authors have used a ConvNet model to classify the handwritten characters of the Tamil language. The model in [1] classifies the characters into a limited number of classes, i.e. 35; whereas there are many more characters present in the language. Data augmentation techniques were used here but were limited to rotation only. Another limitation was the limited number of training samples used for training the model. The model in [3] failed to identify characters that are not an exact replica of the character present in the dataset. The model just recognizes whether the character belongs to the set of Tamil characters or not; there is no classification of the character being done [2]. The authors in [4] have made the use of Residual Neural Network (ResNet) and have managed to achieve an accuracy of 96%; but the model has been trained and tested on a subset of the International Workshop on Frontiers in Handwriting Recognition (IWFHR-10) Tamil Character Dataset (15,000 images). Pragathi et al. [5] have used VGG16, which is a compute-intensive CNN model used for object detection trained on the ImageNet dataset; that is able to achieve an accuracy of 94.52%. Similar to [4] the model was trained on a comparatively smaller dataset, that consisted of 100 images of each character, which sums up to 15,600 images.

In [6], the paper proposes a method to recognize handwritten Tamil characters using Kohonen's Self Organizing Maps. The model is trained and tested on a very small dataset and abnormally written characters are misclassified. In [7, 8], the BanglaLekha-Isolated dataset is used which combined with data from augmentation techniques and characters are classified by CNN [8], Support Vector Machine (SVM) and Long-Term Short Memory (LSTM). Only simple data augmentation techniques, like rotation, position shifting, zoom, etc. In [9], the authors have compared the two main approaches for digitizing handwritten text: word-level classification and character-level classification. The computationally heavy models produced extremely poor accuracies.

A Deep Convolutional Neural Network (DCNN) in [10] is designed for Devanagari Handwritten characters classification. The intensive memory usage of the model is due to the high number of training parameters. A significant number of misclassification errors occurred due to similarities in shape and structure. In [11], the paper tries to automate the recognition of handwritten Hindi isolated characters using multiple classifiers. The models are trained on a very small dataset size of 4,428 handwritten characters. The distorted, similar-looking characters are misclassified. In [12], a data augmentation method that generates synthetic medical images using Generative Adversarial Networks (GANs) is proposed. In [13], the data augmentation is performed for handwritten digits of Latin, Devanagari, Oriya, and Bangla scripts using the generative adversarial network technique. In both cases, it has been observed that the model produced better results after data augmentation.

12.3 Dataset Description

The dataset under consideration is the IWFHR-10 Tamil Character Dataset, curated by HP Labs India [14], which contains approximately contains 500 samples of each of the 156 Tamil characters, written by native Tamil writers including school children, university graduates, and adults from the cities of Bangalore, Karnataka, India, and Salem, Tamil Nadu, India. Total images of characters present in the dataset are 77,553. The dataset has been divided into two sections:

- Train data: This dataset contains approximately 300 samples for each class
- Test data: This dataset contains approximately 170 samples for each class

The dataset contains approximately 500 samples for each character class. We have manually cleaned the dataset by removing noisy images from each character sample. The total number of characters available in the dataset after cleaning is 42,180.

12.4 Methodology and System Architecture

The papers majorly present models that work on images of individual segmented characters rather than images of sentences and entire documents [1–5, 7, 8, 10, 11, 15]. The models presented here were majorly trained on the same dataset–the IWFHR-10 Tamil Character Dataset [1, 3, 4, 15]. Thus, the models were trained on a limited dataset. It has been observed that data augmentation helps in improving the accuracy of the model [7, 12, 13]. Several papers haven't used data augmentation techniques and the ones that used, have applied elementary augmentation techniques such as rotation, skewing, stretching, etc. [1, 3, 4, 7, 9, 10, 12]. The models presented often perform well for a limited number of classes, but often fail to reproduce the same results when the number of classes increases. Thus, these models cannot classify all the characters of the language with high accuracy and hence are not very effective in real-world applications.

We aim to overcome the aforementioned limitations by building a lightweight CNN model [1, 3–5, 7–10] that will be trained on the IWFHR-10 Tamil Character Dataset [14] along with a dataset that will be generated by the GAN [12, 13]. This aims in resolving the issue of the limited dataset, along with the issue of lack of data augmentation. The model proposed works on all the character classes [9] present in the dataset, along with reasonably high accuracy, unlike the aforementioned models, that either worked on a few classes with high accuracy or worked on all the classes but with reduced accuracy.

12.4.1 *Flow of Proposed System*

The user uploads an image of a document containing Tamil characters. Various preprocessing techniques such as rescaling, binarization, noise removal, dilation, deskewing, removal of borders, removal of the transparency layer are applied [2, 6, 16]. The edges of the document in the image are determined, the transformation of perspective is performed and thus the region of interest is retained. Line, word, and character segmentation [2, 6, 16] are done using the contour detection technique. Then the characters are passed to the preprocessing pipeline. The preprocessed characters are then passed through the trained CNN model that recognizes and classifies the characters. The classified characters are reorganized and used for the digitization of documents (Fig. 12.1).

12.4.2 *Training of Model*

The training procedure of the CNN model can be divided into several phases (Fig. 12.2):

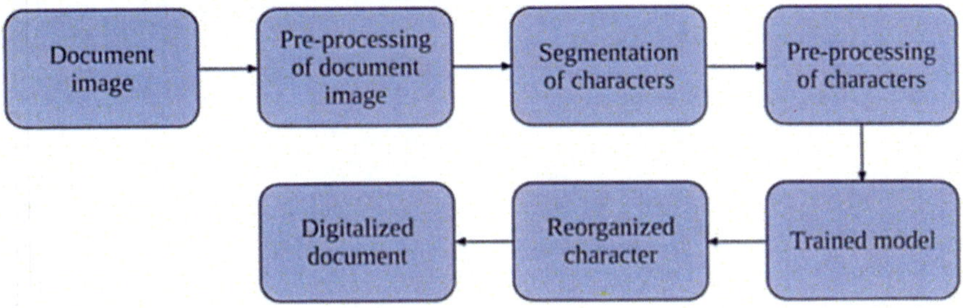

Fig. 12.1 Flow of proposed system

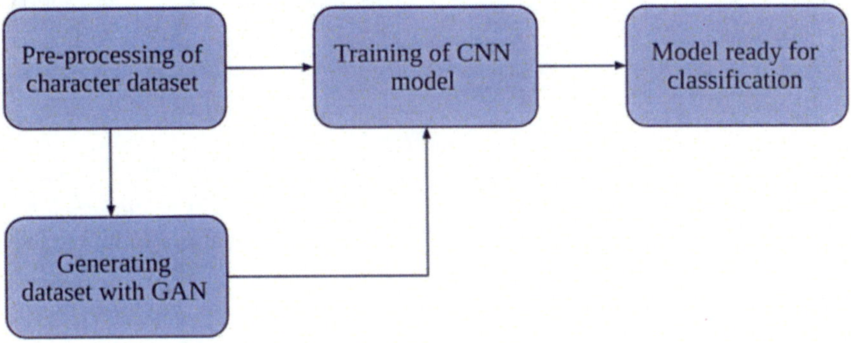

Fig. 12.2 Flowchart for the training of the model

- Preprocessing of characters present in the dataset.
- Generation of images of characters and data augmentation using the preprocessed character dataset, with the help of a GAN.
- Training the CNN model on the expanded dataset, i.e., the original dataset along with the generated data samples from the GAN.

12.4.3 Preprocessing of Characters

In the dataset IWFHR-10 Tamil Character Dataset, we have about 500 images of the 156 unique characters present in the language. The images from the dataset are loaded into the preprocessing pipeline. Initially, the images are converted into grayscale images as we don't need colour information. Then we apply thresholding so that we can convert the image into a binary image. Then colour inversion is applied to the images to make the background black and the character white. Then we find contours to find the bounding box of the character which fully encloses the character. The bounding box is then cropped out. Next, the image will be resized by making the longest side of the image 42 pixels and the other side is calculated with the help

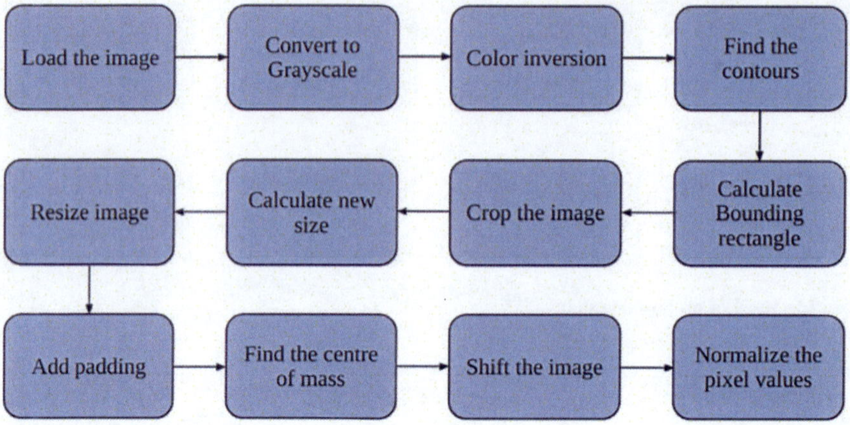

Fig. 12.3 Flowchart for preprocessing of characters

of a scaling factor. Next padding is done to make the image of size 50 × 50. Next, we find the centre of mass of the image. The centre of mass will be closer where the pixel density is higher. Next, the centre of mass of the image is shifted to the centre of the image. Finally, the image is normalized by dividing the pixels by 255 (Fig. 12.3).

12.4.4 Data Augmentation with GAN

The generative adversarial network contains two models which are generator and discriminator [17]. The generator model is used to generate new data from the existing dataset. And the discriminator model is used for classifying whether the generated data is real or fake and they compete against each other. Generative adversarial networks are used here to augment the dataset that generates new data from the existing dataset. This will increase the size and variations in the dataset which can indirectly help to increase the accuracy by training in a larger dataset with variations (Fig. 12.4).

Generator

At the start, the generator which is a neural network takes input as a noise vector and gives the output as a Tamil character. We are using a convolutional neural network as the data with which we are working is image data. As we need to achieve the desired output shape of 64 × 64 px the spatial dimension of input should be increased as it passes through different layers. The first two layers of the network are Dense layers with Rectified Linear Unit (ReLU) activation.

The first dense layer contains 1,024 neurons and the second one contains 32,768 neurons. Then we have the reshape layer. Convolution is to be used afterwards and

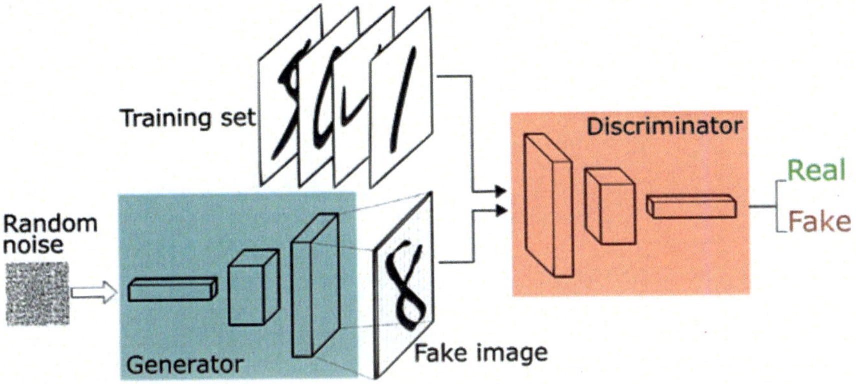

Fig. 12.4 GAN architecture *(Source* https://sthalles.github.io/intro-to-gans/)

hence reshaping is applied, and to apply convolution we need matrix-like entities. The matrix will be upsampled before applying the convolution. (2,2) Upsampling is used which will increase the dimension from 16 × 16 to 32 × 32.

Then we have 2 × 2 convolutional filters, the kernel weights have been initialized according to a normal distribution. LeakyReLU activation has been used for this layer. We again have an upsampling layer and this upsampling layer will give 64 × 64 dimensional matrices as output. Then we have 3 × 3 convolutional filters where LeakyReLU activation has been used and according to normal distribution kernel weights have been initialized. As we only need 1 channel for our grayscale image the last convolutional layer only contains 1 filter (Fig. 12.5).

Discriminator

The discriminator acts as a binary classifier that takes the image as an input and predicts whether the image is fake that is the image formed by the generator or if the image is original. The starting 2 layers of discriminator are convolutional layers which have a stride of 2 and filter size of 5 × 5.

The output of the convolutional layer has to be flattened so that it can be passed to a dense layer. The dense layer is of the size 256 with a 50% dropout and at the end, there is a sigmoid layer. Adam optimizer with a learning rate of 0.0002 is used and the loss used is binary cross-entropy (Fig. 12.6).

In order to train the generator, the loss from the discriminator has to be sent back to the generator but this loss computed from backpropagating discriminator should not affect the weights of the discriminator. For this, the discriminator should not learn anything and hence the discriminator trainable parameter will be set as false and only the weights of the generator will be modified. For this, a new model will be formed by stacking the generator and discriminator. The model will take a noise vector as input and pass it through the generator to create a fake image. This image will then be passed to a discriminator that computes the probability of being an original image. Then we will train the model and the discriminator will not learn anything.

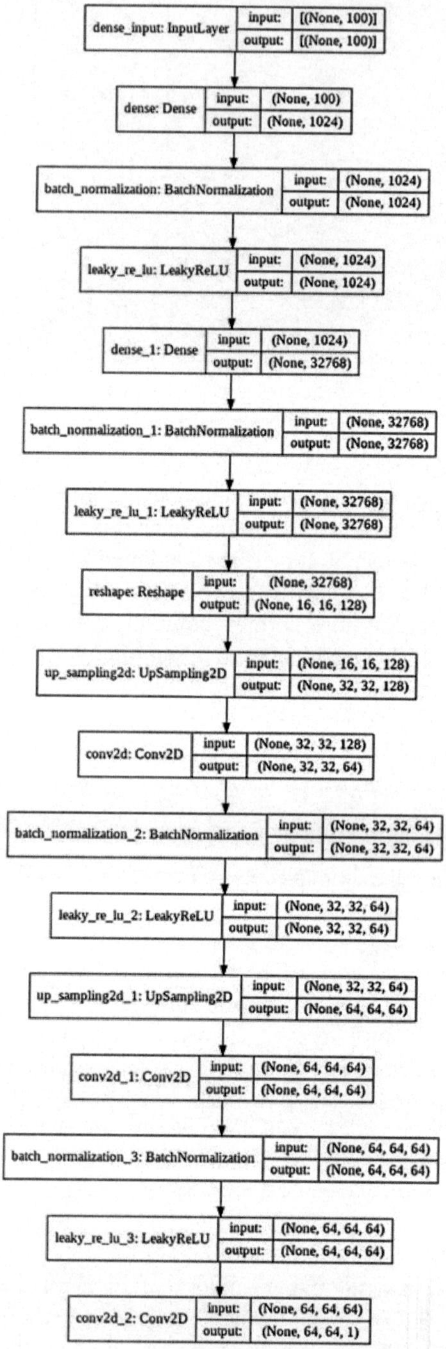

Fig. 12.5 Generator architecture

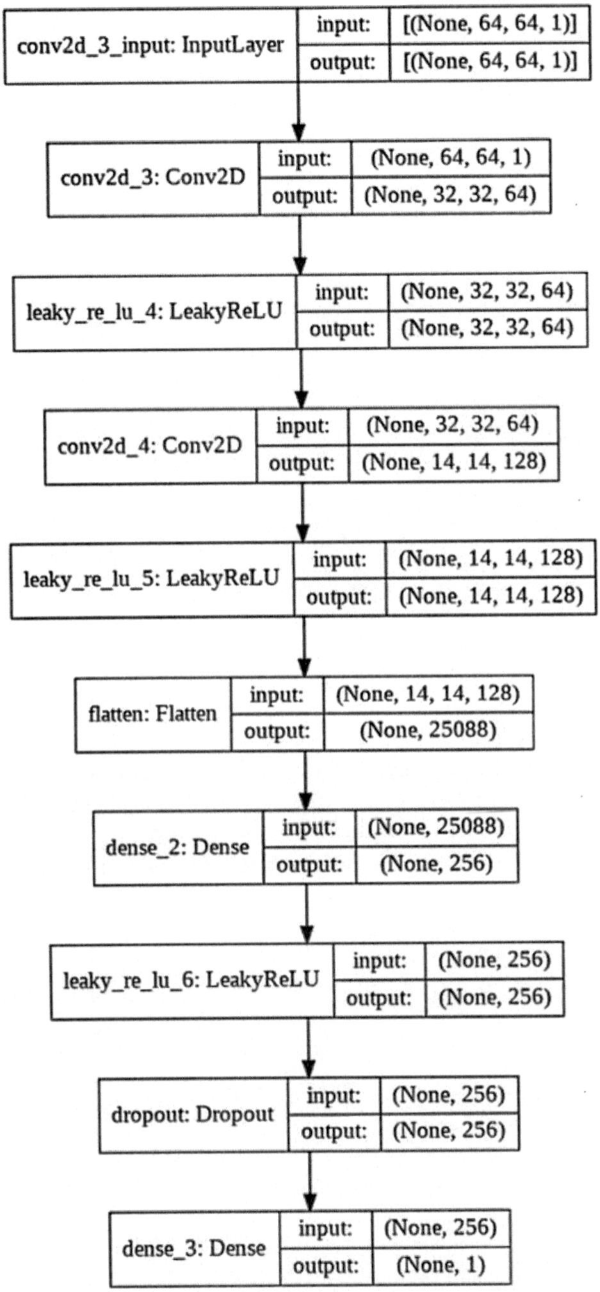

Fig. 12.6 Discriminator architecture

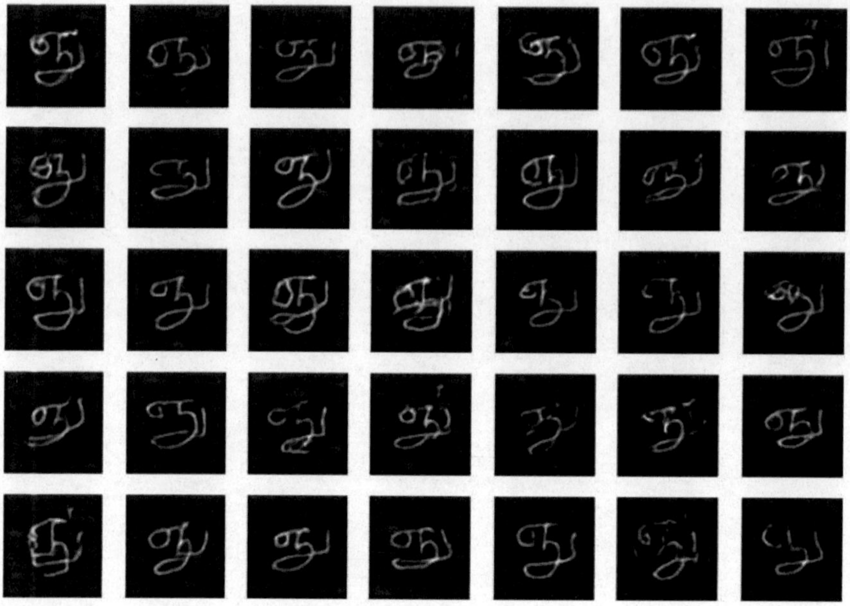

Fig. 12.7 A GAN generated Tamil character: ‘ ஜு’

Image Generation

With the help of GAN, we were able to generate an additional 12,820 images using 42,180 images. So, after combining both, the total number of characters in the dataset is 55,000 images. We have successfully increased the dataset by 30% approximately. This extended dataset has been used to train the CNN model (Fig. 12.7).

12.4.5 CNN Model

A Convolutional Neural Network (ConvNet/CNN) is a deep learning algorithm that can take in an input image, assign importance (learnable weights and biases) to various aspects/objects in the image, and be able to classify them [18]. A CNN can broadly be divided into two parts based on the function that it is performing: feature learning and classification. The feature extraction process is majorly done by a stack of Convolution and Pooling layers. The classification process is usually done by a fully connected layer with n output neurons, where n is the number of object classes to be classified. The role of CNN is to reduce the images into a form that is easier to process, without losing features that are critical for getting a good prediction. The convolution operation is carried out by using a kernel/filter. The output of this convolution operation is called an activation. The objective of the

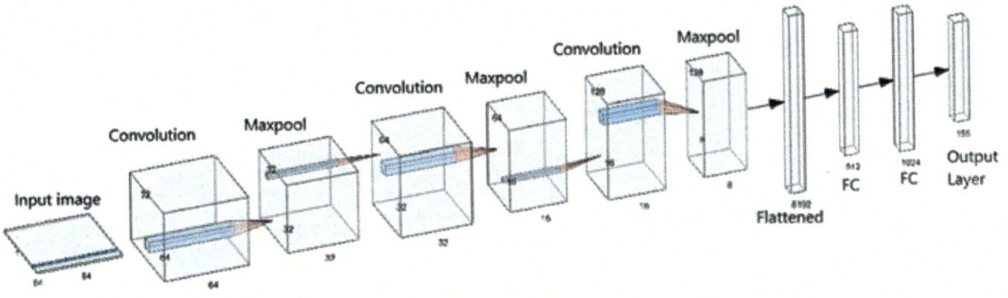

Fig. 12.8 CNN model

convolution operation is to extract the high-level features such as edges, from the input image (Fig. 12.8).

The architecture for Handwritten Tamil character classification consists of nine layers: three convolutional layers, three max-pooling layers, two fully connected layers, and an output dense layer. The network architecture can be described as follows: convolution layer of kernel size (5 × 5) and 32 filters with ReLU activation function and padding same, Max Pooling with pool size (2 × 2) and strides (2 × 2), convolution layer of kernel size (5 × 5), and 64 filters with ReLU activation function and padding same, max-pooling with pool size (2 × 2), convolution layer of kernel size (5 × 5), and 128 filters with ReLU activation function and padding same, max-pooling with pool size (2 × 2), 2 fully connected dense layers of size 512 and 1,024 neurons, and an output layer of 155 classes with a softmax activation function. The input image is of size (64 × 64) and the model classifies it into 155 classes and gives probability values of all classes.

Several hyperparameters needed to be set for the efficient training of the neural network. Hyperparameters such as stride, padding, filters are tuned to make the model better. We have a stride value of 2 to slide two pixels over the input image. We set the parameter padding to "same" to preserve spatial dimensions of the volume such that the output volume size matches the input volume size. We have used glorot_uniform to initialize the kernel weight matrix and the bias vector is initialized to zero. We got the optimum result at 20 epochs. The input images of size 64 × 64 were fed to the network in batches of size 32. The model is compiled setting a few attributes such as optimizer, loss, metrics, and batch size. We have used the Adam optimizer (learning rate = 0.001) as it is used to accelerate the gradient descent algorithm by taking into consideration the exponentially weighted average of the gradients and thus converging towards the minima at a faster pace. We use categorical cross-entropy for our loss function. We use the accuracy metric to see the accuracy score on the testing set when we train the model (Fig. 12.9).

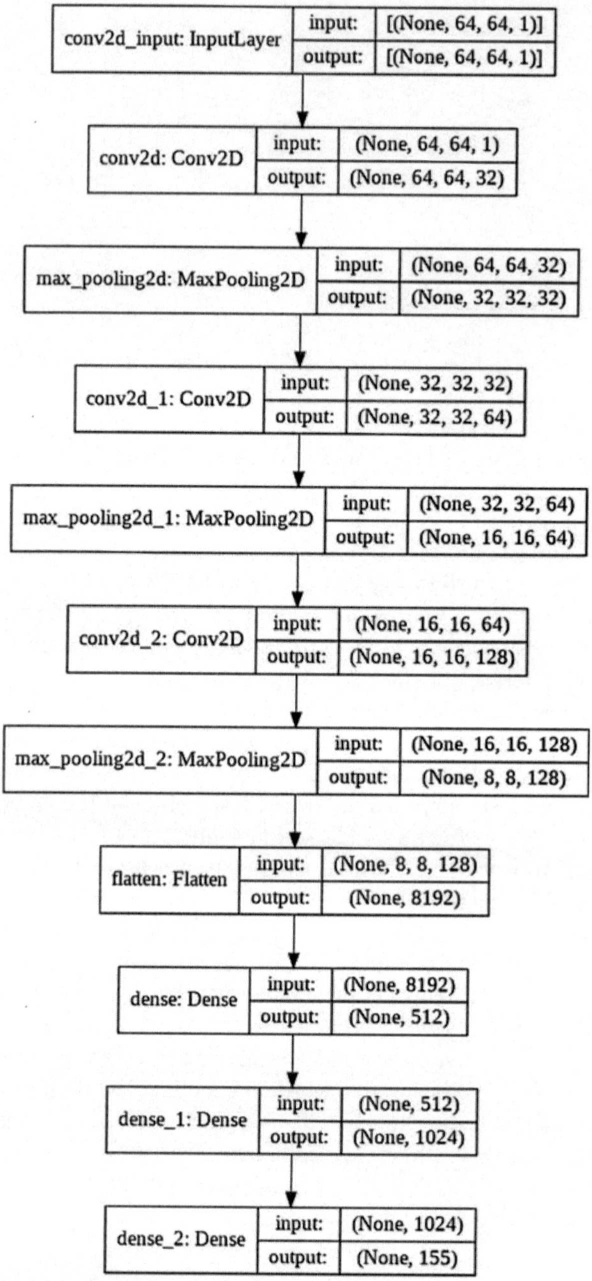

Fig. 12.9 CNN Architecture

12.4.6 Preprocessing of the Document

The input document is subjected to scaling followed by thresholding. Otsu's thresholding method in combination with the binary thresholding method has been used. Several noise-reducing filters such as Gaussian Blur, Median filter, and Bilateral filter have been used to reduce random noise present in the image. A border is applied around the edges of the image. This border is applied to help in segregating the pages when the image contains two pages that meet at the spine of a book. This is followed by Canny edge detection. The image is then subjected to closing, i.e., dilation of the image followed by erosion. This helps in closing the gaps present between the edges. Contour detection is then applied to get the largest convex contour that encompasses the entire handwritten text. Finally, perspective transformation is applied to straighten the input image.

12.4.7 Segmentation

After the initial preprocessing, the edges of the characters are detected using the Sobel filter. This gives us the boundaries of the characters precisely. Then the copy of the input image is subjected to dilation, wherein all the characters in a word are converted to blobs. This is done to segregate the foreground, the region of interest from the background. After converting the words to blobs, contours are detected for these blobs. Bounding boxes are created around these contours (Figs. 12.10 and 12.11).

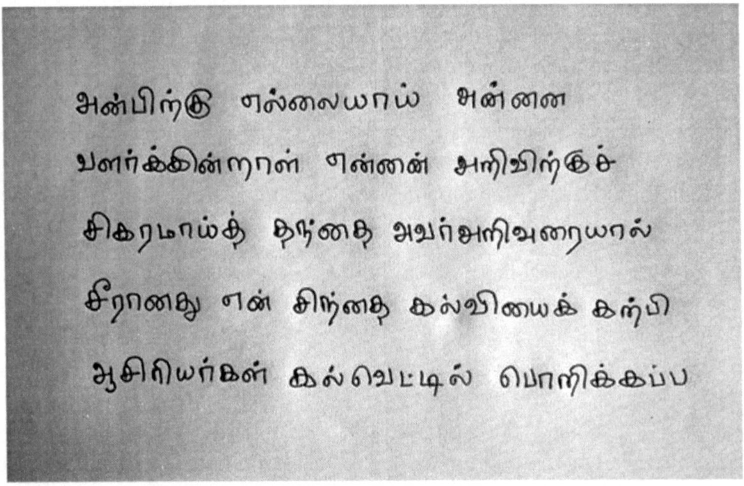

Fig. 12.10 Input image after initial preprocessing

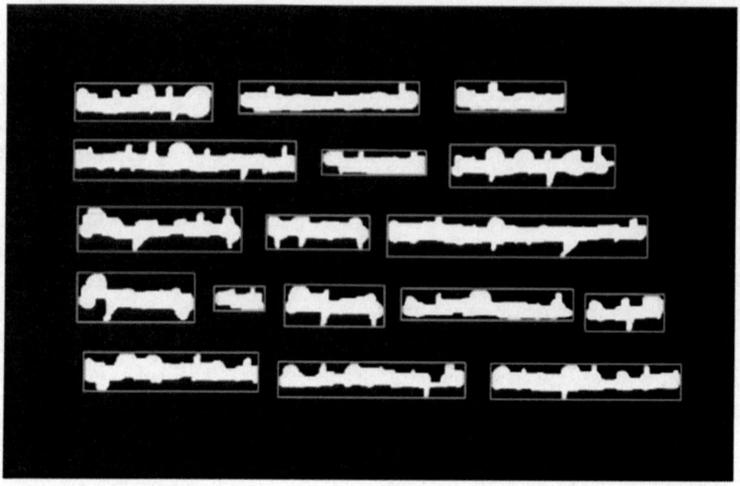

Fig. 12.11 Word segmentation

Now, for each bounding box, the ratio of the number of white pixels to the number of black pixels present in the box, the ratio of the height to the width, and the ratio of the width to the height are calculated. If these ratios satisfy the given constraints, i.e., the values are higher than the calculated threshold, these boxes can be classified as words with certainty. If two or more large bounding boxes intersect, then it is highly probable that these bounding boxes belong to the same word and are hence merged together, thus forming a single bounding box. These bounding boxes are then sorted according to the coordinates and then the preprocessed image is cropped in accordance with them. The images of words are hence obtained. Now for each image of the word, resizing and closing operations are performed. This ensures that characters that are written really close to each other don't appear in a single bounding box. Then the image is thresholded and contours are detected. Bounding boxes for these contours are created and sorted according to their coordinates. As the dot or pulli is not directly a part of the character, it is detected as a separate contour and a separate bounding box is created for the same. This dot and the character are merged into a single bounding box in the following way:

- Firstly, it is checked whether the bounding box above a character is a dot and not random noise. This is done by comparing the contour area to a calculated threshold. If the condition is satisfied, the bounding box certainly contains a dot.
- Then the vertical distance between the dot and its corresponding character is checked.
- If it is within reasonable bounds, the dot is considered to be belonging to the character below it and a bounding box is formed around the character and the dot using the coordinates of the bounding boxes of the two entities.

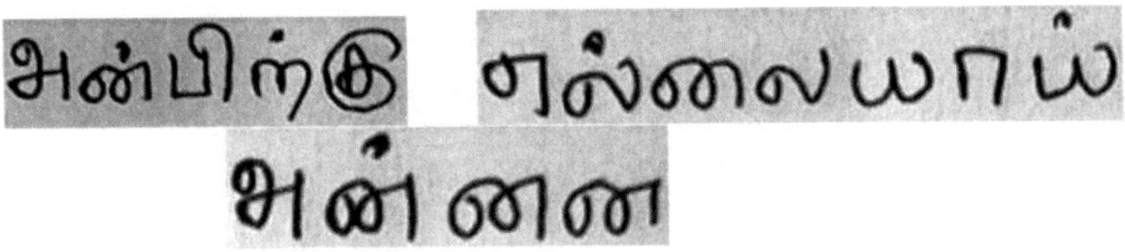

Fig. 12.12 Segmented words

Fig. 12.13 Segmented and preprocessed characters

After the bounding boxes have been established, the image is cropped accordingly. The cropped images are then sent to the preprocessing pipeline (Figs. 12.12 and 12.13.

12.5 Results

12.5.1 Implementation

The model is built using python language. The front end of the website is made using HTML, CSS and JavaScript and Flask is used for the backend. Flask is a small and lightweight python web framework that provides useful tools and features for creating web applications in python. We are loading the saved CNN model in the Flask to classify the characters. We are also performing segmentation in the backend (Fig. 12.14).

The user can upload an image file of the handwritten Tamil document here. The uploaded document is preprocessed, segmented into characters, predicted by the model, converted into an editable text file and then available for download (Figs. 12.15 and 12.16).

12.5.2 Accuracy

The model trained only on the IWFHR-10 dataset gave us an accuracy of about 94.03%. The same model, when trained on the GAN augmented dataset gave us an improved accuracy of 97% (Figs. 12.17 and 12.18 and Table 12.1).

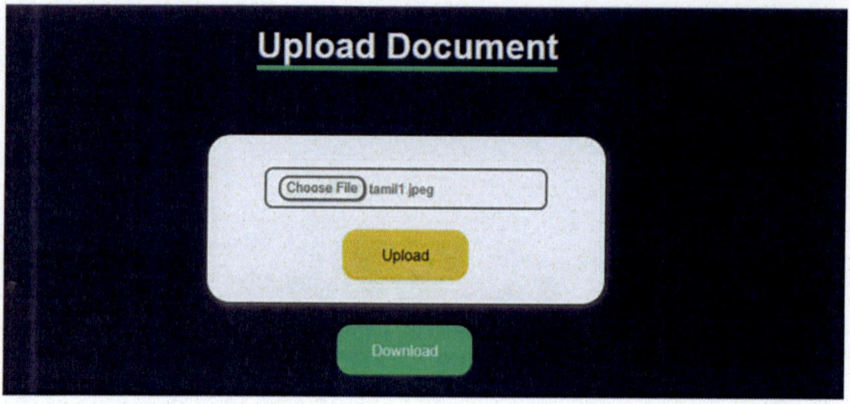

Fig. 12.14 Document digitization website

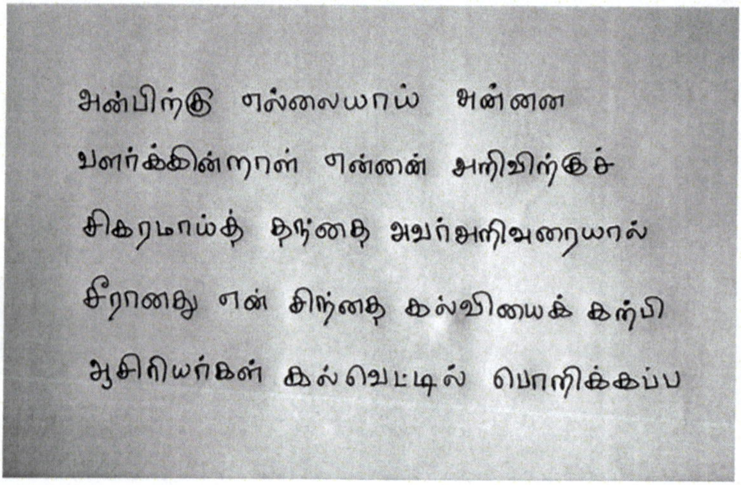

Fig. 12.15 Input handwritten image

12.6 Conclusion

In this paper, we've developed a deep learning system that comprises a combination of GAN and CNN. GAN has been used for expanding the available dataset and data augmentation, and CNN has been used for the task of classification of the Tamil characters. Here, we have presented a lightweight, computationally inexpensive system that classifies all the characters present in the Tamil language, with considerably high accuracy. We have developed a CNN model that achieves an accuracy of 94.03% and about 97% with the help of data augmentation using GAN. The high accuracy of the

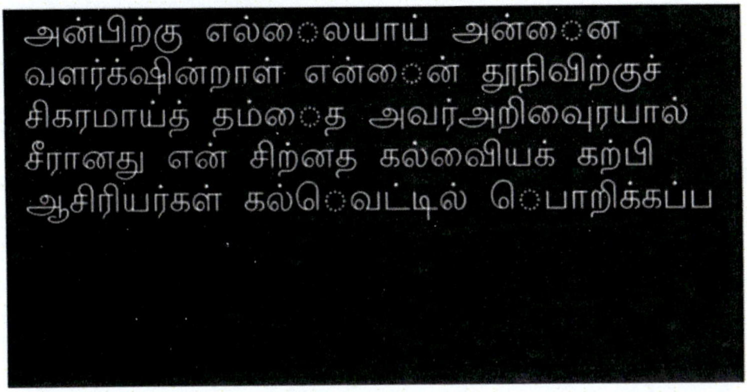

Fig. 12.16 Digitized output of our model

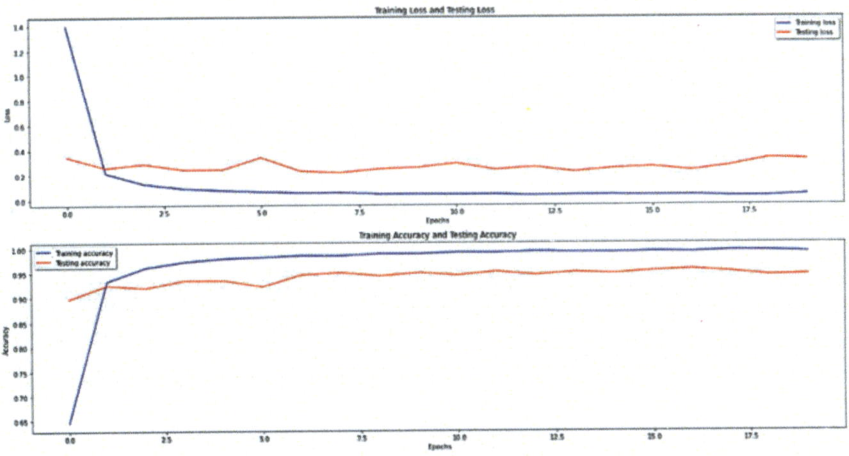

Fig. 12.17 Accuracy of the model (No data augmentation)

model can be attributed to the increase in the number of data points per character in the training dataset and by using a deeper CNN model.

Our system produces a digital, editable word document/PDF from the images of the input document. This system has numerous applications such as digitization of government documents, automated postal addresses and ZIP code digitization, digitization of academic institutional documents and records, data extraction from bank cheques, etc.

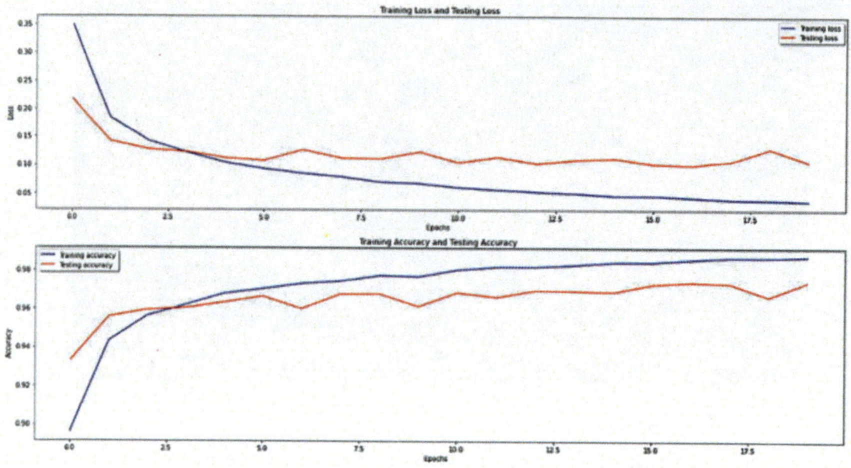

Fig. 12.18 Accuracy of model (With data augmentation)

Table 12.1 Accuracy of existing character recognition models

Model	Dataset	Number of classes	Methodology	Accuracy (%)
Kowsalya et al. [2]	Custom dataset	2	Neural network with Elephant Herding Optimization	93
Banumathi et al. [6]	Custom dataset	8	Kohonen's Self Organizing Feature Map	85
Shanthi et al. [16]	Custom dataset	34	SVM	82.04
Vijayaraghavan et al. [1]	IWFHR-10	35	CNN	94.4
Prakash et al. [3]	IWFHR-10	125	CNN	71.1
Bhattacharya [15]	IWFHR-10	156	Clustering + MLP	87.66
Pragathi et al. [5]	Custom dataset	156	VGG16	94.52
Jayakanthan et al. [4]	Subset of IWFHR-10	156	ResNet	96
Our model	IWFHR-10	156	CNN	94.03
Our model	IWFHR-10 + Data generated using GAN	156	CNN	97

References

1. Vijayaraghavan, P., Sra, M.: Handwritten tamil recognition using a convolutional neural network. In: MIT Media Lab. 2018 International Conference on Information, Communication, Engineering and Technology (ICICET). (2014)
2. Kowsalya, S., Periasamy, P.S.: Recognition of Tamil handwritten character using modified neural network with aid of elephant herding optimization. Multimedia Tools Appl. **78**(17), 25043–25061 (2019)
3. Prakash, A.A., Preethi, S.: Isolated offline tamil handwritten character recognition using deep convolutional neural network. In: 2018 International Conference on Intelligent Computing and Communication for Smart World (I2C2SW), IEEE, (2018)
4. Jayakanthan, R., et al.: Handwritten tamil character recognition using ResNet. Int. J. Res. Eng. Sci. Manag. **3**(3), 133–137 (2020)
5. Pragathi, M. A., et al.: Handwritten tamil character recognition using deep learning. In: 2019 International Conference on Vision Towards Emerging Trends in Communication and Networking (ViTECoN). IEEE, (2019)
6. Banumathi, P., Nasira, G.M.: Handwritten Tamil character recognition using artificial neural networks. In: 2011 International Conference on Process Automation, Control and Computing, IEEE, (2011)
7. Chowdhury, R.R., et al.: Bangla handwritten character recognition using convolutional neural network with data augmentation. In: 2019 Joint 8th International Conference on Informatics, Electronics & Vision (ICIEV) and 2019 3rd International Conference on Imaging, Vision & Pattern Recognition (icIVPR), IEEE, (2019)
8. Purkaystha, B., Datta, T., Islam, M.S.: Bengali handwritten character recognition using deep convolutional neural network. In: 2017 20th International Conference of Computer and Information Technology (ICCIT), IEEE, (2017)
9. Balci, B., Saadati, D., Shiferaw, D.: Handwritten text recognition using deep learning. CS231n: Convolutional Neural Networks for Visual Recognition, Course Project Report, Stanford University, Spring, 752–759 (2017)
10. Deore, S.P., Pravin, A.: Devanagari handwritten character recognition using fine-tuned deep convolutional neural network on trivial dataset. Sadhana **45**(1), 1–13 (2020)
11. Yadav, M., Purwar, R.: Hindi handwritten character recognition using multiple classifiers. In: 2017 7th International Conference on Cloud Computing, Data Science & Engineering-Confluence, IEEE, (2017)
12. Frid-Adar, M., et al.: Synthetic data augmentation using GAN for improved liver lesion classification. In: 2018 IEEE 15th International Symposium on Biomedical Imaging (ISBI 2018), IEEE, (2018)
13. Jha, G., Cecotti, H.: Data augmentation for handwritten digit recognition using generative adversarial networks. Multimedia Tools Appl. 1–14 (2020)
14. HP Isolated Handwritten Tamil Character Dataset: IWFHR-10, http://lipitk.sourceforge.net/datasets/tamilchardata.htm
15. Bhattacharya, U., Ghosh, S.K., Parui, S.: A two stage recognition scheme for handwritten Tamil characters. In: Ninth International Conference on Document Analysis and Recognition (ICDAR 2007), vol. 1. IEEE, (2007)
16. Shanthi, N., Duraiswamy, K.: A novel SVM-based handwritten Tamil character recognition system. Pattern Anal. Appl. **13**(2), 173–180 (2010)
17. Goodfellow, Ian, et al.: Generative adversarial nets. In: Advances in neural information processing systems, p. 27 (2014)
18. Krizhevsky, A., Sutskever, I., Hinton, G.E.: Imagenet classification with deep convolutional neural networks. Adv. Neural. Inf. Process. Syst. **25**, 1097–1105 (2012)

Chapter 13
FactOrFake: Automatic Fact Checking Using Machine Learning Models

V. A. Anusree, K. M. Aarsha Das, P. S. Arya, K. Athira, and S. Shameem

Abstract In today's world, more than traditional news platforms like newspapers, radio and televisions, people are depending on social media platforms to get the latest trending news. These platforms serve as a wonderful way to disseminate important information to the general public in very short amounts of time but at the same time it also allows the spread of fake news to a large section of the population. It is very difficult to verify the integrity of the information contained in the articles, videos and images posted in social media platforms like Twitter, Facebook, etc. Manual methods of identifying a news article as real or fake are very time consuming and require considerable effort. In this project, a machine learning (ML) model has been developed in order to automate this process. Natural Language Processing (NLP) techniques have been used for analyzing the text content of the news and finding patterns that will classify the news as fact or fake. Term Frequency–Inverse Document Frequency (TF–IDF) is the feature extraction technique employed. Two ML classifiers, Logistic Regression and Multinomial Naive Bayes have been used for building the models. Using Logistic Regression, the model was able to achieve about 0.89 accuracy score and precision score of 0.94. This outperformed Multinomial Naive Bayes model which had an accuracy of 0.83 and precision of 0.80 but performed better in terms of recall value of –0.87 compared to that of Logistic Regression with a recall value of 0.83.

13.1 Introduction

The increasing use of social media by people of almost all age groups means that the demand of social media as a source of the latest hot news has also tremendously increased. Nowadays people depend on online social networking sites like Twitter, Facebook, Instagram, etc., to know what's happening around the world. The issue here is that anyone can open an account in these sites and can post stories and there is no immediate way to know the integrity of these claims. More often than not, these

V. A. Anusree (✉) · K. M. Aarsha Das · P. S. Arya · K. Athira · S. Shameem
Vidya Academy of Science and Technology, Thalakottukara, India
e-mail: shameem.s@vidyaacademy.ac.in

J. I.-Z. Chen et al. (eds.), *Machine Learning and Autonomous Systems*, Smart Innovation, Systems and Technologies 269, https://doi.org/10.1007/978-981-16-7996-4_13

are basically just false claims either posted for fun, or to mislead people or to cause community unrest. Not just social networking sites, but such false news are common in news portals and news websites as well.

What does "fake news" actually mean? It is often used interchangeably with rumors. The term was first used in a report published by Buzzfeed to refer to a series of absurd posts in Facebook. Since then it has been used extensively by the media and was selected as the "word of the year" by the American Dialect Society [1]. Some other terms used in place of fake news include satire and parody. In general we can say that fake news refers to articles that contain false information that are intentionally written to mislead readers.

The harm caused by fake news can be severe. Sometimes these are generated by some groups to achieve some political propaganda. Sometimes they target certain communities or personalities. In times of communal riots, such news can add fuel to the fire. They may also be used as a weapon against certain groups to hurt their sentiments.

Also, in times of natural disasters, fake news may send people into panic mode. If a large section of people actually end up believing this fake news, then it can have adverse effects on various fields like business(for example stock prices may drop), government (can lose support), and entertainment.

13.2 Literature Survey

Over the years, a lot of effort has gone into developing effective methods to mitigate the problem of fake news. For example, PoltiFact [2] is a fact checking website, which finds out whether a given piece of news article is fake or real. But since this process is done manually, it takes a lot of time and effort. The emergence of ML and Deep Learning Methods to automate various NLP tasks has given a new ray of hope to an otherwise mundane process.

Despite the fact that the risks associated with fake news had been identified for a long time, the increasing influence of social media in our lives has finally opened the eyes to the adverse effects fake news can have. Nowadays social media sites like Twitter, Facebook, Instagram have become so intertwined with our lives that we have come to trust these sources for the availability of instant news rather than television or newspapers. A piece of news can reach millions around the world within a few seconds of it being first posted. The speed with which the news can spread is exponential. Not only can it shape the public reaction to an event, simply by how the news is worded but it can also have a huge impact on the political foundations of a country.

A very good example of the effect social media platforms and the news generated there can have on society is how Twitter was used by the various political leaders to shape the results of 2016 US Elections. In [3], publicly streaming Twitter data was collected from the Twitter accounts of the five major politicians Hillary Clinton, Donald Trump, Ted Cruz, Bernie Sanders and Paul Ryan. All these individuals possessed active Twitter accounts with more than 2 million followers and engaged in frequent Twitter activity. It was seen that Trump held unique appeal to those espousing hateful ideologies more so than other Republican or Democratic candidates. Even in the 2020 US elections, Twitter was the most used platform by political leaders as well as the common public to make their opinions heard and even going as far as to make false claims about election results [4].

But how can fake news be detected? Manually we humans can use logic and deduce the authenticity by digging a bit deeper, we can look for example, the website in which it is posted, verify using the sources cited in the post. But the need of the hour is to automate this process. But the challenge is to find a way to do this efficiently and effectively on a large scale.

The publication [5] tried identifying fake news by detecting and analyzing fake news features and in turn, identifying the textual and socio-cultural characteristics of fake news. For example, a common feature of fake news that was identified, was the repetitive and excessive use of phrases like "Warnings Issued", "Explosive news report" to capture reader's attention and overuse of quotation marks to make it seem like the information is coming directly from the source although in reality, it might not be. A parsing through the headlines was also carried out to obtain top 20 words used in fake news and real news for a comparison of text features.

Another study that used the news content (text) for fake news detection by analyzing the linguistic features of the news article is [6]. The model is working by capturing the author's style of writing without having access to any kind of metadata. This is a good method to use when user-information or source of the news available but it might fail to work in case the fake news author has tried to copy the style of writing of genuine authors. Also the news articles used to train the model are limited to political and a few other domains.

Another way to identify fake news is by looking at the contents of news articles and echo-chambers (which is a community of users sharing the same opinions on a particular topic) for fake news detection. The same was carried out in [7].

Fake news can also be identified looking at the feedback received by that particular post. In social medias platforms like Twitter and Facebook, if we find a post that looks like it could be fake news, we can always look at the comment section or replies. In 9 out of 10 cases, the reactions of other users to the given post will help us to verify the integrity of the news. A similar concept has been implemented in [8] using User Conversational Tree(UCT) (with four kinds of tags—deny, comment, support and query) to identify fake news patterns. A tool for automatic fake news collection, detection and visualization was developed in [9]. It had mainly used linguistic characteristics of the news articles and the social engagements features for the model training. Since the study is mainly based on Twitter, it would be better to also make use of features like retweets, likes, etc., for the detection task.

Studies have also been carried out to find patterns specific to Fake News. It was seen in [10] that users sharing fake news are more inclined to include hashtags and mentions. Also such tweets had more negative sentiment. This means that sentiment analysis [11] and fact-checking have a direct relation and can be used hand in hand as done in [12]. Monitoring and analyzing negative posts and comments on posts can also help to identify flaming events [13]. This can in turn pave the way for identifying fake news because more often than not, flaming events and fake news spreading goes hand in hand. These processes that we have seen so far can be applied not just in identifying fake "news" but in other areas like identifying fake consumer reviews [14].

Another important research area that can help to identify fake news is bot detection in social media because bots are used immensely to spread false information and panic. One such study that used one class classification approach for bot identification is [15]. This approach uses one class classifiers to extract features of legitimate accounts and compare them to the given account to identify whether it is a bot account or not. The issue with this method is that, nowadays due to the advancements in Artificial Intelligence bot creators are trying to mimic the behavior of legitimate accounts making detection difficult. Hence existing methods need to be improved to keep up with the advancements.

It can be seen that a lot of research papers have been written on the topic of fake news detection and various methods for its identification have also been proposed. The publications on this topic have strictly been increasing over the years which is a good trend since it means that people have finally realized the adverse effects fake news propagation can have on the community. A summary of such works can be found in [16] which is a review article. It selected 87 distinct publications from about 1333 candidates and went through an in-depth analysis of them identifying the various works published over the years, the most active authors, publication venues, the domains in which such works are being applied, the social media in use, commonly used features and methods, and most commonly used datasets. In this study, Twitter was identified as the major social media used for the study of fake news detection. But the interesting fact is that not much research has been done in studying misinformation spreading through other social media like WhatsApp, Instagram and Telegram. It must be because of the complexity of tracking messages because WhatsApp being end-to-end encrypted and Instagram having mostly pictures and videos which are difficult to analyze. Also most of the works are strongly based on text features of the news instead of the available metadata like followers, retweets, multimedia content. The news articles used also seemed to be restricted to any particular domain mostly politics.

Another such study which provides an idea about what fake news is, its characteristics as well as a summary of methods used by prior researchers is [17].

13.3 Proposed Work

In this project, ML models are used to predict whether the news article is fake or real. The proposed model differs from all previous models in the sense that, the data used for training is a combination of different types of news articles pertaining to a variety of topics like world politics, health, etc., from different sources and it tries to identify patterns by looking into their source URLs, as well as content. The combining of

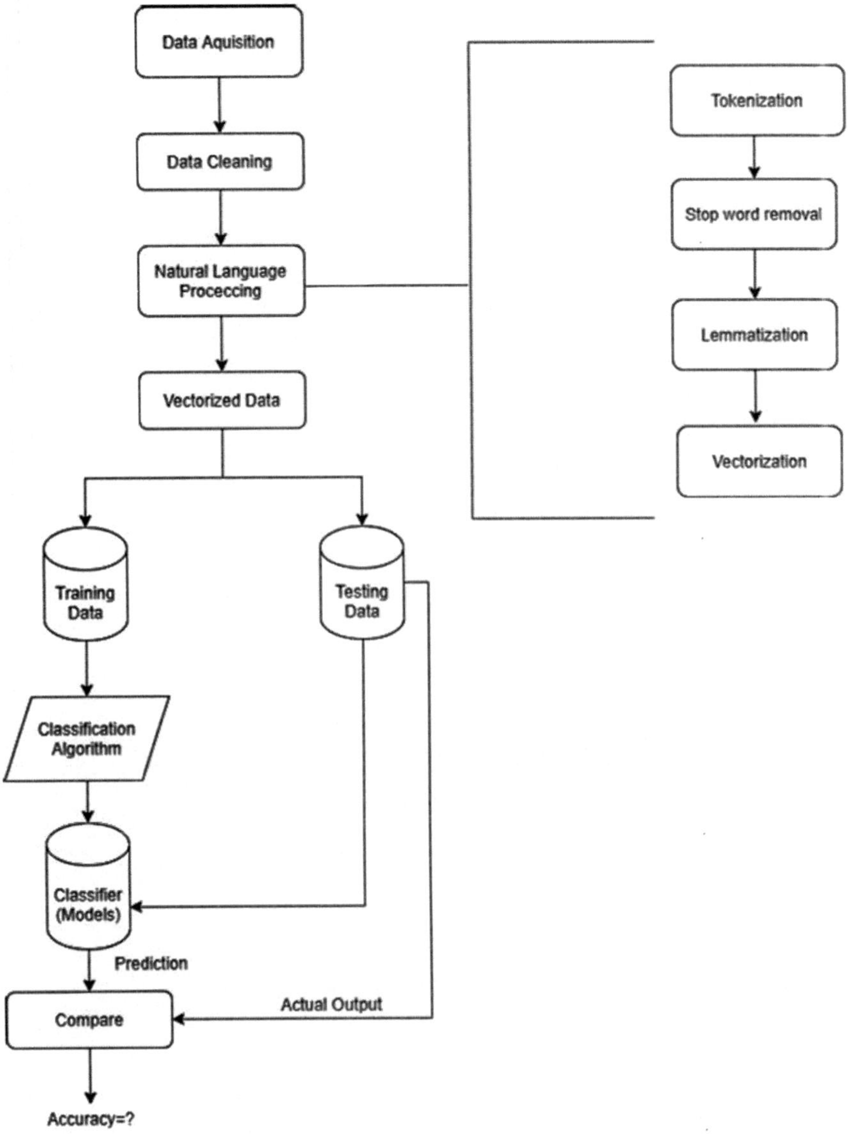

Fig. 13.1 Proposed approach

multiple datasets helps to improve accuracy of the model, since more data is always welcome in ML-based approaches and at the same time the use of different varieties of news topics can make the model perform well in real world scenarios.

The flowchart of the proposed approach is given in (see Fig. 13.1).

Table 13.1 Data count

Poltifact	Gossipcop	Getting real about fake news	Covid-19 fake news dataset
846	20814	12403	3552

Table 13.2 Evaluation measures

Algorithm	Accuracy	Precision	Recall	F1-score	AUC score
Logistic regression	0.889376413	0.941249227	0.834201151	0.884498039	0.892565452
Multinomial Naive Bayes	0.831272437	0.802030457	0.865990682	0.832784293	0.832272201

The first step in any ML-based project is Data Collection. Datasets from a variety of sources were collected (Table 13.1):

- FakeNewsNet [18]:
  - Poltifact
  - Gossipcop
- Kaggle Datasets:
  - Getting real about fake news [19]
  - Covid-19 fake news dataset [20]

The number of samples taken from each of the datasets is given in the table (see Table 13.2). These datasets had different sets of features. So the first task was to clean and combine them so that they would have a common feature set.

After going through all the different available features, the following features were selected:

- News Headline
- Source URL
- News Content

All the rows with null values for any of these features were dropped. Different datasets had different names for the headers of these features. So they were all renamed to "news-URL", "title" and "content" to make the datasets uniform. Since the FakeNewsNet (see Fig. 13.2), didn't have a "content" column, "title" was renamed to "content" in FakeNewsNet and in all other datasets, "title" and "content" were concatenated to form a single feature named "content". A new column "fake" was added which is the target class with binary values (0 for real news and 1 for fake news). Finally, all the datasets were combined into one single dataset (see Fig. 13.3). The next step was to apply NLP techniques to convert text data into vector form. It involved the following steps:

	news_url	title	fake
0	http://www.nfib-sbet.org/	National Federation of Independent Business	0
1	http://www.cq.com/doc/newsmakertranscripts-494...	comments in Fayetteville NC	0
4	https://web.archive.org/web/20070820164107/htt...	Budget of the United States Government, FY 2008	0
5	http://www.politifact.com/truth-o-meter/statem...	Donald Trump exaggerates when he says China ha...	0
6	https://www.law.cornell.edu/constitution/amend...	25th Amendment	0

Fig. 13.2 Poltifact dataset

	news_url	content	fake	total
0	speedtalk.com/forum/viewtopic.php?t=51650	BREAKING: First NFL Team Declares Bankruptcy O...	1	BREAKING: First NFL Team Declares Bankruptcy O...
1	politics2020.info/index.php/2018/03/13/court-o...	Court Orders Obama To Pay $400 Million In Rest...	1	Court Orders Obama To Pay $400 Million In Rest...
2	www.nscdscamps.org/blog/category/parenting/467...	UPDATE: Second Roy Moore Accuser Works For Mic...	1	UPDATE: Second Roy Moore Accuser Works For Mic...
3	https://howafrica.com/oscar-pistorius-attempts...	Oscar Pistorius Attempts To Commit Suicide	1	Oscar Pistorius Attempts To Commit Suicidehttp...
4	http://washingtonsources.org/trump-votes-for-d...	Trump Votes For Death Penalty For Being Gay	1	Trump Votes For Death Penalty For Being Gayhtt...

Fig. 13.3 Combined dataset

	news_url	content	fake	total	text_lemmatized
0	speedtalk.com/forum/viewtopic.php?t=51650	BREAKING: First NFL Team Declares Bankruptcy O...	1	BREAKING: First NFL Team Declares Bankruptcy O...	breaking first nfl team declares bankruptcy o...
1	politics2020.info/index.php/2018/03/13/court-o...	Court Orders Obama To Pay $400 Million In Rest...	1	Court Orders Obama To Pay $400 Million In Rest...	court orders obama to pay 400 million in rest...
2	www.nscdscamps.org/blog/category/parenting/467...	UPDATE: Second Roy Moore Accuser Works For Mic...	1	UPDATE: Second Roy Moore Accuser Works For Mic...	update second roy moore accuser works for mic...
3	https://howafrica.com/oscar-pistorius-attempts...	Oscar Pistorius Attempts To Commit Suicide	1	Oscar Pistorius Attempts To Commit Suicidehttp...	oscar pistorius attempts to commit suicidehtt...
4	http://washingtonsources.org/trump-votes-for-d...	Trump Votes For Death Penalty For Being Gay	1	Trump Votes For Death Penalty For Being Gayhtt...	trump votes for death penalty for being gayht...

Fig. 13.4 Lemmatized dataset

- Combine "content" and "news-URL" columns.
- Remove quotations, stop words, special characters, etc.
- Convert the entire text to lower case.
- Remove stopwords.
- Convert a sentence into individual tokens—Tokenization.
- Convert the tokens into their base form—Lemmatization.
- Finally, convert the lemmatized text into vectors—Vectorization.

After lemmatization, the lemmatized text was added as a new feature to the combined dataframe (see Fig. 13.4).

A word cloud representation of both real and fake news plotted from the lemmatized text is given(see Figs. 13.5, 13.6) respectively.

Next, the lemmatized text was converted to vectors. The vectorization technique used here was TF-IDF implemented using the "TfidfVectorizer" class in the sklearn library of python.

The vector data obtained after vectorization was split into training and testing data (0.8 for training and 0.2 for testing).

Fig. 13.5 Word cloud representation of Real News

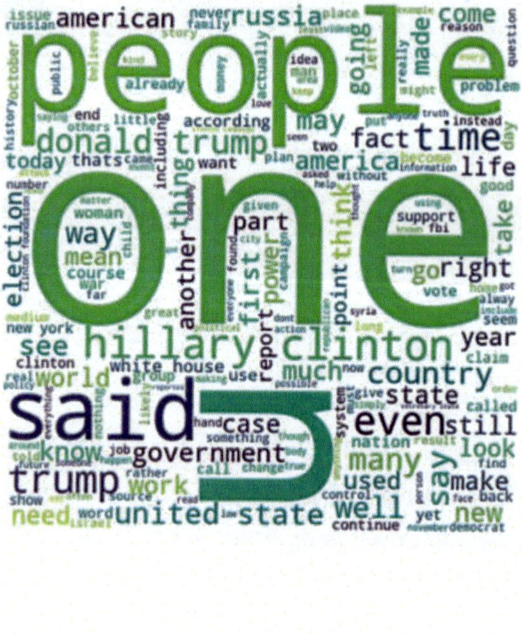

Fig. 13.6 Word cloud representation of Fake News

Two algorithms were used for training:

- Logistic Regression [21]
  Logistic regression is an important member of a class of models called generalized linear models. It can directly predict values that are restricted to (0, 1) interval. Logistic regression predicts the probability y that an instance will belong to a particular category. It finds the log-odds of the probability that you're interested in. In particular, logistic regression assumes that logit(y) is linear in the values of x. It finds the best coefficients to predict y.
- Multinomial Naive Bayes [22]
  Bayesian classifiers are statistical classifiers. They can predict class membership probabilities. Bayesian classification is based on Bayes theorem. Naïve Bayesian classifiers assume that the effect of an attribute value on a given class is independent of the values of the other attributes. This assumption is called class conditional independence.

13.4 Results

The performance measures used to evaluate the models were Accuracy, Precision, Recall, F1-Score,Receiver Operating Characteristic (ROC) Curve and Area Under Curve (AUC) Score (see Table 13.2).

The confusion matrices of the models give us the percentage of TP (True Positives), TN (True Negatives), FP (False Positives) and FN (False Negatives) predicted by the model. From the confusion matrix of Logistic Regression, (see Fig. 13.8) it is clear that the FP percentage is really low, only around 2.68% which has resulted in such a high precision, (see Eq. 13.1). Almost all of the news which are predicted as fake are, in fact, fake. On the other hand, Multinomial Naive Bayes (see Fig. 13.9) has 9.74% of FPs and hence a comparatively lower precision but better recall value since the FNs (7.65%) is lower (see Eq. 13.2) than that of Logistic Regression (9.11%).

$$Precision = \frac{TP}{TP + FP} \tag{13.1}$$

$$Recall = \frac{TP}{TP + FN} \tag{13.2}$$

This difference in the precision and recall values of the two models can clearly be seen in the plot given (see Fig. 13.7). In terms of both accuracy and AUC score, Logistic Regression has performed better than Naive Bayes. But the low FP values in Logistic Regression has brought about the high precision which can be seen by a sudden rise in the line plot at precision score followed by a dip at recall score. On the other hand there is a dip of the Naive Bayes line plot at precision score and a rise at the recall value.

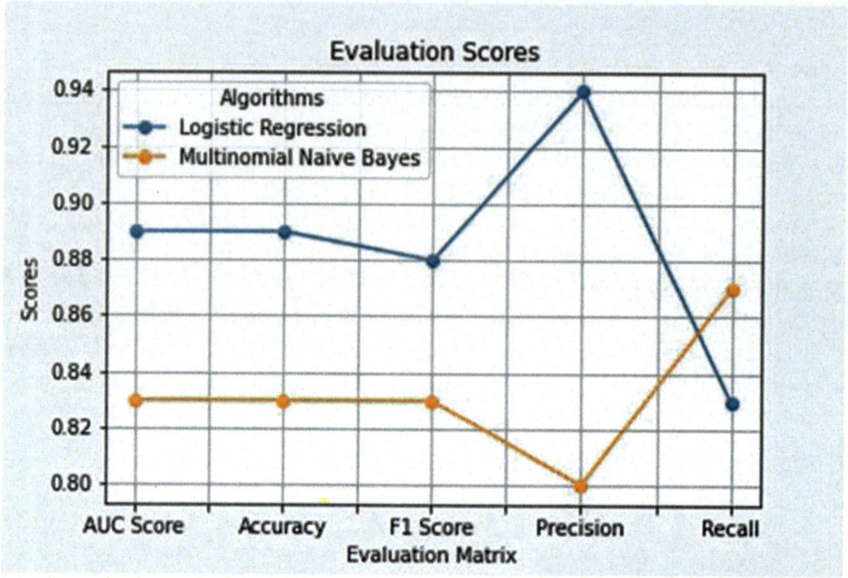

Fig. 13.7 Performance of algorithms

Fig. 13.8 Confusion matrix for logistic regression

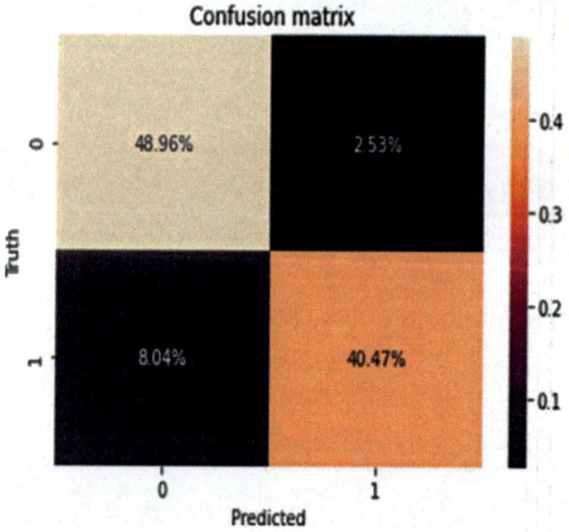

$$F1\ Score = 2 * \frac{(precision * recall)}{(precision + recall)} \tag{13.3}$$

F1 score which conveys the balance between precision and recall (see Eq. 13.3) is also better for Logistic Regression. Hence, we can conclude that the overall performance of Logistic Regression is better than that of Multinomial Naive Bayes.

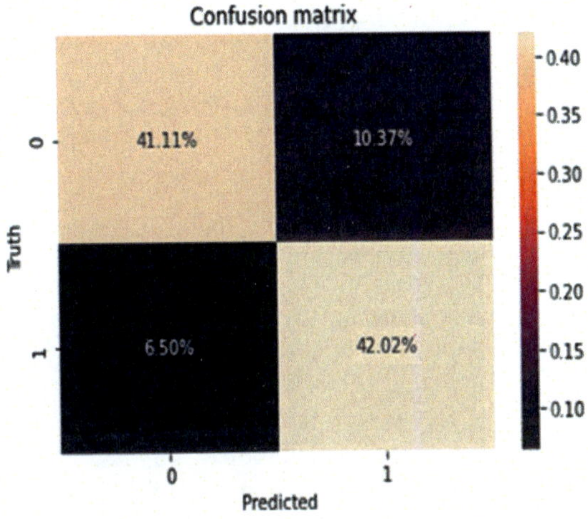

Fig. 13.9 Confusion matrix for multinomial Naive Bayes

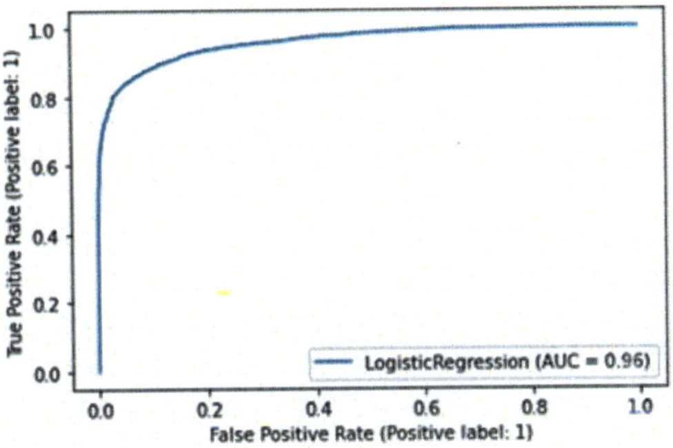

Fig. 13.10 ROC curve for logistic regression

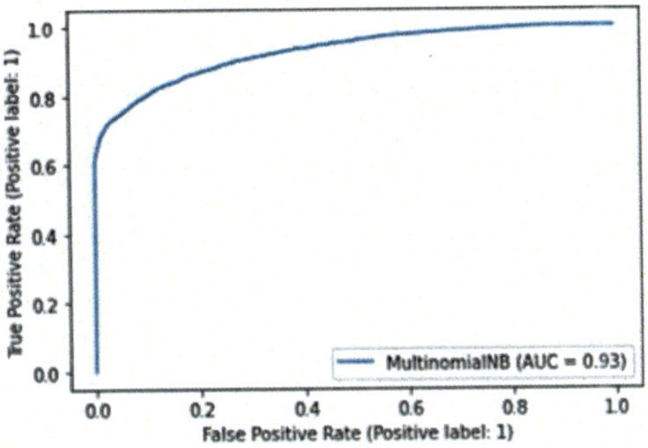

Fig. 13.11 ROC curve for multinomial Naive Bayes

The plot of ROC curves for Logistic Regression and Multinomial Naive Bayes is given in Figs. 13.10, 13.11 respectively. As can be seen from the figures, the curves of both models are close to the top left which means both our models are performing well, Logistic Regression performing better than Multinomial Naive Bayes since it is more closer to the top left. The same can be inferred from the AUC scores also.

13.5 Conclusion and Future Works

Spread of fake news is not a new problem. For centuries, businesses and governments have been using the media to influence people and execute propaganda. But the generation of huge amounts of online content and their fast and vast reach into public is a cause for immediate concern.

As can be seen from the previous section, the Logistic Regression Model was seen to perform better than the Multinomial Naive BayesNaive Bayes Model in terms of all accuracy measures except for Recall. Logistic Regression Model also gave really high precision, around 0.94, which is a desirable behavior in real life. The overall accuracy and other performance measures of both models are also better than what some of the previous works on the same datasets have achieved.

The datasets used in this project are mainly centered around politics, gossip and Covid 19 data. So the performance of the model on any other news topics could be unpredictable. In the future we can try adding more variety of data into the datasets like sports, science, etc. Or we can concentrate on one particular topic and collect news articles only on that topic. This could help the model to recognize interesting patterns and identify fake news from real pertaining to any particular topic. Identifying such specific patterns can really improve the accuracy of a model.

Also, other features like the no.of retweets, likes, etc., can be taken into account. Instead of predicting whether the news article is real or fake, the probability of it being fake or real can also be predicted,since Logistic Regression is already being used and performing well. In social media like Twitter, identification of bots that spread fake news is another improvement that can be made in the existing design. More often than not, fake news is propagated using bots that are designed to retweet tweets containing certain phrases or hashtags. Also certain accounts are found to post articles that cause public unrest to target a section of people. These usually involve false claims, theories or biased accounts of various events. Taking them into account and studying their structures can make the model highly complex, but might improve the accuracy. At the least we might be able to identify interesting patterns.

In short, a lot of efforts and research have gone into the problem of fake news detection. Numerous methods based on both ML, deep learning and even hybrid models have been proposed. Yet, an effective method to differentiate fake news from real news still remains an open challenge.

References

1. Kaur, S., Kumar, P.: Kumaraguru. P, Automating fake news detection system using multi-level voting model. J. Soft Comput. Springer **24**, 9049–9069 (2020)
2. https://www.politifact.com/
3. Sainudiin, R., Yogeeswaran, K., Nash, K., Sahioun, R.: Characterizing the Twitter Network of Prominent Politicians and SPLC-Defined Hate Groups in the 2016 US Presidential Election. Social Network Analysis and Mining. Springer, Berlin (2019)

4. Buzzfeed, https://www.washingtonpost.com/politics/2020/11/05/president-trumps-false-claims-vote-fraud-chronology/
5. Vereshchaka, A., Cosimini, S., Dong, W.: Analyzing and distinguishing fake and real news to mitigate the problem of disinformation. In: Computational and Mathematical Organization Theory, vol. 26, pp. 350–364. Springer, Berlin (2020)
6. Reddy, H., Raj, N., Gala, M., Basava, A.: Text-mining-based fake news detection using ensemble methods. Int. J. Autom. Comput. Springer (2019)
7. Kaliyar, R.K., Goswami, A., Narang, P, Deep, F.: Improving fake news detection using tensor decomposition-based deep neural network. J. Supercomput. Springer, **77**, 1015–1037 (2021)
8. Yavary. A., Sajedi, H., Abadeh, M.S.: Information verification in social networks based on user feedback and news agencies. In: Social Network Analysis and Mining. Springer, Berlin (2019)
9. Shu, K., Mahudeswaran, D., Liu, H.: FakeNewsTracker: a tool for fake news collection, detection, and visualization. Comput. Math. Organ. Theory. Springer, **25**, 60–71 (2019)
10. Kumar, S., Huang, B., Cox, R.A.V. et al.: An anatomical comparison of fake-news and trusted-news sharing pattern on Twitter. Comput Math Organ Theory. Springer, **27**, 109–133 (2021)
11. Dang, N.C., Moreno-García, M.N., De la Prieta, F.: Sentiment analysis based on deep learning: a comparative study. Electronics **9**, 483 (2020)
12. Monica, C., Nagarathna, N.: Detection of fake tweets using sentiment analysis. SN Comput. Sci. **1**, 89 (2020)
13. Rajapaksha, P., Farahbakhsh, R., Crespi, N.: Bruno Defude. Uncovering Flaming Events on News Media in Social Media, IEEE (2019)
14. Hajek, P., Barushka, A., Munk, M.: Fake consumer review detection using deep neural networks integrating word embeddings and emotion mining. Neural Comput. Appl. **32**, 17259–17274 (2020)
15. Rodríguez-Ruiza, J., Mata-Sánchez, J.I., Monroyc, R., Loyola-González, O., López-Cuevas, A.: A one-class classification approach for bot detection on Twitter. In: Computers and Security. Springer (2020)
16. Souza, J., Gomes, J., Jr., Marques, F., de Oliveira Julio, A.M.: A systematic mapping on automatic classification of fake news in social media. In: Social Network Analysis and Mining, vol. 10, p. 48. Springer (2020)
17. Sharma, Sharma, D.K.: Fake news detection: a long way to go. In: 2019 4th International Conference on Information Systems and Computer Networks (ISCON), pp. 816–821. IEEE (2019). https://doi.org/10.1109/ISCON47742.2019.9036221
18. FakeNewsNet, https://github.com/KaiDMML/FakeNewsNet
19. Getting Real about Fake News, https://www.kaggle.com/mrisdal/fake-news
20. COVID-19 Fake News Dataset, https://www.kaggle.com/arashnic/covid19-fake-news
21. Zumel, N., Mount, J.: Practical Data Science with R. Manning Publications
22. Han, J., Kamber, M., Pei, J.: Data Mining Concepts and Techniques, 3rd edn. Elsevier (2011)

Chapter 14
Lane Detection for Autonomous Cars Using Neural Networks

Karishma Vivek Savant, Ghanta Meghana, Gayathri Potnuru, and V Bhavana

Abstract Advancements in lane detection algorithms lead to realizing autonomous driving technology and improving the real-time use of deep learning algorithms that are currently being studied for their various vision-based applications. In this paper, we propose an efficient and accurate algorithm to detect lanes and lane lines for autonomous vehicles. First, an OpenCV lane detection model was built to improve the understanding of lane detection algorithms. Using this knowledge, a CNN solution was adopted for the lane detection algorithm to train more data, work on curved lanes, and improve accuracy. The shortcomings of this model were identified and to solve these shortcomings, a Fully Convolutional Neural Network approach based on SegNet's architecture was adopted. The accuracy of the models implemented was found to be as: OpenCV: 91%, CNN: 94.49%, and FCN: 96.12%.

14.1 Introduction

One of the many goals of 'AI' or machine learning is to describe a scene as precisely as a human being. Recently, the research on autonomous driving technology has increasingly been of many researchers' interest. Technologies for vision-based environment perception tasks have received extensive attention from the automotive industry [1].

Studies in the automotive industry have seen that by providing basic road information, such as the lane structure and the relative position of other vehicles in the lane, the lane detection function can safely guide the self-driving vehicle to the appropriate position within the main lane [2]. The purpose of this work is to develop an efficient and accurate algorithm to detect lanes and lane lines for autonomous vehicles. To achieve this, a model for lane detection for autonomous cars using neural networks is developed.

K. V. Savant (✉) · G. Meghana · G. Potnuru · V. Bhavana
Department of Electronics and Communications Engineering, Amrita School of Engineering, Amrita Vishwa Vidyapeetham, Bengaluru, India

V. Bhavana
e-mail: v_bhavana@blr.amrita.edu

J. I.-Z. Chen et al. (eds.), *Machine Learning and Autonomous Systems*, Smart Innovation, Systems and Technologies 269, https://doi.org/10.1007/978-981-16-7996-4_14

In existing models of lane detection, basic OpenCV is used for straight lane lines [3]. Previously implemented versions of this architecture are found to have a long extraction process, which makes the model consume more time for processing data leading to a slow operation speed. For curved lanes and advanced lane detection implementations, the lane detection models that have been implemented use simple convolutional neural networks that calculate edges in the image of the roads using shape kernels to detect lanes. This type of architecture is found to contain the existence of noise in the outputs of the model, which leads to lower accuracy and increased missed detection, and false detection in lane lines.

In this paper, OpenCV method is first implemented as the basic method of lane detection. The extraction process implemented is modified to consume lesser time and increase accuracy. Next, a Convolutional Neural Network (CNN) model is implemented, which is more robust and faster model than the computer vision-based model. This increases the accuracy and reduces the noise of the lane detection model [4]. The final model implemented is the Fully Convolutional Neural Network (FCN), which omits fully connected layers, leading to a fully convolutional neural network [5], which greatly improves the speed of the algorithm while also causing a reduction in the noise of the outputs [6].

The rest of the paper is organized as follows, Sect. 14.2 discusses few key concepts in related works and previously existing models of lane detection, Sect. 14.3 details the proposed approaches of lane detection implemented and explains the steps involved and its architecture, Sect. 14.4 explains the results and analysis of the three models and compares their performance metrics, Sect. 14.5 concludes the work implemented in this paper, and Sect. 14.6 details the future scope and methods to improve the final model.

14.2 Related Works

The preliminary step involved in the process of training an autonomous vehicle is lane detection. Lane detection is a vital component and the main feature of autonomous vehicles. It is one of the most significant topics of research under the subject and analysis of self-driving vehicles [7]. The process first involves the lane positions being obtained for the vehicle to analyze the road coordinates to avoid the risk of driving off its assigned lane. In the current standard of research related to autonomous vehicles, the main areas of difficulty in lane detection are tackling the problem of accuracy and real-time compatibility simultaneously. This is to ensure the development of an effective and robust lane recognition [8, 9].

Image processing is classified into three main areas. Coding the image handles the compression and computerization of the input image data [10, 11]. This helps in reducing the use of memory space and efficiently uses the channel capacity. Pattern recognition deals with processing the image and recognizing patterns with the help of machine learning algorithms. Restoration of the output image is concerned with the improvement and enhancement of image quality [12].

Furthermore, a short study on the two types of Lane Detection: Preliminary Lane Detection and Enhanced Lane Detection, is analyzed.

14.2.1 Preliminary Lane Detection

In preliminary lane detection, it is assumed that all lanes are straight, as this gives us the simplest coordinates of lane shape to work with. This also makes it easier for the algorithm to identify and detect the lane input. First, lane detection algorithm on the OpenCV model is used to detect the straight lane line edges through the image extraction process, which majorly includes edge detection [13]. Conventional computer vision-based lane detection technology is divided into steps of image processing, which works on extracting vital features of road edges and then takes care of reducing the size of video frames to make the process quicker. Gray-scale implementation on the inputted video file is done next. Then, the Canny algorithm is used to detect edges [14] of the obtained single-channel image and draw out vital features required for lane detection. Finally, the detected lane from the algorithm is superimposed onto the road image [15].

14.2.2 Evolved Lane Detection

The preliminary lane detection method is enhanced by shifting the lane detection approach to the current popular and standard Neural Network for image and video inputs and datasets, known as the Convolutional Neural Network (CNN). The Convolutional Neural Network is an effective model in the field of Image Processing. CNN can recognize and extract simple features, like edges and colors, at the initial layers, and can also extract complex attributes at deeper and more complicated layers [16]. A feature map is created, which helps in learning and predicting the lane lines. The CNN model can learn and predict the lane position even if a few frames are omitted. This helps in improving the accuracy and robustness [17]. Semantic Segmentation method is also used in the field of Image Processing and Object Detection. Semantic Segmentation classifies and labels each pixel, and this helps in accurately predicting the data [18].

Paper	Methodology	Advantage	Disadvantage
(1) Gao et al. 'A Real-Time Lane Detection and Tracking Algorithm" [3]	Gabor filter, Hough transform method, Sobel operator, least squares algorithm	System is real time, efficient and enhances the adaptability for the changing environment of the road scene	The lane detection is based on the assumption that the lane markings are linear, and the road is flat. Some noise is still present

(continued)

(continued)

Paper	Methodology	Advantage	Disadvantage
(2) Yan and Li "A Method of Lane Edge Detection Based on Canny Algorithm' [13]	Canny algorithm, Sobel operator, Hough transform	Can detect linear lanes based on Hough transform	Canny edge extraction process consumes longer time, which causes the slow operation speed
(3) Zang et al. 'Traffic Lane Detection using Fully Convolutional Neural Network' [6]	Substituting the fully connected layers in the classification network with convolution layers. The lane marking is executed by RANSAC	The classification accuracy of different lanes is good	Lane marking algorithms need to be improved to revise the deviation of fitted lanes
(4) Chen et al. 'PointLaneNet: Efficient end-to-end CNNs for Accurate Real-Time Lane Detection' [2]	Inspired by YOLO, this paper proposes a lane detection algorithm for end-to-end training. based on a deep neural network (DNN)	Can simultaneously perform position prediction and lane line classification in a single network	Most of the forward time is spent on the base network. Therefore, the speed could be further improved by using a faster base network
(5) Yang et al. 'Improved Lane Detection With Multilevel Features in Branch Convolutional Neural Networks' [1]	Uses semantic segmentation concepts utilizes multiple level features of the encoder and designs enhanced binary segmentation and reduced pixel embedding branches	Better accuracy performance in low computation. The proposed network, which can help for restoring the occluded lanes makes it robust	Accuracy rate can be increased and the fps rate can be increased as it is lesser than LaneNet

14.3 Proposed Architecture

To design and develop an efficient model for lane detection, three different approaches are implemented and studied. These methods are: (1) OpenCV model for basic lane detection and straight roads, (2) CNN model for curved lanes with a deep learning approach, and (3) FCN for a more robust method of lane detection with improved accuracy.

Approach 1–Lane Detection Using OpenCV

In this approach, the video file is read and decodes the frame, and this frame is converted into gray-scale image. The noise is reduced to eliminate any false edges, and this image is fed into the Canny edge detector, where the edges are detected [3]. The Canny edge detector creates a mask, which is of the same dimension as the road image. This edge detector detects lane lines from the extracted images on the basis of variations in the gradients. It performs the process of masking the canny image and showing the region of interest extracted from the image. The Hough Line Transform algorithm is used to detect the straight lines [19].

Approach 2—Lane Detection using Convolutional Neural Networks (CNN)

The model architecture used in this model uses various popular layers of the CNN architecture such as batch normalization to standardize the inputs and normalize the outputs of the previous layers. This is then followed by four convolutional layers that work to sharpen images and enhance edges. The next layer used is a max pooling layer, which extracts low-level features, a flatten layer that converts the data of the previous layers into a one-dimensional array, and finally, four fully connected layers that perform the decision-making, which, in this algorithm, is the process of lane detection [20]. The final fully connected layer has an output size of six. The Keras-vis library allows to input the trained CNN model into a function using the online Keras tool, for evaluating deep-learning models, and to return the activation maps for the various layers in the model being analyzed for each class. The neural network that is appropriate for working on visual inputs is using Convolutional Neural Networks [21, 22]. Hence, a CNN model is implemented.

Approach 3—Lane Detection using Fully Convolutional Network (FCN)

To further work on consistency issues of the previous CNN model with which parts of the lane image get activated and how highly activated they are, the model can be adapted into a Fully Convolutional Neural Network (FCN). A Fully Convolutional approach is a better approach to image segmentation when it comes to dividing output images into different classes. The FCN architecture is a mirror model, which consists of a sequential arrangement of encoders and decoders. A single encoder consists of batch normalized convolutional layers with ReLU activation function followed by a non-overlapping max pooling layer. A one-dimensional array is the output of the encoder, and it is fed to the decoder [23, 24]. The decoder consists of upsampling layers, which perform the task of unpooling layers used to retain the details of the segmented image. This is followed by deconvolutional layers, also known as transposed convolutional layers, and are used to upsample the low-resolution feature (due to convolution and pooling) back to the original resolution [25–28].

14.3.1 Lane Detection Using OpenCV

The first step involves reading and decoding the inputted video file. The video that is fed into the OpenCV algorithm is captured using the Video Capture object function. This is followed by every frame of the video being decoded. In this stage, the inputted video frames are converted into a succession of image frames. The next step, which is preprocessing involves the frames that exist originally in RGB format being converted to gray-scale. This step is performed since processing of an image existing in a single channel is faster than if the algorithm had to process a three-channel image. As the

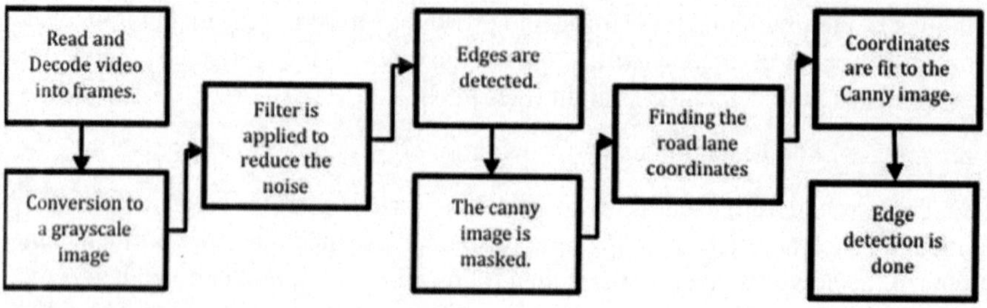

Fig. 14.1 Steps involved in OpenCV lane detection

existence of noise can lead to creating false edges, the process of image smoothening is performed to try to eliminate the noise present in the model. A Gaussian filter is chosen for image smoothening to reduce the noise content. Then, the Canny Edge algorithm is used to evaluate and extract the edges of the lane lines with variation in intensity. The Canny edge detector creates a mask, which is of the same dimension as the road image. This edge detector detects lane lines from the extracted images on the basis of variations in the gradients. The region of interest is detected to take into account only the vital region covered by the lane in the video to be detected. A mask is created in this step, which matches the dimensions of the inputted road image. Next, the bitwise AND operation is executed between the pixels of the canny image and the mask. This leads to the region of interest being identified by the mask. The Hough Line Transform works on straight lines by turning its representation into a 2D array with the parametric representation (rho, theta) where Rho represents the number of rows and theta represents the number of columns. The probabilistic Hough line transform optimizes the Hough Transform by only considering a subset of points from the 2D array required for line detection, this is what returns the result of the identified lane line output. Figure 14.1 represents the steps involved in the OpenCV method of lane detection.

The line that is detected on the image displaying edges is represented by the equation:

$$y = ax + b \tag{14.1}$$

In Eq. (14.1), ‘y’ represents the points in space of the lane line in the y-coordinate of the 2D space, ‘x’ represents the points in space of the lane line in the x-coordinate of the 2D space, ‘a’ is the slope of the line and ‘b’ is the intercept of the line. A singular line on the image displaying edges produces a point on the Hough Space. This is since a line can be represented by its slope ‘a’ and its intercept ‘b’ whereas, a point on the image displaying edges. In Eq. (14.2), ‘x_i’ represents the point on the image representing the edge of the lane line in the x-coordinate, ‘y_i’ represents the point on the image representing the edge of the lane line in the y-coordinate (x_i, y_i) can possess an infinite number of lines passing through it. This makes a way for a point on the image displaying edges creating a line in the Hough Space represented by the equation:

$$b = ax_i + y_i \tag{14.2}$$

Figure 14.2 represents the gray-scale canny image after converting the image from the RGB channel into a single channel for efficient extraction. Figure 14.3 represents the line image obtained by the algorithm after the road coordinates are identified. Figure 14.4 represents the final output of the OpenCV model showing the line image of the detected lane lines merged with the road image of the video input into the model.

Fig. 14.2 Gray-scale Canny Image obtained from the Canny filter

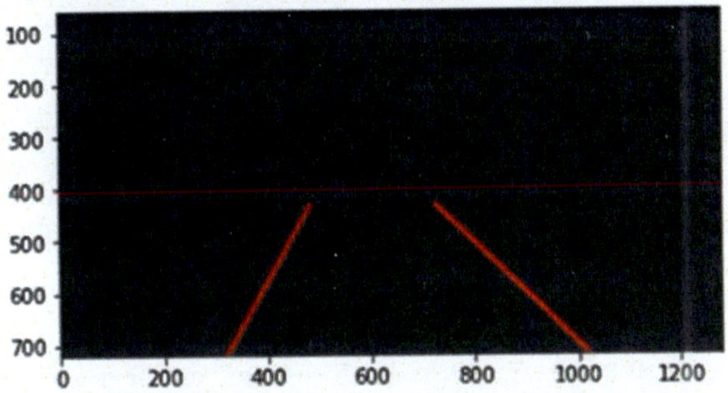

Fig. 14.3 Line image obtained from OpenCV lane detection model

Fig. 14.4 Line image merged onto Road image displayed as output of OpenCV model

14.3.2 Lane Detection Using CNN

A model CNN model is created that can take in a road image, this image is taken from a video file, which does not depend on perspective transformation. In the preprocessing step of CNN, the aspect ratio of the input image is changed to 80 × 160 × 3 (from 90 × 160 × 3) because the model has pooling layers of 2 × 2 dimension. The model architecture used here includes batch normalization to standardize the inputs and normalize the outputs of the previous layers. This is then followed by four convolutional layers that work to sharpen images and enhance edges. The next layer used is a max pooling layer, which extracts low-level features, a flatten layer that converts the data of the previous layers into a one-dimensional array, and finally, four fully connected layers that perform the decision making. The last fully connected layer has an output size of six obtained from the six-lane coefficients of the line equation $ax^2 + bx + c$ for either side of the road, making it a total of 6 coefficients being predicted. Figure 14.5 represents the lane drawn by the CNN model in the 'G' channel, and Fig. 14.6 represents the final output of the CNN model, which merges the detected lane of the algorithm onto the road image of the inputted video.

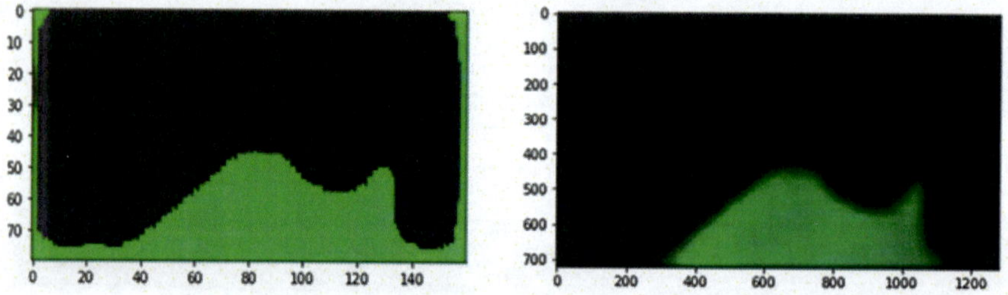

(a) Lane to be drawn from the model in G color

(b) RGB image of the lane drawn

Fig. 14.5 Images in the channel 'Green' of the lane drawn by using the algorithm designed to detect the lane using CNN model

Fig. 14.6 Lane detected using CNN algorithm merged onto road image

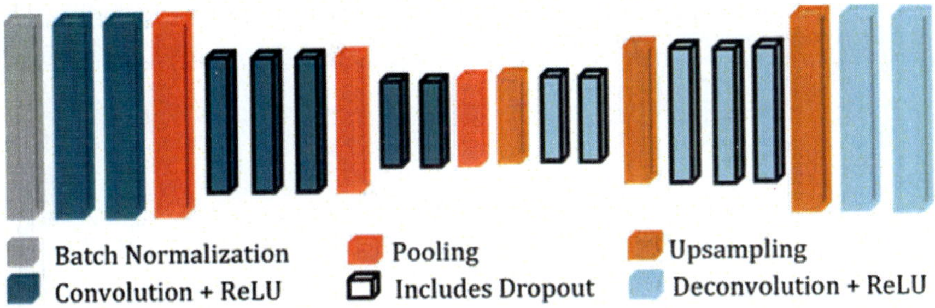

Fig. 14.7 Fully Convolutional Network Architecture

14.3.3 Lane Detection Using FCN

It was found that where lanes could get re-drawn in wrong or changing perspectives, changing the current model to reflect the architecture of Segnet could solve the identified consistency issues. This is possible as the SegNet architecture neural network can allow for the model to re-draw the predicted lane on its own. The process of imputing the road images into the model will remain the same, which is a video and then is taken frame-wise, and the lane drawing obtained from the previous coefficient labels will be viewed as the output image.

The lanes that the model detected were in the channel green, hence, the output 'filter' of the model is selected to be the 'G' channel from RGB. The two blank filters 'R' and 'B' can be stacked alongside it to make the subsequent images merge in with the original road image. Modifying the model to fit this new architecture allows for the ease of flipping the road image and the lane image label during the pre-processing of the data simultaneously. The rotated images during pre-processing aid in procuring additional labels while ensuring to maintain coincidence of the road images in the model and its various labels (Fig. 14.7).

Coming to these conclusions, SegNet's architecture was used as a guideline for the next approach for the lane detection model. Using Keras made sure the process was not complicated. Max pooling layer will select the maximum element from the region of the feature map covered by the filter. This reduces the feature maps by half on each dimension and results in an output that is one quarter the area of the input. The Convolutional layers present in the architecture perform downsampling by applying each filter across the input images. This results in an output feature map that is smaller. The up-sampling layer works by repeating the rows and columns of the input. In the first half of the model, down sampling the spatial resolution of the image helps developing complex feature mappings. In each convolution step, further detailed information of the inputted image is obtained. At this stage, a highly efficient discrimination between different classes is extracted. The last layers have lesser resolution and up-sampling helps in producing better details in that last layer and also reduces the number of computations.

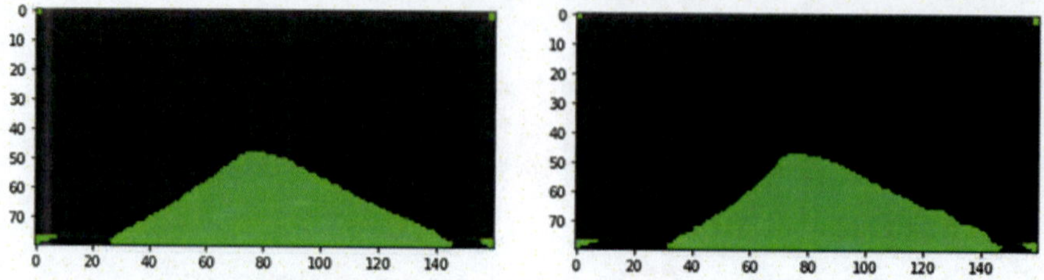

Fig. 14.8 Lane drawn in G color after the FCN model prediction has been done

The deconvolutional layers introduced had to match up the math correctly with what the model was trying to do. For this, a mirror model is implemented. In the preprocessing step, the aspect ratio of the input image is changed to 80 × 160 × 3 (from 90 × 160 × 3) because the model has pooling layers of 2 × 2 dimension, the change in the aspect ratio helps to minimize the issues in imitating the encoder half of the model. A decoder is added after the final pooling layer. The dropout is used in the inner layers, like the encoder. A one-dimensional array is the output of the encoder, and it is fed to the decoder. The decoder upsamples the low-resolution image to retain the details of the segmented image. The last deconvolutional outputs a filter with 80 × 160. The lane image labels were normalized by dividing by it so that the labels were from 0 to 1 for the 'G' pixel values. This improved the time taken by the model for evaluation and running as well as the overall quality and results of the final outputs. Figure 14.8 represents the lane detected by the FCN model in the 'G' channel.

14.4 Results and Analysis

In this section, the results of the FCN model are discussed. The calculation and the comparison of the performance of all the three methods implemented are tabulated.

Figure 14.9 represents the FCN model's 'G' colored image converted into its

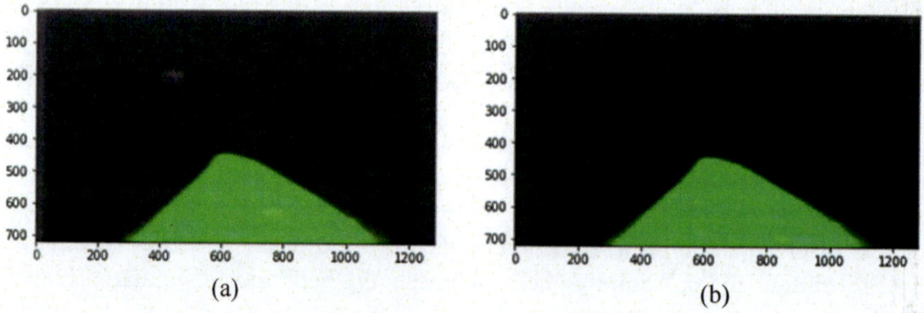

Fig. 14.9 G color image converted to RGB image

Fig. 14.10 RGB image merged with the road image

RGB format before merging it onto the image of the lane extracted from the inputted video file. The final FCN model implemented can identify the lane line coordinates with more precision than the previous CNN model. This is represented by the output obtained by the FCN model as shown in Fig. 14.10.

The performance metrics are calculated for all three models. Accuracy is the measure of how often the prediction made by the algorithm is correct. This is calculated with the help of Eq. (14.3).

$$Accuracy = \frac{numberofaccuratelypredictedrecords}{totalnumberofrecords} \tag{14.3}$$

Mean Square Error is used to calculate the loss as mentioned in Eq. (14.4).

$$Loss = square(trueclassification - predictedclassification) \tag{14.4}$$

F1 Score is the harmonic mean of precision and recall to and measures the cases that are incorrectly classified.

$$F1Score = 2 * \frac{precision * recall}{precision + recall} \tag{14.5}$$

Precision is the ratio of correctly predicted positive observations, i.e., True Positives, and total positively predicted observations, i.e., sum of True Positives and False Positives.

$$Precision = \frac{TruePositive(TP)}{TruePositive(TP) + FalsePositive(FP)} \tag{14.6}$$

Recall is the ratio of correctly predicted positive observations, i.e., True Positives, and total actual positive observations, i.e., sum of True Positives and False Negatives.

$$Recall = \frac{TruePositive(TP)}{TruePositive(TP) + FalseNegative(FN)} \tag{14.7}$$

Table 14.1 Performance comparison between different lane detection implementations

Lane Detection Implementation	Accuracy (%)	Loss
OpenCV	91	–
CNN (Fully Connected Layers)	94.49	3.26%
FCN (Fully Convolutional Network)	96.12	0.48%

Table 14.2 FCN model performance analysis

Precision	Recall	F1 Score
96.34%	95.37%	95.85%

The performance metrics for OpenCV were calculated manually and for CNN and FCN, Keras metrics were used to calculate the performance metrics.

Table 14.1 shows a detailed summary of performance measures between the final FCN model implemented and the previous lane detection models, i.e., the OpenCV model and the CNN model (fully connected model). Table 14.2 represents the precision, recall, and F1 score of the FCN model. The results are analyzed by performance measures such as accuracy, loss, precision, recall, and F1 score. The dataset used by the final model for testing consisted of over 12,000 images.

As displayed by the tabulated performance statistics, the first two models using FCN and CNN displayed a relatively better performance in being able to classify the coordinates of the road and detect the lane efficiently. In comparison, the OpenCV model was not as efficient as the two neural network models. The poorer performance of the OpenCV model is due to its use of the Canny edge detector, which is used to compute the gradient and mask. There is a probability that it is difficult to detect the weak edges that are not marked well enough for the algorithm to pick up on. Hough transform may classify a binary pixel to an actual lane marker, or it is noise. This results in an increase in incorrect detection rate. In the CNN implementation, the activation may get triggered due to the rotations and flips of the images, which causes misdetection and reduces the accuracy. The proposed FCN model eliminates the concept of perspective transform and fully connected layers, which helps increase the accuracy and robustness. Figure 14.11 represents the accuracy and loss graphs of the FCN model.

14.5 Conclusion

In this paper, a Fully Convolutional Neural Network approach was proposed to detect the lane lines. The utilization of the FCN approach creates a model that fixed the problem of robustness for curved lines that were seen in the OpenCV model. The problems of uniformity in the processing of the algorithm with regard to which areas of the road were marked as highly activated and the percentage of activation, as seen in the CNN model, are also solved with the Fully Convolutional Neural

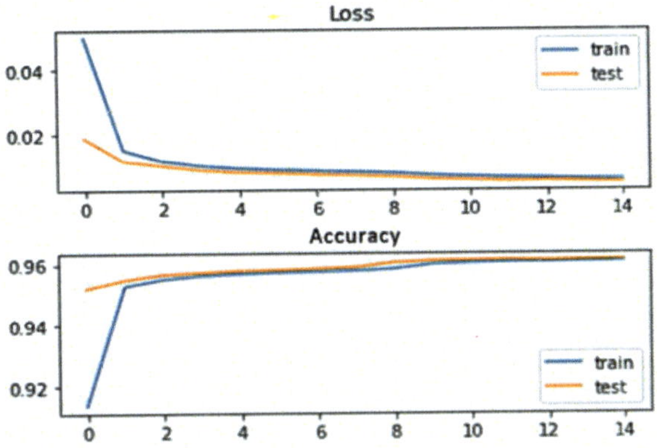

Fig. 14.11 Accuracy and Loss graph of FCN Model

Network model. The FCN model architecture improves on previously implemented lane detection algorithms by reducing noise, improving the speed of processing, and greatly increasing accuracy and other performance parameters. This final FCN model fixes activation problems in the processed images present in other models, leading to accurate lane detection in various conditions and different datasets. The FCN model gives a good result in terms of accuracy of about 96.12% when compared with OpenCV, which has an accuracy of 91%, and the CNN model that gives an accuracy of 94.49%.

14.6 Future Scope

There are ways the results of the current model can be modified to further improve the robustness and efficiency of the algorithm by making use of larger and more versatile data sets, which will enhance the model's results. Feeding data with transitions between light and shadow and data without lane lines or only one line to train the model will increase the robustness of the deep learning model. This further allows the model to be used in more situations and environments where lane lines are not well defined. A recurrent neural network (RNN) can be used, which has the ability to use past information for the next prediction, and can increase the model's memory and subsequently its results.

References

1. Yang, W., Cheng, Y., Chung, P.: Improved lane detection with multilevel features in branch convolutional neural networks. IEEE Access **7**, 173148–173156 (2019)
2. Chen, Z., Liu, Q., Lian, C.: PointLaneNet: efficient end-to-end CNNs for accurate real-time lane detection. In: 2019 IEEE Intelligent Vehicles Symposium (IV), Paris, France, 2019
3. Gao, Q., Feng, Y., Wang, L.: A real-time lane detection and tracking algorithm. In: IEEE Information Technology, Networking, Electronic and Automation Control Conference (2017)
4. Gurghian, A., Koduri, T., Bailur, S.V., Carey, K.J., Murali, V.N.: DeepLanes: End-to-end Lane position estimation using deep neural networks. In: Proceedings of the IEEE Conference Computer Vision and Pattern Recognition Workshops (CVPRW), Las Vegas, NV, USA, pp. 38–45. (2016)
5. Bhavya sree, B., Yashwanth Bharadwaj, V., Neelima, N.: An inter-comparative survey on state-of-the-art detectors- R-CNN, YOLO and SSD. In: Smart Innovation, systems and technologies, vol. 213, pp. 475–483. (2021)
6. Zang, J., Zhou, W., Zhang, G., Duan, Z.: Traffic lane detection using fully convolutional neural network. In: APSIPA Annual Summit and Conference (2018)
7. Redmon, J., et al.: You only look once: unified, real-time object detection. In: Proceedings of the IEEE conference on computer vision and pattern recognition. (2017)
8. Wang, Z., Ren, W., Qiu, Q.: LaneNet: real-time lane detection networks for autonomous driving. arXiv preprint arXiv:1807.01726 (2018)
9. Kheyrollahi, A., Breckon, T.P.: Automatic real-time road marking recognition using a feature driven approach. Mach. Vis. Appl. **23**(1), 123–133 (2012)
10. Bhavana, V., Krishnappa, H.K.: Multi-Modality Medical Image Fusion using Discrete Wavelet Transform. In: 4th International Conference on Eco-friendly Computing and Communication Systems, ICECCS 2015, ELSEVIER Procedia Computer Science, pp.625–631
11. Bhavana, V., Krishnappa, H.K.: Fusion of MRI and PET images using DWT and adaptive histogram equalization. In: 2016 International Conference on Communication and Signal Processing (ICCSP)
12. Greenhalgh, J., Mirmehdi, M.: Detection and recognition of painted road surface markings. In: Proceedings of the 4th International Conference Pattern Recognition Applications and Methods, Lisbon, Portugal, pp. 130–138. (2015)
13. Yan, X., Li, Y.: A method of lane edge detection based on Canny algorithm. In: Chinese Autmation Congress (CAC). (2017)
14. Raju, J.V.V.S.N., Rakesh, P., Neelima, N.: Driver drowsiness monitoring system. In: Smart Innovation, Systems and Technologies, (SIST vol.169), pp. 675–683 (2019)
15. Aminuddin, N.S.: A new approach to highway lane detection by using Hough transform technique. J. Inform. Commun. Technol. **16**(2), 244–260
16. Zou, Q.: Robust lane detection from continuous driving scenes using deep neural networks. IEEE Trans. Veh. Technol. (2019)
17. Qin, B., Liu, W., Shen, X., Chong, Z.J., Bandyopadhyay, T., Ang, M.H., Frazzoli, E., Rus, D.: A general framework for road marking detection and analysis. In: Proceedings 16th International IEEE Conference Intelligent Transportation Systems-(ITSC), The Hague, Netherlands, pp. 619–625 (2013)
18. Neven, D., et al.: Towards end-to-end lane detection: an instance segmentation approach. arXiv preprint arXiv:1802.05591 (2018)
19. Li, X., Wu, Q.X., Kou, Y., Hou, L., Yang, H.: Lane detection based on spiking neural network and Hough transform. In Proceedings of the 8th International Congress Image and Signal Processing (CISP), Shenyang, China, pp. 626–630 (2015)
20. Li, J., Mei, X., Prokhorov, D., Tao, D.C.: Deep neural network for structural prediction and lane detection in traffic scene. IEEE Trans. Neural Netw. Learn. Syst. **28**(3), 690–703 (2017)
21. Kim, J., Kim, J., Jang, G.J., Lee, M.: Fast learning method for convolutional neural networks using extreme learning machine and its application to lane detection. Neural Netw. **87**, 109–121 (2017)

22. Oliveira, G.L., Burgard, W., Brox, T.: Efficient deep models for monocular road segmentation. In 2016 IEEE/RSJ International Conference on Intelligent Robots and Systems (IROS), Daejeon, South Korea, pp.4885−4891. (2016)
23. Hur, J., Kang, S.-N., Seo, S.-W.: Multi- Lane detection in urban driving environments using conditional random fields. In: IEEE Intelligent Vehicles Symposium (IV), pp. 1297–1302. (2013)
24. Lim, K.L., Drage, T., Bräunl, T.: Implementation of semantic segmentation for road and lane detection on an autonomous ground vehicle with LIDAR. IEEE Int Conf Multisens Fus Integ Intell Syst (MFI) **2017**, 429–434 (2017). https://doi.org/10.1109/MFI.2017.8170358
25. Veni, S., Jyothirmayie, P.G., Reddy, O., Vineela, G., Praveenkumar, S., Srihrsha, VLV.: Lane detection and departure warning system. In: International Conference on Recent Trends in Engineering and Technology & Mechanics, Simulation and Control (ICRTET 2013). pp. 125–130. (2013)
26. Vats, A., Nair, B.B.: An improved driver assistance system for detection of lane departure under urban and highway driving conditions. In: Information, Communication and Computing Technology, Singapore, vol. 835, pp. 31–43. (2019)
27. Vinothkanna, R.: A survey on novel estimation approach of motion controllers for self-driving cars. J. Electron. **2**(04), 211–219. (2020)
28. Manoharan, S.: An improved safety algorithm for artificial intelligence enabled processors in self driving cars. J. Artif. Intell. **1**(02), 95–104 (2019)

Chapter 15
Role of Swarm Intelligence and Artificial Neural Network Methods in Intelligent Traffic Management

Umesh Kumar Lilhore, Sarita Simaiya, Pinaki Ghosh, Atul Garg, Naresh Kumar Trivedi, and Abhineet Anand

Abstract An increase in the volume of vehicles in metropolitan areas increases various issues, i.e., traffic jams, pollution, parking issues, noise, security, and accidental causes. Heavy traffic harm the social system because a large amount of time is squandered on it and then responsible for controlling the vehicle congestion problems is essential. An IoV (internet of Vehicles) innovation had also enhanced a wide range of applications associated with ITM. In this work, we are presenting an Intelligent traffic management system based on swarm intelligence and neural network. In the first phase, the proposed system applies a swarm intelligence method PSO for smart traffic light control. PSO detects the accurate cycle program for the management of traffic lights. In the second phase, an artificial neural network method will be applied to accurately detect and count the number of vehicles over the road. This recognition process mainly uses the roadway visual data and begins by extracting relevant the image mostly from the subsequent video sequence. As once context is recognized, consequent images have been used to identify objects moving through the object computation. The final phase applies the combined approach of phase-1 and phase-2 to achieves the best outcomes in terms of less congestion, less travel time, and better average speed.

U. K. Lilhore · S. Simaiya (✉) · P. Ghosh · A. Garg · N. K. Trivedi · A. Anand
Chitkara University Institute of Engineering and Technology, Chitkara University, Punjab, India
e-mail: sarita.simaiya@chitkara.edu.in

U. K. Lilhore
e-mail: umesh.lilhore@chitkara.edu.in

P. Ghosh
e-mail: pinaki.ghosh@chitkara.edu.in

A. Garg
e-mail: atul.garg@chitkara.edu.in

N. K. Trivedi
e-mail: nareshk.trivedi@chitkara.edu.in

A. Anand
e-mail: abhineet.anand@chitkara.edu.in

J. I.-Z. Chen et al. (eds.), *Machine Learning and Autonomous Systems*, Smart Innovation, Systems and Technologies 269, https://doi.org/10.1007/978-981-16-7996-4_15

15.1 Introduction

Intelligent traffic management has been one of the critical issues in certain metropolitan areas. The significant problem is to categorize traffic congestion circumstances to establish effective regulation management and strategic planning. With its rapid growth, the Internet of Vehicles (IoV) has enhanced a broad range of applications. IoV enhances communication strengths among automobiles in the transport network, which might prevent collision accidents, tends to result in a significant boost in accident prevention. Vehicle-to-vehicle communication does provide true traffic flow details [1].

To control the traffic system [2] presented a vehicular frequency control system depending on "Action Heuristic Dynamic Programming". It's designed specifically for the traffic signal model. In research article [3] suggested a new signalized intersection approach based upon this cellular network's electron orbiting selection, but also examined the reliability and receptiveness of the methodology. In the research article [4] A series of traffic signal scheduling techniques relying on the DNN were also developed to achieve traffic application effectiveness.

Video cameras have been implanted throughout most public transportation services systems to track vehicle traffic. Heavy traffic improvements are monitored using video surveillance, and suitable actions are implemented as just a result. The above surveillance cameras can be used for vigilance and in the event of an emergency [5]. Traffic categorization is performed using artificial neural networks (ANNs), which are accurate prediction techniques. Camera and video surveillance are a comfortable way to collect traffic flow that ANN will categorize [6].

In this work, we are presenting an Intelligent traffic management system based on swarm intelligence and neural network. The complete article is divided into various sections which covers the introduction part, related work, proposed model, implementation and finally covers results and discussion; conclusions, and future aspect.

15.2 Related Work

The majority of relevant strategies to our current topic will be described in the whole section. The whole section consists of two subcategories because our methodology is classified into two primary sections: The role of (a) Swarm intelligence into ITM and (b) ANN into ITM.

15.2.1 Swarm Intelligence in ITM

Swarm intelligence has been the collaborative pattern of behavior of self-organized structures, distributed, naturally, or artificially. The principle is used in machine learning research. In the research article [7] an innovative neuronal ant-based method associated with smart traffic lamps premised on city traffic and road load status has also been suggested. The current technique divides the traffic network further into individual cells, with each automobile navigating via the least congested route to its desired location using Aco for each compartment.

In the research article [8] the application of ACO toward the dilemma of traffic control has been explored in this investigation. The said algorithm's decentralized, collaborative, probabilistic, and self-organization characteristics are well-suited to road network traffic. The research article [9] figured out that instead of altering the moment of traffic signals, it is possible to efficiently prevent congestion problems.

All the above-discussed methods mainly concentrated on controlling traffic lights and controlling vehicle flow to avoid congestion and accidents in ITM. However, the existing methods encounter some weak points which need immediate action:

- The research mainly managed a limited number of vehicles with a fixed number of traffic controllers and other important entities, i.e., number of vehicles, roads, direction, etc. To resolve the issue our proposed method utilizes the swarm intelligence method PSO. The PSO method can able to detect the scenario, which has an optimum cycle path for higher volume traffic elements.
- Most of the works are related to a particular scenario. Few of the works are related to the density of traffic. To resolve the issue our proposed method utilizes the property of ANN. The proposed approach will be able to detect the number of the vehicle, approximate counts, and possible traffic diversion routes.
- In existing research works only a few algorithms have been applied for Comparision with existing work. To resolve this issue the proposed hybrid swarm intelligence (PSO) and ANN-based method will be compared with the random search method and ant colony method.

15.2.2 Neural Network in ITM

It is a set of methodologies that attempts to identify relationships between two sets of data using a method that imitates how the biological brain works. In the research article [10] the idea for creating a road traffic monitoring system using neural networks and the IoT has been suggested in terms of reducing the queuing for automobiles at stoplights [11]. In the article [12] surveillance footage data has been used in the article to classify traffic flow just using a Deep Convolutional Neural network. While particularly in comparison to other different classifiers, DCNN requires the least amount of filtering and is well-known for all of its precision [13].

In the research article [14] a traffic management method is proposed based on a neural network. Regulating traffic and providing smart transportation guidelines are the main goals of a traffic flow prediction system (TFFS). The study of TFFS is an interaction among automobiles, road users, as well as transportation systems (including roadways and traffic controls) well to identify and construct an ideal transport system with cost-effective traffic flow and negligible road traffic issues.

The research article [15] discusses two distinct artificial neural frameworks. That was evolved for interconnection into an ATMS bulletin board architectural design. The very first framework has been a Perceptron neural network-based automatic traffic lamps regulator, whereas the other one is a backpropagation prototype trained to anticipate road vehicular traffic problems.

15.3 Proposed Hybrid Model for ITM

The proposed system is based on a combined approach of swarm intelligence and ANN. In the first phase, the proposed system applies a swarm intelligence-based PSO methodology for smart traffic light control. In the second phase, an artificial neural network method will be applied to accurately detect and count the number of vehicles over the road. The working of the complete system is as follows.

15.3.1 Traffic Model

The proposed hybrid model is mainly based on the "vehicle-to-traffic" light balance framework concept. Which can be used to optimize traffic signals and find the best route for automobiles. To provide a suitable and predefined period, smart traffic signals will apply specific red and green beam durations to different roads in a crossroads. In this article, the term "ideal route" refers to a route with light density and a sensible length to the desired location [16].

Figure 15.1 is representing the proposed traffic model. The proposed model maintains two queues for vehicle information. To maintain the interaction details and number of vehicles. On traffic lights and automobiles, wireless sensors have been installed. As just a result, automobiles within the transmission range of traffic signals will deliver a notification to the traffic signal next to the front. The vehicle's position, speed, position, and obtained signal are included in this particular message.

Based on everyone's locations and received signal data, traffic lights will evaluate the furthest vehicle for each road as well as transmit a connection query to companions. Inability to answer to the traffic signal query message and furthest automobiles broadcast a query signal in their data transmission band. The whole signal data will

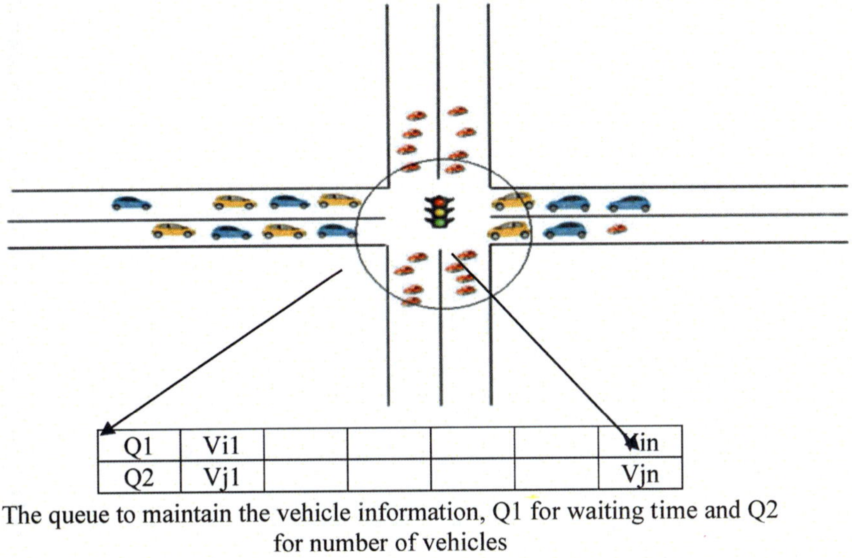

Fig. 15.1 Proposed traffic light controlling model

be received by automobiles in the transmission range behind the furthest automobile. An overall amount of information received has been sent to the traffic signal via the automobile that is the furthest away.

15.3.2 Phase-1: Traffic Light Controller (TLC) with PSO

Inside this phase one, we introduce a new distributed optimization technique based on the real traffic relevant information utilizing vehicular networks as well as the PSO algorithm. Motivated first by foraging birds, in this phase we consider a frequency choice strategy of one crossing the particle and also the selections of the first complete traffic flows channel as just a connectivity search space.

The next step performs the mapping of the consecutive signal processes within each crossroads toward the incremental upgrades of the speed and direction of the correlating virtualization particle triggered by a PSO method [17]. Each particle does have its corresponding feature and during the whole vehicle traffic flow, all the particles have a connection with other certain particles.

A PSO algorithm relies mostly on groups, looking to move the individual or group to just a nice location as per the wellness of its ecosystem. Even so, it doesn't use a genetic operator too for the entity, and thus wants to treat almost every entity as just a density-free particle throughout the D-dimensional solution space. Which moves within solution space till a certain constant speed depends on its flying Knowledge and companion flying experience.

Algorithm: Algorithm for Particle swarm optimization method

1. ***Initialization of the swarm variable, perform the following for each particle:***

1.1 Firstly do the Initialization for the position of each of the particles.

j= 1,.....................,n

1.1.1 Let uniformly distribution

1.1.2 as-Pj [0] ~ U [UB, LB] ------(1)

Where UB= upper bond, LB = Lower bond, for a particular search space

1.1.3 Now set the particle best P_{best}*, towards the search space.*

P_{best} *[j, 0] = P[j(0)]* ------(2)

1.1.4 Now set the swarm best G_{best}*, towards the search space.*

G_{best} *[0] = Min f (P[j(0)])* ------(3)

1.1.5 Now set the velocity value:

$V_j \sim U[(-|UB-LB|), (|UB-LB|)]$ ------(4)

2. ***Repeats the step untill a satisfactory termination condition not found -***

2.1 for each of the particles, j= 1,......................,N_p

do, the following for each particles

2.1.1 Select a random number

$(R_1, R_2) \sim U[(0,1)]$ ------(5)

2.1.2 Update the velocity of partciles by formula -

$V^j_{g+1} = [W.\ V^j_g + \Omega+1]$ ------(6)

2.1.3 Update the positions of partciles by form ula-

$X^j_{g+1} = [X^j_g + V^j_{g+1}]$ ------(7)

2.1.4 Check the best fit-

If $P_j[t] < f[P_{best}(j, t)]$*, repeat the following steps*

2.1.4.1 Upgrade the particle's absolute best place:

If (J : $P_{best}[(j,t)]$*)*

Upgrade the swarms finnest position

$G_{best}[t] = P_i[t]$ ------(8)

2.1.4.2 Increment the value t, t=t+1;

3. The final output will be held by the G_{best}*.*

15.3.3 Phase:-2 Traffic Density and Count by ANN

In just this phase-2, an ANN methodology is presented to trace and count several vehicles over the road. A background has been extracted differently from multiple frames during the first stage of the development of the methodology. By utilizing the Sobel operator the background image, as well as the latest image, has been detected from moving automobiles on the road. An edge item's dimension has been reduced next, and features have been obtained besides processing by the ANN for vehicle classification and finding the counts [18].

Figure 15.2 is representing the phase-2 scenario for finding the density and count of the vehicles over the road. Figure 15.2 is showing the 3-scenarios for traffic lights and the density of the vehicles. The first scenario has red light where all the vehicles are stopped and waiting for the green light and in scenario 2 the light is green and vehicles are moving over the road and in case-3 vehicles are moving without a signal. All these scenarios will generate a different density and count for vehicles [19].

Figure 15.3 is representing the various steps that apply for finding the road traffic entities and the count and density of vehicles. There are five main steps to

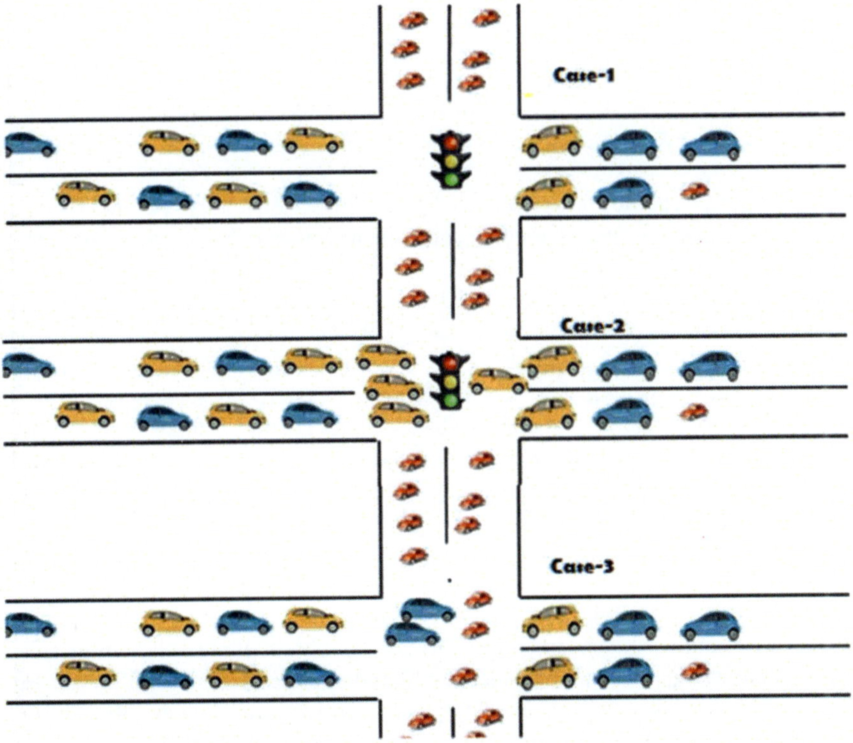

Fig. 15.2 Vehicle count and density by Phase-2 ANN method

Fig. 15.3 Steps in Phase-2 ANN method

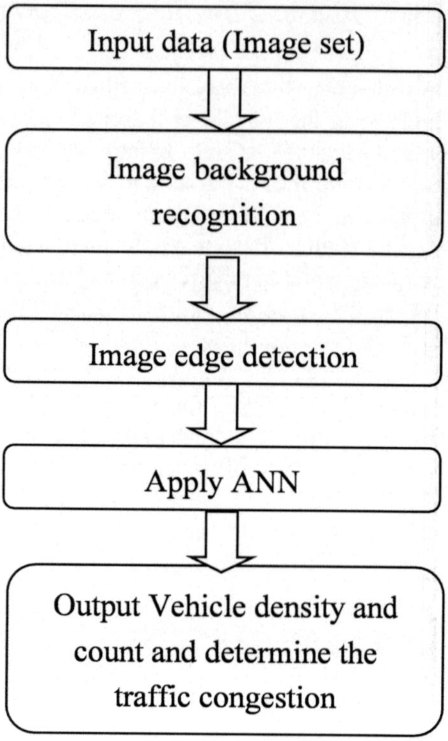

be followed in this phase. The picture background has been retrieved during the first stage. Once the background of the image has been recognized, a notification will be also required To use the recognized background picture as well as existing frames, the next stage creates a feature vector edge of moving vehicles. Each edge image, which was collected previously from road traffic, has almost a similar resolution value, equivalent to the actual image sets.

For the forecasting of the flow of traffic, a multilayer perceptron network has been used in an investigation. The ANN could be used after the horizontally and vertically estimates have been measured. Estimated images will be utilized in strengths e for the network attributes. We must formulate an ANN predicated on such a vector with recognized outputs and inputs, which necessitates the creation "supervised learning-based neural network called SNN". The creation of an SNN can be defined in a variety of manners. The Backpropagation method has been applied for implementing supervised techniques [20].

After determining the traffic congestion the proposed method will help in route diversion and setting the traffic signal. It will improve the road traffic flow without congestion and reduce the chances of collision. Also, reduce the overall travel time from source to destination for vehicles. The following traffic parameters have been used by the ANN method to count the vehicle, density, and congestion.

15.3.3.1 Traffic Density

The percentage of automobiles occupying a particular length of a major road is known as "traffic density" [21]. This can be composed as

$$T_{density} = TN_{vehicles}/R_{length} \tag{15.9}$$

where $T_{density}$ = *traffic density,* $TN_{vehicles}$ = *Total number of vehicles, and* R_{length} = *Length of road*

15.3.3.2 Average Speed

It can be calculated by dividing the total count of vehicles by the sum of their speeds. The speed of a vehicle is defined as "The maximum distance traveled by the motor vehicle per time unit" [22].

$$\text{Savg} = \sum_{i=0}^{n} \text{Speed(i)/Tn(i)}$$

where Savg = average speed of a vehicle, Speed (i) = speed of a vehicle, Tn (i) = total number of vehicles.

Figure 15.4 is representing the ANN model for phase-2. The hidden layer in ANN, which is made up of many neurons, conducts arithmetic on input data and sends the findings to the next output neuron. An ANN can choose any quantity of hidden neurons and we'll only take into account one hidden unit with 5 neurons in this instance. An output division is in charge of delivering data toward the outside universe. The proposed hybrid model has a total of 3 neurons, one of which correlates

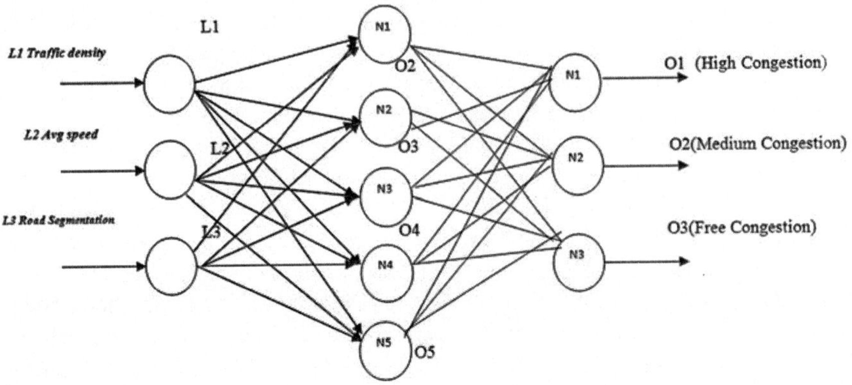

Fig. 15.4 ANN model for phase-2

to a distinct congestion region, since this research divides congestion regions further into the 3 groups: high-level congestion (HLC), medium level congestion (MLC), and congestion-free flow (CFF).

15.3.4 Final Module: Hybrid Strategy

In this work, we are presenting an Intelligent traffic management system based on swarm intelligence and neural network. In the first phase, the proposed system applies a swarm intelligence method PSO for smart traffic light control. PSO detects the accurate cycle program for the management of traffic lights. In the second phase, an artificial neural network method will be applied to accurately detect and count the number of vehicles over the road. This recognition process mainly uses the roadway visual data and begins by extracting relevant the image mostly from the subsequent video sequence. As once context is recognized, consequent images have been used to identify objects moving through the object computation.

The final phase applies the combined approach of phase-1 and phase-2 to achieve the best outcomes in terms of less congestion, less travel time, and vehicle detection ratio. The idea behind applying the combined strategy is to detect the vehicle density and volume by ANN and control the traffic signal accordingly by phase-1. This will help in improving road safety by reducing accidents. Also will reduce the overall travel time.

15.4 Results and Discussion

On MATLAB, we simulate the proposed hybrid model relying on a transport network with four crossings. The dataset is collected from the online Kaggle dataset [23]. This data set contains 4060 traffic image files. A few artificial neural networks frameworks have been proposed and provided with training on the Kaggle traffic dataset. To adapt the edge weights throughout the network, each of these frameworks has been trained at multiple frequencies. To possibly compare with traditional fixed-time direct control, we fixed almost the same input values and components for the proposed method. We originally estimated each mobility as vacant. In every mobility, the highest passing capability has been 450 automobiles per 60 min.

Figure 15.5 is representing the vehicle images collected over the road to measure the density and count. Image Edges have been identified of a passing vehicle. The Sobel transform has been applied now to identify vertically and horizontally outer edges of the image. Just after edges have been discovered, individuals have been loaded and processed and get more considerable pixel values. We distinguish an entrance real-time traffic circulation to view the efficiency of the classifier under diverse circumstances. Whereas if outflow numbers are much lesser than just the inflow frequency, the situation can be defined as unsaturated, and consequently, this

(a) Frontal image of vehicle (b) Lateral image of Vehicle

Fig. 15.5 Road Vehicle Images captured by cameras

was delineated as saturated. The proposed hybrid model and existing random search method and ant colony method have been implemented on MATLAB simulator and various performance comparison parameters have been calculated. Figures 15.6 and 15.7 are representing the simulation outcomes of the proposed method. The following experimental results have been calculated.

Figure 15.8 is representing the simulation result for Avg. speed and congestion level. This data has been calculated among phase-2 method (ANN) and hybrid method (PSO with ANN). The experimental results are showing better outcomes for the proposed hybrid method. Similiar in Fig. 15.9 simulation results for are representing the simulation results of queuing length has been calculated. The results are showing outcomes of phase-1 PSO method, existing ant colony [3], and random search [4] method, with hybrid (PSO with ANN). The experimental results are showing better outcomes for the proposed hybrid method.

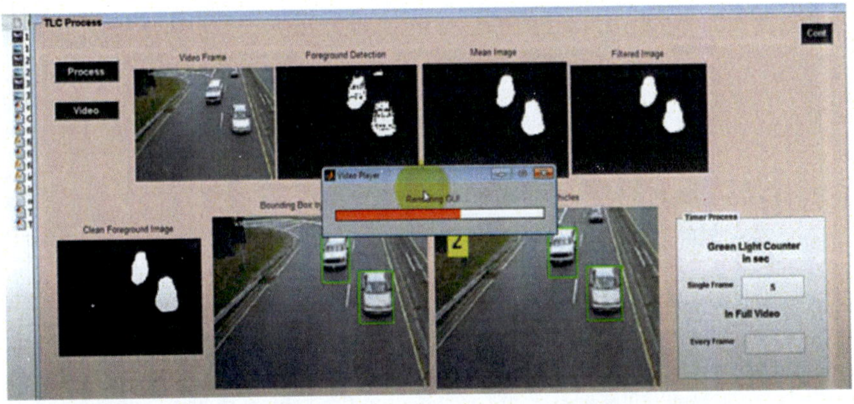

Fig. 15.6 Simulation results of ANN for counting the vehicle and measure the density

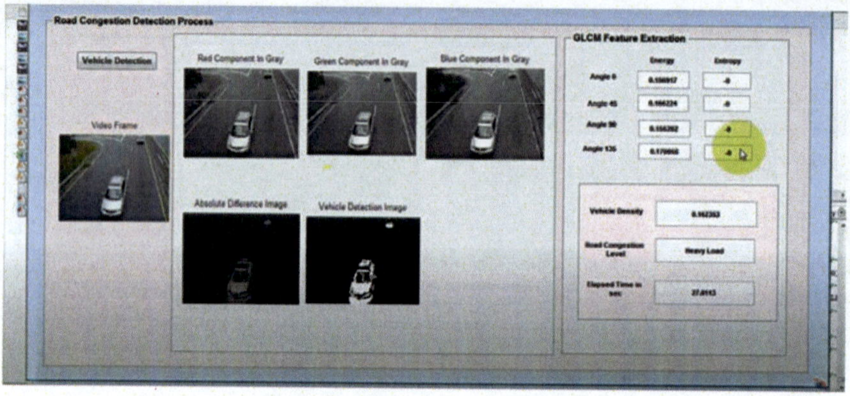

Fig. 15.7 Simulation result of the Proposed model to set the traffic signal to handle the congestion

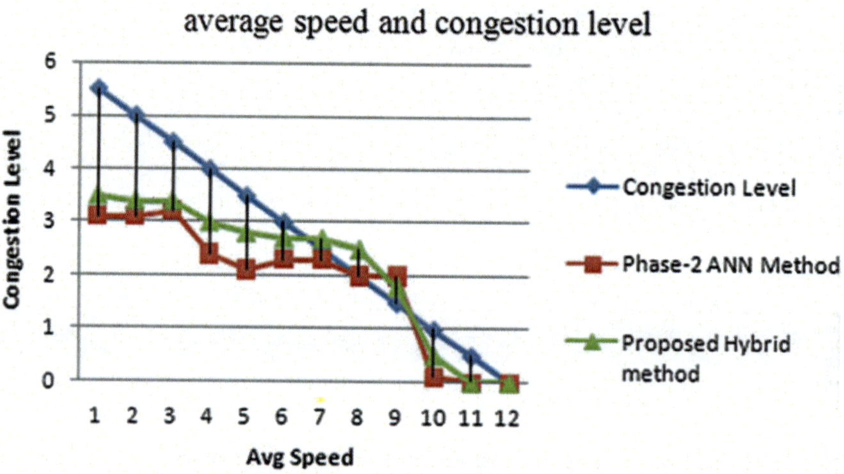

Fig. 15.8 Simulation results for Avg. speed and congestion level

15.5 Conclusions

This research article is presented a hybrid model for Intelligent transport management. We utilize a swarm intelligence-inspired PSO method in phase-1 to discover the sensible ratio of traffic signal lighting fixtures as well as map the sequence of traffic control efforts and also to enhance the status of a vehicular intersection. Throughout phase-2 extends the ANN approach for detecting the vehicle density and count the vehicle. We modulate the four-intersection scenario throughout this article. As even the findings demonstrate, the proposed hybrid methodology has quite a significant progress under various situations.

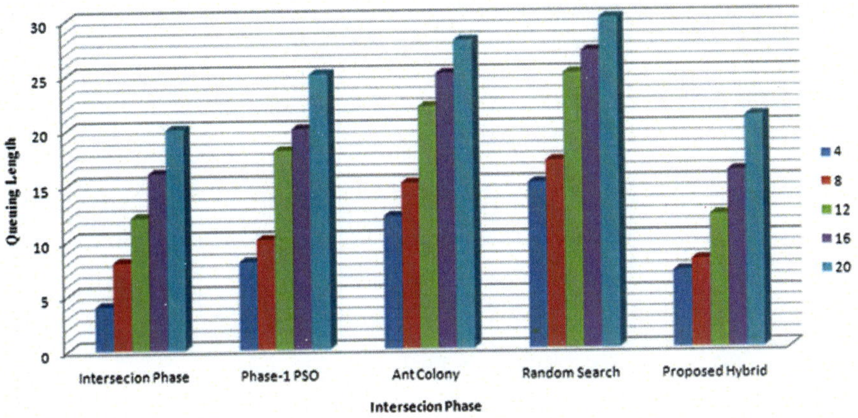

Fig. 15.9 Simulation results for comparison of queuing length

References

1. Agafonov, A.A.: Short-term traffic data forecasting: a deep learning approach. Optic. Mem. Neural Netw. **30**(1), 1–10 (2021)
2. Alkhalidi, N.A., Yaseen, F.A.: FDPHI: Fast deep packet header inspection for data traffic classification and management. Int. J. Intell. Eng. Syst. **14**(4), 373–383 (2021)
3. ASCE Task Committee on Application of Artificial Neural Networks in Hydrology.: Artificial neural networks in hydrology. II: hydrologic applications. J. Hydrol. Eng. **5**(2), 124–137 (2000)
4. Ata, A., Khan, M.A., Abbas, S., Ahmad, G., Fatima, A.: Modelling smart road traffic congestion control system using machine learning techniques. Neural Netw. World **29**(2), 99–110 (2019)
5. Chamoso, P., de La Prieta, F.: Swarm-based smart city platform: a traffic application. Adcaij Adv. Distrib. Comp. Artif. Intell. J. **4**(2), 89 (2015)
6. da Silva, D.L., Seijas, L.M., Bastos-Filho, C.J.A.: Artificial Bee colony optimization for feature selection of traffic sign recognition. Int. J. Swarm Intell. Res. **8**(2), 50–66 (2017)
7. Guillen-Perez, A., Cano, M.-D.: Intelligent IoT systems for traffic management: a practical application. IET Intell. Transport Syst. **15**(2), 273–285 (2021)
8. Hu, W., Wang, H., Yan, L., Du, B.: A swarm intelligent method for traffic light scheduling: application to real urban traffic networks. Appl. Intell. **44**(1), 208–231 (2016)
9. Kesuma, P.A., Rohman, M.A., Arif Prastyanto, C.: Analysing the root cause of the damage risk in heavy vehicles to reduce traffic congestion. Int. J. Construct. Manag. 1–12 (2020)
10. Kotsi, A., Mitsakis, E., Psonis, V.: Coordinated provision of C-ITS services for dynamic traffic management. In: 2020 IEEE 23rd International Conference on Intelligent Transportation Systems (ITSC). IEEE (2020)
11. Kyriakopoulos, C., Nicopolitidis, P., Papadimitriou, G., Varvarigos, E.: Exploiting IP-layer traffic prediction analytics to allocate spectrum resources using swarm intelligence. Int. J. Commun. Syst. **33**(14), e4516 (2020)
12. Li, X., Yi, P., Jiang, Y., Lu, X.: Traffic anomaly detection algorithm based on improved salp swarm optimal density peak clustering. In: 2021 4th International Conference on Artificial Intelligence and Big Data (ICAIBD), IEEE (2021)
13. Lilhore, U.K., Saurabh, P., Verma, B.: A new approach to overcome problem of congestion in wireless networks. In: Advances in Intelligent Systems and Computing, pp. 499–506. Springer Berlin Heidelberg, Berlin, Heidelberg. (2013)
14. Lilhore, U.K., Simaiya, S., Guleria, K., Prasad, D.: An efficient load balancing method by using machine learning-based VM distribution and dynamic resource mapping. J. Comput. Theor. Nanosci. **17**(6), 2545–2551 (2020)

15. Mao, Y., Qin, G., Ni, P., Liu, Q.: Analysis of road traffic speed in Kunming plateau mountains: a fusion PSO-LSTM algorithm. Int. J. Urban Sci. 1–21 (2021)
16. Raghuwanshi, V., Lilhore, U.: Neighbor trust algorithm (NTA) to protect VANET from denial of service attack (DoS). Int. J. Comput. Appl. **140**(8), 8–12 (2016)
17. Sergeev, S., Bozhuk, S., Pletneva, N., Evdokimov, K., Klochkov, Y.: Dynamic analysis of the traffic of vehicles used in car sharing. Int. J. Math. Eng. Manag. Sci. **6**(3), 847–859 (2021)
18. Shan, Q., Mostaghim, S.: Achieving task allocation in swarm intelligence with bi-objective embodied evolution. Swarm Intell. (2021). https://doi.org/10.1007/s11721-021-00198-2
19. Siddique, M.S.A., Hoque, S.: Predicting the daily traffic volume from hourly traffic data using artificial neural network. Neural Netw. World **27**(3), 283–294 (2017)
20. Sun, X., Chen, X., Qi, Y., Mao, B., Yu, L., Tang, P.: Effects of advanced traffic signal status warning systems on vehicle emission reductions at signalized intersections. J. Air Waste Manag. Assoc. **69**(4), 391–401 (2019)
21. Wang, R., Kan, Z., Cui, Y., Wu, D., Zhen, Y.: Cooperative caching strategy with content request prediction in internet of vehicles. IEEE Internet Things J. **8**(11), 8964–8975 (2021)
22. Yang, S., Yang, B., Kang, Z., Deng, L.: IHG-MA: inductive heterogeneous graph multi-agent reinforcement learning for multi-intersection traffic signal control. Neural Netw. Offic. J. Int. Neural Netw. Soc. **139**, 265–277 (2021)
23. Kaggle List traffic dataset.: (15–06–2021). Available from: https://www.kaggle.com/mbornoe/lisa-traffic-light-dataset (2019)

Chapter 16
Identification of Tamil Characters Using Deep Learning

S. Akashkumar, Atreya Niranjan Dyaram, and M. Anand

Abstract Optical character recognition is a key field that attempts to convert printed text into a digital representation. In this digital era, more printed materials are available, and they are mostly unclassified. When dealing with large data, any unclassified data are useless. The study of categorizing digital messages into some intelligible structure is known as natural language processing (NLP). Also, the input should be in digital format. OCR mitigates the burden of translating printed text into digital text. Even though OCR has become popular for English scripts and many other foreign languages, there has been not much development in OCR for Indian regional languages like Tamil, Telugu, Malayalam, etc. The major reason is that the existing literature on regional language contains a large number of characters, and several methods are proposed to combine them in order to make words. Tamil has 247 characters, which include alphabets (12), consonants (18), and alphabets + consonants (216). The combination in character makes the language difficult in converting to digital format. This research work has introduced a method to label each class with a particular number by reducing the class length to 108. This makes it much easier to process due to less number of classes. The usage of Convolutional Neural Network (CNN) made the processing smoother. The proposed CNN architecture was self-made with various hit and trial methods, which will also be discussed in this paper.

16.1 Introduction

Everything is becoming automated in this age of technology and internet. When the world is shrinking, tasks are becoming shorter, and only data are growing exponentially. Image data are becoming increasingly large as a result of the developments in computer science and image processing. There are several studies in the field of image classification. There are also several studies in the field of text classification.

S. Akashkumar (✉) · A. N. Dyaram · M. Anand
Vellore Institute of Technology, Vellore, Tamilnadu 632014, India

J. I.-Z. Chen et al. (eds.), *Machine Learning and Autonomous Systems*, Smart Innovation, Systems and Technologies 269, https://doi.org/10.1007/978-981-16-7996-4_16

Among these are number recognition, English character recognition, Devanagri script recognition, and many others. However, there is a scarcity of research work in Tamil character recognition. Tamil is one of the oldest languages in the world. One of the South-Indian languages, Tamil is predominantly spoken in Tamil Nadu, an Indian state. Tamil also has its own script, which is unique in itself. Thus, the paper aims to carry the research work in the domain of Tamil character recognition.

This era is an era of Machine Learning, Deep Learning, Data science and IOT. The use of Data Science has grown to a higher level in recent days. Basically, Machine Learning uses Statistical Analysis for predicting outcomes for a particular instance. Machine Learning can be classified as two major classes: Supervised Learning and Unsupervised Learning. There is another section called Reinforcement learning. But this paper focuses on Supervised Learning, as the data are labeled. The paper compares different models for the identification of Tamil characters. One of the proposed models uses KNN and SVM for classification purposes. But the model doesn't work well.

This paper uses the concept of Deep Learning for the Recognition of Tamil characters. Deep learning is a branch of Artificial Intelligence, which learns the model automatically upon feeding the data. When the data are complicated enough to be trained by a Machine Learning model, the answer is Deep Learning. The next two models use Neural Networks for character classification. The models perform much better than the previous model and give satisfactory results. The remainder of the paper is divided into the following sections: Literature Review, Research Gap, Proposed Methodology, Algorithms Novelty and Process, Implementation, Testing verification and validation, Comparative Analysis, Future Improvements and Conclusion.

16.2 Literature Review

In [1], the authors use multi-layer feedforward neural network for handwritten digit recognition. The use of extended Kalman filter in their methodology gives an edge over normal neutral networks as it reduces the network size and also speeds up the learning. Also, the suggested method is better than monotonic classifiers. Sunil Kumar Khatri, Shivali Dutta and Prashant Johri use LVQ (Learning Vectorized Quantization) and integrate ANN and supervised learning to classify digits though the approach fails to overcome the challenges of various writing styles, shapes and sizes [2]. Pre-trained layers using sparse autoencoders with predefined sequences of training process and rounds are used to train the net in [3], and the reconstructed images during the training process of autoencoders using different parameters have been used to expand the training set. But the proposed method takes reasonable training time, and also feature extraction and representation can be improved. Handwritten Arabic Numeral Recognition using Deep Learning Neural Networks [4] is one of the first works focusing on recognizing Arabic digits. The proposed method has a record accuracy of 97.4 for Arabic digits recognition.

In [5], the authors have used unsupervised pre-training along with autoencoder and deep Convolutional Neural Network for recognizing handwritten Bangla digits. The proposed approach achieves 99.50% accuracy, which is the best for recognizing handwritten Bangla digits. But the paper limits itself to Bangla. Second, pre-training should be tried on larger datasets as well to check if better results can be achieved. Reference [6] deals with comparison of Logistic Regression and ANN for Handwritten Digit recognition and uses vectorization for the optimization of model. The design of the model is flexible enough to incorporate different data according to the application. This work could be faster if integrated along with a Computer Architecture made specifically for systems as such. In [7], the authors suggest OCR in handwritten Arabic languages with a novel approach for identifying handwritten isolated Arabic characters. They use encoded Freeman chain code and recognize the importance of OCR in Arabic text. For [8], the authors created the dataset by actually collecting samples from different people. The proposed method uses convolutional neural network to extract features. It uses a method different from the older method that requires handcrafted features and needs to be used for finding features in the text. But they have not applied to all the characters as complete dataset is in the creation process.

A Comparative Study on Handwriting Digit Recognition Using Neural Networks [9] discusses Neural Network (NN) approaches. The famous NN approaches are deep neural network (DNN), deep belief network (DBN) and convolutional neural network (CNN). The three NN approaches are compared and evaluated in terms of accuracy and performance. Reference [10] tells about creating a model that uses MNIST dataset to train with the digits. The model comprises four layers, each with different activation functions for better efficiency. It mimics the learning capability of a full architecture CNN to better recognition accuracy in the MNIST handwritten digit recognition problem. But, due to variation in size and shape of handwritten scripts, the efficiency of the output cannot be as high as expected. The authors of [11] make use of ensemble neural networks with ensemble decision tree is new and unexplored. The proposed method performs better than ensemble neural networks and single neural networks. But the accuracy of 84% can be improved. Vincent Christlein, Martin Gropp and Stefan Fiel make use of CNN activation features in a simple yet effective manner [12]. Training CNN in an unsupervised manner and the use of surrogate classes is something new the authors have approached. Also, activation features in unsupervised manner are better than the ones with supervision.

In [13], BP neural network is used in order to obtain higher recognition rate, with a simple model. The paper studies the double hidden layer BP neural network with different feature extractions. This helps in getting higher recognition rate. Gaber feature extraction has been used to increase efficiency. Simple CNN has been used, hence the time of the network training is longer with the increase of the convolution layer. The network will be difficult for training because of its many layers. Handwritten Digits Recognition with Artificial Neural Network [14] implements a multi-layer fully connected neural network with one hidden layer for handwritten digits recognition. The multi-layer ANN helps in increasing the accuracy up to 99.6%. But the model is limited to MINST dataset and cannot be expanded with other larger dataset unless we change the ANN model. Sneha Shitole and Savitri Jadhav worked on Recognition of Handwritten Devanagari Characters using LDA [15]. They compare between LDA and PCA for which yields better results. Finally, LDA has proven to be better than PDA. Dataset for Devanagari characters is created for testing out the model. The authors, Kumar and R.K Agarwal show the use of unsupervised training using convolutional autoencoder with deep ConvNet in order to detect handwritten Devanagari digits in [16].

The authors apply deep convolution neural network for recognition of Bangla digits recognition in Bangla Handwritten Digit Recognition Using an Improved Deep Convolutional Neural Network Architecture. The model uses seven-layered D-CNN containing three convolution layers, three average pool layers and one fully connected layer for recognizing handwritten digits [17]. Very high accuracy (up to 99.9% on training data and 97.6% on test data) has been obtained by the usage of seven-layered neural network. But using data augmentation for increasing the amount of data and a deeper network may increase the performance significantly. OCR on cursive handwriting is a difficult task since they differ from person to person and detecting handwriting is difficult. There is no dataset available for Urdu language, hence 500 different person's handwritings were taken, and model has been trained with the collected data [18]. Since the dataset is not very huge, the model did not train well, and it has the error rate of 6.04–7.93%. Handwriting is also a biometric feature of a person, and neural network model can make use of it to solve criminal records and can be used for other useful purposes and [19] aims in identifying the person who has written the scripts based on his handwriting. The advantage of this proposed system is to extend the designed framework for the other Indian scripts like Bengali, Gujarati, Tamil, Gurumukhi and so on.

K. Manjusha, M. Anand Kumar and K. P. Soman make use of scattering transform wavelet features for the first layer of CNN in [20]. The proposed hybrid model performs better than the standard CNN model, which has equivalent self-learning architecture, but it can be improved to implement self-learning architecture as well as the proven filter. In [21], Mehdi Rizvi, Hasnain Raza, Shahab Tahzeeb and Shan Jaffry use two subclasses of computer division together: OCR and Object Recognition. The proposed approach reduces the workload after examination due to the initial work and is also time efficient. This work can be extended toward the recognition of from digits to characters for academic purposes. Ali et al. [22] uses DL4J framework and records a good accuracy of 99.21 A complete review of different writer identification methods at the same time comparing different feature selection methods is done in [23]. The authors do classification or analysis of different languages with respect to different script structures. But there is a need of employing modern techniques like fine-tuning and transfer learning.

16.3 Research Gap

A lot of research work on optical character recognition using convolutional neural networks/cross-correlation methods/template matching have been already carried out. However, most of them suppose the text in the document in English. Although some reports are purported for Tamil, they demand organic text without casted shadows and dimensional variation. In another front, various other techniques have been implemented to detect Tamil, few of which have been successful too.

However, the aim of this project is to improve reliability on high-frequency words overall. The previous approaches had used zoning, SVM and KNN classifiers for detecting and converting Tamil scripts. We have tried out those methods as well as Convolutional Neural Network approach for this research work. The ultimate aim is to have the highest possible accuracy and convert text with a considerable amount of noise

16.4 Proposed Methodology

Images are large in size and contain a lot of details. They can be split into three layers most of the times; viz. Red, Green and Blue. Every image is converted into an array of pixels, which is multi-dimensional. For e.g., A 256 by 256 image will be converted into 256*256*3 by 1 vector. In this way, features are extracted from the image. But looking into the size of the vector, one can notice how tedious the job of analyzing and pre-processing the vector is. This is when Deep Learning comes into the picture. Deep Learning maps many nets among the many layers created from the image dataset. CNN, DNN, RNN,etc. are some of the well-known deep learning algorithms.

This paper aims in using ConvNets for digitalizing and recognizing the Tamil characters. This is supposed to help adapt models for western scripts and test their viability for Tamil scripts. There is no defined dataset for Tamil scripts. So we have created our own dataset manually for all the Tamil characters with their respective dictionary values to use it in our model.

The project gives alternatives and additions to the existing methods, leading to a custom-defined model. The new model is used for examination, detection and digitalization of Tamil text on any surface with any texture or color with significant noise.

16.5 Algorithms, Novelty and Process

The novelty of this paper's work is that it considers the OCR for printed Tamil characters rather than handwritten characters. Many previous works have been done on handwritten data. But little work has been done in the domain of printed text. The data have been collected from text books [24]. There are a total of 10386 images for all of the 108 characters. Among the total images, the number of images for each label is approximately 90–100. The initial pre-processing of the data contains few steps. The images of Tamil characters are first converted to numpy array, further removing noise and shadow from it. This numpy array is saved as npy file (X).

We tried a method followed by [19] where each character undergoes Zoning for extracting the feature from that image. Zoning is a process in which the image is divided into zones where each zone has an image in itself. For any particular zone, the number of ones is counted, and this works as a feature. So, for e.g.: if an image is divided into 16 zones, every image in the dataset has 16 features. The output or the labels of the images are converted into one-hot encoding format. The features are further fed into SVM model and K Nearest Neighbor (KNN) model for scaling the observed features. From those features, the weights were collected and tested with some other sample images for testing. Unfortunately, this model didn't give good accuracy as expected. The reason could either be due to high variance, the data might have been over-fitted or maybe the image was not under similar classification.

Due to the failure of this model, ConvNet was used for feature extracting and classification of data. The classifier output (y) is denoted in y list, which is one-hot encoded for feeding into ConvNet. Again this list is converted into numpy array and saved in npy format. This X is used as input layer for our ConvNet model, and y is given as the label for training. This input layer was used to test few models for extracting the features and classifying the Tamil characters. The model contains two layers containing Relu (Rectified Linear Unit) as activation layer and used max-pooling layer along with each layer. The final layer contains Softmax activation for determining integer values. We used mini-batches symmetrically with 71 images in each batch. This helped in reducing the time consumed for each epoch. The next model proposed in the paper is a deeper network than the previous one. This neural network architecture has a similar architecture as the previous one, but with more dense layers in the end. Also, the image dimensions are reshaped to improve the efficiency, i.e. the image dimensions were powers of 2 for improved performance. Also, the images used for training are grayscale. This is done with the intention of reducing noise and indirectly reducing the dimensionality of input. Figures f1 and f2 show the architecture used by the models.

16.6 Implementation

16.6.1 Pre-processing

Pre-processing forms an important part of predictive analysis. The major steps of pre-processing done for the data are:

- Getting the images and labels from the images folder
- Reading the image and storing in suitable data structure(numpy array).

- Denoising the image
- Converting to grayscale (second model)
- Normalizing the image data
- One hot encoding the labels

16.6.2 Training

For the first model, we have implemented a Convolutional Neural Network model with two Conv-2D layers, each followed by Maxpool layer. The model was implemented on Google Colaboratory. The data was divided into epochs, with the model taking as input mini-batches of images (i.e. Training data). Figure 16.2 shows the graph of cost after every iteration for the first model.

For the second model, we don't use the images in RGB format. We converted the image to grayscale. Also, the dimensions of the images were different. We implemented a CNN model with two Conv-2D layers, each followed by a Maxpool layer. These layers are followed by a flatten layer and three dense layers. For all the layers, the activation function used is a Rectified Linear Unit (ReLU), except the last dense layer where the activation is SoftMax as the output of the task is multi-class classification. Figure 16.3 shows the graph of change in accuracy after every iteration for the second model.

16.6.3 Model Architecture

Figure 16.1 shows the architecture used for the Deep Neural Network model. The first layer is the convolutional layer, which is used to extract features from an image. It helps to retain the relationship between pixels while reducing the image size at the

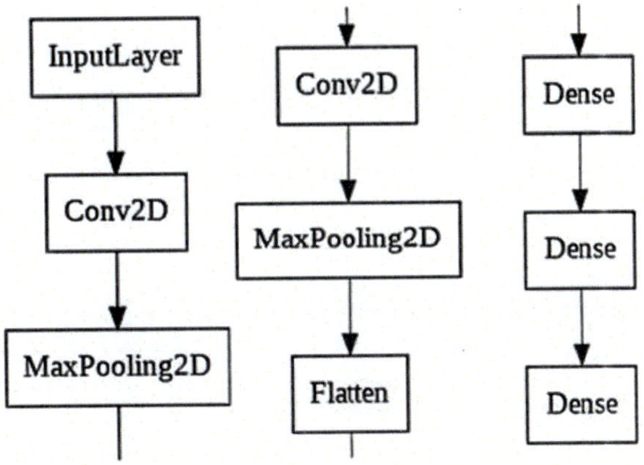

Fig. 16.1 Architecture for second model

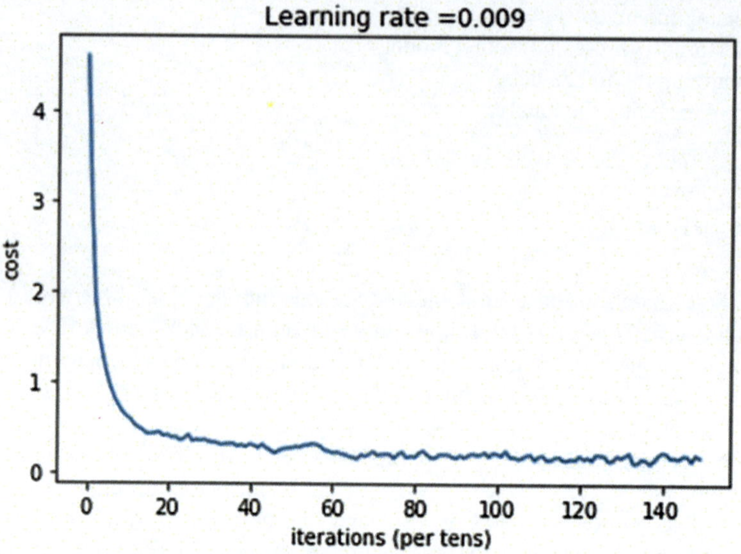

Fig. 16.2 Result of training the data with a CNN (first model) (loss)

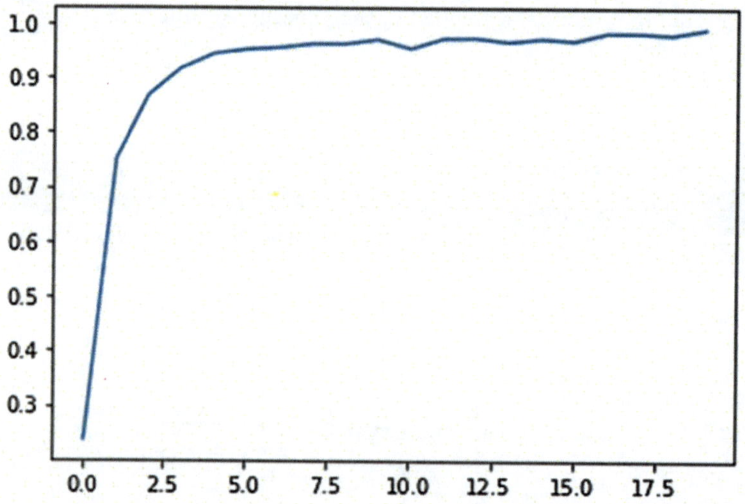

Fig. 16.3 Result of training the data with a CNN (second model) (accuracy)

same time. Next layer in the architecture is max-pooling layer, which is a pooling layer. This layer takes the maximum value from a block it chooses, which helps in retaining the most significant feature while moving to the next layer.

This is followed by another set of convolution and max-pooling layer. Using convolution layers more than once helps in learning hierarchical representations and forms more abstract representations. For example, the first layer may classify edges but the second layer will use the edges to form shapes.

After these four layers, we are flattening the output in order to convert the received input to one-dimensional array so that it can be used by the next layers. After flattening, follow the dense layers. The dense layer uses the features extracted and does the classification. The reason for adding multiple dense layers is that while extracting features, convolutional layers work in a differentiable manner. Thus adding dense layers helps the network classify the extracted features in a better way.

Let us see the different types of layers in detail.

Convolutional Layer It is a layer that has a matrix of features, which looks like a grid of neurons. In this layer, convolution operation is performed upon the input received from the previous layer, and the result is passed to the next layer. Convolutional layer has a kernel, which performs the same operation on every location it slides over. Thus, it transforms the matrix to another matrix with different set of features.

We have used convolution layer at two places. The convolution layer reduces the image size while also bringing the similar features into a single pixel. As the input image is a grayscale image, we are using kernel of size $3 \times 3 \times 1$. In general, a neuron layer of size $N \times N$, and filter of size $m \times m$, gives output of size $(N - m + 1) \times (N - m + 1)$ While propagating forward, for a filter w, data(x) passed to the activation function is:

$$x_{ij}^{l} = \sum_{a=0}^{m-1} \sum_{b=0}^{m-1} w_{ab} y_{(i+a)(j+b)}^{l-1} \tag{16.1}$$

where y_{ij}^{l-1} is the neuron from previous layer. The value is then passed to the activation function.

$$y_{ij}^{l} = \sigma(x_{ij}^{l}) \tag{16.2}$$

Max-pooling layer It is a pooling layer. It takes the output of convolutional layer and performs pooling operation to produce a single output for a chosen grid depending on the size of the grid. There are different types of pooling like max pooling, average pooling or some other type. In our case, we are choosing max pooling. In pooling layers, there is no learning. In general, a kernel size of N/2 is chosen for max-pooling layer so as to reduce the size of the input by 2. Hence, following the same convention, we have chosen the size of the kernel in the same way.

In general, a neuron layer of size $N \times N$, and some region of size $m \times m$ is chosen. For the above mentioned parameters, a layer of size $\frac{N}{m} \times \frac{N}{m}$ is the output.

Dense Layer This is the most commonly used layer. For dense layer size, most people consider powers of 2 like 32, 64 and 128. Sometimes, it is just trial and error method as this is also considered a part of hyper-parameter tuning. We have also followed the same convention, starting with powers of 2, and trying out various combinations to achieve the best result. The final dense layer though has a size of 108 as the number of labels is 108. We are using SoftMax activation as the classification problem in our case is multi-label classification.

When we evaluate our model, we need to optimize it based on the error it gives after each epoch. We need to correct the error and make the neural network correct the error it made. This process should be done iteratively for each epoch with a standard function. This function is called as loss function, and we need to choose a strong loss function, which makes our model's learning curve to flatten after specified number of epochs. This loss function along with accuracy can also be used to determine or validate how efficient our model is and how we can improve it. The loss function we are using for our model is Categorical Cross-Entropy loss function, given by Eq. (16.3). Since we are using multi-class classification, this loss function is more suitable.

$$Loss = -\sum_{i=1}^{outputsize} y_i \cdot \log \hat{y}_i \tag{16.3}$$

Here, y_i is the target value and $\hat{y}_i$ is the (ith) scalar value in the model output This loss function gives very good measure of probability distribution across discrete classes and helps in distinguishing between the classes, which is required for multi-class classification.

16.7 Testing, Verification and Validation

As mentioned before, the zoning method did not show noteworthy results.

Next, we consider the first model used for CNN. The data were split into training and testing datasets, in the ratio of 2:1. For validation, there are many approaches and metrics like cross-validation of model, confusion matrix, ROC curve, accuracy, Cohen's K score, etc. These above-mentioned methods are useful in various situations, according to the type of classification, type of data, data distribution and many other factors. The validation technique used for this model is accuracy. Since it is a multi-label classification model, accuracy can be used. Accuracy is an easy to use and a reasonable metric for both binary as well as multi-class classification model. This technique can be used when the dataset we use in the model is well balanced. Here in our dataset, we have 100 images classified for each class, which makes it appropriate for this metric. Accuracy can be calculated with a simple formula given by Eq. (16.4), which is sum of true positives and true negatives divided by total number of cases examined.

$$Accuracy = \frac{TP + TN}{TP + FP + TN + FN} \tag{16.4}$$

Table 16.1 Table for epochs v/s accuracy

No. of epochs	Accuracy
10	88%
20	92%
25	90%
30	90%

where,

$$TP = TruePostive,$$

$$TN = TrueNegative,$$

$$FP = FalsePositive,$$

$$FN = FalseNegative$$

The first CNN model was trained for over 140 epochs. This model showed a training accuracy of approximately 93% and testing accuracy of approximately 72%. Coming to second CNN model, the results of the model are pretty satisfactory. We trained the model with different number of epochs. Table 16.1 shows the number of epochs v/s accuracy. All the observations have better results compared to the first model. As it can be seen, the model with 20 epochs has the highest test accuracy of approximately 92%.

The models were tested with single data also. When an image was given as an input to the model for predicting the label, the model gave an output of one-hot encoded array, where the index corresponding to label would be the correct label.

16.8 Comparative Analysis

As it is visible from the results, the results of model 2 are far better than results of model 1. The reason can be attributed to the grayscaling of images in the second case. Also, in the second case, the model is somewhat deeper. One observation from the results is that for the first model to get better results, we need to grayscale the image, and more data can add to its accuracy.

Now, among the many observations of the second model for different number of epochs, 20 epochs seem to be perfect. Figure 16.3 shows the graph of accuracy for the second model run for 20 epochs. Now, when the number of epochs was increased, the model started giving lesser accuracy. The reason for the slide in accuracy is due to over-fitting. Over-fitting is the production of an analysis that corresponds too closely

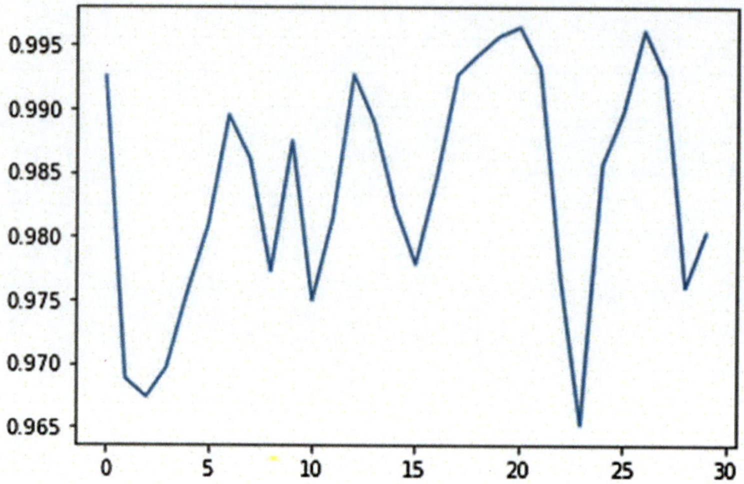

Fig. 16.4 Training for 30 epochs

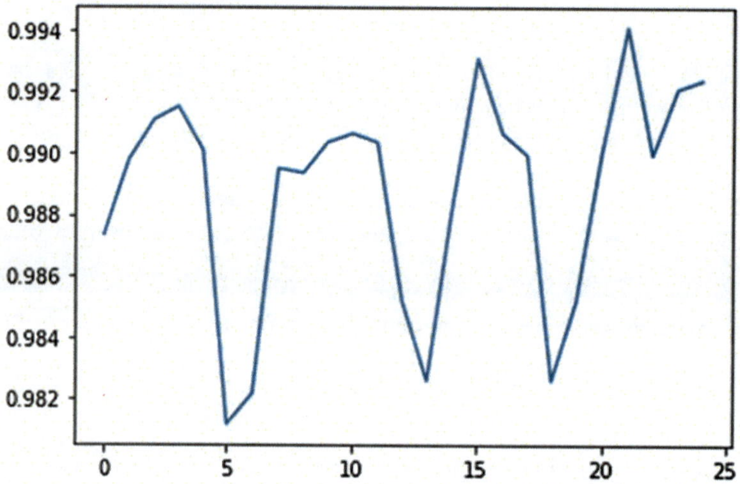

Fig. 16.5 Training for 25 epochs

or exactly to a particular set of data and may, therefore, fail to fit additional data or predict future observations reliably. As we can observe from Figs. 16.4 and 16.5, the graphs of accuracy for 30 epochs and 25 epochs are not monotonically increasing and clearly show that the model is over-fitting. But if we observe Fig. 16.3, the accuracy is monotonically increasing for 20 epochs.

16.9 Future Improvements

92% is not a bad accuracy, but not the best either. There is scope for improvement. But increasing the epochs is not the solution as it will only increase the training time, but finally end up overfitting. Thus, we need to have more data. In case, additional data are not possible, we can enhance the model by using transfer learning where the first few layers(shallow layers) are pretrained and the remaining layers (deeper layers) are trained and the weights are learned.

16.10 Conclusion

Optical character Recognition (OCR) is a wide field of study and always has potential to be improved. In the modern world, Natural Language Processing (NLP) has found its way through various appliances, i.e. mobile voice detection, Alexa, Google Home, Smart TV etc. OCR plays a vital role in the development of NLP. Tamil being a regional language in southern India, there are very less researches done in Tamil OCR. We initiated a possible way to create our own dataset and make a model which is good enough to predict Tamil characters with 92% accuracy. This has a wider scope in terms of improvement and development of our CNN. Our first result with our model was coming out to be 72%. We tried manipulating the model with various layers and weight parameters. We went through a lot of hit and trial method before concluding our final CNN architecture. The proposed method is a scalable system and still has the scope of enhancements. However, it has already outperformed most conventional techniques, what remains is to break its own feat. The project out of subject also seeks technological advancements for regional sects, its first step to achieve that being an advanced OCR system for Tamil. It will constantly be under improvement and drafts, only to make it better.

References

1. Wong, K.W., Leung, C.S., Chang, S.J.: Handwritten digit recognition using multilayer feedforward neural networks with periodic and monotonic activation functions. In: Object Recognition Supported by User Interaction for Service Robots, vol. 3, pp. 106–109. IEEE (2002)
2. Khatri, S.K., Dutta, S., Johri, P.: Recognizing images of handwritten digits using learning vector quantization artificial neural network. In: 2015 4th International Conference on Reliability, Infocom Technologies and Optimization (ICRITO)(Trends and Future Directions), pp. 1–4. IEEE (2015)
3. Saabni, R.: Recognizing handwritten single digits and digit strings using deep architecture of neural networks. In: 2016 Third International Conference on Artificial Intelligence and Pattern Recognition (AIPR), pp. 1–6. IEEE (2016)

4. Ashiquzzaman, A., Tushar, A.K.: Handwritten arabic numeral recognition using deep learning neural networks. In: 2017 IEEE International Conference on Imaging, Vision and Pattern Recognition (icIVPR), pp. 1–4. IEEE (2017)
5. Shopon, M., Mohammed, N., Abedin, M.A.: Bangla handwritten digit recognition using autoencoder and deep convolutional neural network. In: 2016 International Workshop on Computational Intelligence (IWCI), pp. 64–68. IEEE (2016)
6. Farooq, F., Tandon, S., Parashar, P., Sengar, P.: Vectorized code implementation of logistic regression and artificial neural networks to recognize handwritten digit. In: 2016 IEEE 1st International Conference on Power Electronics, Intelligent Control and Energy Systems (ICPEICES), pp. 1–5. IEEE (2016)
7. Althobaiti, H., Lu, C.: A survey on arabic optical character recognition and an isolated handwritten arabic character recognition algorithm using encoded freeman chain code. In: 2017 51st Annual Conference on Information Sciences and Systems (CISS), pp. 1–6. IEEE (2017)
8. Nair, P.P., James, A., Saravanan, C.: Malayalam handwritten character recognition using convolutional neural network. In: 2017 International Conference on Inventive Communication and Computational Technologies (ICICCT), pp. 278–281. IEEE (2017)
9. Ghosh, M.M.A., Maghari, A.Y.: A comparative study on handwriting digit recognition using neural networks. In: 2017 International Conference on Promising Electronic Technologies (ICPET), pp. 77–81. IEEE (2017)
10. Teow, M.Y.: A minimal convolutional neural network for handwritten digit recognition. In: 2017 7th IEEE International Conference on System Engineering and Technology (ICSET), pp. 171–176. IEEE (2017)
11. Larasati, R., KeungLam, H.: Handwritten digits recognition using ensemble neural networks and ensemble decision tree. In: 2017 International Conference on Smart Cities, Automation and Intelligent Computing Systems (ICON-SONICS), pp. 99–104. IEEE (2017)
12. Christlein, V., Gropp, M., Fiel, S., Maier, A.: Unsupervised feature learning for writer identification and writer retrieval. In: 2017 14th IAPR International Conference on Document Analysis and Recognition (ICDAR), vol. 1, pp. 991–997. IEEE (2017)
13. Hou, Y., Zhao, H.: Handwritten digit recognition based on depth neural network. In: 2017 International Conference on Intelligent Informatics and Biomedical Sciences (ICIIBMS), pp. 35–38. IEEE (2017)
14. Islam, K.T., Mujtaba, G., Raj, R.G., Nweke, H.F.: Handwritten digits recognition with artificial neural network. In: 2017 International Conference on Engineering Technology and Technopreneurship (ICE2T), pp. 1–4. IEEE (2017)
15. Shitole, S., Jadhav, S.: Recognition of handwritten Devanagari characters using linear discriminant analysis. In: 2018 2nd International Conference on Inventive Systems and Control (ICISC), pp. 100–103. IEEE (2018)
16. Kumar, S., Aggarwal, R.: Augmented handwritten Devanagari digit recognition using convolutional autoencoder. In: 2018 International Conference on Inventive Research in Computing Applications (ICIRCA), pp. 574–580. IEEE (2018)
17. Saha, C., Faisal, R.H., Rahman, M.M.: Bangla handwritten digit recognition using an improved deep convolutional neural network architecture. In: 2019 International Conference on Electrical, Computer and Communication Engineering (ECCE), pp. 1–6. IEEE (2019)
18. Ahmed, S.B., Naz, S., Swati, S., Razzak, M.I.: Handwritten urdu character recognition using one-dimensional blstm classifier. Neural Comput. Appl. **31**(4), 1143–1151 (2019)
19. Dargan, S., Kumar, M., Garg, A., Thakur, K.: Writer identification system for pre-segmented offline handwritten Devanagari characters using k-nn and svm. Soft Comput. 1–12 (2019)
20. Manjusha, K., Kumar, M.A., Soman, K.: Integrating scattering feature maps with convolutional neural networks for malayalam handwritten character recognition. Int. J. Docum. Anal. Recognit. (IJDAR) **21**(3), 187–198 (2018)
21. Rizvi, M., Raza, H., Tahzeeb, S., Jaffry, S.: Optical character recognition based intelligent database management system for examination process control. In: 2019 16th International Bhurban Conference on Applied Sciences and Technology (IBCAST), pp. 500–507. IEEE (2019)

22. Ali, S., Shaukat, Z., Azeem, M., Sakhawat, Z., Mahmood, T., Rehman, K.: An efficient and improved scheme for handwritten digit recognition based on convolutional neural network. SN Appl. Sci. **1**(9), 1125 (2019)
23. Rehman, A., Naz, S., Razzak, M.I.: Writer identification using machine learning approaches: a comprehensive review. Multimedia Tools Appl. **78**(8), 10889–10931 (2019)
24. Senthil, A.: https://github.com/Akashkumarsenthil/Tamil_OCR_

Chapter 17
Vocal Eyes Communication System

S. Gayathri, Anirudh Chandroth, K. Riya Ramesh, R. N. Sindhya Shree, and Surojeet Banerjee

Abstract Communication is one of the most vital elements for a human being. Amyotrophic lateral sclerosis also known as ALS is one of the progressive neurodegenerative diseases. This affects nerve cells in the brain and spinal cord. A-myotrophic is from Greek, it means "No muscle nourishment". Whenever a muscle has no nourishment, it becomes weak. The nerve cells present in person's spinal cord control the muscles, whose location is identified by lateral. As a result, the person might lose the ability to speak as there are almost 100 muscles that are responsible for speech. The main objective of the proposed work is to give voice to voiceless. Even in severe cases of ALS, the patient can still move his/her eyes. This ability is utilized to create an application in which eye movements of the patient can be used to select the letters displayed on virtual keyboard. Then this will eventually be converted into speech. The entire work is based on OpenCV and image processing algorithms.

17.1 Introduction

Human beings are naturally enabled with the ability to do their work on their own and to communicate. Unfortunately, some people do not have the ability to perform these basic tasks due to some diseases. ALS is one such disease. ALS is a condition where the patient might lose complete control over their movements and ability to communicate. Hence, a person suffering from ALS will be able to control their eye movements only.

There is no known cure for ALS till date. Spine stimulation, human-comp interface and exoskeleton are a few technologies that were developed as an alternate. Unfortunately, these technologies require the implantation of electro devices, which

S. Gayathri · A. Chandroth (✉) · K. R. Ramesh · R. N. S. Shree · S. Banerjee
ECE Department, Sri Jayachamarajendra College of Engineering, JSS Science and Technology University, Mysore, India

S. Gayathri
e-mail: sgmurthy_65@sjce.ac.in

J. I.-Z. Chen et al. (eds.), *Machine Learning and Autonomous Systems*, Smart Innovation, Systems and Technologies 269, https://doi.org/10.1007/978-981-16-7996-4_17

is very difficult and expensive. Therefore, these technologies were not as efficient as expected.

Many methods have been developed for eye tracking. Occulography is a method for eye tracking. Eye movements are necessary to efficiently scan a scene in order to get visual information. Various image processing algorithms can be implemented to get the visual information. Various libraries such as Dlib, OpenCV, etc., are used to locate the facial landmarks and obtain the eye movements.

In the proposed system, alphabets are detected based on the patient's eye movements and are used to form a word and then it is converted into speech. Hence the patient can communicate by himself with the world.

17.2 Research Gap and Problem Formulation

The existing systems emphasize the need for a system that is useful for paralyzed patients. Their main aim is to provide a system that is useful for patients suffering from diseases like ALS in such a way that they can communicate without having to use their limbs to make any gestures.

However, there is an inevitable gap in the research being conducted in the field of these types of communication methods. Almost all of these systems give importance to only detecting the eye movements or gestures and finding the word or phrase. The research being conducted in this field gives less emphasis to how the detected data is conveyed as speech.

Also, in most of the pre-existent systems which involve eye movements, too many eye movements are needed to select a single letter or entity, which makes the whole communication process tedious for the patient.

This shows that a lot of focus is given on just finding a method of communication which is suitable for patients, rather than finding a method which is convenient enough for the patient to use for his/her day-to-day communication.

Since the research in the field of communication systems for patients suffering from diseases like ALS mainly focuses on just the methodology to be implemented, rather than aspects like comfort of the patient, etc. There are other aspects that are not given due importance. One such factor is the cost of implementation of the system.

These communication systems are supposed to be available to almost all the patients irrespective of their economic status. This is because being able to communicate is a basic human right. Unfortunately, the existing systems are quite expensive too. Hence, all the patients may not be able to afford the same.

Apart from this, these systems also suffer from other technical problems such as decreased user autonomy, slowed inputs, decreased quality of communication and poor communication partner engagement.

Therefore, we can conclude that a communication system for patients suffering from diseases like ALS is necessary. Also, the communication system must be convenient to use, efficient and affordable for the patient. The system must help the patient in relaying the information to the person that they are speaking with, without any difficulties.

17.3 Related Work

Following papers were referenced to design the proposed system architecture.

Paper [1] presents a system called E-ACE which is a smartphone-based system, which utilizes nine directions of eye movement rather than the traditional four directions used by similar systems. This system allows its users to select the four eye movements that they are comfortable with from the set of nine eye movements.

Paper [2] demonstrates a smartphone-based system GAZE speak is an eye gesture communication system that can interpret eye movements in real-time, analyzes the movement and form the relevant word from the gestures made. It is a cost effective, robust and easy to learn as compared to an e-Tran board.

In Paper [3] the drawbacks like communication quality, autonomy and many other issues have been discussed, an AACrobat system is designed which is used to overcome all these drawbacks. Initially, qualitative finding on issues with gaze-based AAC has been presented. A system called AACrobat is designed which embodies all the guidelines and introduces novel interactions by extending these devices with mobile companion app.

Paper [4] contains an overview of the project and the techniques used to solve the patient's problem. Limitation and description of algorithm are comparing the researched technique. The smart solution is obtained using python programming with openCV.

Erica is a workstation-based system that helps patients with motor disabilities. This workstation consists of imaging hardware along with its software which is used to record a digital portrait of the patient's eye. This system used eye gaze technology and is described in paper [5].

Paper [6] proposes a method that utilizes centroid analysis technique for estimation of eye position and direction. This paper represents a methodology for the detection of eye blinking robustly in a real-time environment. OpenCV platform is used to implement this methodology.

In Paper [7], a method to detect eye region has been proposed. This is using haar classifier method by means of modifying the direction of sliding window. With this method, computation time can be decreased and an accuracy rate of 92.4% has been achieved.

Paper [8] is about a device that allows visually impaired people to interact with computers by using their voice. This technique utilizes the concept of Optical character recognition and text to speech synthesizers in Raspberry pi. It is an effective and cost-beneficial technique that enables patients to hear the contents of text files rather than reading them.

Paper [9] is about a text to speech synthesizer that converts text into synthesized speech which is read out to the patient. This synthesized speech can be saved as a mp3 file. The paper talks about the development of such applications and how they can help visually impaired patients.

In paper [10], the relationship between the emotion of a patient and eye movements is presented. This paper gives an insight into emotional relevant features extracted from eye tracking along with the current issues faced in this field.

Paper highlights the use of blinking of eyes of a patient to analyze the intention of the patient. It uses image processing to detect the blinking of the eyes and then select the set of intentions that are displayed on the screen.

Paper focuses on a communication support interface for patients who suffer from motor paralysis and are unable to speak. It uses the EOG technique to realize the eight directions of eye movement with high efficiency.

Paper focuses on how Eye tracking can be used as an indirect means of communication for patients with ALS, and hence can be used to perform cognitive tests on ALS patients to evaluate their condition. This method is reliable and easy to administer to patients with motor disabilities.

Paper uses eye "writing" technique for communication. The technique employs the method where the patient rotates their eyeballs to trace a certain pattern which is then converted to its corresponding symbol. The eye movement is detected by EOG technique.

In paper, EOG-based eye-writing method is used for alternative inputs for the human–computer interface and can be implemented for new communication methods such as in augmented reality. For "eye-writing" the system uses traces using EOG technique and each trace is mapped to a symbol. The system is trained with a set of eye traces and matches the input eye trace from the set to return the correct symbol.

17.4 Summary of Literature Survey

The existing solutions are described by the tabular column mentioned above. Basically, there are two factors to be considered, the former being the eye movement detection and the subsequent recognition of the letter, and the latter is the text to speech conversion (Table 17.1).

Various approaches are available for the detection of eye movement. It could be either sensor-based or it could be computer vision-based. The sensor-based approach makes use of two pairs of electrodes placed around one eye to detect the electric signals, and this approach is known as electroocculogram (EOG). This approach is more expensive than the computer vision-based approach. The computer vision-based approach could either be real-time detection, wherein the eye movements are captured by a camera in real-time and the eye movements are detected, or, it could be the one where a video of the eye movements is taken and then it is used to detect

Table 17.1 Summary of literature survey

Sl no	Title of the paper	Drawbacks	Future scope
1	Eye gesture-based communication for people with motor disabilities in developing nations	In order to select a letter, at least three eye movements are compulsory, which is a major inconvenience for the patient. Also, the patient has to remember the letters lying in a certain range in order to select a letter, which can make it very difficult for the user to learn to use the system	The option selected by the patient can be played as speech Also, the patient should be able to confirm if the selected option was the one that they needed
2	Smartphone-based gaze gesture communication for people with motor disabilities	Gaze Speak mainly focuses on implementation in iOS devices and does not have a separate hardware device to implement the same	A separate low cost and portable hardware device can be implemented, so that the system is portable and caters to specific needs
3	AACrobat: using mobile devices to lower communication barriers and provide anatomy with gaze-based AAC	During face-to-face conversations, these devices have a problem of slow input, which causes communication breakdowns	Efficiency of the system can be improved. Also, specific, low cost and portable hardware can be designed for the same
4	Review on Smart Eye Blink Solution for MND Patient using Python	Here, the patient's eye movement is recorded using a camera and then, the eye movements are detected. But this is not real-time detection, as it makes use of a video of the eye movement	Real-time eye movement could be more effective for such systems, thus enabling more efficient communication
5	Human–computer interaction using eye gaze input	A near-infrared light source, a light-intensity imaging board, etc., additional hardware has been used in the technology used here. This is not really feasible for a communication system for paralyzed people	Here, the system is limited to prosthetic applications
6	Real-time eye blinking detection and tracking using OpenCV	Although this system is useful in detection of drowsiness, it does not form a large part of the eye movement-based communication system	This system is useful in detecting drowsiness in drivers. Detection of more eye movements can be included to form a more efficient system

(continued)

Table 17.1 (continued)

Sl no	Title of the paper	Drawbacks	Future scope
7	Image processing for rapid eye detection based on robust haar sliding window	By using haar conventional methods some texting images cannot be detected. Also, images with different eye gaze cannot be detected and here lighting can also influence the detection of eye movements in an adverse manner	The detection of eye can be improved efficiently by using the scheme of changing position of scanning window. The computational speed can also be improved
8	Text to speech Conversion	The image of handwritten text is converted into speech in this system. Although this might have a lot of other useful applications, it is not specifically useful for providing a communication system for people who are paralyzed	The detection of text can be made more efficient. Also, this system can be very useful for visually impaired people
9	Design and implementation of text to speech conversion for visually impaired people	This system only concentrates on the detection of text and its conversion to speech. Also, it does not concentrate on the detection of the text from real-time user inputs	This system can be implemented in many other platforms such as ATM machines, video games, etc. Also, the system can be made to work with local languages
10	Emotion recognition using eye tracking: Taxonomy, review and current challenges	Although the proposed system may have a lot of useful applications, the identification of emotions by using eye tracking alone may not be practical in real-life situations	Emotion detection can be further expanded to the analysis of the changes in any other facial expressions other than eye tracking alone
11	A Communication system for ALS patients using eye blink	The intentions of patients are visible on the screen and they should blink their eye to select it. The time taken is just 0.8 s	The system is very complicated and costly. It is not affordable
12	Development of EOG-based communication system controlled by eight-directional eye movements	The system is mainly based on measuring horizontal and vertical occulograms using electrodes and then interpreting the movement. This process increases efficiency of the system	The system is mainly based on threshold. So, if the threshold is not selected properly the whole system goes wrong. Also, the system is quite complex

(continued)

Table 17.1 (continued)

Sl no	Title of the paper	Drawbacks	Future scope
13	Eye tracking control to assess cognitive functions in patients with amyotrophic lateral sclerosis	This system is based on cognitive tests. The properties of stimuli are used to assess cognitive functioning	The assessment of cognitive status is quite difficult
14	Eye-writing Communication for patients with amyotrophic lateral sclerosis	Here the patient need not select anything, instead he can trace any symbols he needs to with the help of his eyeballs	Only symbols can be traced, it would be efficient if one can trace letters or words
15	Real-time "eye-writing" recognition using electrooculogram (EOG)	Inputs are taken initially from EOG waveforms and system reconstructs it. It is compared with the trained images and the output ID obtained. The accuracy is more if the trained images are more	If the trained images are less in number, then the accuracy decreases and the interpretation of letters will be wrong

the sequence of eye movements which is used for the subsequent detection of eye movement. The former approach is more useful in real-life scenarios. In the proposed system we are using Dlib library to detect the eye movements. Dlib is used to locate the facial landmarks as it is necessary to detect face before detecting eyes. Using Dlib initially face and then eyes are detected.

After the detection of eye movements, the next factor to be considered is the detection of letters corresponding to the eye movements. Most of the approaches involve the use of software which is often installed in a mobile device. Although this can bring in the aspect of mobility in the proposed system, specific hardware is necessary because these devices are specifically designed for patients who are paralyzed and hence, they need a device that specifically caters to their communication. Many systems have used cognitive analysis but, in those systems, it is difficult to assess all the stimuli, when EOG is used for this purpose, much data is necessary for training. OpenCV is used as its processing speed is more compare to other methods. Count non-zero thresholding is used to select the letters.

The next aspect to be considered is text to speech conversion. In the existing systems, Digital Signal Processing (DSP) techniques are used to convert the identified text data to speech/audio data. Also, there are systems that make use of Natural Language Processing (NLP) to identify the text as well. Also, Text to Speech Synthesizer (TTS) is also used to convert text to speech. In proposed system, gTTs is used to convert text into speech. It is a simple CLI interface tool which is used to convert text to speech.

17.5 Software Requirements

Python: It is a high-level language. The entire system is based on image processing, and this is implemented using python.

OpenCV: It is a library of programming functions. These functions are mainly aimed at real-time computer vision. OpenCV is used to process the image that appears in the webcam.

Pyglet: It is a multimedia library for python. This is used to get better clarity in visuals.

NumPy: It is a python programming language library. This provides additional support for multi-dimensional arrays and matrices. It allows user to write fast programs whenever most of the operations work on arrays or matrices instead of scalars.

Dlib: It is a library written in C++ programming language. This library is used to detect the facial landmarks in order to detect eyes.

gTTs: It is a CLI tool interface and also a python library with Google Translates text to speech API. This library is used to convert the word or sentence in a text format to speech.

17.6 Proposed Work

The proposed system is explained in this section. The eye movements are taken as input through a camera and the corresponding alphabets are selected by deciding the grid location using image processing. This process continues till the sentence is completed and is stopped once completed. The detected sentence or phrase is then converted to speech audio. The speech audio is the output.

17.6.1 Detection of Facial Landmarks

In order to detect eyes, initially face has to be recognized.

17.6.2 Display Grid

A display grid that contains letters to be selected is prepared accordingly.

17.6.3 Selection of Letters

The required letters are selected. Count non-zero thresholding method is used for this purpose.

17.6.4 Text to Speech Conversion

A word or a sentence is formed which is then converted into speech.

17.7 Implementation

The eye movements are taken as input through a camera and the corresponding alphabets are selected by deciding the grid location using image processing. This process continues till the sentence is completed and is stopped once completed. The detected sentence or phrase is then converted to speech audio. The speech audio is the output. Figure 17.1 shows the block diagram of the proposed system (Fig. 17.2).

In order to move forward to process the image, we need to first find eyes. In order to find eyes, we need to find a face first. The facial key point detector takes a rectangular object of Dlib module as input, which is nothing but coordinates of the face. In order to find faces, we can use Dlib, which is an inbuilt frontal face detector.

Initially we will be using the "dlib. get_frontal_face_detector()" to detect the four corner coordinates of the face. It is basically a pre-trained model which makes use of HOG and Linear SVM techniques. HOG (Histogram of oriented Gradients) is used

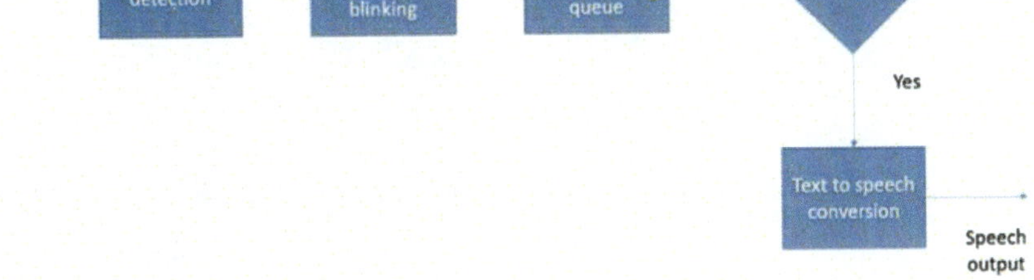

Fig. 17.1 Block diagram of proposed system

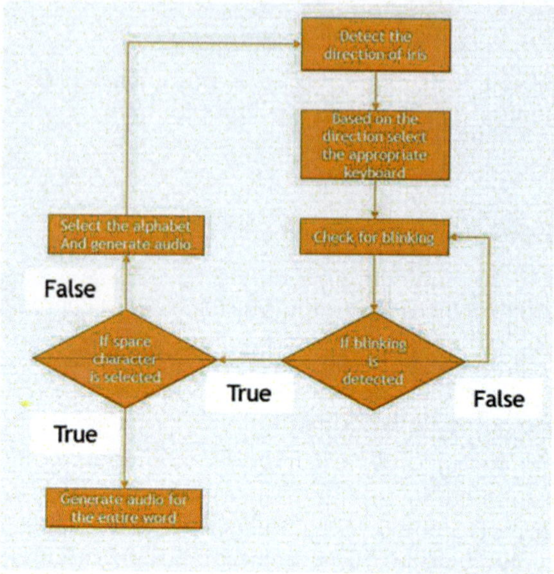

Fig. 17.2 Flow chart of the proposed system

primarily as an object detector. It provides information about pixel orientation and direction. The Linear SVM method is used to differentiate and group similar objects together. The dlib provides both of these methods combined for face detection. The image shown below is obtained by joining the coordinates obtained from the function (Fig. 17.3).

The next task is to detect the eyes. For this we will be making use of the trained model file, "shape_predictor_68_face_landmarks.dat", and pass it to dlib.shape_predictor() function. This will return 68 points on the face as shown

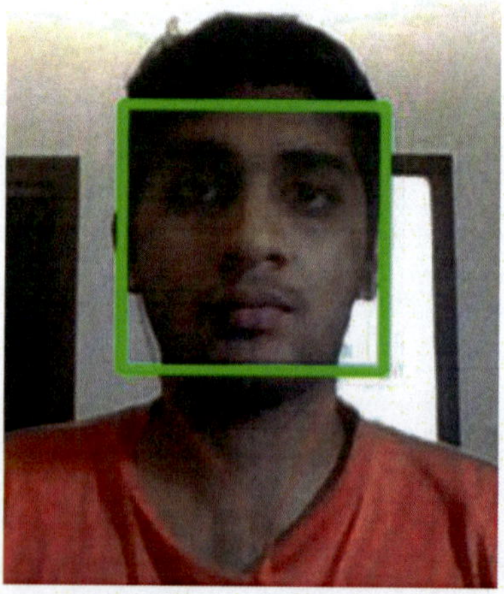

Fig. 17.3 Face detection

below. We are interested in points 36–47 as these points are around the eyes. In this way we are able to isolate the eyes from the face (Fig. 17.4).

The grid used by the users consists of letters. The main menu has the option to select either left or right side of the display. The letters are divided equally on the left- and right-hand sides of the display. There is a back button on each side of the display which helps the user to go back to the main menu if any wrong side is selected. The right side of the display has an additional space frame that is used for two purposes. One is to provide space and another one is to play audio of the word or sentence whenever it is encountered (Figs. 17.5, 17.6 and 17.7).

Using a camera, the eye movements of the person is observed and the grid location is selected based on the position of the iris. User can select any side based on the letter needed by moving their Iris either to left or right. Based on the white space in the eye, the side would be detected using countNonZero() function. Basically, the entire eye is divided into two regions, left and right. countNonZero() is applied to both the regions. White is represented by one and Black is represented by zero. The function counts the number of ones and returns the value. Now the total number of ones in both sides is compared. The side where there are less number of ones is the side where the iris is and vice versa. Hence the gaze can be determined using

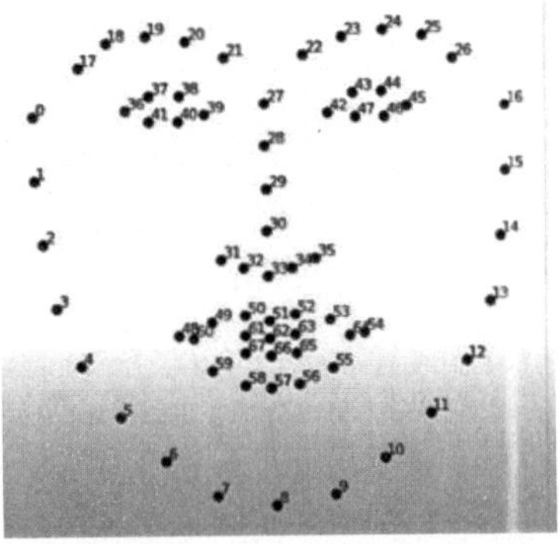

Fig. 17.4 Facial landmarks

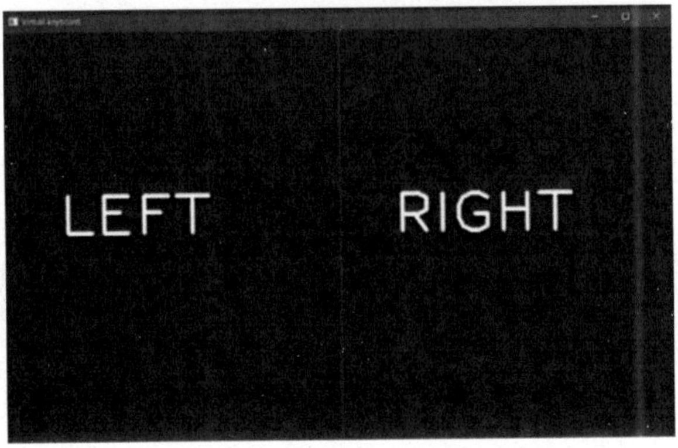

Fig. 17.5 Main menu of the display

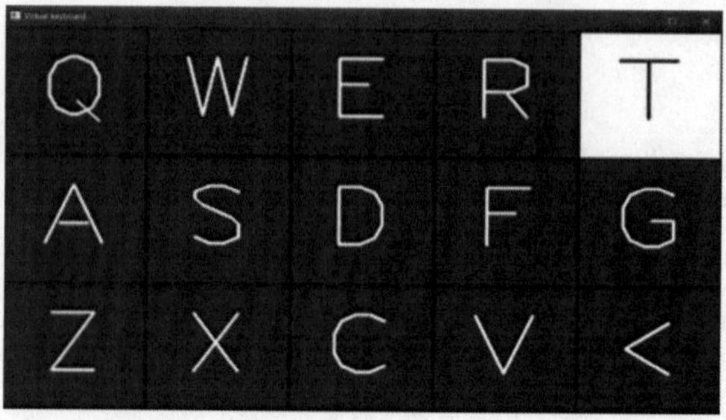

Fig. 17.6 Letters on the left-hand side of the display

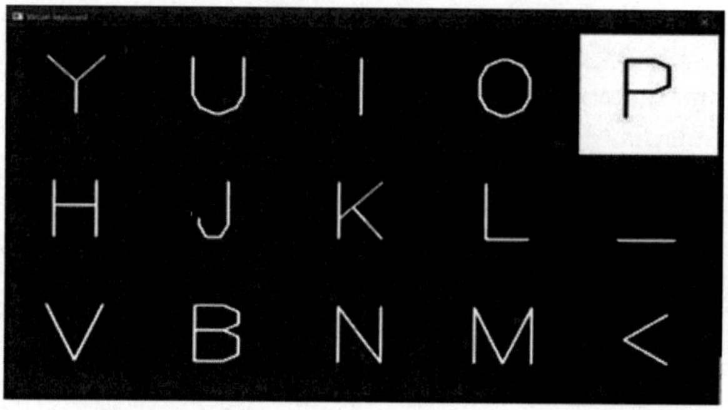

Fig. 17.7 Letters on the right-hand side of the display

this method. Once the side is selected, user can select the letters by closing the eyes when the blink is at a particular letter. Once the space is encountered, the audio of the word will be played and the same procedure is to be followed to form a sentence (Fig. 17.8).

Each time a letter is selected by the user, audio of the letter selected is produced. Once the selection of letters is completed, space should be selected, as soon as the space is encountered, audio of the entire word is produced. Same procedure is to be followed to form a sentence. Once the user has completed the sentence, it

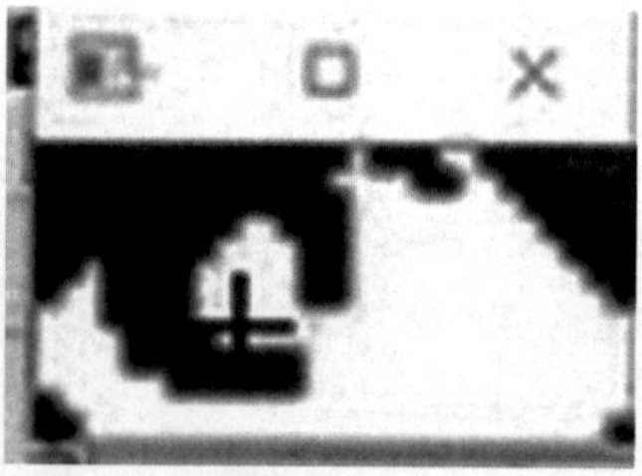

Fig. 17.8 Thresholding (finding the location of iris)

Fig. 17.9 Example output–1

is converted into complete audio data and is produced through the speakers. This process is achieved using gTTs library.

gTTs (Google Text to speech) is a python library with gTTs API, which converts the text to speech. Here the text variable is nothing but a string that is used to store user's input. We have created a thread which is called each time when the space character is encountered. We will be passing the text to the gTTs function and this will be running parallel with the main process.

17.8 Results and Discussion

Various letters, words and sentences are tested on the system. Initially, only letters were tested after which words were tested. Upon successfully testing words, sentences of different complexity were tested. The desired output of getting the audio for each alphabet, then word and finally the entire sentence was observed (Figs. 17.9 and 17.10).

The time taken for generating the desired output was measured with various inputs and is tabulated as shown below (Table 17.2).

17.9 Novelty of Proposed Work

As mentioned earlier, the communication system needed for the patients must be comfortable, efficient and affordable. Here, only two eye movements, i.e., left and right, and blink are needed to identify letters. This makes it much more convenient for the patient to choose letters. The main purpose of this project is to devise a

Fig. 17.10 Example output–2

Table 17.2 Time analysis for different inputs

Complexity	Time taken
Alphabet	5–10 s
Word	1–2 min
Phrase	2–4 min
Sentence (five to ten words)	4–7 min

methodology through which eye movements can be converted into text and hence into speech. The project largely depends on techniques of digital image processing. The language that has been chosen to implement this is python as it is highly compatible with image processing techniques and supports various libraries such as OpenCV. The detected text is then converted to speech. This is implemented using the google text to speech conversion library.

17.10 Conclusion

Human beings are naturally enabled with the ability to be self-productive and to communicate. Unfortunately, some people do not have the ability to perform these basic tasks due to some diseases. ALS is one such disease, in which a person can lose complete control over their movements. In the proposed system, alphabets are detected based on the patient's eye movements and are used to form a word and then it is converted into speech. Hence the patient can communicate by himself with the world. OpenCV, Dlib, gTTs and many other libraries are used to implement this. The future work is, adding backspace, minimizing the time taken to select letters to form words and sentences.

Acknowledgements The authors of the paper acknowledge and thank our Institution JSS Science and Technology University (Mysore), all the faculty and supporting staff of the Institution for the constant support.

References

1. Pai, S., Bhardwaj, A.: Eye gesture based communication for people with motor disabilities in developing nations. In: IJCNN 2019. International Joint Conference on Neural Networks. Budapest, Hungary 14–19 July 2019
2. Zhang, X., Kulkarni, H., Morris, M.R.: Smartphone-based gaze gesture communication for people with motor disabilities. In: ACM Press the 2017 CHI Conference-Denver, Colorado, USA (2017.05.06–2017.05.11), 6–11 May 2017
3. Fiannaca, A., Paradiso, A., Shah, M., Morris, M.R.: AACrobat: using mobile devices to lower communication barriers and provide autonomy with gaze-based AAC, CSCW 2017, Portland, OR, USA, February 25–March 1, 2017
4. Tambe Samreen, M., Rajeshwari, P.: Review on smart eye blink solution for MND patient using Python. IJSDR **4**(3) (2019)
5. Hutchinson, T.E., Jr. White, K.P., Senior Member, IEEE, Martin, W.N., Reichert, K.C., Frey, L.A.: Human-computer interaction using eye-gaze input. IEEE Trans. Syst. Man Cybern. **19**(6) (November/December 1989)
6. Pimplaskar, D., Nagmode, M.S., Borkar, A.: Real time eye blinking detection and tracking using OpenCV. Int. J. Eng. Res. Appl. **3**(5), 1780–1787, (Sept–Oct 2013). ISSN: 2248-9622
7. Utaminingrum, F., Primaswara, R., Sari, Y.A.: Image processing for rapidly eye detection based on robust haar sliding window. Int. J. Electr. Comput. Eng. (IJECE) **7**(2) (April 2017)
8. Venkateswarlu, S., Kamesh, D.B.K., Sastry, J.K.R., Rani, R.: Text to speech conversion. Indian J. Sci. Technol. **9**(38) (October 2016). https://doi.org/10.17485/ijst/2016/v9i38/102967
9. Isewon, I., Oyelade, J., Oladipupo, O.: Design and implementation of text to speech conversion for visually impaired people. Int. J. Appl. Inf. Syst. (IJAIS), Foundation of Computer Science FCS, New York, USA, **7**(2) (April 2014). ISSN: 2249-0868
10. Lim, J.Z., Mountstephens, J., Teo, J.: Emotion recognition using eye-tracking: taxonomy, review and current challenges. Sensors **20**(8), 2384 (2020). https://doi.org/10.3390/s20082384. (22 April 2020)

Chapter 18
A Smart Driver Assistance System for Accident Prevention

Tarush Singh, Faaiza Sheikh, Ashish Sharma, Rahul Pandya, and Arpit Singh

Abstract The proposed method comprises a computer vision-based system highly applicable in the detection of a possible accident while driving. A stereo depth camera is placed in front of the vehicle at a point through which all crucial viewing angles are captured. The camera will be covering a viewing angle of 180°. This idea proposes an efficient framework for accurate object detection using CNN along with an object tracking algorithm. This system will determine the probability of an accident based on the three parameters, which are: object approaching towards the line of movement, relative distance between the object and vehicle, and relative speed of the object with respect to the vehicle. It will provide a reliable technique to attain an efficient rate of detection with less erroneous indication of an accident. Since it covers 180°, it can accurately detect any object/vehicle coming from either side. By judging the speed of the vehicle and the trajectory of the object, the system will calculate the expected time of collision and in order to avoid the impact, an automatic braking mechanism will take place to cautiously reduce the speed of the vehicle in a calculated manner. In conclusion, this computer vision-based system will be a game-changer in accident detection and will make driving safer, easier, and more efficient.

18.1 Introduction

With the evolution in the fields of automobiles and technology, the advancements in the areas of accident prevention are limited. All the vehicles, majorly four-wheelers, are highly equipped with safety and precautionary measures for its driver and the co-passengers. The automobile units undergo several safety tests, drop tests, and many more. There is also an introduction of airbags, seat belts, suspension units, and hydraulic brakes. However, these are just limited to the safety of people inside the vehicle. There is a significant probability that even with these parameters, the accident shall be prevented and the lives are not risked.

T. Singh (✉) · F. Sheikh · A. Sharma · R. Pandya · A. Singh
Manjara Charitable Trust's Rajiv Gandhi Institute of Technology, Mumbai, Maharashtra 400053, India
e-mail: lncs@springer.com

J. I.-Z. Chen et al. (eds.), *Machine Learning and Autonomous Systems*, Smart Innovation, Systems and Technologies 269, https://doi.org/10.1007/978-981-16-7996-4_18

Over the decades, many safety precautions have been introduced in various fields to ensure safe driving. They include dual-front airbags, rear parking sensors, seat belt pretensioners, speed-sensing door locks, reinforced B-pillar, and anti-lock braking system (ABS) with paired Electronic Brake-force Distribution (EBD). These are the most important and fundamental safety measures a vehicle must-have. The latest technology that is currently being introduced is the auto-pilot system. Despite these measures being implemented, there is no significant reduction in the number of road accidents per annum. As per the report metrics, there are approximately 1.35 million fatalities each year with the majority coming from the driver's error and lack of responsibility [1]. This gives us the idea that there might be errors regarding road safety measures (road signs, potholes, speed breakers, sharp turns, an inclination of road height, etc.), public inattentiveness while crossing roads, and many more such factors. Yet the majority is caused by the negligence of the driver itself. These are inclusive of alertness and overspeeding among general irresponsibility.

Therefore, to increase road and vehicular safety, special designs and special adaptive, methodological, and smart technologies need to be implemented to ensure the focus of the driver to minimize the damage caused by accidents both inside and outside the vehicle. There have been introductions of several technical advancements in cars, like auto-pilot systems in cars designed by Tesla, automated sensors at the rear and front of vehicles for parking systems, etc. This paper proposes a similar smart and effective approach that could be included in a vehicle, which aims to prevent and hence reduce the number of accidents. The proposed system aims to detect the likelihood of collisions between the vehicle and obstacle and to take actions accordingly to prevent accidents. This can be done by the following steps:

- Increasing the field of vision by placing a wide-angle stereo camera at the front of the vehicle.
- Tracking and classifying objects and calculating their respective distances and relative speeds and hence identifying possible collisions.
- Applying breaks in a stage-wise manner by implementing automated advanced braking systems.

18.2 Literature Survey

18.2.1 Related Work

The motivation behind the proposed work is to make the driving experience more convenient and safe for the driver. since India is one of the busiest countries in terms of road traffic. In 2019 alone, the figure for road accidents across India adds up to 4,49,000. Every year, about three to five percent of the GDP of India was used up in road accidents. The country has about one percent of the global vehicle population but it accounted for six percent of the world's road traffic accidents. India observed over 1,37,000 fatal road accidents across the country in 2019. Road accidents are a major concern across the Indian subcontinent. In 2019 alone, due to road accidents, the country reported over 151,000 fatalities. Therefore, the proposed system can be applied in vehicles and it can help prevent a number of road accidents and the damages caused by vehicle-vehicle accidents.

In paper [2], the author proposes a braking system, assisted by computer vision. The computer vision system aided by a camera placed at the bumper of the car detects the lane of the car and calculates the amount of error to give a reference angle for the braking system. The paper suggests placing an ultrasonic camera at the front of the car to detect obstacles in the vehicle's path. In addition, the model estimates the angular relation between the steering wheel and the tires using gravitational sensors. This is done to effectively steer the car away from an obstacle if applying brakes is not safe.

In this paper [3] the author proposes a vision-based system along with an automatic braking system. Using image and video processing they have made an attempt to detect any pedestrians or vehicles. "Braking is done automatically when the driver gets distracted or when his visibility is poor. Automatic braking is applied by the output of algorithms using decision logic. The system is also designed for easy future augmentation for, e.g., night driving conditions. This system shows improved robustness with respect to the algorithms that are fast enough to be suitable for on-line operation."

Saito et al.'s [4] paper describes a method for velocity detection of vehicles using stereo cameras and describes an equation for the same. Furthermore, they derive an equation to detect vehicle speed in bad weather, or when one of the cameras is obscured using various parameters like stereo reliability, stereo weight, normalize parameter, and bias gain parameter.

Yang et al.'s [5] paper proposes a method to estimate the speed of vehicles using binocular stereovision. Their system does stereo matching on license plates only, as it is a part of any vehicle and is easier to detect. The speed can be measured by taking the ratio of distance between two selected 3D points by frame intervals. "Vehicle trajectory is depicted by connecting multiple selected 3D points."

Tourani et al. [6] put forward "Motion-based vehicle speed measurement for intelligent transportation systems" and this paper proposes a novel stereovision-based system for speed measurement. "The important feature of this system is the use of an optimized single shot multibox detector network that can efficiently detect license plates in the captured two view stereo videos."

18.2.2 *Stereo Depth Vision*

We have used a stereo depth-enabled camera for distance measurement. It consists of two image sensors separated on the same baseline at a distance L. In this system, the images are in the same plane. In stereovision, the concept of parallax plays a significant role it is defined as "Parallax is a displacement or difference in the apparent position of an object viewed along two different lines of sight and is measured by the angle or semi-angle of inclination between those two lines" [7].

This is the main idea behind the functioning of stereo cameras. The distance D between the object and the camera is a function of the disparity d which is calculated by a microcontroller unit, baseline distance L, focal length f, and sensor pitch w [4].

$$D = \frac{L \times f}{w \times d} \tag{18.1}$$

While there are multiple methods to calculate the distance from an object, the calculation of velocity has remained the same for a while. The widespread method to calculate speed is the ratio of difference per unit time, but this will not be accurate in adverse weather conditions.

One method to calculate the relative velocity of the object from the vehicle can be given by the formula. The result will be relative because the distance is calculated from the vehicle.

$$V = \frac{\Delta D}{\Delta t} \tag{18.2}$$

An alert will be sent when the relative velocity is a high positive value, or the distance is small but the velocity is significant.

18.2.3 Types of Braking System

The braking systems that exist today include a number of mechanical as well as autonomous braking systems that are highly efficient but have their own limitations. One can categorize them majorly into different models. Firstly, service brake or foot-oriented braking involves the driver's manual input; however it completely relies on the driver's attention. In case of failure, one can apply hand brakes, however, this does not overcome the limitation of reliance on the driver's reflexes. Moreover, hand brakes are too sudden, creating an abrupt jerk, and they may in turn damage the vehicle or harm the passengers. An Automatic Emergency Braking System (AEB) scans the road ahead for obstacles and warns the driver of possible collisions by employing different signals like lights and sounds. If the driver does not respond to these warnings by slowing down the vehicle or steering away from the object, a sudden automatic brake is applied. However, this still endangers. The complete safety of the vehicle and the passengers. Our proposed model employs the benefits of a number of braking systems while eliminating each one's limitations. Cascaded braking aims to prevent sudden jerks caused by AEB. If the vehicle is in imminent danger and the driver doesn't respond to the warnings, it smoothly applies brakes in a stage-wise manner, slowly reducing the speed of the vehicle to avoid any casualties.

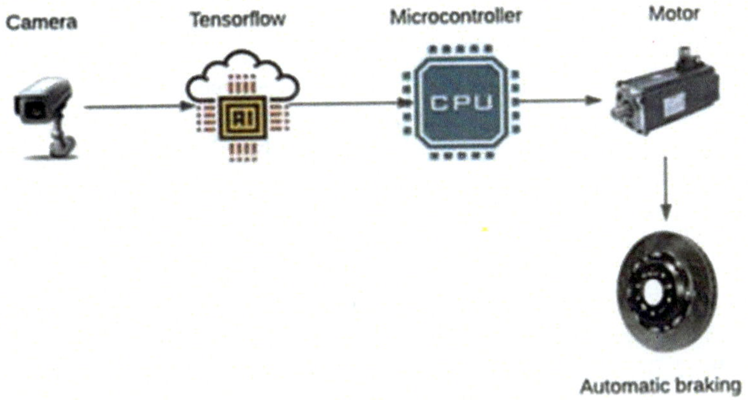

Fig. 18.1 Flow of working [3]

18.3 Proposed System

18.3.1 Methodology

This paper proposes a vision-based system in which a high-resolution stereo depth camera is placed at a point at the front of the vehicle. This camera will be covering a certain viewing angle necessary to plot any threat along the line of movement. By using TensorFlow's Object Detection API we were able to detect any vehicle within the visible range of the camera. If a vehicle is in the line of movement, our system will iteratively calculate its speed and distance, and when the likelihood of a collision crosses a certain threshold value X, our system will show a WARNING alert to the driver and will automatically apply brakes in a stage-wise manner to avoid sudden jerks.

Using the microcontroller and the Braking system the speed of the vehicle will be gradually decreased. As shown in Fig. 18.1 a way of using our system, the collision could be prevented.

18.3.2 Algorithm and Approach

The algorithm consists of four main functions. Detection for identifying the vehicles based on the size of the Boxes, calculating distance from the obstacle, calculating relative velocity, tracking for the progression, and decision for application of brakes. The working is represented in the form of a flowchart in Fig. 18.1.

For object detection the actual model used was/frozen_inference_graph.pb. Using this model different vehicles were detected. After loading this model into the system memory, it will start identifying the objects based on the size of the array. And will further iterate the boxes for supplementary analysis [3] (Fig. 18.2).

By classification we can distinguish different Boxes according to the size of the array, for example, a car will be classes [0] [i] == 3 and a bus will be classes [0] [i] = = 6. Also, when the car will be closer it will appear bigger yet it will not give an erroneous identification as the distance will be constantly calculated by the stereo camera and thus the classification will be nearly accurate.

Now the most important part is the tracking of objects. If an object is approaching too close to the vehicle there is a chance of collision. To avoid the accident, a threshold is calculated. This threshold is the ratio of distance with respect to relative velocity, and it should be greater than a certain value X. If the threshold value drops below a certain value X, the chance of accident will be more than 80% and the system will alert the driver. Now, one can argue that there will be a delay in the reaction time

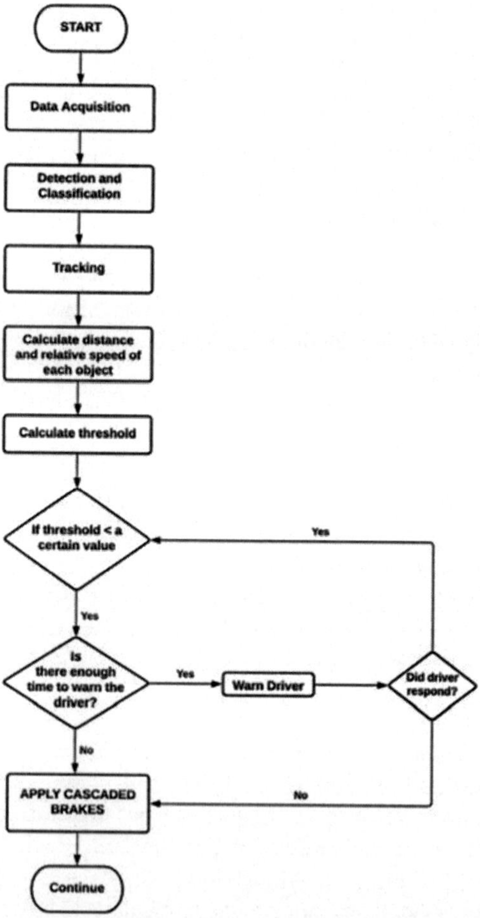

Fig. 18.2 Flowchart of proposed system

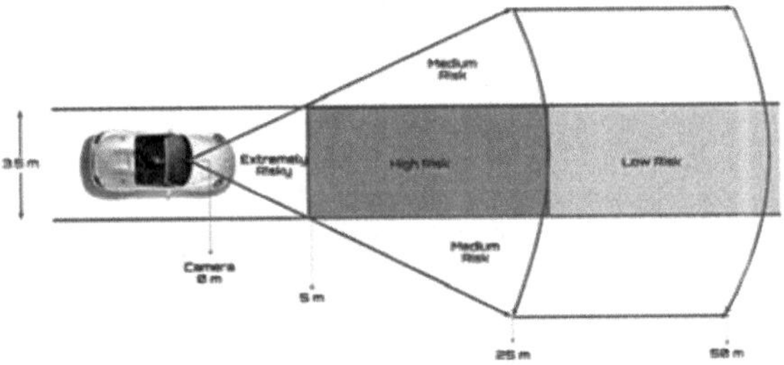

Fig. 18.3 Cascaded braking with respect to distance [3]

of the driver to take any actions to avoid the collision. This delay in reaction time (T_{REACT}) combined with velocity/deceleration (V/A_{DRIVER}) of the vehicle will give the T_{FCW} (Forward collision time). This can be represented as:

$$T_{FCW} = T_{REACT} + T_{STOP} = T_{REACT} + \frac{V}{ADRIVER} \tag{18.3}$$

This delay in the driver's reaction can lead to an accident. Therefore, our automatic braking system comes into picture. If the driver is taking too long to respond to the warnings, our system sends a signal to the microcontroller. The microcontroller is programmed in such a way that the input signal from the detection algorithm and the distance from the stereovision camera is compared with certain threshold values. If the distance is less than the threshold distance the microcontroller sends an output signal to apply Cascaded brakes as shown in Fig. 18.3.

The figure illustrates the typical risky areas to be tackled by an Automatic braking system. The main motive is determination of the distance parameter or the range of the object accordingly with the speed of the vehicle. This can be categorized into four zones of collision:

- Low Risk: This region lies in the range of 25 m or more from the driver unit. The Low-Risk region is the farthest region from the driving unit. The vehicle can be accelerated or decelerated very smoothly when the object or obstacle lies in this region depending on the user as well as relative speed of both vehicles in case the other is moving. It may also contain pedestrians crossing the streets with no danger of imminent collision but still need to be detected by the system to decelerate or change the path accordingly as they are in the line of path of motion.
- Medium Risk: This region is an arc region lying in the range of 5–25 m on either side of the line of motion as shown in Fig. 18.3. Pedestrians can cross streets even in the medium-risk zone with a slightly higher risk as compared to the Low-Risk region. Still, the system needs to infer and detect the obstacle approaching its line of motion and regulate the braking system accordingly.
- High Risk: This region is right in front of the driving unit ranging from 5 to 25 m. Since it is directly in the line of motion, any approaching vehicle or stationary object in this region carries a high risk of collision. This also depends on the speed of the driving unit and the rate of deceleration or acceleration. The detection needs

to be very precise and sharp in this region with the braking system coming into immediate effect or the driver changing the direction of the vehicle more quickly.

- Extreme: This region is the region lying in the range of 5 m from the driving unit in the line of motion. This region carries the highest risk of collision and can lead to severe damage to the pedestrian or the driver and may even be fatal. This still does not count when the vehicle is in traffic or moving at a very low speed. The automatic braking system should come into immediate effect in case the approach is relatively at a higher speed and the airbags should be blown to prevent the driver from getting subjected to injury from the sudden jerk from inertia.

In regular conditions, the vehicle stopping distance is about 5 m at 30 km/h, increasing up to 12 m at 50 km/h, thus the systems must intelligently focus their techniques on the danger of detecting a pedestrian in the above-mentioned regions of risk.

18.3.3 Working

The proposed system constantly calculates the likelihood of a collision by object detection API. The speed of the Ego vehicle is updated as an input to the system in real-time. This paper has presented the "Detection" algorithm, in which the moving vehicles in the frames are detected. This is done by the frozen model used in Tensorflow followed by the classification of the vehicles. In this model different vehicles of different sizes are detected by setting the boxes and classes in the program. The WARNING sign in the result section is an indication that the front vehicle is in the region of interest and there is a chance of collision. Even though automobiles are not interfering in the line of movement, that is they are in the medium-risk zone, yet it is in close range to the ego vehicle. In this case the detection algorithm shows the chance of collision in percentage anyway, if there is any sudden movement from either side towards the line of movement, the automatic braking system will get activated.

Using a tracking function, it provides continuity to the system. As the detection algorithm works one frame at a time thus, tracking gives constant processing of the frames. The detection function might not necessarily detect every information in a frame always and due to which the classification might not be able to classify the boxes on the vehicles at that moment. Therefore, a tracking function balances this out by giving it a robust continuity [8] (Fig. 18.4).

Furthermore, the decision function compares the threshold and the ratio of distance between the two vehicles to the deceleration of two vehicles. In this comparison if the value of X is greater than the threshold value, and the distance between the ego and front vehicle is less than 5 m, the braking system will get turned on. Our system typically has to do with Cascaded Braking which involves multi-stage braking in three stages. When the system shows the "Forward Collision Warning" alert to the driver that means the distance between ego and the front vehicle is reducing at a rate if not stopped will result in a collision. Therefore, to avoid this collision our braking

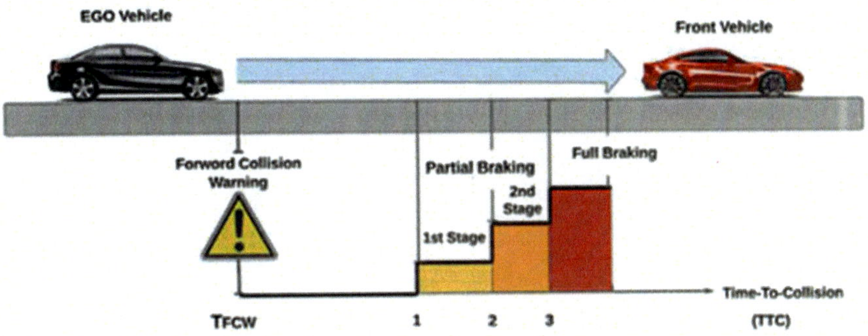

Fig. 18.4 Braking with respect to time to collision

system using its servo motor applies partial braking in two steps. If the distance is about 50 m at 60 km/h the system will apply partial break reducing the speed up to 20 km/h, if the distance is about 25 m at 45 km/h then it will move into stage two and will have to reduce the speed up to 10 km/h (Fig. 18.5).

This plot of Time to collision vs Stopping time indicates a correlation between time to collision and the stopping times for the Forward Collision Warning (FCW). PB1 time = time taken to reach the first stage of partial braking, PB2 time = time taken to reach the second stage and FB time = time at which full braking is applied [8] (Fig. 18.6).

This plot indicates the acceleration of the ego car. As you can observe the drop in the acceleration of the car decreased from 2.3 s and then decelerated till full brakes were applied. This decrease in the acceleration will act as a strong force in the prevention of the collision (Fig. 18.7).

In this plot the velocity of the ego car is shown. From the graph, it can be observed that the velocity of the vehicle decreases with a slope from around 2.2–2.3 s when the partial brakes were applied. Hence it follows the multi-stage braking and reduces

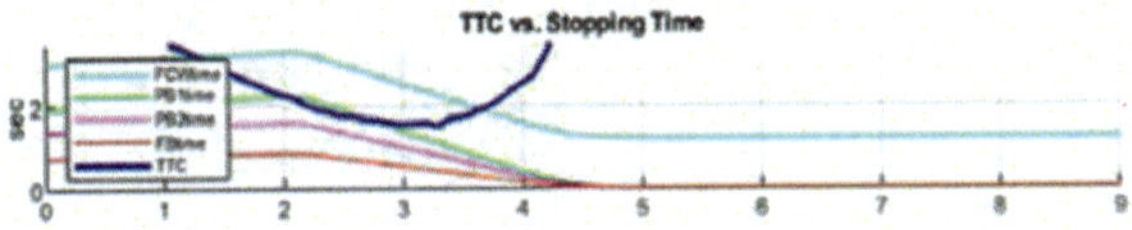

Fig. 18.5 TTC versus stopping time

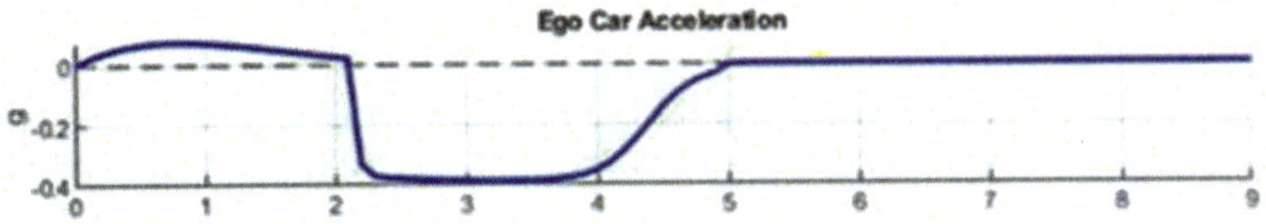

Fig. 18.6 Ego car acceleration

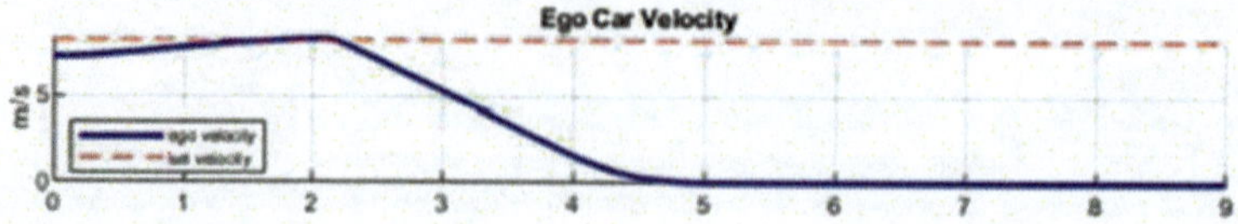

Fig. 18.7 Ego car velocity

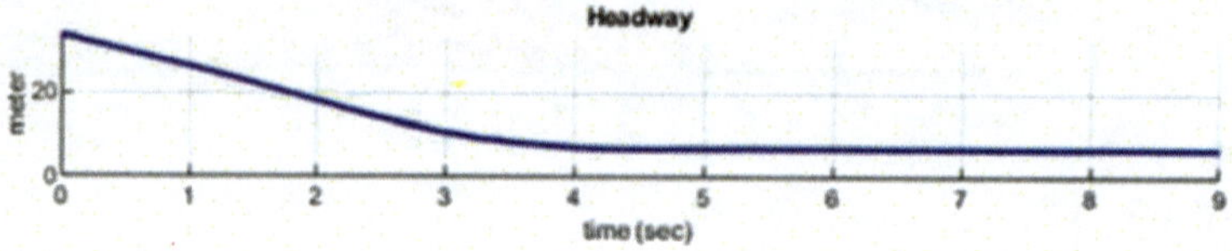

Fig. 18.8 Headway between two cars

the velocity of the vehicle in a gradually decreasing manner. However, if one would try to reduce the speed from a high speed to a very low speed, the driver may avoid the collision with the front vehicle but he/she might take some serious damage due to the sudden stopping of the vehicle. This damage can be harmful to the vehicle as well as to the driver [8] (Fig. 18.8).

Here, Headway is the time interval between the ego car and the front car moving in the same direction or same path. From these plots we can observe that for the first 2 s the vehicle's speed was at a set velocity. At 2.3 s our detected algorithm detects a very high chance of collision with the front object. Promptly after the detection, a WARNING alert is initiated.

After a fraction of a second at 2.4 s, the first stage of the partial brake is applied and the velocity of the vehicle is reduced, as the vehicle decelerate the second stage is applied at 2.5 s where the speed of the ego vehicle is further reduced to an extent that full brakes can be applied without any major harm to the driver.

Finally, when the vehicle slows down to a very low speed at 3.9 s. The distance between the ego vehicle and the detected object/front vehicle is 2.3 m. In summary, our system prevented full collision of the vehicles. These graphs were generated using MATLAB software.

18.4 Results

After the implementation of the above-stated smart driver assistance system, we were able to generate a fair accuracy in successfully predicting and detecting the vehicles, pedestrians, and obstacles. The features were implemented via a game and the model designed was successfully able to identify the target values accurately [9].

In Fig. 18.9, we can clearly observe that the moving cars in the other lanes are being detected with their percentage of nearness and threat based on distance. It can

Fig. 18.9 Detection of three cars and pedestrians

Fig. 18.10 Detection of vehicle in side lane

also be observed that the speed is checked via clock and the pedestrians are also identified in case of possible road crossing. While in Fig. 18.10 our system detected the vehicle while the vehicle was still approaching in the side lane signifying the field of vision of our system and detect the presence of a vehicle accurately (Fig. 18.11).

Fig. 18.11 Detection of vehicle in extreme range

Fig. 18.12 Warning sign in extreme region

The vehicle in Fig. 18.10 lies in the Extreme threat region while it is still passing by the standstill car of the driver unit. It detects that while moving away and generates a WARNING sign to alert the driver. Similarly, in Fig. 18.12 we can observe that the approaching vehicle though on a low speed when entering the Extreme threat region that is, within the range of 5 m, while moving towards the driving unit produces a WARNING sign.

However, in Fig. 18.13 it can be observed that the system produced a warning sign both for the vehicle in the line of motion of the driving unit and also for the vehicle that is in the medium-risk region as both of them lie within the threshold value of 25 m range.

The model while at a crossroad junction can easily identify the traffic lights indicator, the threats or possibility of mishap due to the vehicles in the same or the other lane and also the pedestrians that might be crossing the street.

It can be observed in Fig. 18.14 that our system has successfully detected the presence of multiple units at the very same instantaneous time while the driving unit is still approaching them at a speed of approximately 60 kmph [9].

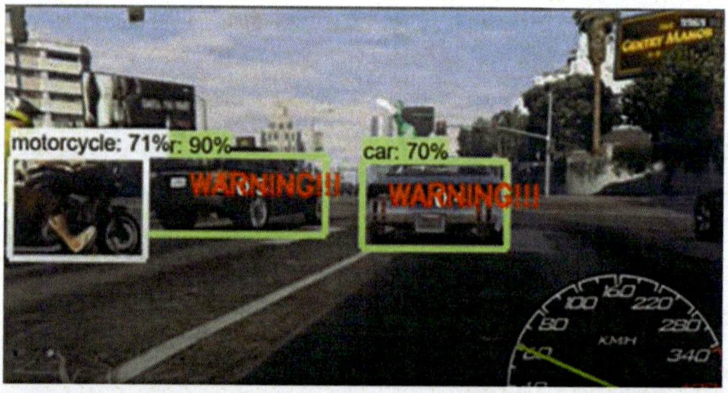

Fig. 18.13 Warning sign in high-risk region

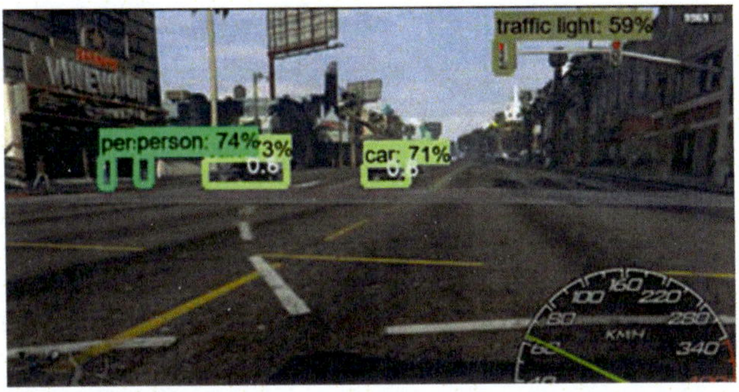

Fig. 18.14 Detection of traffic lights and other objects

In comparison to other existing models cited in this paper, our model works on stereo depth vision giving it an extra edge over the existing models. Since other models employ ultrasonic sensors, which is efficient for short-range detection but will not be efficient to prevent collisions in case of vehicles fast approaching from a distance. In addition, ultrasonic sensors calculate only distance between vehicles and hence will raise false alarms to the driver. Our system overcomes this by calculating relative speed between the vehicle and the obstacle and only raising an alarm when the driver's attention is crucial.

In regards to object detection, existing models majorly detect license plates. Our model detects various other objects like traffic lights, pedestrians, and based on the size of the vehicle it classifies the type or class of the vehicle. Our model not only prevents collisions between two vehicles but also between ego vehicles and the objects stated above.

18.5 Conclusion

In this study, we proposed a system for Automatic Emergency braking. Previous models of automatic braking detect the proximity of an obstacle and apply sudden brakes if an obstacle is closer than a certain distance. These models may not be entirely effective as they would warn the driver even if the vehicle and obstacle are stationary. In addition, these models apply very sudden brakes putting the vehicle and the passengers in potential danger. Our model overcomes both these limitations, by calculating a certain threshold value instead of the distance between vehicle and obstacle. The threshold is a function of relative velocity and distance between the vehicle and the obstacle, and if it drops below a certain value, the driver is warned by flashing dim lights or sounds. If the driver does not respond to the signals, automatic cascaded brakes are applied. The safety of the vehicle and the passengers is significantly improved as compared to the previous systems with the introduction of cascaded brakes, due to the absence of sudden jerks. The risk of passengers in other vehicles and pedestrians is also reduced. But this also depends on the up-gradation of the road safety norms and people to follow it with discipline.

However, our system may not guarantee total prevention of accidents as there are possible chances of personal lapses and system failures. This project is to a certain level inclusive of the driver's alertness and may not guarantee proper results during bad weather when vision is compromised. In the future, we aim to explore other methods of calculating relative speed and improve this model for bad weather. Moreover, this model can be improved by integrating stereovision with ultrasonic sensors to accurately calculate proximity for obstacles that are too close to the vehicle.

18.6 Future Scope and Applications

The proposed collective system can not only be implemented in self-driving cars but also in trucks and other heavy motor vehicles. This system when installed and configured to operate in multiple vehicles can generate enhanced results. It can be also used to generate the metric data of possible threats due to other factors like weather conditions and varied terrains, if the system is updated with further advancements.

Configuration with Traffic Control Systems: Upon configuring the installed vehicular to interact with the traffic control systems, the system can warn the driver regarding the upcoming signal and duration to it and also the traffic present in lane. This will significantly reduce signal breaches and prevent crossroad collision accidents.

Edge Computing: Linking of the entire sector containing numerous networks to the cloud after pre-processing of the data in the fog layers and edge layers can significantly reduce the amount of data load on the systems and learning models with improvisation in prediction [10].

Minimal use of Traffic Personnel: The system can be configured with the control rooms of the traffic department to identify breaches in traffic rules easily. The traffic personnel would not have to roam around firing speed guns and chase vehicles. Rather the linking of the system will alert the control room at the instantaneous moment, track the vehicular moment and the guilty party can be questioned.

Pothole and Speed Breaker Detection: The vehicle system can be further developed by integrating ultrasonic sound waves to detect uneven roadways. Detecting potholes and speed breakers and taking some action beforehand can keep the ride smooth and even.

References

1. Road Accidents per Annum Worldwide. https://www.who.int/news-room/fact-sheets/detail/road-traffic-injuries#:~:text=Approximately%201.35%20million%20people%20die,of%20their%20gross%20domestic%20product

2. Joukhadar, A., Issa, H., Kalaji, Y.: Design and implementation of auto car driving system with collision avoidance. Cogent Eng **5**(1), 1485458 (2018). https://doi.org/10.1080/23311916.2018.1485458
3. Detection System Available at: Murugan, S., Bose, N (2013) Automatic braking system assisted by image & video processing for pedestrian and vehicle detection using matlab. Int. J. Inf. Technol. Comput. Sci. **2**, 361– 369
4. Saito, T., Okubo, T., Takahashi, N.: Robust and accurate object velocity detection by stereo camera for autonomous driving. In: 2020 IEEE Intelligent Vehicles Symposium (IV), pp. 1242–1248 (2020). https://doi.org/10.1109/IV47402.2020.9304742. http://www.ncbi.nlm.nih.gov
5. Yang, L., Li, M., Song, X., Xiong, Z., Hou, C., Qu, B.: Vehicle speed measurement based on binocular stereovision system. IEEE Access, 1–1 (2019). https://doi.org/10.1109/access.2019.2932120
6. Tourani, A., Shahbahrami, A., Akoushideh, A., Khazaee, S., Suen. C.Y.: Motion-based vehicle speed measurement for intelligent transportation systems. Int. J. Image, Graphics Signal Process **11**(4), 2019
7. To find more information on parallax visit the wikipedia page at https://en.wikipedia.org/wiki/Parallax
8. Autonomous Braking System. https://in.mathworks.com/help/driving/ug/autonomous-emergency-braking-with-sensor-fusion.html
9. Simulation of software in game. https://pythonprogramming.net/detecting-distances-self-driving-car/
10. Edge Computing. Kim, K., Hong, Y.: Autonomous network traffic control system based on intelligent edge computing. In: 2019 21st International Conference on Advanced Communication Technology (ICACT), pp. 164–167 (2019). https://doi.org/10.23919/ICACT.2019.8701939

Chapter 19
A Comparative Study of Algorithms for Intelligent Traffic Signal Control

Hrishit Chaudhuri, Vibha Masti, Vishruth Veerendranath, and S. Natarajan

Abstract In this paper, methods have been explored to effectively optimize traffic signal control to minimize waiting times and queue lengths, thereby increasing traffic flow. The traffic intersection was first defined as a Markov Decision Process, and a state representation, actions and rewards were chosen. Simulation of Urban MObility (SUMO) was used to simulate an intersection and then compare a Round Robin Scheduler, a Feedback Control mechanism and two Reinforcement Learning techniques—Deep Q-Network (DQN) and Advantage Actor-Critic (A2C), as the policy for the traffic signal in the simulation under different scenarios. Finally, the methods were tested on a simulation of a real-world intersection in Bengaluru, India.

19.1 Introduction

The objective of this paper is to provide a comparative analysis between four policies used to manage queue build-up in traffic signals. This work aims to identify the relative strengths and weaknesses of the current state-of-the-art Reinforcement Learning (henceforth, RL) approaches to this problem, against those of simpler non-RL-based approaches. This work is necessitated to determine if the computational intensity of RL algorithms is viable and justified. The two RL techniques chosen are the Deep Q-Network (DQN) and the Advantage Actor-Critic (A2C) techniques, compared against a Round Robin (RR) scheduler and a Feedback Control Mechanism (FCM). All traffic data used in these simulations have been randomly generated, according

Chaudhuri, Masti and Veerendranath: These authors contributed equally to this work.

H. Chaudhuri · V. Masti (✉) · V. Veerendranath · S. Natarajan
Department of Computer Science and Engineering, PES University, Bengaluru, India
e-mail: vibha@pesu.pes.edu

V. Veerendranath
e-mail: vishruth@pesu.pes.edu

S. Natarajan
e-mail: natarajan@pes.edu

J. I.-Z. Chen et al. (eds.), *Machine Learning and Autonomous Systems*, Smart Innovation, Systems and Technologies 269, https://doi.org/10.1007/978-981-16-7996-4_19

to a Weibull generation. Simulations have been performed using Eclipse's traffic simulator, Simulation of Urban MObility (SUMO).

19.2 Related Work

The first course of action was to identify the possible formalisms that could be adopted towards the problem. This is a well-known issue in tackling the signal-timing problem, and individual approaches taken by authors include Wey's Mixed-Integer Linear Programming (MILP) model [1], as well as the Tournament-Based Elimination Pairing system treatment utilized by Eriskin et al. [2]. However, the intuitive notion of each traffic signal behaving like a selfish agent intending to minimize the average waiting time of its lane allows us to formalize the intersection as a stochastic repeated game, an assumption that leads to the natural formulation of the problem as a Markov Decision Process (henceforth, MDP).

Existing research tends to differ on the fundamental definition of the MDP. In particular, authors have not settled on a universal definition for the state, action, probability, or reward spaces that comprise the MDP. As an example, Abdoos et al. [3] describe the environment in terms of the queue length (an approach also used by this project), whereas another popular description involves visualizing the environment in terms of the average waiting time [4]. Upon settling on a single model for experimentation, the question of choosing an appropriate learning algorithm becomes crucial. The simplest way to approach this is to use tabular methods such as Q-learning over a discretized state space, which employs a simple search-for-maximum-reward mechanism over each epoch.

Signal-timing algorithms have traditionally been divided into *fixed-time control*, where the signal-timing plan is deduced a priori, without any feedback from the relevant environment, and the availability of real-time data does not affect the mechanism in any way, and *actuated methods*, where the algorithm also deduces a posteriori, with a set of rules provided whose violations can lead to triggering an update in the working of the algorithm.

Actuated methods have also been studied in the context of wireless sensor networks. Pandit et al. [15] discussed the use of the Oldest Arrival First (OAF) algorithm, a derivative of the Oldest Job First (OJF) scheduling algorithm, and the usage of vehicular ad hoc networks (VANET) to implement adaptive traffic signal control. More recently, work by Hilmani et al. [14] has explored the use of wireless sensor networks to manage traffic signals in detail.

RL approaches have also been examined [5, 6, 13] with two crucial algorithms being benchmarked against the efficacies of a fixed-time control and an actuated mechanism. The first of these is a deep Q-network, based on the work by Vidali et al. [7], where a deep neural network is trained to learn the state-action function based on the generic Q-learning equation. The second is an Actor-Critic mechanism: The Advantage Actor-Critic (A2C), which stems from its asynchronous variant, Asynchronous Advantage Actor-Critic A3C [8].

Several methods for obtaining traffic data have also been explored, with three main approaches looked at: real-time traffic data from surveillance cameras [11], real-world data obtained from Google Maps and road cameras [12] and synthetic data generated according to a Weibull distribution [7]. The authors of this paper have used synthetic data generated according to a Weibull distribution for the experiments.

19.3 Experimental Setting

All the simulations involved in the undertaking of this experiment have been executed using the Eclipse Foundation's Simulation of Urban MObility (SUMO) traffic simulator [9]. Interfacing with the Python RL scripts has been performed using the Traffic Control Interface (TraCI), and data has been generated according to SUMO's internal data dumps.

The problem as defined by Vidali et al. [7, pp.44] is: '*given the state of the intersection, what is the traffic light phase that the agent should choose, selected from a fixed set of predefined actions, in order to maximize the reward and consequently optimize the traffic efficiency of the intersection.*'

Here, the state of the intersection defines the state space of the MDP, with varying definitions according to the algorithm chosen. The action space consists of choosing a particular arm of the intersection to signal green at any time. The reward function accordingly varies from algorithm to algorithm. The intersection used for experimentation is a four-way single-lane intersection with a single traffic light controller in the middle as shown in Fig. 19.1. The length of the road in each direction is fixed at 242.8 m. The state representation for the intersection (environment) is chosen suitably based on the signal-timing algorithm employed.

The action set (1) is the set of possible actions that the agent can take from any given state s_t of the environment. For our RL approaches, the action set (1) represents the possible green phases that the agent—the traffic signal—can activate.

$$A = \{S, E, N, W\} \tag{19.1}$$

Every action of set (1) belonging to A represents the incoming lane (arm) for which the green phase of the traffic light is activated. For instance, the action E represents

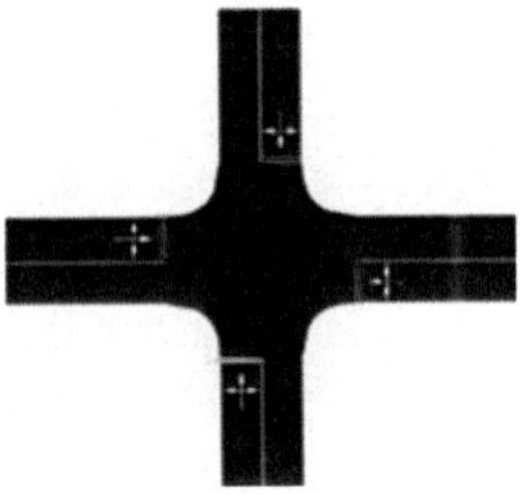

Fig. 19.1 Four-way single lane intersection (environment)

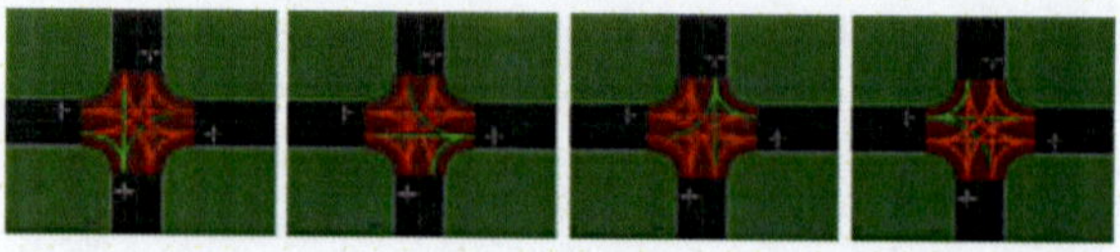

Fig. 19.2 The four green phase actions (*S, E, N and W*)

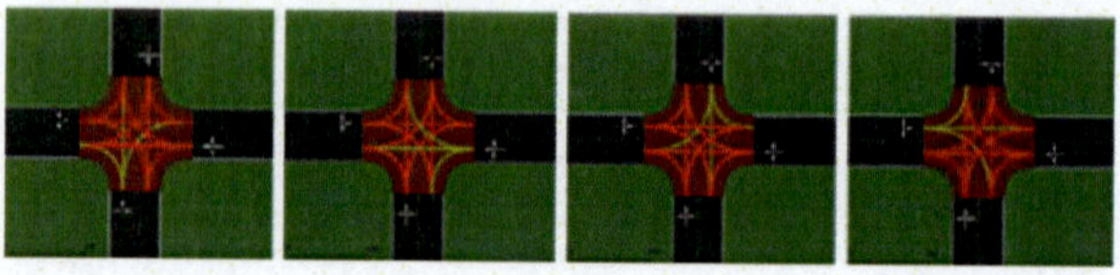

Fig. 19.3 The four intermediate yellow phase actions

the agent activating the green phase for the east (E2T) lane and deactivating the green phase for the remaining three incoming lanes—north (N2T), south (S2T) and west (W2T) (Fig. 19.2).

It must be noted that when the agent takes an action a_t and it differs from the previously taken action a_{t-1}, an intermediate action of yellow phase activation must be taken. For instance, if the previously taken action was E and the next action to be taken is N then an intermediate action E_{yellow} must be taken where the yellow phase for the incoming east lane (E2T) is activated (Fig. 19.3).

Traffic route files used for testing and training are generated with vehicular density per lane plotted according to a Weibull distribution, to ease loading over time. Three different traffic scenarios are tested for each signal scheduling algorithm, with the traffic conditions under each instance referred to by a specific SCENario (SCEN) number, as done by Vidali et al. [7]. In the first case, traffic is distributed uniformly between all arms (SCEN-1). In the second, the density of traffic on the N2T-S2T arms is heavier—90% of the vehicles travel on the N2T-S2T arms—(SCEN-2). Finally, a heavier traffic density is placed on the E2T-W2T arms (SCEN-3). A summary is shown in Table 19.1.

In the case of our Feedback Control approach, the action space is defined to be a range of numbers from min_time to max_time. Here, the order in which green phases are activated is fixed and only the green signal timing varies based on traffic density. In this approach, an intermediate yellow phase action is taken after every green phase action (in other words, 50% of the actions are yellow phase actions).

19.3.1 Round Robin Scheduling

In this method, each agent is assigned a fixed quantum of time during which its corresponding arm is signalled green. This quantum remains invariant under all traffic conditions, including empty arms. Each quantum consists of 30 timesteps of

Table 19.1 Traffic probability for a single car

Direction	Source	Destination	Probability in SCEN-1	Probability in SCEN-2	Probability in SCEN-3
Straight	North	South	0.1875	0.3375	0.0375
	South	North	0.1875	0.3375	0.0375
	East	West	0.1875	0.0375	0.3375
	West	East	0.1875	0.0375	0.3375
Turn (Left or Right)	North	East	0.03125	0.05625	0.00625
	North	West	0.03125	0.05625	0.00625
	South	West	0.03125	0.05625	0.00625
	South	East	0.03125	0.05625	0.00625
	East	North	0.03125	0.00625	0.05625
	East	South	0.03125	0.00625	0.05625
	West	South	0.03125	0.00625	0.05625
	West	North	0.03125	0.00625	0.05625

signalling green, followed by three timesteps of signalling yellow. No distinction is made between right-turning, straight-going and left-turning traffic. The experiment is repeated thrice, with each of the traffic scenarios as described above (SCEN-1, SCEN-2 and SCEN-3).

Round Robin (RR) was explored and compared against other methods in this paper since RR is the most prevalent traffic signalling algorithm used in most traffic intersections around the world. Using RR as a baseline enables us to effectively evaluate the relative performance improvements offered by other methods (Fig. 19.4).

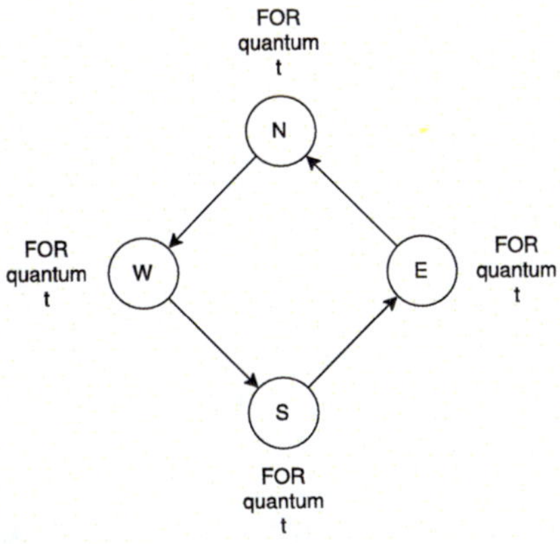

Fig. 19.4 Round Robin Scheduling for time quantum *t*

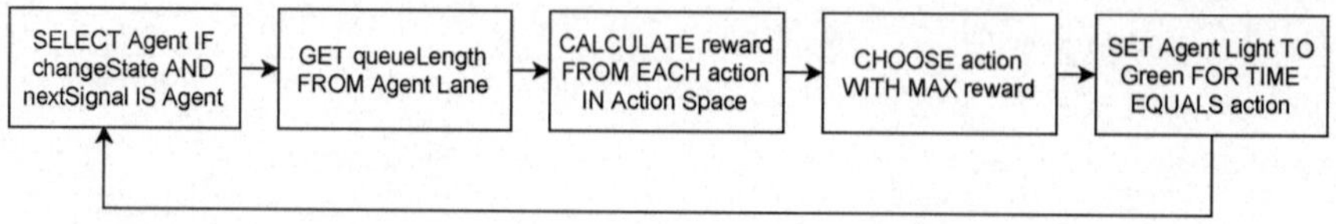

Fig. 19.5 Feedback control loop

19.3.2 Feedback Control Mechanism

This is a slight modification of RR. The primary idea of this approach is to enable the signal to perform some basic 'learning' by reading inputs from the environment and updating its actions based on those inputs. For this reason, the action space is selected as an array of discrete-time intervals which the traffic agent can select from. The action represents the length of time the agent signals green.

This model uses a game-playing strategy, where each agent is given a 'turn' in RR fashion, similar to players in most board games, such as Monopoly. For this reason, this mechanism will be referred to as MONOPOLY, henceforth. The MONOPOLY reward function searches overall actions in the agent's reward space and calculates the reward for each possible action. The reward function is defined in (19.2), where s_t represents the state of the corresponding lane at time t. Here, the state

$$R(s_t,\ a) = |\, s_t - v_t a \,| \tag{19.2}$$

of the lane only refers to the length of the queue at time t. v_t refers to the average speed of the cars in the junction at the last timestep, and a is the action currently being compared. The action generating the maximum value of R implies that the signal neither overshoots nor undershoots but stays green for exactly the time required to clear out the queues (Fig. 19.5).

19.3.3 Deep Q-Learning

In a reinforcement learning model, an agent is placed in an environment and is made to observe its state s_t. The agent then proceeds to choose an action a_t from a given set of actions according to a Markov Decision Process (MDP), transitioning the environment from the current state s_t to the next state s_{t+1}. Upon making the transition, the agent receives a reward r_{t+1} based on the previous action a_t. The goal of the agent is to choose a sequence of actions in the environment that maximizes the reward.

Q-learning is a reinforcement learning technique where at every step the value function (known as the Q-function) is updated following an equation (typically the Bellman Equation). In regular Q-learning, a table of Q-values $Q(s, a)$ is maintained to estimate the optimal Q-function $Q^*(s, a)$ and the entries of the table are updated as the agent learns. As the state space gets larger, maintaining a table becomes infeasible. Deep Q-learning makes use of a deep neural network to estimate the value of the optimal Q-function $Q^*(s, a)$. In this paper, the Q-function used for Deep Q-learning (3) is the one proposed by Vidali et al. [7].

$$Q(s_t, a_t) = r_{t+1} + \gamma \cdot max_A Q'(s_{t+1}, a_{t+1}) \tag{19.3}$$

$Q'(s_{t+1}, a_{t+1})$ represents the Q-value of taking the action a_{t+1} in the next state s_{t+1}, the term r_{t+1} represents the reward obtained by taking the action a_t from state s_t, the term $Q(s_t, a_t)$ represents the Q-value of the action a_t taken from the state s_t. The term max_A refers to the action a_t out of all possible actions taken in state s_{t+1} that maximizes the value of $Q'(s_{t+1}, a_{t+1})$. The value γ represents a discount factor between 0 and 1 that diminishes the weightage of future rewards and emphasizes the immediate reward. The value α represents a learning rate between 0 and 1 that updates the current Q-value with a discounted learned value. As the agent learns and updates the Q-table, it acts as a better estimate for the optimal Q-function $Q^*(s, a)$.

The state describes a snapshot of the environment (in this case, the four-way intersection serves as the environment) at a given timestep t and is represented as s_t. The state should give an accurate description of the number of cars in each lane at a given timestamp t. The state representation used is similar to the one described by Vidali et al. [7]. Each incoming road is divided into cells of varying size, where the cells are smaller and more concentrated closer to the traffic light and get larger as the distance from the traffic light increases (Fig. 19.6).

Each lane consists of nine cells leading to a total of 36 cells in the environment of four incoming lanes. The state can be represented as a boolean vector of length 36 where the value of each cell in the vector is one if a vehicle is present in the cell and zero if the cell is empty. Using this state representation, a total of 2^{36} states exist in the state space. Computationally, having a boolean state space vector is much simpler than simply having a state vector of length four, where each entry is the number of vehicles in a given lane.

The model used to train the agent for Deep Q-learning is a neural network of one input layer, five hidden dense layers and one output layer (Fig. 19.7). The input to the network is a vector of length 36 and the output from the network is a vector of

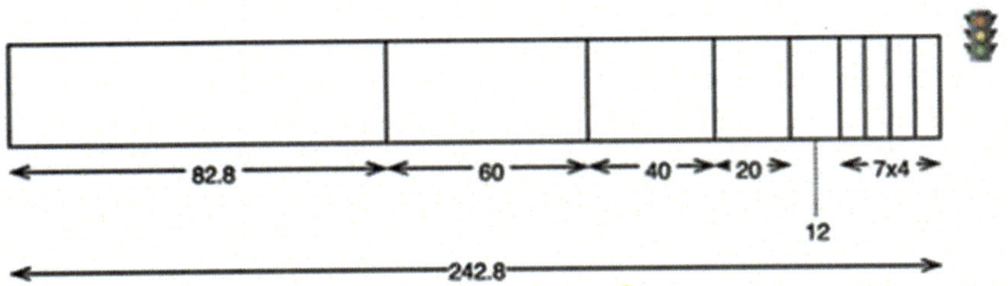

Fig. 19.6 State Representation (nine cells)

Fig. 19.7 Neural Network (DQN) Architecture

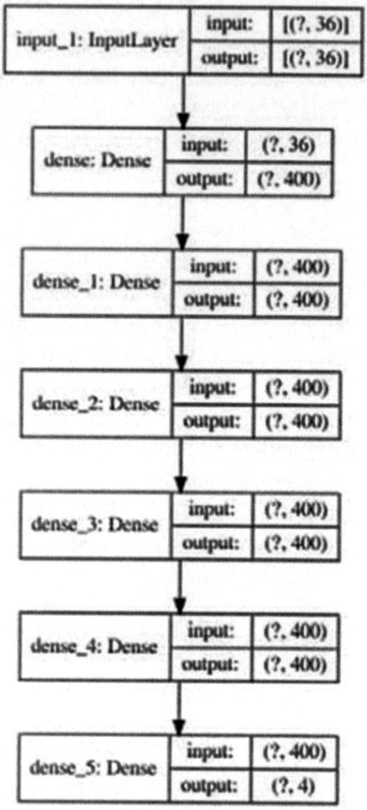

length four, representing the four possible actions that can be taken at every stage. The green phase duration for every arm is a multiple of five timesteps. At every step, the agent is trained to decide which arm of the signal to activate the green phase for.

The agent is trained for 100 episodes, alternating between the three traffic scenarios, and in each episode, the experience replay is sampled from 800 times. The value of γ is set at 0.75 and a learning rate of 0.001 is used.

19.3.4 Advantage Actor-Critic (A2C)

The Advantage Actor-Critic mechanism belongs to a family of RL algorithms collectively known as Actor-Critic algorithms. These algorithms offer a method to bifurcate the processes of learning and evaluating strategies or actions and their relative effects on the environment. Two separate processes, termed the 'actor' and 'critic' processes, are employed in this case. The actor method is responsible for *choosing* a particular strategy. The critic method then evaluates the effects of that action (by calculating

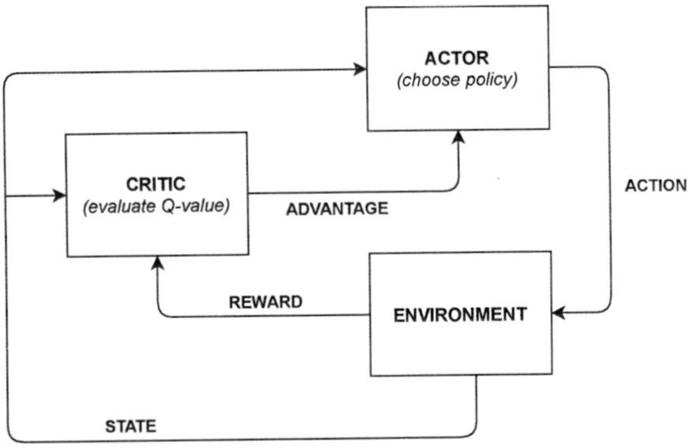

Fig. 19.8 A2C architecture

something akin to a Q-value in the DQN approach), measures the regret attached to that action and sends this feedback to the actor process.

The Advantage function in A2C is a modification to this Actor-Critic architecture making use of Temporal Difference Learning (TDL). In TDL, the actor now tries to *predict* rewards. It assumes that the current reward also includes some discounted future reward that it may achieve by moving into a particular *next state*. The critic's work is to calculate this next state reward valuation. An error, termed the temporal difference error, is now introduced as a parameter: it is the difference between the predicted value of all future rewards from the current state and the current value of the state, i.e. the extra reward gained by taking that particular action, over the average reward that would be gained from the other actions. This is the *advantage* of the state. The actor can now learn by measuring its advantage: if the advantage is greater than zero, it implies that the action leading to this advantage is good for the agent, and therefore the agent should continue choosing this action (Fig. 19.8).

The Actor-Critic method also provides an opportunity to incorporate parallelism into Deep Reinforcement Learning as described in Minh et al. [8]. This is achieved by using multiple agents that learn on their independent instances of the environment in parallel. An agent-environment pair is referred to as a worker. Since the environment for each agent is independent, the experiences of the agents differ, enabling the model to learn on a larger set of states in less time. This decreases training time significantly, also enabling the training of Deep Q-Learning networks on a CPU rather than a GPU.

A2C synchronously updates the global neural network, by averaging the weights of all the workers, while in A3C the workers asynchronously update the network by themselves. The agent is trained on the three scenarios described earlier using SUMO-RL [10] and the three scenarios are tested on it.

19.4 Results

The results of testing the different traffic light agents (RR, MONOPOLY, DQN and A2C) in the three traffic scenarios (SCEN-1, SCEN-2 and SCEN-3) for a time period of 5,400 timesteps are compared. The main parameters of interest are the peak total queueing length of vehicles waiting in all four incoming arms of the intersection, the time spent above half queue length and the similarity in the graphs of all traffic scenarios.

19.4.1 Round Robin Scheduling

The agent is made to perform in the three traffic scenarios and the total queueing lengths are observed in Fig. 19.9 (queue length vs timestep). The peak total queue length in SCEN-1 is around 88 vehicles, whereas for SCEN-2 and SCEN-3 the peak total queue length is around 57 vehicles. The agent performs well in uniform traffic scenarios (SCEN-1) and the queue length is above half only 33% of the time, but in non-uniform scenarios (SCEN-2 and SCEN-3), it is above half ~80% of the time. The shapes of the uniform and non-uniform graphs are not similar.

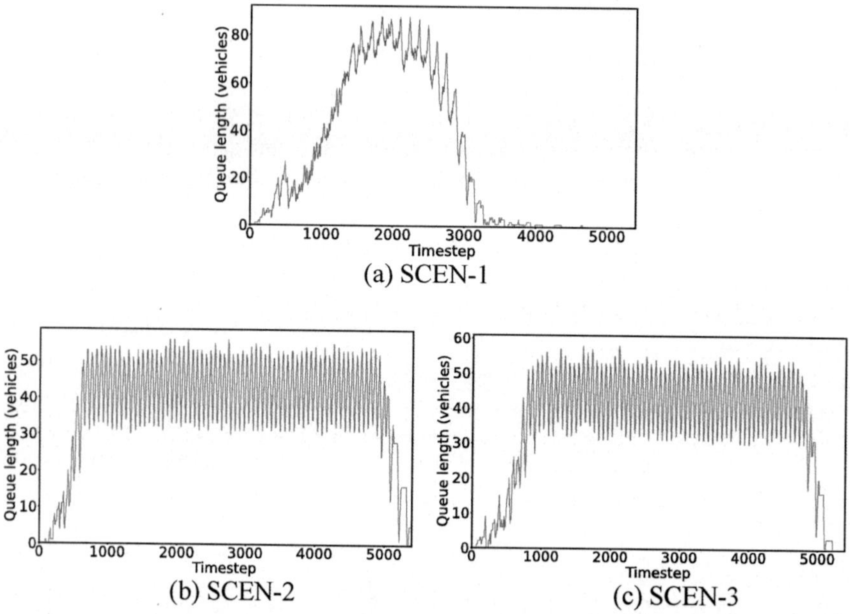

Fig. 19.9 Total queue lengths (y-axis) versus timestep (x-axis) using Round Robin (RR)

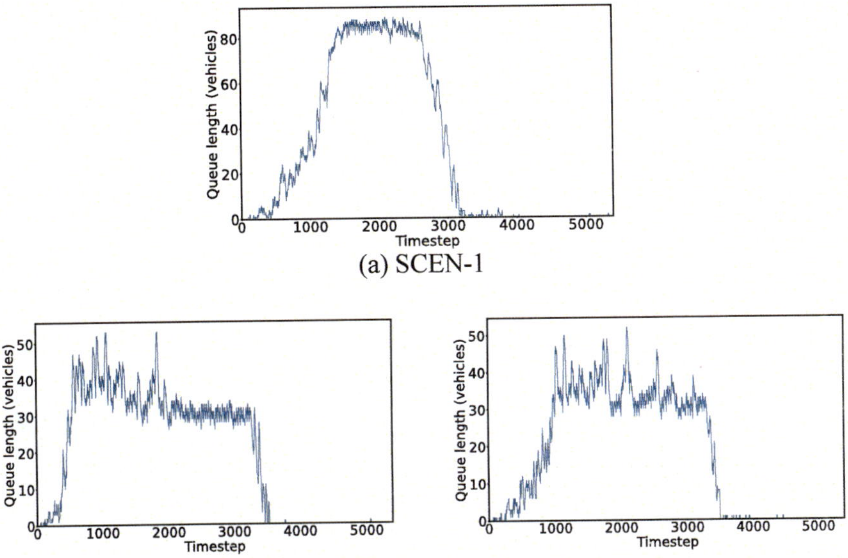

Fig. 19.10 Total queue lengths (y-axis) versus timestep (x-axis) using Feedback Control

19.4.2 Feedback Control Mechanism

MONOPOLY was tested against SCEN-1, SCEN-2 and SCEN-3 traffic density conditions, with the plots in Fig. 19.10 (queue length vs timestep).

The performance of MONOPOLY in uniform traffic scenarios (SCEN-1) displays a similarity in behaviour to RR. However, under non-uniform traffic densities, the agent performs significantly better than the RR agent. The time spent with a queue length above half is 50% because green signalling time is not wasted unnecessarily on empty lanes. The shapes of the graphs in uniform and non-uniform traffic are quite similar. The peak total queue length under uniform traffic is around 89 vehicles and for non-uniform traffic, it is around 53 vehicles.

A significant disadvantage of MONOPOLY is the fact that under high traffic loads, the agents act selfishly, with each agent signalling green for the maximum permissible time, causing the mechanism to revert to RR.

19.4.3 Deep Q-Networks (DQN)

The DQN agent was tested against SCEN-1, SCEN-2 and SCEN-3 traffic scenarios, and the corresponding queuing lengths are shown in Fig. 19.11 (queue length vs timestep). The results of testing the DQN agent on the uniform traffic agent show no

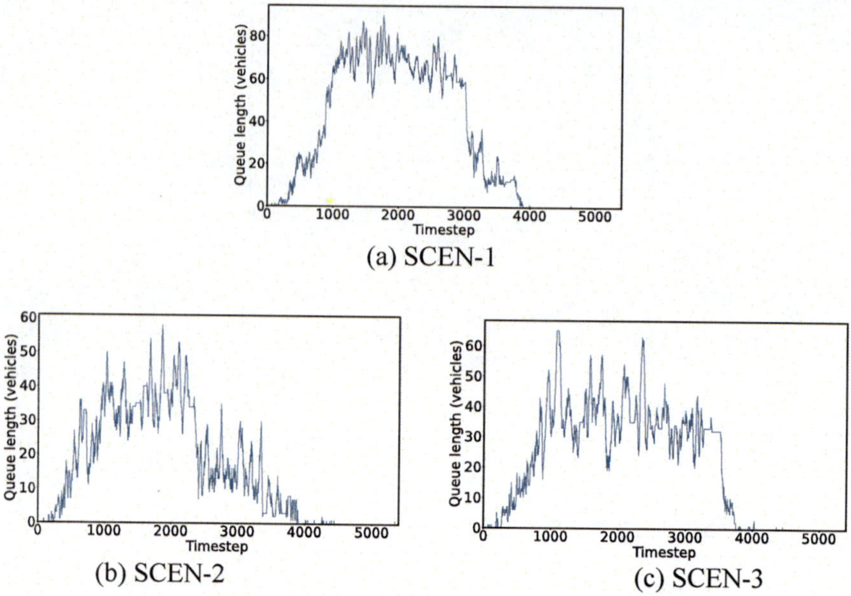

Fig. 19.11 Total queue lengths (y-axis) versus timestep (x-axis) using DQN

major improvement over RR and MONOPOLY. However, its performance in non-uniform traffic densities is significantly improved. The queues of traffic in all four arms of the intersection eventually clear out to completion or near completion by around 4,000 timesteps, and the graphs of all three traffic scenarios are quite similarly shaped. The peak total queue length under uniform traffic is around 90 vehicles and for non-uniform traffic, it is around 60 vehicles.

An important advantage of the DQN agent is in the symmetry of its performance in the three traffic scenarios. The uniform and non-uniform plots are close to indistinguishable, with around 30%–40% of the time spent with a queue length above half. However, the training process takes a long time and as the networks get more complicated, the training time is expected to increase drastically.

19.4.4 Advantage Actor-Critic (A2C)

The A2C agent was tested against the same three traffic scenarios. The plots of queue length over time (Fig. 19.12) were observed to be similar to DQN in all three scenarios when the number of processes (workers) was set to one since the underlying mechanism of A2C is a Multilayer Perceptron (MLP) Neural Network. Though the queue lengths were the same, the training time was significantly lesser than DQN.

A performance improvement was observed when the number of processes (N-CPU) was increased. The performance measure chosen for the comparison was the

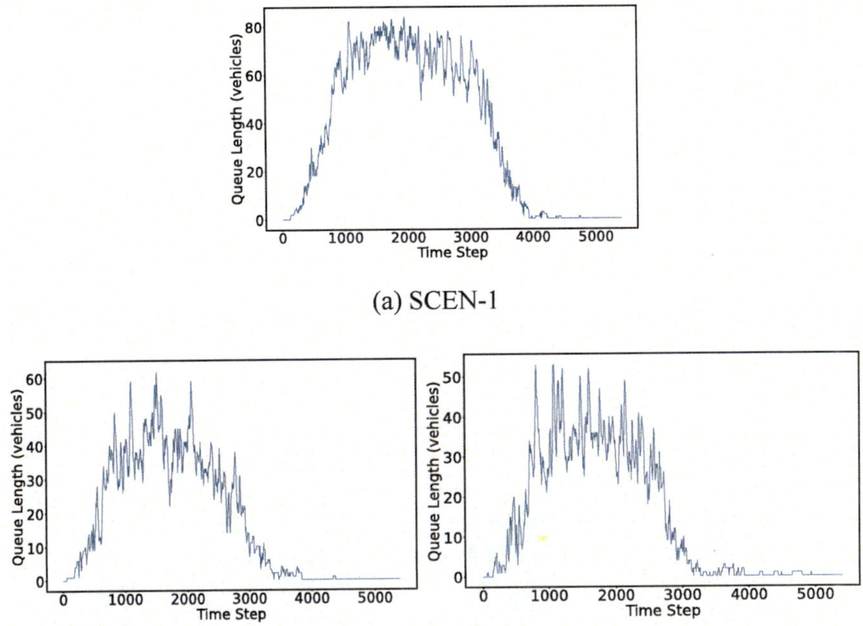

(a) SCEN-1

(b) SCEN-2

(c) SCEN-3

Fig. 19.12 Total queue lengths (y-axis) versus timestep (x-axis) using A2C

total wait time, instead of the queue lengths. The results of the comparison between the different number of processes are shown in Figs. 19.13, 19.14 and 19.15 (waiting time vs timestep and waiting time vs no. of processes).

The major advantage of A2C is the performance improvement (reduction in total wait times) achieved with lower training times, as the number of processes was increased. This is due to the parallelism that is incorporated in A2C, as described in Mnih et al. [8].

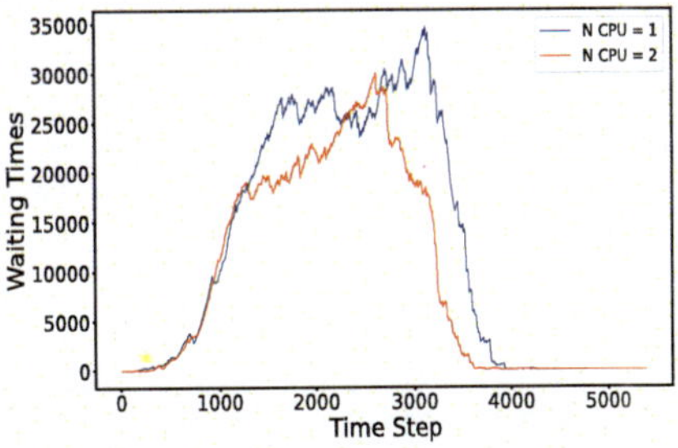

Fig. 19.13 N-CPU = 1 versus N-CPU = 2

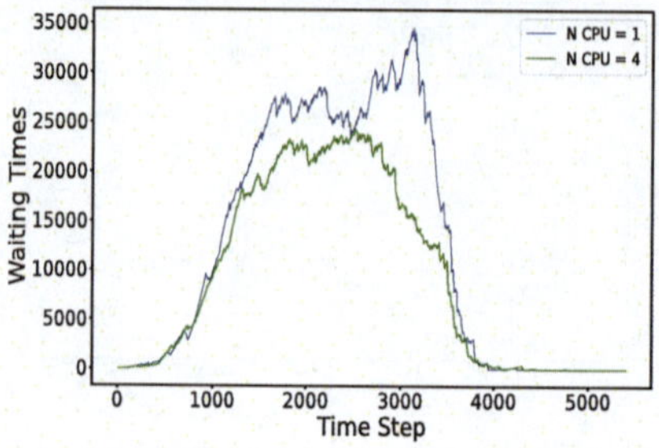

Fig. 19.14 N-CPU = 1 versus N-CPU = 4

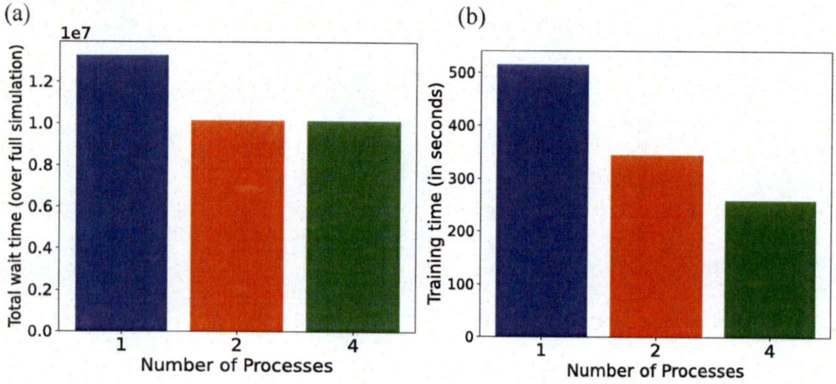

Fig. 19.15 Comparison of N-CPU = 1, 2 and 4. **a** Total waiting times versus no. of processes **b** Training times versus no. of processes

19.4.5 Summarized Results

A comparison of the parameters as a result of testing the different traffic light agents (RR, MONOPOLY, DQN and A2C) in the three traffic scenarios (SCEN-1, SCEN-2 and SCEN-3) are summarized in Table 19.2.

Table 19.2 Summarized results

Algorithm	Peak queue Length (vehicles)			Fraction of time spent at above half queue length		
	SCEN-1	SCEN-2	SCEN-3	SCEN-1	SCEN-2	SCEN-3
RR	88	56	58	0.33	0.84	0.77
MONOPOLY	89	53	52	0.33	0.51	0.45
DQN	90	58	65	0.40	0.26	0.34
A2C	84	62	53	0.47	0.29	0.31

19.4.6 Real-World Intersection

The performance of the A2C agent was observed to be the most optimal and scalable and was therefore tested on a simulation of a real-world intersection (JP Nagar Sarakki Signal in Bangalore, Karnataka, India. *Coordinates: 12.906468, 77.573254*). The network was obtained from OpenStreetMaps (OSM) and cleaned up in netedit as shown in Fig. 19.16. The traffic data was generated as per the Weibull distribution mentioned earlier.

The queue lengths over time were observed for the RR agent (Fig. 19.17) and the A2C agent (Fig. 19.18). The peak queue length formed by the A2C agent (35 vehicles) was approximately half of the peak formed by the RR agent (60 vehicles).

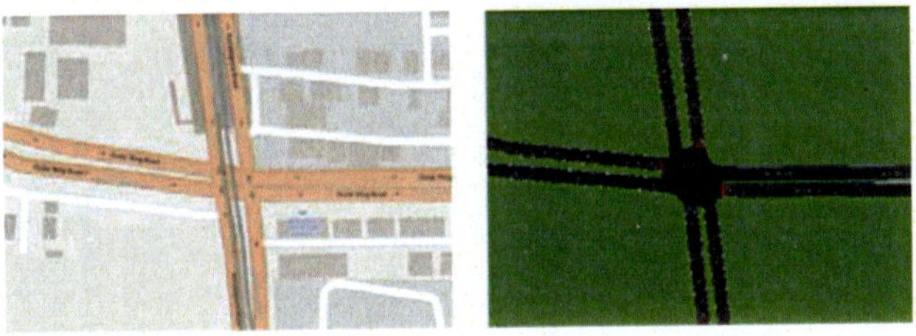

Fig. 19.16 Real-world intersection in OSM (left) and in Netedit (right)

Fig. 19.17 Total queue length versus timestep RR

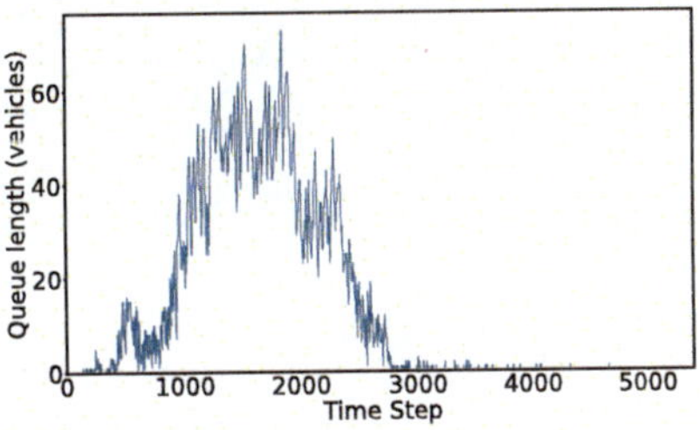

Fig. 19.18 Total queue length versus timestep A2C

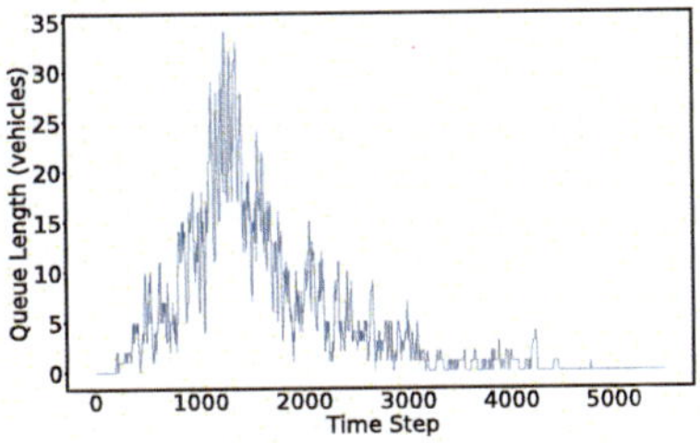

19.5 Conclusion

In this paper, a comprehensive comparative study of some of the techniques and algorithms that can be used for Intelligent Traffic Signal Control (ITSC) has been presented. While Round Robin (RR) is the most prevalent, it is inefficient in clearing out traffic, especially in non-uniform traffic conditions. RR is advantageous as it is computationally simple and can be used in low-density intersections. Feedback Control (MONOPOLY) presents an improvement in non-uniform traffic conditions while staying computationally efficient as well. It can be considered the most efficient when looking at the *performance-computation trade-off*. Deep Q-Networks (DQN) is a modern Reinforcement Learning method that yields impressive performance but is computationally intensive. Advantage Actor-Critic (A2C) is an optimization of DQN that performs just as well, while reducing the computation/training time required, by incorporating parallelism in Reinforcement Learning. A2C and other Actor-Critic methods can be best applied in high-density intersections if performance is paramount and computational resources are available.

References

1. Wey, W.M.: Model formulation and solution algorithm of traffic signal control in an urban network. Comput. Environ. Urban Syst. **24**(4), 355–378 (2000)
2. Eriskin, E., Karahancer, S., Terzi, S., Saltan, M.: Optimization of traffic signal timing at oversaturated intersections using elimination pairing system. In: Procedia Engineering, vol. 187, pp. 295–300 (2017)
3. Abdoos, M., Mozayani, N., Bazzan, A.: Traffic light control in non-stationary environments based on multi agent Q-learning. In: 14th International IEEE Conference on Intelligent Transportation Systems (ITSC) (2011)
4. Brys, T., Pham, T., Taylor, M.: Distributed learning and multi-objectivity in traffic light control. Connect. Sci. **26**(1), 65–83 (2014)
5. Gregurić, M., Vujić, M., Alexopoulos, C., Miletić, M.: Application of deep reinforcement learning in traffic signal control: an overview and impact of open traffic data. Appl. Sci. **10**(11), 4011 (2020)
6. Rasheed, F., Yau, K.L.A., Noor, R.M., Wu, C., Low, Y.C.: Deep reinforcement learning for traffic signal control: a review. IEEE Access (2020)
7. Vidali, A., Crociani, L., Vizzari, G., Bandini, S.: A deep reinforcement learning approach to adaptive traffic lights management. In: WOA, pp. 42–50 (2019)
8. Mnih, V., Badia, A., Mirza, M., Graves, A., Lillicrap, T., Harley, T., Silver, D., Kavukcuoglu, K.: Asynchronous methods for deep reinforcement learning. In: International Conference on Machine Learning, PMLR (2016)
9. Krajzewicz, D., Hertkorn, G., Rössel, C., Wagner, P.: SUMO (Simulation of Urban MObility)-an open-source traffic simulation. In: Proceedings of the 4th Middle East Symposium on Simulation and Modelling, MESM20002, pp. 183–187 (2002)
10. Alegre, L.: SUMO-RL. GitHub, github.com/LucasAlegre/sumo-rl (2019)
11. Wei, H., Zheng, G., Yao, H., Li, Z.: IntelliLight: A reinforcement learning approach for intelligent traffic light control. In: KDD (2018)
12. Muthupalaniappan, A., Nair, B.S., Rajan, R.A., Krishnan, R.K.: Dynamic control of traffic signals using traffic data from google maps and road cameras. Int. J. Recent Technol. Eng. (2019)

13. Zheng, G., Xiong, Y., Zang, X., Feng, J., Wei, H., Zhang, H., Li, Y., Xu, K., Li, Z.: Learning phase completion for traffic signal control. In: CKIM, International Conference on Information and Knowledge Management (2019)
14. Hilmani, A., Maizate, A., Hassouni, L.: Automated real-time intelligent traffic signal control system for smart cities using wireless sensor networks. In: Wireless Communications and Mobile Computing (2020)
15. Pandit, K., Ghosal, D., Zhang, H. M., Chuah, C.: Adaptive traffic signal control with vehicular ad hoc networks. IEEE Trans. Veh. Technol. **62**(4), 1459–1471 (2013)

Chapter 20
A Comparative Study of Deep Learning Neural Networks in Sentiment Classification from Texts

Tanha Tahseen and Mir Md. Jahangir Kabir

Abstract Nowadays, text data has become the dominating medium for expressing views and opinions on various issues on online platforms. The necessity of mining information and polarity from public opinion is outgrowing. The analysis of sentiment on social networks, such as Twitter, Facebook and E-commerce websites is an effective way of learning about the users' opinions and reviews. In recent years, it has been experimentally shown in some research works that deep learning neural networks are a promising solution to the challenges of Natural Language Processing. This research conducts a comparative study among some neural networks such as LSTM, Bi-LSTM, GRU-CNN and the combination of CNN and RNN. Two datasets from different domains have been used: The IMDB Movie Review (50 k), which is a benchmark dataset and an Amazon Product Review (10 k) dataset. Amazon is also a standard platform for collecting datasets. Analyzing the experimental results, we realize that CNNs can sometimes outperform RNNs. Also the comparative performance may differ in different domains and datasets.

20.1 Introduction

This is the age of internet and people express their emotions and opinions on different issues on different social media and e-commerce platforms via text. Today e-commerce websites altogether are a large platform for buying and selling products, and those have a review section there where the buyers state their opinion about the quality of the products they have bought. Viewing those reviews, buyers can decide whether or not to buy a particular product. Similarly, the businesspersons know to what extent they have to improve their quality of products. Integrating sentiment analysis with recommender systems can be helpful to generate efficient recommendations [1]. The problem of extracting features and opinions from text data and classifying them falls in the field of Natural Language Processing (NLP). The sources of data for text classification are the social media and e-commerce platforms.

T. Tahseen (✉) · M. Md. J. Kabir
Department of Computer Science and Engineering, Rajshahi University of Engineering & Technology, Rajshahi, Bangladesh

J. I.-Z. Chen et al. (eds.), *Machine Learning and Autonomous Systems*, Smart Innovation, Systems and Technologies 269, https://doi.org/10.1007/978-981-16-7996-4_20

For e-commerce websites, the main concern lies within determining the polarity of customer reviews, i.e. whether majority of the reviews are positive or negative. In our research, we work on sentiment polarity classification, to determine whether a text/review expresses positivity or negativity with deep learning methods and make a comparative analysis among the performance of these methods. We perform sentence-level classification here.

Though Natural Language Processing is a difficult and challenging task, it has been experimentally shown in recent years that deep learning neural networks can provide promising performance in the field of NLP [2]. This is why sentiment analysis with deep learning neural networks is a growing research topic. Both machine learning and deep learning techniques fall within the category of supervised learning. Traditional machine learning algorithms such as Linear regression, Logistic regression, Naïve Bayes, SVM, Maximum Entropy Classifiers have been used in sentiment analysis in several research articles and the Naïve Bayes [3], SVM [3, 4] and Maximum Entropy classifier have been shown to perform very well, with varying performance metrics depending on datasets. Yet various Deep Learning Neural Networks are coming to light because of their remarkable performance [2]. Hybrid and customized neural networks can be expected to give more promising results [5, 6]. Studies are going on to improve the performances of deep learning neural networks on texts to overcome the challenges of text classification.

The growing importance of sentiment classification and deep learning neural networks' growing remarkable performances in text classification have been the main motivations behind this study. The major limitation of the existing models is their inconsistency in performance. Which neural network will have the best performance on a new dataset cannot be foretold without experiment. To achieve a deep learning model to generate admirable results, existing models must be customized. For that purpose, a comparative study on the existing models should be conducted. How each type of neural network works and what level of performance they provide on a particular benchmark dataset requires attention.

Section 20.2 in this paper introduces a brief research history related with our work. Section 20.3 discusses theoretical background of the deep learning models used in this study. Section 20.4 represents our experimental workflow, describes our datasets, architecture of our deep learning models and the criteria that we have considered to evaluate the models. Finally, Sect. 20.5 analyses the obtained results and compares the performances of the deep learning models used in this study.

All the experimental works have been done on Google Colaboratory with GPU support.

20.2 Related Work

Till now many neural networks have been used in sentiment classification and those are LSTM, Bi-LSTM, Tree-LSTM, Discourse-LSTM, simple RNN (SRN), Recursive Neural Network, CNN, GRU, Bi-GRU, Deep Belief Network (DBN) and Deep

Neural Network (DNN). Lin Li applied SRN, LSTM and CNN on the IMDB Movie Review (50 k reviews) Dataset and made a comparative analysis considering different dataset size and review length and there the CNN (with 87.42% accuracy) outperformed LSTM and SRN, LSTM achieved the second-highest accuracy (85.82%) and SRN the least (78.10%) in 2018 [7]. Lina Maria shows in her review paper that on the movies reviews corpus (Pang & Lee 2005) which contains 10,662 reviews, 78.6% accuracy by LSTM and 79% by Bi-LSTM was achieved by Li in 2015 [10]. In 2018, Alharbi did a comparative experiment on the Naïve Bayes, Support Vector Machine (SVM), J48 (an algorithm to generate a decision tree), K-Nearest Neighbors (KNN), LSTM and CNN using two Twitter datasets published by SemEval-2016. In that experiment, CNN and LSTM outperformed the machine learning algorithms, with CNN having the highest accuracy [11]. But that cannot draw the conclusion that Deep Learning networks perform way better in sentiment analysis than machine learning algorithms. Monika Kabir and Shuxiang Xu achieved 94.18% and 91.02% using maximum entropy classifier and 90.78% and 90.27% using Boosting method on Amazon (10 k reviews) and IMDB (10 k reviews) datasets in 2019 [4]. Jennifer and Smys proposed a hybrid model composed of the combination of Naive Bayes and SVM for early detection of depression from text data, applied it on a Twitter dataset and achieved 91% accuracy in 2021. Whereas SVM, Naive Bayes, Random Forest and Decision Tree classifier achieved 84%, 70%, 80% and 65%, respectively, on that dataset [21]. Hybrid networks composed of the combination of both CNN and RNN are seen to achieve better performance than CNNs or RNNs alone [5, 6, 12–14].

Comparative performances of different deep learning methods from some more literatures of recent years are depicted in Table 20.1.

20.3 Deep Learning Neural Networks

Deep learning neural networks are complicated structures built of artificial neurons arranged in multiple layers. In this section, brief descriptions of the neural networks used in this study are discussed.

Long Short-Term Memory (LSTM): LTSM is a type of Recurrent Neural Network (RNN) that is capable of learning order dependence in sequence prediction problems. Short-term memory is a problem for Recurrent Neural Networks. LSTM was designed to overcome the vanishing gradient problem and gradient exploding problem which were major drawbacks of simple RNNs [15].

Bidirectional LSTM (BiLSTM): BiLSTM is an LSTM variant. It is a sequence processing model with two LSTMs per layer: one takes the input sequence in a forward direction, and the other receives the same input sequence in a backward direction [17]. Thus BiLSTMs effectively expand the amount of information available to the network.

Table 20.1 Comparative performance (in terms of testing accuracy/AUROC/f1-score) of deep learning neural networks from literatures throughout the years 2016–2021

Paper Year & Ref	Datasets	Train-test split and k-folds	Methodologies applied	Accuracy/AUROC/f1-score	Embedding
(2021) [24]	112,412 restaurant reviews from Yelp.com posted during January–June 2020	80%–20%	Simple Embedding + Average Pooling BiLSTM	0.90 (F1-score)	Glove
				0.92(F1-score)	
(2020) [23]	Amazon pet supplies	50%–20%–30% (train- val-test)	1- layer CNN	0.8215	–
			Vanilla RNN	0.7481	
			LSTM	0.8527	
			Bi-LSTM	0.8529	
			GRU	0.8546	
			Bi-GRU	0.8351	
				(AUROC)	
(2020) [6]	17,289 Turkish tweets between 2011 and 2017 collected from shared user tweets about a GSM operator in Turkey	80%–20%	CNN	78.06%	Word2Vec
			LSTM	79.07%	
			Bi-LSTM	79.18%	
			GRU	79.59%	
			CNN-NiLSTM	82.14%	Character + FastTex
(2020) [2]	IMDB(50 k) from Stanford University	k = 10	DNN	85.512%	TF-IDF
			CNN	80.628%	
			RNN	58.72%	
			DNN	80.25%	Word2Vec

(continued)

Table 20.1 (continued)

Paper Year & Ref	Datasets	Train-test split and k-folds	Methodologies applied	Accuracy/AUROC/f1-score	Embedding
			CNN	82.62%	
			RNN	86.69%	
(2019) [13]	IMDB movie reviews from http://rottento--matoes.com	80%–20%	CNN	85.1%	Word2Vec
			LSTM	85.5%	
			CNN-LSTM	91%	
(2019) [12]	STSC(4 k), STSC(1.5 k), SS-TDS, HCR-TDS	–	Bi-LSTM	76.52%	Random initialization
			GRU	75.75%	
			Bi-GRU	78.21%	
			CNN	68.23%	
			Bi-LSTM-CNN	81.99%	
			GRU-CNN	83.24%	
			Bi-GRU-CNN	84.23%	
				(Average accuracy On 4 datasets)	
(2019) [8]	IMDB Movie Reviews	–	Tree-LSTM (N-ary)	84.7%	–
			Discourse-LSTM (N-ary)	84.7%	
	movie reviews from Rotten Tomatoes (20 k)	K = 10	Tree-LSTM (Child- sum)	74.0%	–
			Discourse-LSTM (Child-sum)	55.0%	

(continued)

Table 20.1 (continued)

Paper Year & Ref	Datasets	Train-test split and k-folds	Methodologies applied	Accuracy/AUROC/f1-score	Embedding
(2017) [9]	SSTb (Stanford Sentiment Treebank)	80%–20%	CNN non-static	87.2%	Word2Vec
			CNN multichannel	88.1%	
2016) [22]	Stanford Sentiment Treebank from http://nlp.stanford.edu/sentiment/	80%–20%	CNN-GRU	89.95%	Word2Vec
			CNN-LSTM	89.56%	
			Avg GRU	89.61%	
			CNN-GRU	86.64%	Random initialization
			CNN-LSTM	86.80%	

Gated Recurrent Unit (GRU): GRU is a variant of RNN. Just like the LSTM, GRU was developed to overcome the short-term memory problem of simple RNN. GRU is called the younger sibling of LSTM [18].

GRU has only two types of gates—reset gate and update gate. As a result, GRU is simpler in structure and computation than LSTM and way more fast than LSTM [6, 18].

The relationship between the gates of a node of GRU and their functions can be explained by the following equations [17]:

$$\text{gate}_{\text{reset}} = \sigma\left(W_{\text{inputreset}} \cdot x_t + W_{\text{hiddenreset}} \cdot h_{t-1}\right) \tag{20.1}$$

$$r = \tanh\left(\text{gate}_{\text{reset}} \odot \left(W_{h1} \cdot h_{t-1}\right) + W_{x1} \cdot x_t\right) \tag{20.2}$$

$$\text{gate}_{\text{update}} = \sigma\left(W_{\text{inputupdate}} \cdot x_t + W_{\text{hiddenupdate}} \cdot h_{t-1}\right) \tag{20.3}$$

$$u = \text{gate}_{\text{update}} \odot h_{t-1} \tag{20.4}$$

where r = output of reset gate and u = output of update gate (Fig. 20.1).

Convolutional Neural Network (CNN): CNN or Convolutional Neural Network is also an effective deep learning model for classifying texts. Though it is mainly used in image classification, satisfactory results have been derived using CNN in sentiment classification in recent years [2, 7, 9, 11]. For text classification, 1-dimensional CNNs are used. After feature extraction by convolutional layers with filters and having the dimensionality of the extracted features reduced by pooling layers, the outputs are passed to the fully connected layer to train the texts with their respective class labels [6] (Fig. 20.2).

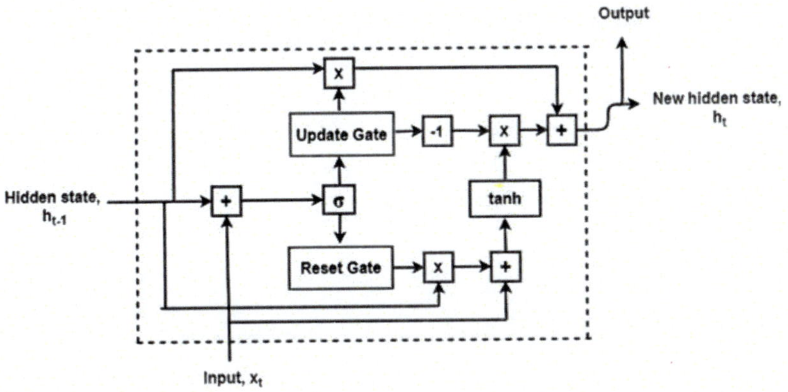

Fig. 20.1 Internal architecture of a GRU node

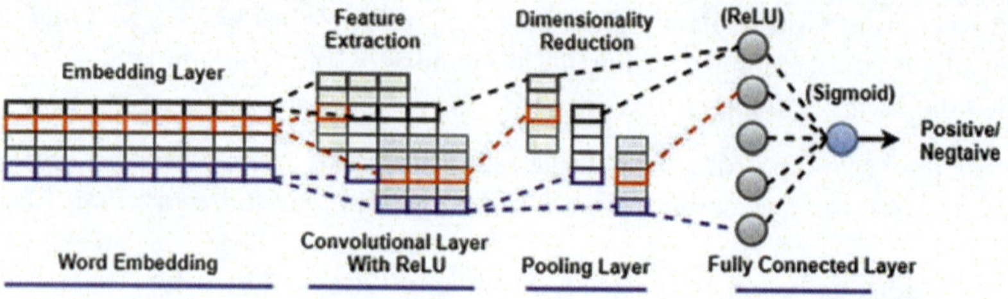

Fig. 20.2 1D CNN architecture for text classification

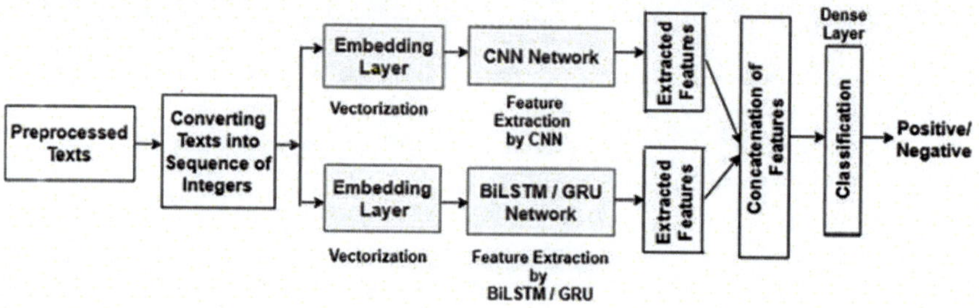

Fig. 20.3 A demonstration of the combination of a CNN and an RNN

CNN + BiLSTM / CNN + GRU: This type of architecture can be implemented by passing the features extracted by one or two convolutional layers into an RNN network (LSTM/BiLSTM/GRU/ SRN) [19], or by merging two separate CNN and RNN networks [5, 6]. In our study, we have implemented the latter. This method is said to combine the strengths of both CNN and RNN. Figure 20.3 pictures the working model of the ensemblement of a CNN and an RNN architecture.

20.4 Experimental Setup

In this section, the entire workflow of our study, the datasets we have used, the way we preprocessed our data, how we built our models, the criteria that we considered to evaluate our models and how overfitting was handled will be discussed. Figure 20.6 represents our entire workflow (Fig. 20.4).

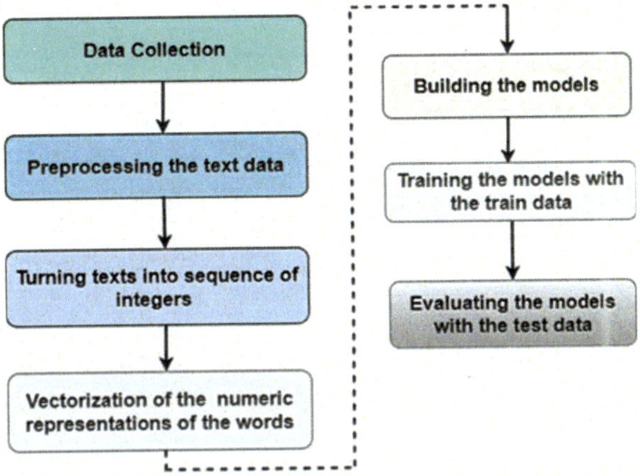

Fig. 20.4 The steps of our experiment

Table 20.2 A brief description of our datasets

Dataset	Contents	Number of positive reviews	Number of negative reviews
IMDB	Movie reviews	25,000	25,000
Amazon	Reviews of DVD, video game, books and clothing	4903	5097

Table 20.3 Dataset split in our every experiment

Train data	Test data	Validation data
75%	25%	20%

20.4.1 Datasets

We have worked on the following two binary labelled (positive and negative) datasets in our study: (1) The IMDB Movie Review Dataset (50 k Reviews) [25], (2) An Amazon Product Review Dataset (10 k Reviews) [26] (Tables 20.2 and 20.3).

20.4.2 Text Preprocessing

Text preprocessing is a crucial step in Natural Language Processing. It is required to remove non-important features from texts to make room for the important features in the word dictionary. Preprocessing tasks serve this purpose. Fig. 20.5 depicts our whole preprocessing task.

WordNetLemmatizer from nltk library was used to lemmatize the words of the texts.

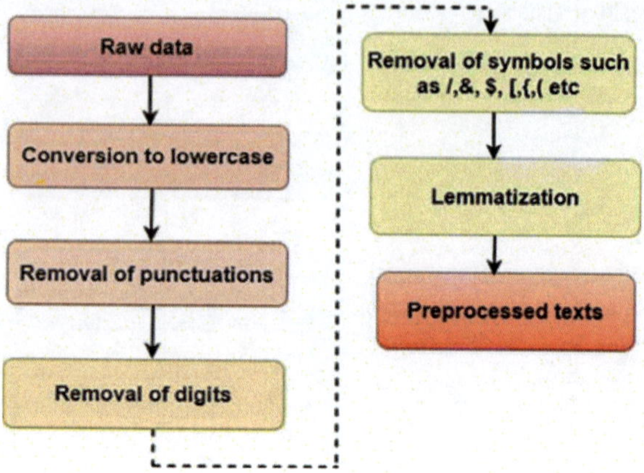

Fig. 20.5 Text preprocessing in this study

20.4.3 *Word Vectorization*

Deep learning or machine learning algorithms cannot process raw text or image data, since computer can work only with numeric data. That's why the text data must be converted into numeric form in the task of NLP. After each sentence is converted into a sequence of integer numbers, each word in a sentence is to be converted into an n-dimensional vector, to add features to the word, hence making the words distinguishable from each other. We simply used the Tokenizer from keras to turn the text data into numeric sequences, and finally the embedding layer to vectorize the integers. No pre-trained word embedding was used.

20.4.4 *Building the Models*

In this sub-section, descriptions of our model architectures are given in tabular representations (Table 20.4).

The number and names of the layers along with the number of nodes within the layers, embedding dimension in the embedding layer and dropout percentage used in the dropout layer are presented in Table 20.5. In both LSTM and GRU, ReLU was

Table 20.4 Vocabulary size, number of epochs and batch size in our every network

Vocabulary size for IMDB Dataset	Vocabulary size for Amazon Dataset	Epochs	Batch size	Optimizer
20,000	6,000	21 for LSTM on IMDB 25 for the rest	200 except for LSTM on the IMDB dataset, where the batch size was 380	Adam

Table 20.5 Description of our LSTM (left) and GRU (right) architecture

Layer(Type)	Number of nodes	Embedding dimension	Dropout percentage	Layer (Type)	Number of nodes	Embedding dimension	Dropout percentage	Activation functions
Embedding		64		Embedding		64		
LSTM	64			GRU	80			
LSTM	32			GRU	40			
Dense	64			Dense	64			ReLU
Dropout			40%	Dropout			40%	
Dense Sigmoid	1			Dense	1			sigmoid

used in the first fully connected (dense) layer and Sigmoid was used in the final dense layer. We built our BiLSTM model similarly to the LSTM. Only difference is that we used bidirectional LSTM nodes instead of unidirectional LSTM nodes (Table 20.7).

In our CNN and CNN-RNN architectures, ReLU was used in every Conv1d layer and the 1st dense layer and Sigmoid was used in the final dense layer.

Table 20.6 Description of our CNN architecture

Layer (Type)	Number of nodes/ filters	Embedding dimension	Dropout percentage	Activation functions
Embedding		64		
Conv1D	64 (filters)			ReLU
Maximum pooling				
Conv1D	32 (filters)			ReLU
Maximum Pooling				
Flatten				
Dense	64			ReLU
Dropout			40%	
Dense	1			Sigmoid

Table 20.7 Description of our CNN-BiLSTM and CNN-GRU architecture

Our CNN + BiLSTM architecture	Our CNN + GRU architecture
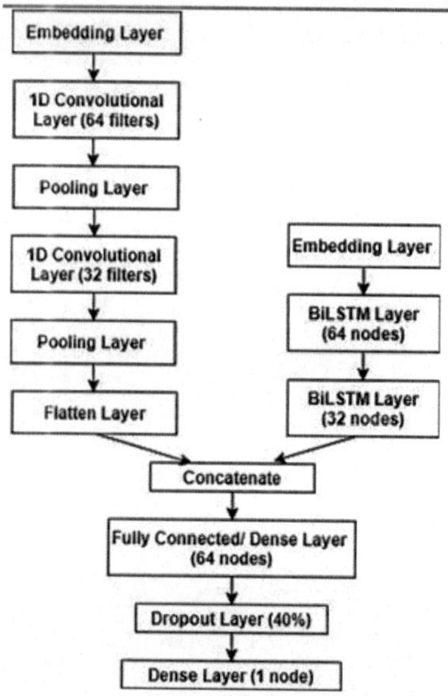	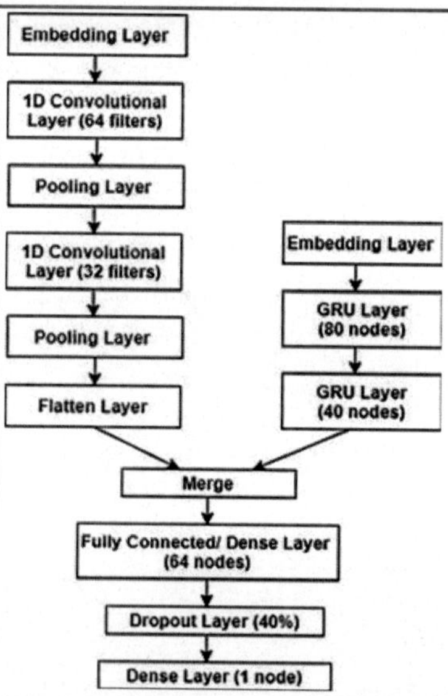

20.4.5 Evaluation Metrics or Evalutation Criteria

Testing accuracy: Testing accuracy is determined from the following formula, where TP, TN, FP and FN are true positives, true negatives, false positives and false negatives determined from confusion matrix:

$$\text{Testing accuracy} = \frac{\text{TP} + \text{TN}}{\text{TP} + \text{TN} + \text{FP} + \text{FN}}(\%) \quad (20.5)$$

Precision, recall and f-score: Precision, recall and f-score are determined from the following formulae:

$$\text{Precision} = \frac{\text{TP}}{\text{TP} + \text{FP}} \quad (20.6)$$

$$\text{Recall} = \frac{\text{TP}}{\text{TP} + \text{FN}} \quad (20.7)$$

$$\text{f} - \text{score} = \frac{2{*}\text{precision}{*}\text{recall}}{\text{precision} + \text{recall}} \quad (20.8)$$

Matthews Correlation Coefficient (MCC): The formula for calculating the MCC score is as follows:

$$\text{MCC} = \frac{\text{TP.TN} - \text{FP.FN}}{\sqrt{(\text{TP} + \text{FP}).(\text{TP} + \text{FN}).(\text{TN} + \text{FP}).(\text{TN} + \text{FN})}}$$

The range of MCC values lies between -1 and + 1. A model with a score + 1 is a perfect model and a model with a score -1 is a poor model. MCC is the only binary classification rate that produces a high score only if the binary predictor correctly predicts the majority of positive and negative data instances [20].

20.4.6 Handling Overfitting

One of the most challenging tasks of training a neural network is to eliminate overfitting. It occurs when a network is too specialized, or in other words, contains too much parameters for the classification task. By checking the gap between training loss curve and validation loss curve, and training accuracy curve and validation accuracy curve, it can be determined whether a model is overfitted or underfitted or properly fitted.

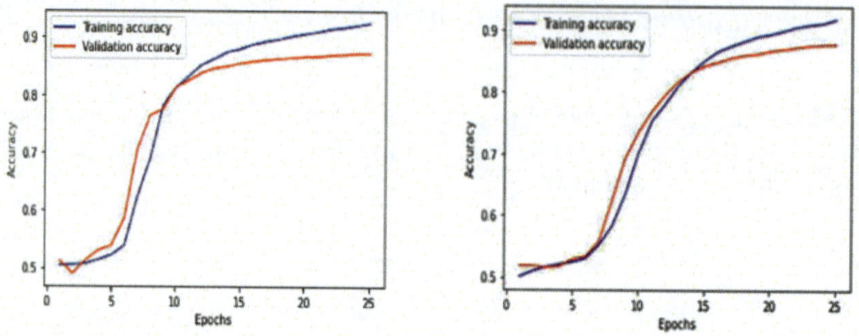

Fig. 20.6 Training accuracy curves of CNN-BiLSTM (left) and CNN (right) on the IMDB dataset

A little overfit might be acceptable (below 10%), but if the difference of the training accuracy and validation accuracy, or training loss and validation loss is greater than that then it must be reduced.

We used regularizers (ranging within 0.0001–0.0002), dropout layers and adjusted the learning rates (ranging within 0.00001–0.000025) to handle overfitting (Fig. 20.6).

20.5 Experimental Results and Performance Analysis

In this section, we display the experimental results in tabular representation, and compare the models based on the different performance metrics. The highest scores are bolded in the tables (Table 20.8).

The table exhibits that on the IMDB dataset, CNN has achieved the highest accuracy (88.18%) and LSTM has achieved the 2nd highest accuracy (88.06%). The combination of CNN and BiLSTM stands in 5th position in this case, and the combination of CNN and GRU has the least performance (86.26%).

Table 20.8 Testing accuracies on IMDB (left) and Amazon (right) datasets

IMDB		Amazon	
Model	Accuracy (%)	Model	Accuracy (%)
LSTM	88.06	LSTM	80.88
BiLSTM	87.55	BiLSTM	80.32
CNN	**88.18**	CNN	81.56
GRU	87.80	GRU	79.96
CNN-BiLSTM	87.74	CNN-BiLSTM	**81.64**
CNN-GRU	86.26	CNN-GRU	78.76

Table 20.9 Precision, recall and f-score values on IMDB (left) and Amaon (right) datasets

IMDB				Amazon			
Model score	Precision	Recall	F-score	Model score	Precision	Recall	F-score
LSTM	0.88	0.87	0.87	LSTM	0.81	0.80	0.80
BiLSTM	0.88	0.87	0.87	BiLSTM	0.80	0.80	0.80
CNN	**0.89**	0.87	**0.88**	CNN	0.81	**0.81**	**0.81**
GRU	**0.89**	0.86	0.87	GRU	0.80	0.78	0.79
CNN-BiLSTM	0.87	**0.88**	0.87	CNN-BiLSTM	**0.84**	0.77	0.80
CNN-GRU	0.85	0.87	0.86	CNN-GRU	0.80	0.75	0.77

Table 20.10 MCC scores on IMDB (left) and Amaon (right) datasets

IMDB		Amazon	
Model	MCC score	Model	MCC score
LSTM	0.76109	LSTM	0.61721
BiLSTM	0.75097	BiLSTM	0.60617
CNN	**0.76375**	CNN	0.63087
GRU	0.75626	GRU	0.59879
CNN-BiLSTM	0.75476	CNN-BiLSTM	**0.63383**
CNN-GRU	0.72538	CNN-GRU	0.57515

On the Amazon dataset, the combination of CNN and BiLSTM has outperformed all the other models (81.64%). It could not beat CNN at a significant level. Here also, the combination of CNN and GRU has the least accuracy score (78.76%).

Table 20.9. shows that the CNN and GRU have the highest precision score (0.89), the ensemblement of CNN and BiLSTM has the highest recall score (0.88) and the CNN has the highest f-score value (0.88) on the IMDB dataset. The combination of CNN and BiLSTM outperforms others in the case of recall value here. The ensemblement of CNN and BiLSTM has the highest precision score (0.84), and CNN has the highest recall value and f-score (0.81) on the Amazon dataset.

Table 20.10. shows that CNN is the best performing classifier on the IMDB dataset, and the combination of CNN and BiLSTM is the best classifier for the Amazon dataset based on the MCC scores.

20.6 Conclusion and Future Work

We have conducted a comparative experiment on different deep learning neural networks and analyzed the results. In our experiments, we have seen that CNN outperformed RNNs on our both datasets. But the combination of CNN and RNN should have performed better than CNN by a significant percentage. One of our ensembled models (CNN-BiLSTM) has outperformed CNN on the Amazon dataset slightly, but the difference should have been higher because as we have seen in a number of research papers that the combination of CNN and RNN (LSTM or BiLSTM) generally always outperforms CNN significantly.

In future work, we will use pre-trained word embedding models and check how our hybrid models perform in that case.

Acknowledgements We would like to thank IMDB and Amazon for making the datasets available.

References

1. Hung, B.T.: Integrating sentiment analysis in recommender systems. ın: reliability and statistical computing: modeling, methods and applications. In: Pham, H. (ed.), pp. 127–137. Springer International Publishing, Cham (2020)
2. Dang, N.C., Moreno-García, M.N., De la Prieta, F.: Sentiment analysis based on deep learning: a comparative study. Electronics **9**(3), Art. no. 3 (Mar. 2020). https://doi.org/10.3390/electronics9030483
3. Jagdale, R.S., Shirsat, V.S., Deshmukh, S.N.: Sentiment analysis on product reviews using machine learning techniques. In: Cognitive Informatics and Soft Computing, Singapore, pp. 639–647. https://doi.org/10.1007/978-981-13-0617-4_61(2019)
4. Kabir, M., Kabir, M., Xu, S., Badhon, B.: An empirical research on sentiment analysis using machine learning approaches. Int. J. Comput. Appl. (2019). https://doi.org/10.1080/1206212X.2019.1643584
5. Jang, B., Kim, M., Harerimana, G., Kang, S., Kim, J.: Bi-LSTM model to increase accuracy in text classification: combining Word2vec CNN and attention mechanism. Appl. Sci. **10**, 5841 (2020). https://doi.org/10.3390/app10175841
6. Salur, M.U., Aydin, I.: A novel hybrid deep learning model for sentiment classification. IEEE Access **8**, 58080–58093 (2020). https://doi.org/10.1109/ACCESS.2020.2982538
7. How textual quality of online reviews affect classification performance: a case of deep learning sentiment analysis SpringerLink. https://link.springer.com/article/10.1007/s00521-018-3865
8. Kraus, M., Feuerriegel, S.: Sentiment analysis based on rhetorical structure theory: learning deep neural networks from discourse trees. Expert Syst. Appl. **118**, 65–79 (2019). https://doi.org/10.1016/j.eswa.2018.10.002
9. Ganda, R., Mahmood, A.: Deep Learning Approach for Sentiment Analysis of Short Texts. pp. 705–710 (2017). https://doi.org/10.1109/ICCAR.2017.7942788
10. Rojas-Barahona, L.M.: Deep learning for sentiment analysis. Lang. Linguistics Compass **10**(12), 701–719 (2016). https://doi.org/10.1111/lnc3.12228
11. Alharbi, S.M., de Doncker, E.: Twitter sentiment analysis with a deep neural network: An enhanced approach using user behavioral information. Cognit. Syst. 63 Res. **54**, 50–61 (2019). https://doi.org/10.1016/j.cogsys.2018.10.001
12. Abid, F., Alam, M., Yasir, M., Li, C.: Sentiment analysis through recurrent variants latterly on convolutional neural network of Twitter. Futur. Gener. Comput. Syst. **95**, 292–308 (2019). https://doi.org/10.1016/j.future.2018.12.018
13. Rehman, U., Malik, A.K., Raza, B., Ali, W.: A hybrid CNN-LSTM model for improving accuracy of movie reviews sentiment analysis. Multimed. Tools Appl. **78**(18), 26597–26613 (2019). https://doi.org/10.1007/s11042-019-07788-7
14. Yenter, A., Verma, A.: Deep CNN-LSTM with combined kernels from multiple branches for IMDb review sentiment analysis. In: 2017 IEEE 8th Annual Ubiquitous Computing, Electronics and Mobile Communication Conference (UEMCON), pp. 540–546 (2017). https://doi.org/10.1109/UEMCON.2017.8249013

15. J. Brownlee: A Gentle Introduction to Long Short-Term Memory Networks by the Experts. Machine Learning Mastery (May 23, 2017) https://machinelearningmastery.com/gentle-introduction-long-short-term-memory-networks-experts/
16. Papers with Code-BiLSTM Explained. https://paperswithcode.com/method/bilstm
17. Gated Recurrent Unit (GRU) With PyTorch, FloydHub Blog (2019) https://blog.floydhub.com/gru-with-pytorch/
18. Chollet, F.: Deep Learning With Python. Manning, Shelter, Island (2017)
19. Zhou, C., Sun, C., Liu, Z., Lau, F.C.M.: A C-LSTM neural network for text classification. arXiv:1511.08630 [cs] (2015). http://arxiv.org/abs/1511.08630.
20. Chicco, D., Jurman, G.: The advantages of the Matthews correlation coefficient (MCC) over F1 score and accuracy in binary classification evaluation. BMC Genomics 21(1), 6 (2020). https://doi.org/10.1186/s12864-019-6413-7
21. Smys, S., Raj, J.S.: Analysis of deep learning techniques for early detection of depression on social media network-A comparative study. J. Trends Comput. Sci. Smart Technol (TCSST) **3**(01), 24–39 (2021)
22. Wang, X., Jiang, W., Luo, Z.: Combination of convolutional and recurrent neural network for sentiment analysis of short texts. In Proceedings of the COLING 2016, the 26th International Conference on Computational Linguistics, Osaka, Japan, pp. 2428–2437 (2016)
23. Seo, S., Kim, C., Kim, H., Mo, K., Kang, P.: Comparative study of deep learning-based sentiment classification. IEEE Access **8**, 6861–6875 (2020). https://doi.org/10.1109/ACCESS.2019.2963426
24. Luo, Y., Xu, X.: Comparative study of deep learning models for analyzing online restaurant reviews in the era of the COVID-19 pandemic. Int. J. Hospitality Manag. **94**, 102849 (2021)https://doi.org/10.1016/j.ijhm.2020.102849
25. Available at https://www.kaggle.com/lakshmi25npathi/imdb-dataset-of-50k-movie-reviews
26. Available at https://github.com/Gunjitbedi/Text-Classification/blob/master/corpus.csv

Chapter 21
An Efficient Employee Retention Prediction Model for Manufacturing Industries Using Machine Learning Approach

S. Radhika, S. Umamaheswari, R. Ranjith, and A. Chandrasekar

Abstract This paper explores the advantage of machine learning algorithm in the prediction of employee retention in manufacturing industries in India. The effect of different work-related and personal features on the retention of employees is discussed in detail. Further the main features that contribute to the retention of employees are also found. Using the best machine learning algorithm, prediction model to predict the retention of employees is proposed. Real-time data set collected from Ceramic Manufacturing sector in India is used to analyse the employee retention. Statistical parameters such as percentage analysis and correlation matrix are used to obtain the correlation and contribution of each feature on employee retention. The parameters such as accuracy, F1 score, precision and recall were used to identify the best prediction model. Results indicate that Support Vector Machine (SVM) is the best machine learning algorithm that can be used to build the employee retention prediction model that can help the organization to take required steps and decisions to retain their employees.

21.1 Introduction

Machine Learning has penetrated into each and every sector that includes medicine, astronomy, production, etc. [1–3]. Nowadays the use of machine learning algorithms for Human Resource (HR) Management has brought in several advantages like increase in efficiency, improvement in decision making, etc. It is a known fact that

S. Radhika (✉)
Department of EEE, Sathyabama Institute of Science and Technology, Chennai, India

S. Umamaheswari
Department of Management Studies, Sathyabama Institute of Science and Technology, Chennai, India

R. Ranjith · A. Chandrasekar
St.Joseph's College of Engineering, Chennai, India
e-mail: ranjithr@hotmail.co.in

J. I.-Z. Chen et al. (eds.), *Machine Learning and Autonomous Systems*, Smart Innovation, Systems and Technologies 269, https://doi.org/10.1007/978-981-16-7996-4_21

the growth of any organization largely depends on the employees or human resources as their quality and efficiency decide the strength of the organization [4–7].

Generally company needs employees to perform specific jobs and their high retention rate is essential for a good company. This is particularly important in manufacturing industries where skilled workforce is inevitable [8]. When an employee leaves an organization, it is not only the present loss in terms of money, time and energy spend on the employee, but it can even act as a recurring loss as the organization has to look for new employees which in turn results in a lot of investment for recruitment, training and other activities. Not only that, the organization may face problems in meeting the quality and demands of the customers. This may lead to loss of customers and even to a major setback in future business prospects. Hence employee retention is one of the key aspects in the growth of any organization [9] and it is very much essential as the employees are the real assets for an organization. Employee retention is the systematic procedure used by the organizations to prevent the leaving of employees. Usually the HR team gets a lot of data from the employees; however, only less amount of information is efficiently utilized. Machine learning algorithm opens up new avenues for the retention of employees in manufacturing industries. The application of Machine Learning in the field of employee retention involves utilization of data to explore useful information and make a suitable prediction of their retention levels which can aid in taking suitable measures to retain the employees and even improve the HR activities.

A lot of research is being done to prove the efficacy of machine learning algorithms in the field of HR management to assist the HR managers to improve production, expand business as well as employee retention levels. In [10, 11] different machine learning algorithms were attempted to predict the attrition of employees. From the analysis it is proved that machine learning algorithms were effective in prediction of attrition level of employees. Logistic regression is use in [12] to predict employee attrition rate. The results yield 84% accuracy which clearly indicates that machine learning is a suitable tool for the prediction of employee retention. In [13] the factor affecting attrition rate using Weka is discussed. All the above analysis makes use of IBM data set. Various classification models were attempted in [14] to predict the performance of employees and it also yields promising results. Thus from the literatures it is clearly found that machine learning algorithms can be effectively used in predicting the retention level of employees in an organization. It can also aid the HRM team in providing the path way which can help them to prevent the attrition level of employees. Till now, to the authors' knowledge there is no attempt made to address the prediction of employee retention in Ceramic Manufacturing industries in India using machine learning techniques.

In this work, a detailed analysis features that affect the retention of employees is made. Using the real data set collected from Ceramic Manufacturing sector in India, the main features that contribute to employee retention are identified. The data set collected includes 17 features and 416 samples. Using percentage analysis, the characteristics of employee retention based on personal details are discussed. Using the correlation, the highly correlated features for employee retention are identified. Further, detailed associations among the features are also studied. Finally by using feature selection, top-performing features are identified and are used for further analysis. Using training and test data split in the ratio 70:30, prediction is performed. Accuracy, F1 score, Precision and recall were used to select the best predictor. Using all the above, a model to build the prediction of employees is built. Thus model built using machine learning algorithm can be used to predict whether the employee will retain or not and can support the HR department to adopt several measures so that skilled employees are retained. This paper is organized as follows: Section 2 describes the methodology adopted. It includes the description of dataset, data preprocessing and feature selection; Model building is discussed in Sect. 3. Results and discussion are made in Sect. 4. Conclusions are given in Sect. 5.

21.2 Methodology

This section is devoted to describe the methodology used in the proposed work. The steps involved in the proposed approach are illustrated in Fig. 21.1. As seen from Fig. 21.1 the data set collection is the first step. This is followed by preprocessing and feature selection process. Then classification of the employees as retained or not retained is done. Finally using the different performance metrics, the best model is made for the prediction of employee retention.

21.2.1 Collection of Data Set

The data has been collected from Ceramic sanitary ware factories in India. The manufacturing units of the top ten brands of sanitary wares were identified and the request was made to the factory managers to cooperate for the research survey. Five factories agreed to cooperate with some restrictions including keeping anonymity of employee's identity and name of the factory. Thus five sanitary ware factories located in different places have been fixed for collecting primary data for this study, and named A, B, C, D and E for study purpose. A total of 550 samples were collected and the distribution of data collected among the five factories is shown in Table 21.1 [15].

The data set includes two types of features namely personal features and work-related features. A total of 17 features were collected and they are tabulated as follows in Table 21.2.

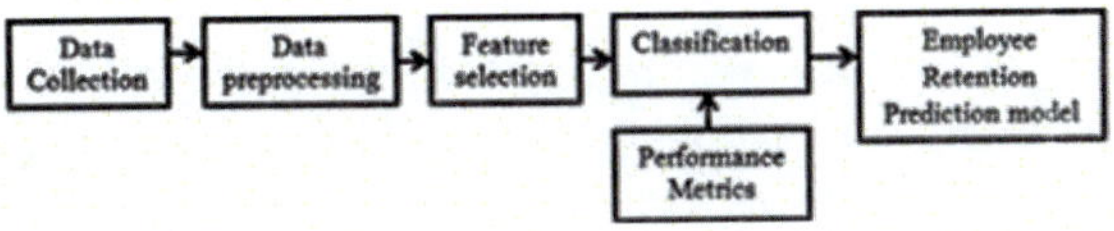

Fig. 21.1 Methodology for employee retention prediction

Table 21.1 Data set collection distribution

S. no	Name of the factory	Total employees	Sample size
1	A	450	115
2	B	350	90
3	C	600	153
4	D	350	90
5	E	400	102
	Total	2150	550

Table 21.2 Features set

Age	Work environment
Gender	Training
Education qualification	Flexible working hours
Income	Encouragement
Experience	Promotion
Designation	Manager support
Department of working	Goodwill
Employee retention	Satisfaction
Marital status	

The target feature is employee retention. It consists of a response which is a 'yes' for the employees who have retained and a 'no' for the employees who left the organization.

21.2.2 Data Preprocessing

The next step in the application of machine learning algorithm is to apply preprocessing stage. In this step the redundant and duplicate data sets were removed. Unfilled rows in the data set are also removed as their inclusion may produce inaccurate prediction. After removing the redundant and null samples, the total samples with no null were found to be 416 and it is further used for analysis and classification. The qualitative and categorical features were converted to numbers so that further processing using machine learning can be made. For instance, the department in the data set includes four departments namely Finance, HR, production and marketing. They were transformed into numerical values by assigning integers ranging from 1 to n. Therefore the department feature is transformed as follows: 1-Finance, 2-h, 3-Marketing and 4-Production. In the similar way other features such as age, gender, marital status, income and educational qualification were also converted into numerical values for further analysis.

21.2.3 *Percentage Analysis*

Percentage analysis is one of the statistical measures used to describe the characteristics of the sample. Percentage analysis involves computing measures of variables selected and their finding gives easy interpretation. In the present data set with 416 samples, the total employees retained is found to be 304 which is 73% and those left is 112 employees. Percentage analysis is done for age, gender, education, income, experience, department and designation of the respondents and interpretations are also suitability made.

Table 21.3 reveals the percentage analysis of employee retention with respect to age of employees. From Table 21.3 as well as from the line plot in Fig. 21.2, it is evident that the retention level is worst in below 25 years age group and high retention is seen among employees whose age is above 45 years old of age. Gender wise retention using percentage analysis is shown in Table 21.4 as well as Fig. 21.3. From the Fig. 21.3 it is found that the retention level is more among male employees than female employees. Marital-based percentage retention of employees is illustrated in Fig. 21.4 and Table 21.5. From that, it can be inferred that employees' retention is found to be affected by marriage also. Retention level is found to be 76% among married employees which is 14% more than unmarried employees.

Table 21.6 explains the retention level of employees based on their educational qualification. The percentage analysis is represented as line plot in Fig. 21.5. It

Table 21.3 Employee retention by age

S. no	Age (Years)	No of samples	Employees retained	Employees not retained	% of Employees retained
1	Below 25	34	13	21	47
2	Between 26 and 35	193	83	110	69
3	Between 36 and 45	146	68	78	77
4	Above 45	43	27	16	95

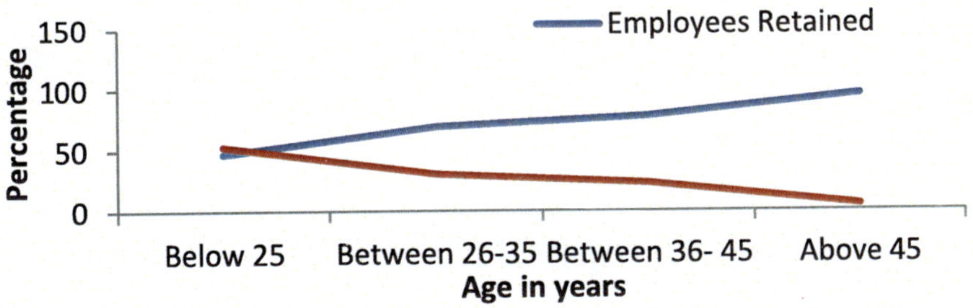

Fig. 21.2 Percentage analysis of employee retention by Age

Table 21.4 Employee retention by gender

S. no	Gender	No of samples	Employees retained	Employees not retained	% of Employees retained
1	Male	386	285	101	74
2	Female	30	19	11	63

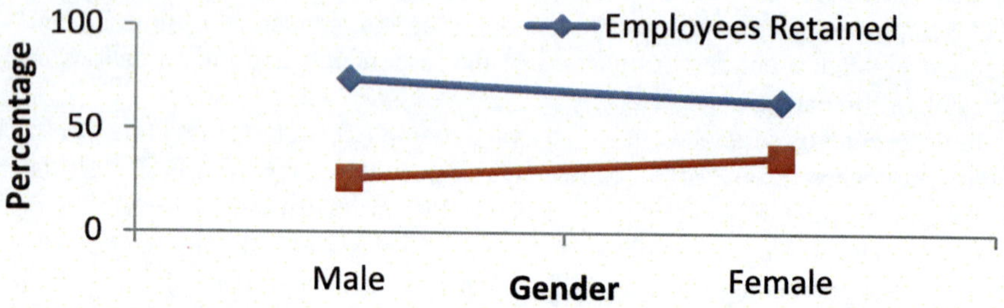

Fig. 21.3 Percentage analysis of employee retention by gender

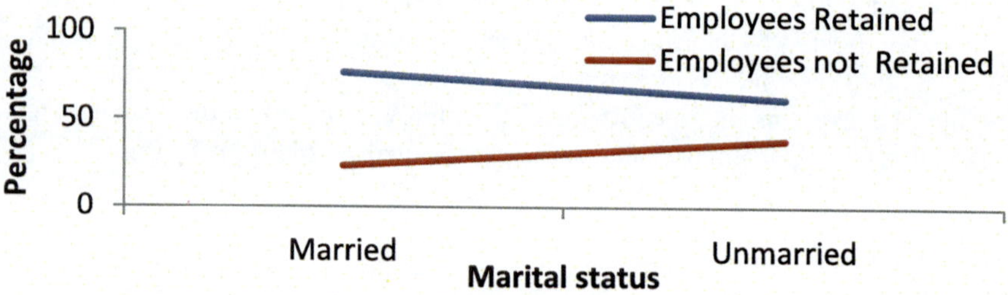

Fig. 21.4 Percentage analysis of employee retention by Marital status

Table 21.5 Employee retention by marital status

S. no	Marital status	No of samples	Employees retained	Employees not retained	% of Employees retained
1	Married	335	256	79	76
2	Unmarried	86	53	33	62

Table 21.6 Employee retention by education

S. no	Education group	No of samples	Employees retained	Employees not retained	% of Employees retained
1	UG	85	64	21	75
2	PG	155	105	50	68
3	Professional	147	127	20	86
4	Others	29	19	10	66

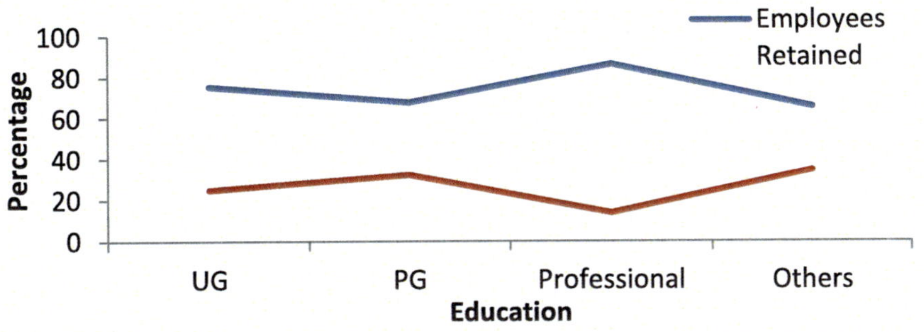

Fig. 21.5 Percentage analysis of employee retention by educational qualification

is inferred that employees having professional degree has high retention level of 86% followed by 75% contributed by UG degree employees. Thus the organization can enjoy the benefits of educated employees in implementing their target. Higher qualification increases marketability and for effective retention importance needs to give to the factors satisfying employees (Fig. 21.6).

Experience is also considered as an essential aspect for the study, since the duration of association of employees with the company exerts an impact on their retention. From Table 21.7 it is found that 83% of employees are having experience of more

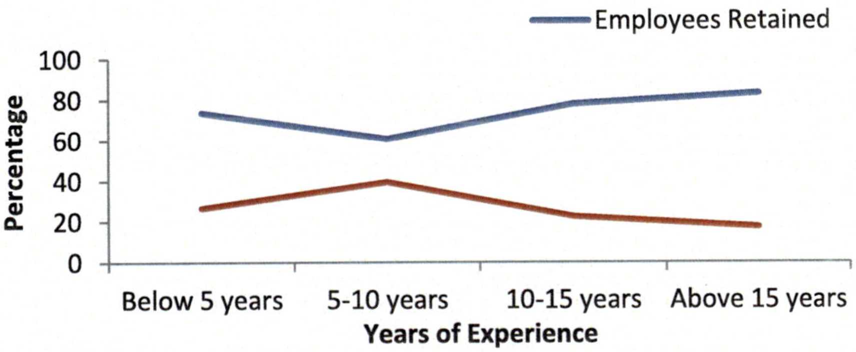

Fig. 21.6 Percentage analysis of employee retention by experience

Table 21.7 Employee retention by experience

S. no	Experience group	No of samples	Employees retained	Employees not retained	% of Employees retained
1	Below 5 years	83	61	22	73
2	6–10 years	122	74	48	61
3	11–15 years	117	91	26	78
4	Above 15 years	94	78	16	83

than 15 years and have shown highest retention. The retention level is lowest among six to ten years which is 61% followed by 11–15 years of experience employees with 78% retention level. From Table 21.8 and Fig. 21.7, income-wise employee retention can be understood. Below 20,000 income contributes to 61% and above 40,000 comes to 94%. The middle income group ranging from 20,000 to 30,000 contributes to 71% while 30,000–40,000 accounts for 76%. This can be concluded that more income can improve the retention level of employees.

From Fig. 21.8 it is found that employees of marketing department are predominant and their contribution is 84% followed by production with 76%. The tabular illustration is given in Table 21.9. Since ceramic sanitary ware products are consumed by a wide range of households, reasonable numbers of employees are required for

Table 21.8 Employee retention by income

S. no	Income group	No of samples	Employees retained	Employees not retained	% of Employees retained
1	Below 20,000	128	78	49	61
2	20,000–30,000	120	85	35	71
3	30,000–40,000	102	78	24	76
4	Above 40,000	66	62	4	94

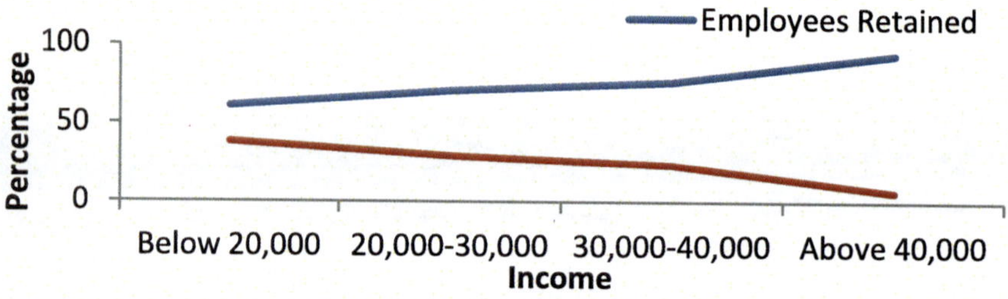

Fig. 21.7 Percentage analysis of employee retention by Income

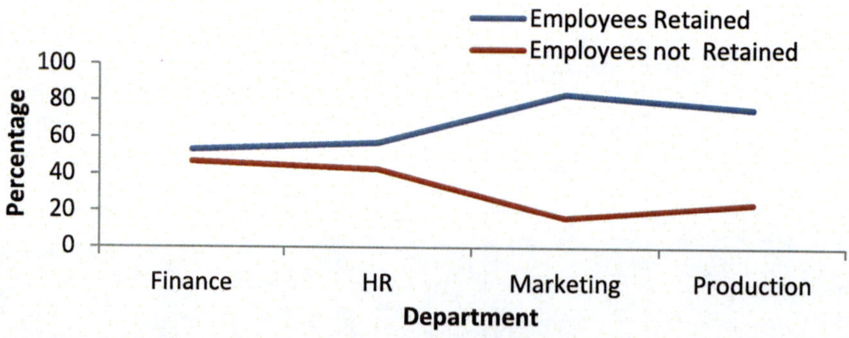

Fig. 21.8 Percentage analysis of employee retention by Department

Table 21.9 Employee retention by department

S. no	Department group	No of samples	Employees retained	Employees not retained	% of employees retained
1	Finance	32	17	15	53
2	HR	56	32	24	57
3	Marketing	80	67	13	84
4	Production	248	188	60	76

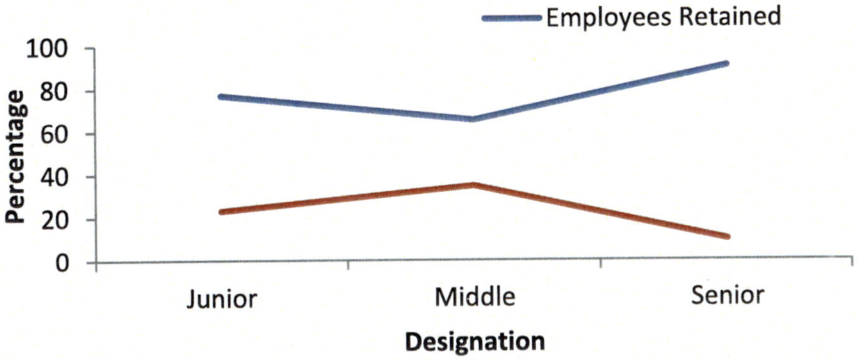

Fig. 21.9 Percentage analysis of employee retention by designation

Table 21.10 Employee retention by designation

S. No	Designation group	No of samples	Employees retained	Employees not retained	% of employees retained
1	Junior	121	93	28	77
2	Middle	222	145	77	65
3	Senior	73	66	7	90

marketing the products. Participants of HR and finance department contributed to 57% and 53%, respectively. Figure 21.9 and Table 21.10 show retention of employees by their designation. "Senior" dominates to the tune of 90%, followed by juniors 77% and middle designated employees contribution is 65%.

21.2.4 *Correlation Analysis*

The next step is to perform feature selection on the data to identify the key features that contribute to employee retention. For that purpose, correlation between all features and the target feature is performed and the results are tabulated in Fig. 21.10 as well as in Table 21.11. The correlation ranges from −0.136 to 0.614. The following

Table 21.11 Correlation matrix

	Employee retention	Age	Gender	Qualification	Experience	Monthly Income	Department	Marital status	Designation	Work Condition	Training	Flexible working hours	Encouragement	Promotion	Manager Support	Goodwill	Satisfaction
Employee retention	1.000	0.236	−0.061	0.108	0.123	0.233	−0.117	−0.136	0.065	0.586	0.537	0.523	0.504	0.582	0.533	0.606	0.614
Age	0.236	1.000	−0.062	0.102	0.292	0.298	−0.041	−0.208	0.158	0.135	0.094	0.069	0.179	0.144	0.177	0.179	0.103
Gender	−0.061	-0.062	1.000	−0.146	−0.098	−0.023	0.114	−0.026	0.006	0.043	0.048	0.059	0.096	0.067	0.115	0.047	0.130
Qualification	0.108	0.102	−0.146	1.000	0.238	0.195	−0.131	−0.093	0.061	0.124	0.061	0.076	0.035	0.041	0.057	0.064	0.024
Experience	0.123	0.292	−0.098	0.238	1.000	**0.760**	0.006	−0.571	**0.612**	0.036	0.063	0.070	0.080	0.102	-0.008	0.113	0.010
Monthly income	0.233	0.298	−0.023	0.195	**0.760**	1.000	0.088	−0.493	**0.628**	0.218	0.259	0.196	0.217	0.236	0.178	0.263	0.197
Department	−0.117	−0.041	0.114	−0.131	0.006	0.088	1.000	−0.157	0.028	−0.200	−0.184	-0.152	-0.187	-0.146	-0.133	-0.156	-0.188
Marital status	−0.136	−0.208	−0.026	−0.093	−0.571	−0.493	−0.157	1.000	−0.436	−0.026	−0.087	-0.116	-0.148	-0.123	0.034	-0.176	-0.077
Designation	0.065	0.158	0.006	0.061	0.612	0.628	0.028	−0.436	1.000	0.081	0.138	0.106	0.090	0.092	0.047	0.138	0.081
Work condition	**0.586**	0.135	0.043	0.124	0.036	0.218	−0.200	−0.026	0.081	1.000	0.735	0.599	0.661	0.631	0.554	0.672	0.748
Training	**0.537**	0.094	0.048	0.061	0.063	0.259	−0.184	−0.087	0.138	**0.735**	1.000	0.792	0.692	0.619	0.585	0.718	0.742
Flexible working hours	**0.523**	0.069	0.059	0.076	0.070	0.196	−0.152	−0.116	0.106	**0.599**	**0.792**	1.000	0.694	0.617	0.528	0.702	0.781
Encouragement	**0.504**	0.179	0.096	0.035	0.080	0.217	−0.187	−0.148	0.090	**0.661**	**0.692**	**0.694**	1.000	0.667	0.609	0.754	0.739
Promotion	**0.582**	0.144	0.067	0.041	0.102	0.236	−0.146	−0.123	0.092	**0.631**	**0.619**	**0.617**	**0.667**	1.000	0.716	0.695	0.722
Manager support	**0.533**	0.177	0.115	0.057	−0.008	0.178	−0.133	0.034	0.047	**0.554**	**0.585**	**0.528**	**0.609**	**0.716**	1.000	0.653	0.699
Goodwill	**0.606**	0.179	0.047	0.064	0.113	0.263	−0.156	−0.176	0.138	**0.672**	**0.718**	**0.702**	**0.754**	**0.695**	**0.653**	1.000	0.758
Satisfaction	**0.614**	0.103	0.130	0.024	0.010	0.197	−0.188	−0.077	0.081	**0.748**	**0.742**	**0.781**	**0.739**	**0.722**	**0.699**	**0.758**	1.000

characteristics are identified from the correlation Table 21.11.

Highest positive correlation exists between employee retention and satisfaction. This indicates that retention can be maintained by giving priority to satisfaction of employees.

High positive correlation is seen between satisfaction and working condition and flexible working hours which clearly indicates that importance should be given by the HR team to these features for employee retention.

Negative correlation exists between employee retention and marital status. This indicates that employees leave the organization when they are unmarried.

21.3 Model Building

In this step, the various machine learning algorithms are used for prediction analysis. The machine learning algorithms used in this paper are decision tree, Naive Bayes, AdaBoost, Logistic regression, K-Nearest Neighbour (KNN) and Support Vector Machine (SVM). The entire process is simulated in Python 2.7 using Orange tool on a Windows 10 operating system. Here the aim used is to find the best model that makes accurate prediction of retention of employees. In order to perform predictions, it is necessary to train the models using the data set. It is also required to test the model using the test data set. There is one single real-time data set available, the single data set is divided into training data and testing data set in the ratio of 70:30 [8]. In order to perform model validation, Stratified K-Fold approach is used as it provides better results than k-fold cross-validation technique.

Feature selection using decision tree classifier is performed to identify the features that are more important for employee retention. The results are shown in Fig. 21.11. Hence from Fig. 21.11, the top three features namely satisfaction, promotion and goodwill towards organizations are taken for further processing using the different prediction algorithms as they have highest value of correlation greater than 0.6. The parameters used for evaluating the performance of the prediction algorithms are accuracy, area under Region of convergence, F1, precision and recall. Table 21.12 compares the various algorithms in terms of these parameters.

As SVM has the highest value of F1 score together with accuracy, it is found to be the best choice for prediction of employee retention. Therefore the confusion matrix for SVM is shown in Table 21.13 as follows. As seen from Table 21.13 it can be found that the actual employees who retained are 304 and who left is 112. Using SVM the correctly predicted value of employees retained is 303 and those who left the organization is 113, which is quite nearer to the actual.

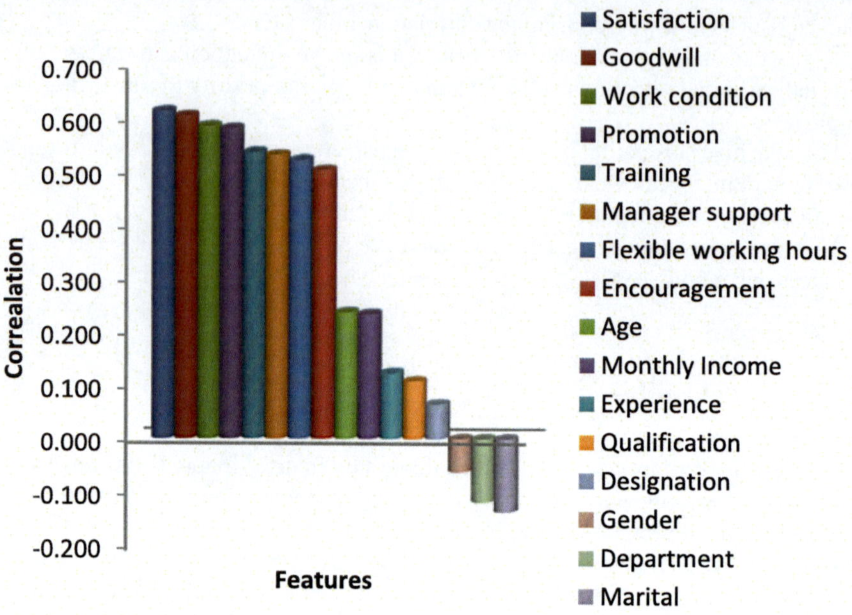

Fig. 21.10 Correlation analysis

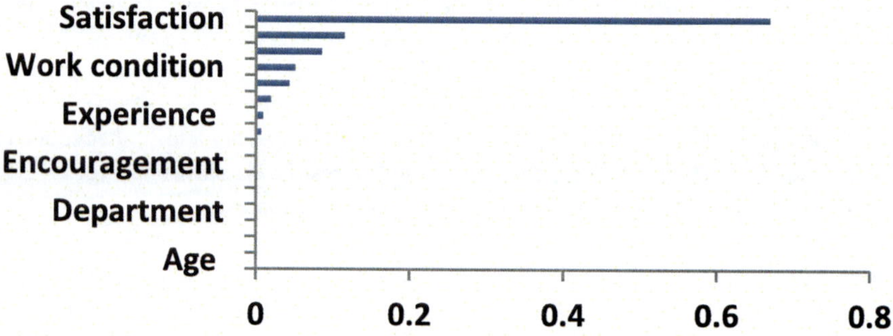

Fig. 21.11 Feature Selection by decision tree classifier

Table 21.12 Comparison of prediction algorithms

Model	AUC	CA	F1	Precision	Recall
SVM	0.9984	0.9832	0.9832	0.9832	0.9832
AdaBoost	0.9784	0.9808	0.9808	0.9809	0.9808
Decision tree	0.8912	0.9375	0.9367	0.9369	0.9375
Naive bayes	0.9896	0.9351	0.9365	0.9414	0.9351
Logistic regression	0.9487	0.8654	0.8520	0.8752	0.8654
KNN	0.8928	0.8462	0.8433	0.8421	0.8462

Table 21.13 Confusion matrix of SVM

		Predicted		
		0	1	Σ
Actual	**0**	109	3	**112**
	1	4	300	**304**
	Σ	**113**	**303**	**416**

21.4 Result and Discussion

This work presents prediction of employees in ceramic manufacturing industries in India. The percentage analysis made for the personal details related features indicates that the retention level increases with age and marriage. More male employees are found to be retained than female employees. More retention is seen among employees with more than 15 years of experience and with salary above 40,000 who are mostly senior designation groups which proved in the correlation Table 21.11 also. Retention is lower for employees with PG qualification than with UG. Highest retention is seen among marketing employees and lowest among finance department employees. The top three features that determine the retention of employees were satisfaction, promotion and goodwill. Employee satisfaction is the highest determining factor for employee retention. The SVM classifier is found to produce highest accuracy and F1 score than other classifiers. Hence it can be used to build the employee retention prediction model.

21.5 Conclusions

This paper presents a machine learning approach to predict the employee retention in ceramic manufacturing industries in India. The factor that contributed to the retention of employees was identified. Also the prediction of likelihood of an employee to leave the organization is also discussed. To perform all these, firstly data preprocessing is performed to remove redundant and missing values. Then percentage analysis is performed to identify the impact of personal factors on the retention of employees. Then feature selection is performed to select the best feature that aid in the prediction of employee retention. In order to validate the training and test phase stratified cross-validation is performed. Finally various machine learning algorithms were used for the prediction of employee retention. The results indicate that using satisfactory level, promotion and goodwill as features and SVM as classifier, it is possible to predict whether an employee will retain or not. Thus our proposed model can be used by HR managers to select suitable measures to retain the employees. In future, this work can be further extended by giving more emphasis to feature selection with inclusion of more features such as market value, societal impact and environmental impact. Also research towards time of retention of employees can also done.

References

1. Thilaka, B., Sivasankaran, J., Udayabaskaran, S.: Optimal time for withdrawal of voluntary retirement scheme with a probability of acceptance of retirement request. J. Inf. Technol. **2**(04), 201–206 (2020)
2. Chakrabarty, N.: A regression approach to distribution and trend analysis of quarterly foreign tourist arrivals in India. J. Soft Comput. Paradig (JSCP) **2**(01), 57–82 (2020)
3. Koresh, M.H., Deva, J.: Analysis of soil nutrients based on potential productivity tests with balanced minerals for maize-chickpea crop. J. Electron. **3**(01), 23–33 (2021)
4. Garg, S., Sinha, S., Kar, A. K., Mani, M.: A review of machine learning applications in human resource management. Int. J. Product. Perform Manage (2021)
5. Mikalef, P., Gupta, M.: Artificial intelligence capability: conceptualization, measurement calibration, and empirical study on its impact on organizational creativity and firm performance. Inf. Manag. **58**(3), 103434 (2021)
6. Young, A.G., Majchrzak, A., Kane, G.C.: Organizing workers and machine learning tools for a less oppressive workplace. Int J Inf Manag **59**, 102353 (2021)
7. Kakulapati, V., Chaitanya, K.K., Chaitanya, K.V.G., Akshay, P.: Predictive analytics of HR-A machine learning approach. J. Stat. Manag. Syst. **23**(6), 959–969 (2020)
8. Umamaheswari, S., Krishnan, J.: Retention factors and their relative significance in ceramic manufacturing industries in India. Asian Soc. Sci. **11**(13), 260 (2015)
9. Jain, N., Tomar, A., Jana, P.K.: A novel scheme for employee churn problem using multi-attribute decision making approach and machine learning. J. Intell. Inf. Syst. **56**(2), 279–302 (2021)
10. Qutub, A., Al-Mehmadi, A., Al-Hssan, M., Aljohani, R., Alghamdi, H.S.: Prediction of employee attrition using machine learning and ensemble methods. Int. J. Mach. Learn. Comput **11** (2021)
11. Fallucchi, F., Coladangelo, M., Giuliano, R., William De Luca, E.: Predicting employee attrition using machine learning techniques. Computers **9**(4), 86 (2020)
12. Ponnuru, S., Merugumala, G., Padigala, S., Vanga, R., Kantapalli, B.: Employee attrition prediction using logistic regression. Int. J. Res. Appl. Sci. Eng. Technol. **8**, 2871–2875 (2020)
13. Alao, D.A.B.A., Adeyemo, A.B.: Analyzing employee attrition using decision tree algorithms. Comput. Inf. Syst. Dev. Inf. Allied Res. J. **4**(1), 17–28 (2013)
14. Al-Radaideh, Q.A., Al Nagi, E.: Using data mining techniques to build a classification model for predicting employees performance. Int. J. Adv. Comput. Sci. Appl. **3**(2) (2012)
15. Umamaheswari, S., Krishnan, J.: Work force retention: role of work environment, organization commitment, supervisor support and training & development in ceramic sanitary ware industries in India. J. Indust. Eng. Manag. (JIEM) **9**(3), 612–633 (2016)

Chapter 22
Drowsiness Detection with Alert & Notification System

Souvik Sarkar, Rohit Kumar, Sidhant Singh, and Debraj Chatterjee

Abstract India is considered to be a developing country, along with development in transportation sectors but road accidents are a major hurdle. According to the Transport Research Wing of the Ministry of Road Transport and Highways, 2019, India ranks first in the number of road accident deaths across the 199 countries and accounts for almost 11% of the accident-related deaths in the World. Road accidents occur mainly due to poor road conditions, careless driving, and sometimes due to fatigue-related causes like drowsiness while driving. This study mainly focuses on drowsiness detection while driving and a way to prevent accidents. The National Highway Safety Administration reported that drowsy driving results in 1,550 deaths, 71,000 injuries, and more than 100,000 accidents per year. The majority of drivers reported being falling asleep in less than an hour while driving. This could be prevented by monitoring the eye-lid positions of the driver and if it closes for a certain interval of time, an alert system will be invoked alerting the driver to be watchful and along with this emergency contacts of the driver will be notified of this situation.

22.1 Introduction

Machine learning [1] is a technology used for getting computers to learn and behave like artificial humans along with spontaneous auto-improvement by feeding data and information to them in form of observations and real-world interactions. Machine learning algorithms are designed as programs that can learn data itself and improve from experiences, without the need for any human involvement. It is the future,

S. Sarkar (✉) · R. Kumar · S. Singh · D. Chatterjee
Techno International New Town, Kolkata, India

R. Kumar
e-mail: rohit.kumar.2018@cse.tict.edu.in

S. Singh
e-mail: sidhant.singh.2018@cse.tict.edu.in

D. Chatterjee
e-mail: debraj.chatterjee@tict.edu.in

J. I.-Z. Chen et al. (eds.), *Machine Learning and Autonomous Systems*, Smart Innovation, Systems and Technologies 269, https://doi.org/10.1007/978-981-16-7996-4_22

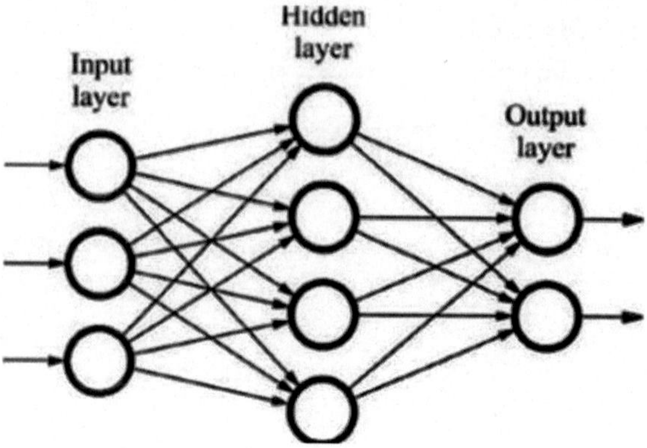

Fig. 22.1 Layers of a neural network

implemented in a wide range of fields like image recognition, speech recognition, traffic prediction, language translation, drowsiness detection, etc. There are three types: supervised learning, unsupervised learning, and reinforcement learning. This study is based on supervised learning. Supervised learning can be described as using labeled training data to learn the mapping function from the input variables (X) to the output variable (Y).

$$Y = f(X) \tag{22.1}$$

A neural network [2] is a connection of layers of nodes that are connected in such a way that the output of one neuron acts as the input for the neuron in the next layer. A neural network comprises an input layer and output layer and hidden layers all the layers are connected in such a way that all the layers act as a single unit to produce a result. Figure 22.1 represents a neural network with different layers.

CNN (Convolutional Neural Network) [3] is a classification algorithm where images are taken as input, objects/aspects are assigned with weights/biases, and then it is differentiated from other objects. It is mostly applied to analyze visual images. Pre-processing required in CNN is much lower as compared to other classification algorithms. CNN can be visualized as shown in Fig. 22.2.

Each neuron takes an input and after passing from a function of weight and biases, it enters an activation function. Activation functions [4] decides if a neuron will fire or not. In the machine learning models, there is a lot of activation function, for example, sigmoid, ReLU, softmax, tanh, etc. But for our model, we had used softmax and ReLU as activation functions. ReLU is one of the commonly used activation functions in the machine learning world and can be expressed mathematically as in Eq. (22.2):

$$f(x) = \max(0, x) \tag{22.2}$$

So, any input less than 0 will simply be 0 and the rest will return the number itself. It can be expressed easily as in Fig. 22.3.

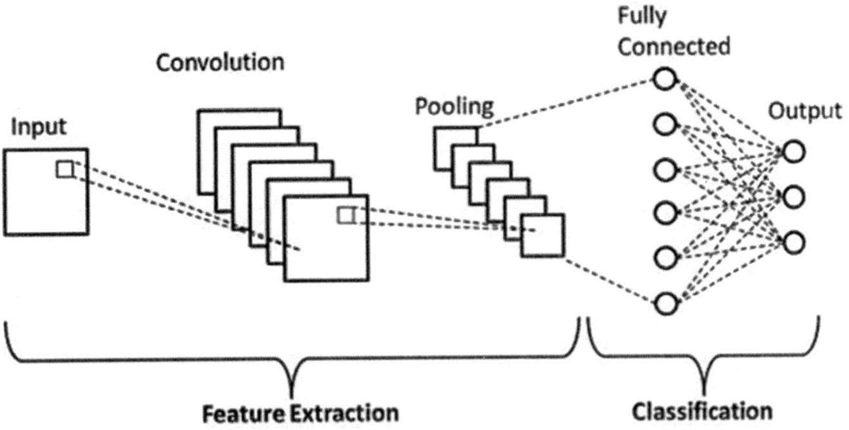

Fig. 22.2 A convolutional neural network

Another activation function that has been used in the model in the output layer is the softmax function, it represents the argmax of the likelihood of all the possible outcomes and is mathematically represented by Eq. (22.3)

$$f(x)_i = \frac{e^{x_i}}{\sum_{j=1}^{k} e^{x_j}} \tag{22.3}$$

Here, x is array i, and j is the indices of the array f(x) that will generate an array.

Drowsiness [5] is the state of sleepiness. Major causes for this state are insufficient sleep, medications, and drinking alcohol. During such a condition operating machines, driving could be risky and might cause serious casualties. Some modules used in this study for this purpose are Computer Vision, OpenCV, Tensor Flow with Keras, and Pygame.

Computer Vision [6] is studying and understanding 2D images and reconstructing a 3D scene from them, in accordance with the structure and their properties present. It comprises modeling and duplicating visions of humans using computers and various devices.

OpenCV [7] is a python library used to develop real-time computer vision applications. Mainly it emphasizes video capturing, image processing, and analysis comprising of features like object detection, face detection, etc.

TensorFlow [8] is also a python library, which is generally used for various tasks such as training of neural network based on dataflow and different layers.

Keras is an API that is written in python and runs on top of the machine learning platform Tensorflow. It is used to make the implementation of neural networks a feasible way.

Using computer vision, the face of the driver will be monitored via video stream and eyes will be extracted from the facial image using OpenCV. A model has been prepared using Tensorflow with Keras to detect open and closed eyes. Extracted eye

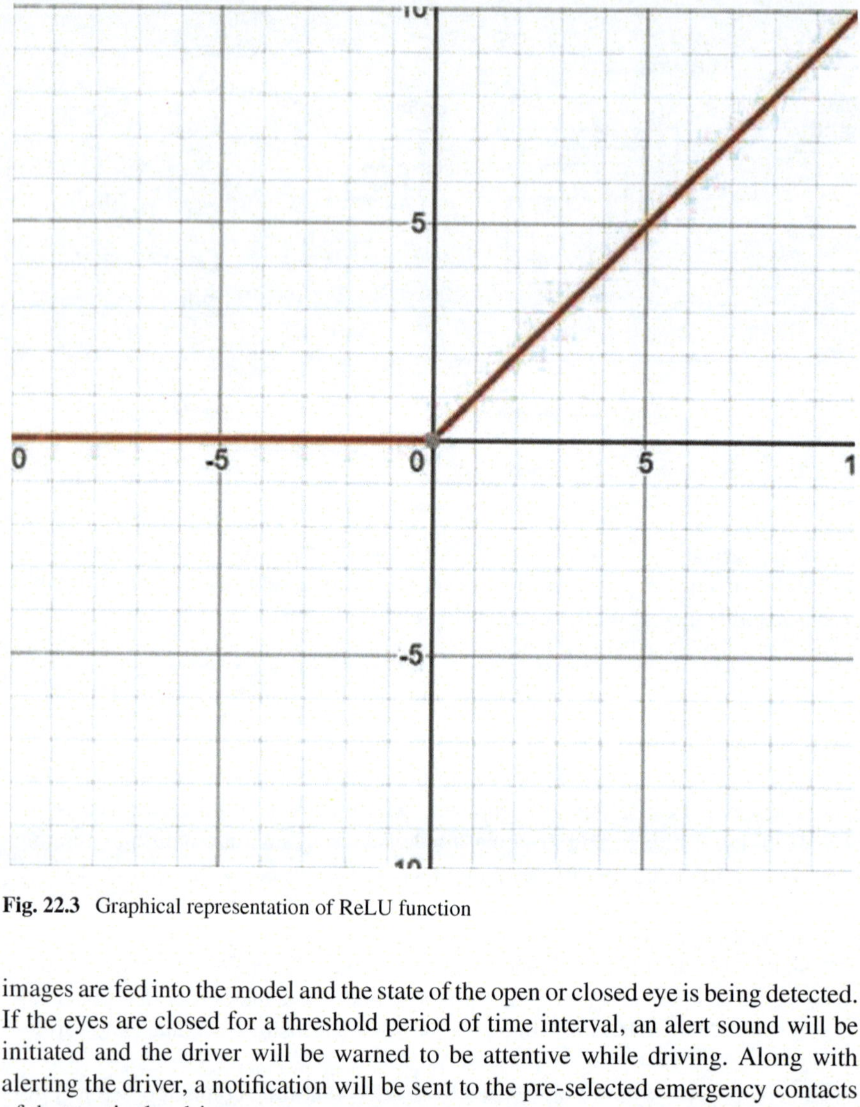

Fig. 22.3 Graphical representation of ReLU function

images are fed into the model and the state of the open or closed eye is being detected. If the eyes are closed for a threshold period of time interval, an alert sound will be initiated and the driver will be warned to be attentive while driving. Along with alerting the driver, a notification will be sent to the pre-selected emergency contacts of that particular driver.

There are various studies on this topic and most of them had stopped after the feature extraction or has used various sophisticated devices that can really increase the cost as well as can cause a problem for the driver such as obstructing driver vision or has done the detection part but didn't propose any work on the alert system to aware the driver.

In our work we have used the neural network to detect and classify the extracted feature. Moreover, this model can be implemented on a mobile device as well as a very low hardware cost. This work also deals with the alertness of the drowsy driver as discussed in the methodology section.

22.2 Methodology

22.2.1 Planning

In this segment, we are going to define a proper methodology to detect drowsiness and prevent accidents. At first, we will define our dataset and method for pre-processing of that data. After that, we will build a model to train the data and validate it with test data. In Fig. 22.4 the steps for training the data and validation is shown.

OpenCV captures the face from the input video and extracts the eye from it. It is then fed into the trained model and the probability of the eye being closed is carried out. If the eye is found to be in a closed position for a certain time then an alert system is triggered (Fig. 22.5).

The whole method discussed in this paper can be implemented at a very low cost as well as this model is efficient in different lighting conditions with minimal errors. Moreover, for the feature extraction, even if the model is unable to detect the eyes due to some reason then certain areas of the face will still be extracted from the video stream for analysis. This work can be easily implemented on a website or on an app running on any platform easily. It can also be implemented as a dedicated device as the work is not highly hardware-dependent.

22.2.2 Data Pre-Processing

The dataset used for training the model is from the eyes-open-closed-dataset [9] of kaggle, this dataset contains a lot of open and closed eye images in different lighting

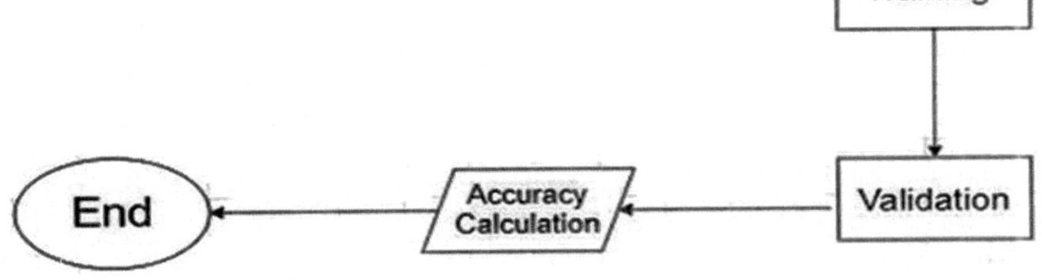

Fig. 22.4 Flow chart showing Training of Model

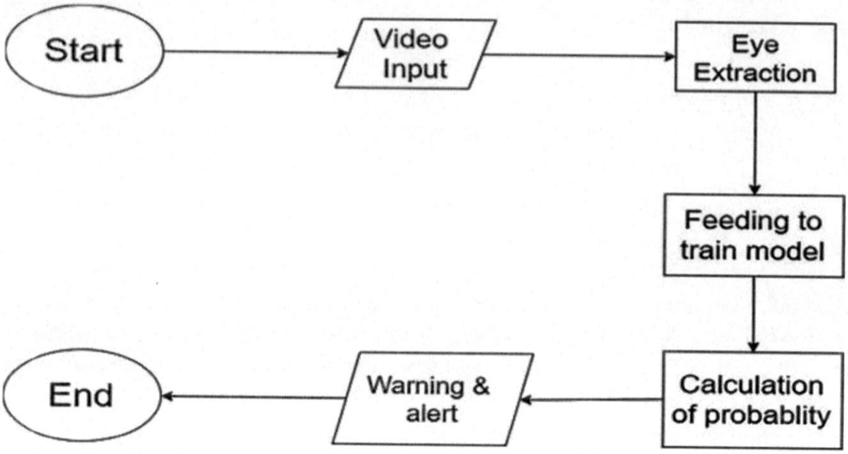

Fig. 22.5 Flow chart showing the overall mechanism

conditions which is ideal for our work as the condition of the image is not always ideal in a journey. The dataset is first separated into two categories of open-eye and closed-eye manually for the training purpose. Then, the separated images are pre-processed [10] so that each of the images in the dataset will have the same dimension and corrupt images will be rejected. After the standardization of the images, the images are shuffled to limit the chances of our model getting biased. The images present in the dataset has an initial size of 82 × 82 pixels but the computation of 13,031 such images will be expensive so to reduce the computation cost-effectively such that the reduced size will not affect the accuracy of the model we have reduced the size to 50 × 50 and converted to Grayscale channel. Figure 22.6 represents the scaling of the image.

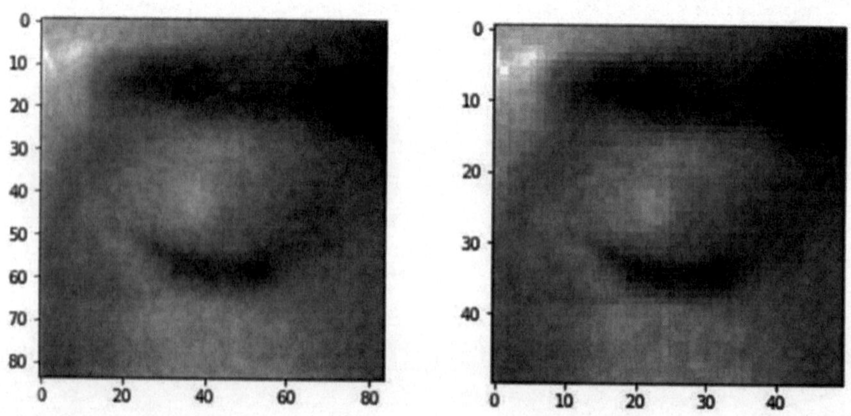

Fig. 22.6 Image compression

The prepared data is then divided into an 8:2 ratio for training and testing data, respectively.

22.2.3 Model Preparation

The training data is then fed to a sequential model so, the input layer will be a matrix of 32 × 50 × 50, 32 represents the batch size. Further, the model comprises convolutional, pooling, dense layers attached with appropriate activation functions. Specifically, the model consists of two convolutional layers with 256 filters with a kernel having size 3. Further computed by ReLU function, followed by Maxpooling layer, followed by a dense layer of 64 nodes and two nodes, respectively. The final layer is later on computed to a softmax activation function giving the result as 0 or 1 representing close or open, respectively. Figure 22.7 represents the model and its layers.

The model is trained with a learning rate of 10e-3 with Adam optimizer and with a loss function of binary cross entropy for 20 epochs. The model is saved for further use.

Model: "sequential"

Layer (type)	Output Shape	Param #
conv2d (Conv2D)	(None, 48, 48, 256)	2560
max_pooling2d (MaxPooling2D)	(None, 24, 24, 256)	0
conv2d_1 (Conv2D)	(None, 22, 22, 256)	590080
max_pooling2d_1 (MaxPooling2	(None, 11, 11, 256)	0
flatten (Flatten)	(None, 30976)	0
dense (Dense)	(None, 64)	1982528
dense_1 (Dense)	(None, 2)	130
activation (Activation)	(None, 2)	0

Total params: 2,575,298
Trainable params: 2,575,298
Non-trainable params: 0

Fig. 22.7 Different layers of the model

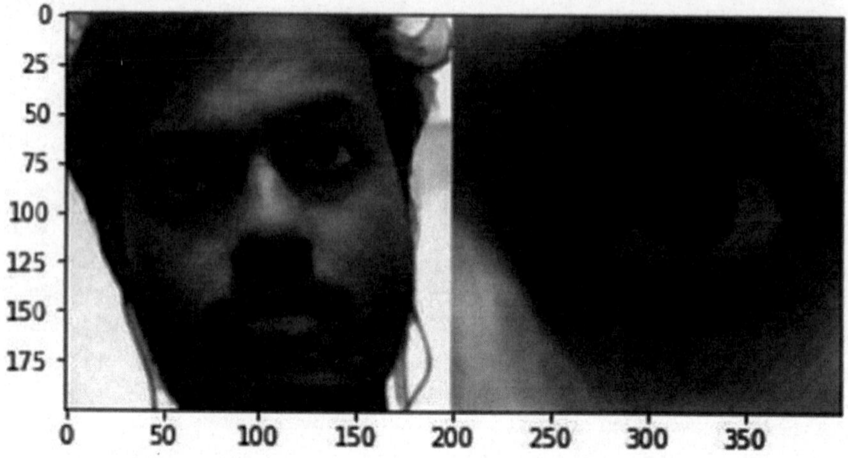

Fig. 22.8 Feature extraction

22.2.4 *Feature Extraction*

OpenCV is used in this paper for continuous video input then it is converted to an image. Facial features were extracted from the video input. Figure 22.8 shows a small demonstration of the extraction of features.

But the image that has been extracted is in a large dimension, our model in the input layer can take input of image of size 50 × 50 but this image is of size 200 × 200. So we have to scale the image down to the expected size so after re-sizing the image, the face and feature look like Fig. 22.8.

Figure 22.9a, b represents the extraction of eye from face image in close and open state, respectively. The eye extracted image is then passed to the trained model to predict if the extracted eye image is in open or close state and the duration of eye in a close state is calculated.

22.2.5 *Alert Mechanism*

If the calculated value exceeds a threshold value then an alert system is provoked. The alert system first plays warning sounds to the driver if the driver's state did not improve after continuous warning then his preferred emergency contacts will be notified. For the notification purpose the whole system must be connected to an internet-enabled device and beforehand emergency contacts must be added by the person, so that the added contacts will get the live footage and location of the person and can act accordingly to avoid any serious accidents. The whole system will be connected to a web server and can be accessed by website or an app. The

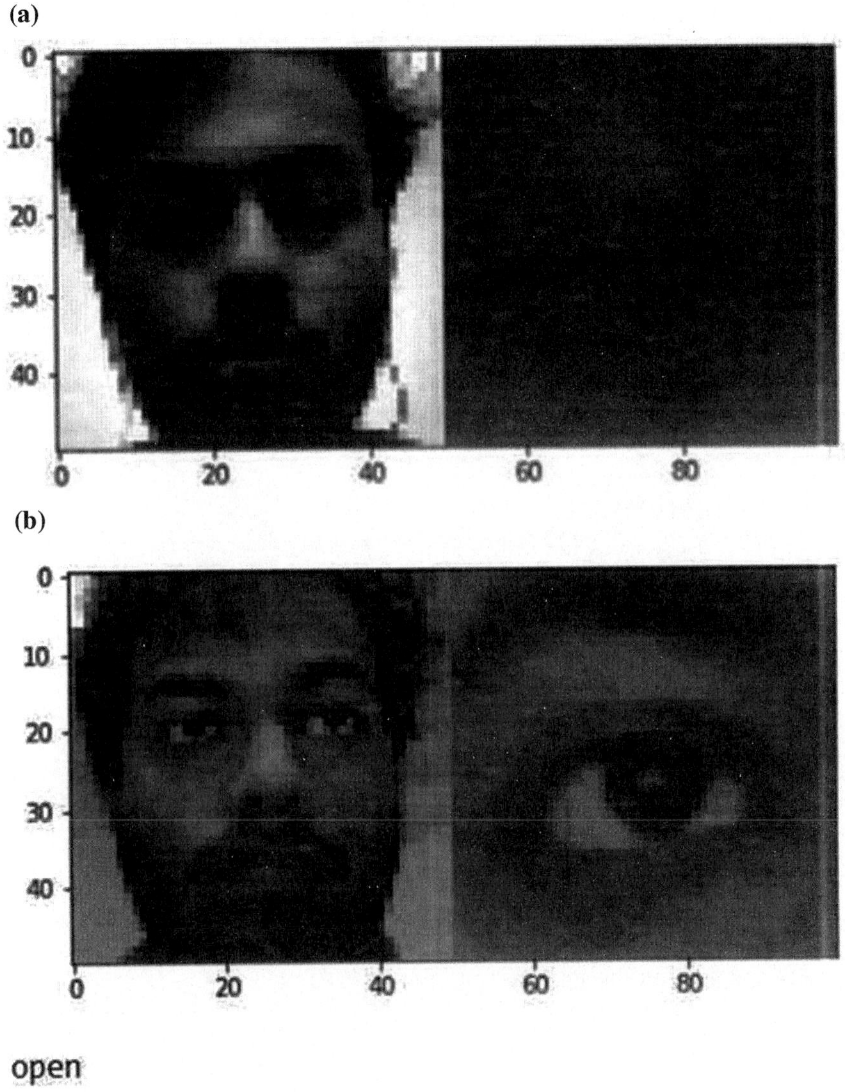

Fig. 22.9 **a** Close state of eye. **b** Open state of eye

website could be built over the Django framework [11] along with Javascript, Django channels, WebRTC [12], and WebSockets [13]. For connection between two peers in the network, they have to set up a peer-to-peer connection for which they have to share some information with one another, i.e. SDP (Session Description Protocols) [14] which will be collected when a user inputs their emergency contacts. So, the SDP will be shared using the intermediary server called Signaling. In our proposal, Django channels will act as the Signaling server and the connection to the Signaling

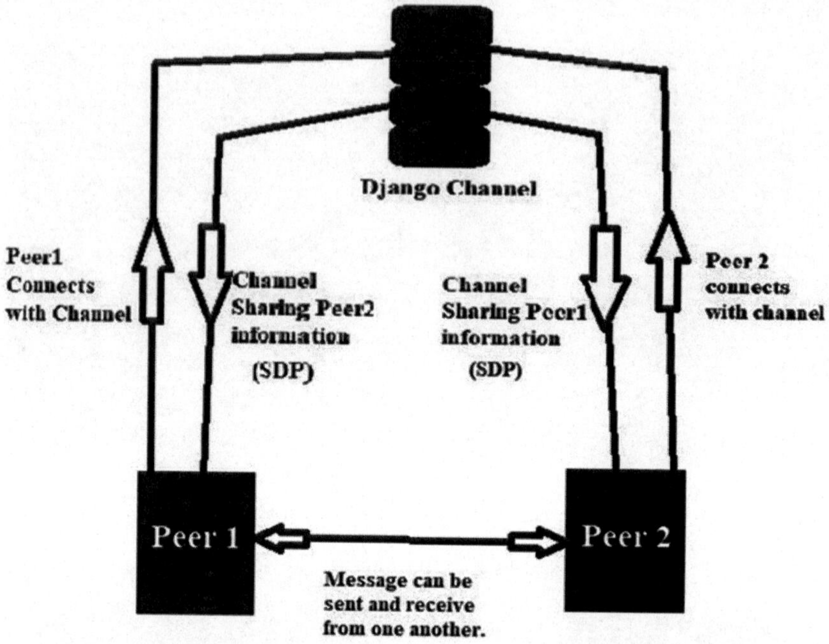

Fig. 22.10 Notification system illustration

server is done using the WebSockets. Once both the peers have their SDP, the peer-to-peer connection is established. This can be illustrated using the image below in Fig. 22.10.

After setting up the connection, both the peers are ready to share information, i.e. with the established connection whenever there is a need the system will send the live footage as well as the current location of the driver. As both the image streams as well as the geographical coordinate can be converted to python objects, they can be easily sent using the channels.

22.3 Result

22.3.1 Eye: Close-Open State

The results from the model as well as extraction of required features from the face are quite good with high accuracy. The model can easily differentiate between the open and close eye. Figure 22.11 shows the results of some of the validation dataset.

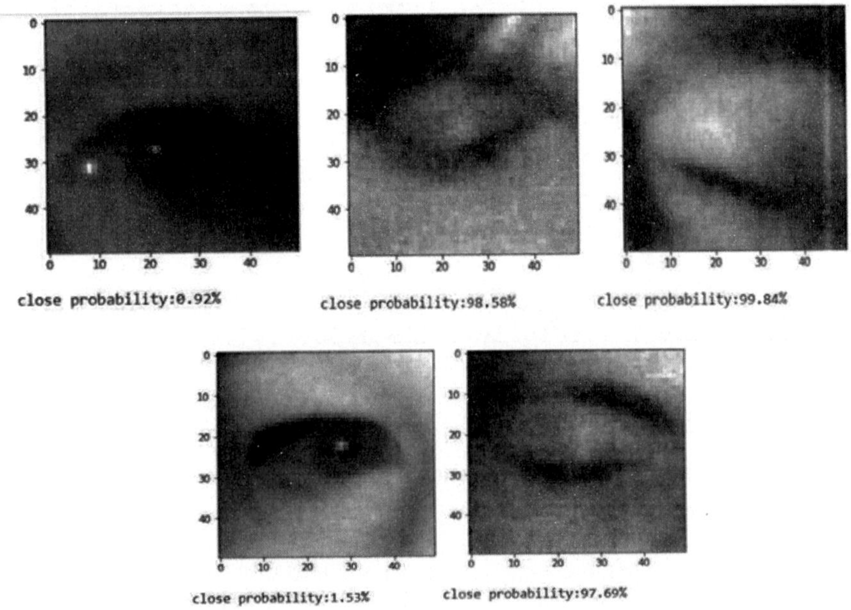

Fig. 22.11 Model detecting state of eye

22.3.2 *Model Efficiency*

The accuracy and loss that has been achieved while training this model for both validation and training data is good and can be visualized by the below figure generated by the Tensorboard [15] which is a module Tensorflow.

Figure 22.12 represents the variation of epoch accuracy in the y-axis with the number of epochs [16] on the x-axis. Both the curves represent the value of accuracy

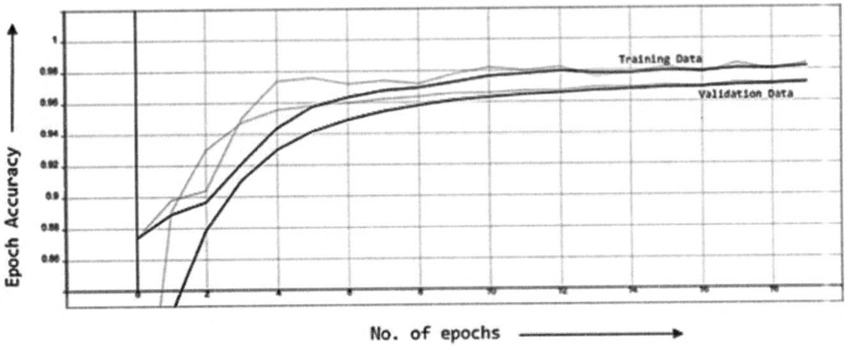

Fig. 22.12 Graph showing Model accuracy

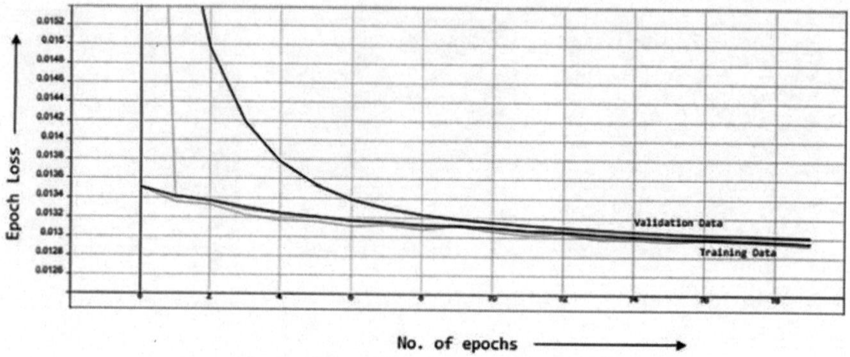

Fig. 22.13 Graph showing epoch-loss in the model

for validation (bottom) and testing (topmost) datasets, respectively. As we can see from the figure that training and validation dataset has achieved a good accuracy of 96% and 98.9%, respectively.

Figure 22.13 represents the variation of epoch-loss in the y-axis with the number of epochs on the x-axis. For both the training and validation dataset, the loss decreases from 12.2% and 13.7% to 1.3% and 1.32%, respectively. In the above figure, the green and gray represent validation and testing dataset, respectively.

22.3.3 Graphical Representation of Eye Pattern

The images from Fig. 22.9 then fed into this trained model to identify the state of the eye and if it finds any abnormality in the eye for a duration of four seconds the warning system activates.

Figure 22.14 is the representation of an open-eye and a close-eye for a period of 30 s, the x-axis represents the time and the y-axis represents the probability of a closed state of the eye. The black area represents the amount of time the driver has passed the threshold limit. As we can see that for a period of three to eight seconds (approximately) during that period it crosses the time limit of four seconds, which is the threshold value. The model has detected that the probability of eye in a close state is higher so, the machine will trigger the alarm system and continues for a while. If the driver did not respond during that period the system will notify and send warnings to the emergency contacts.

Figure 22.15 represents the pattern of a normal person passing the threshold as we can see the duration of a person passing the threshold limit never exceeds the four seconds mark so the alarm will not trigger.

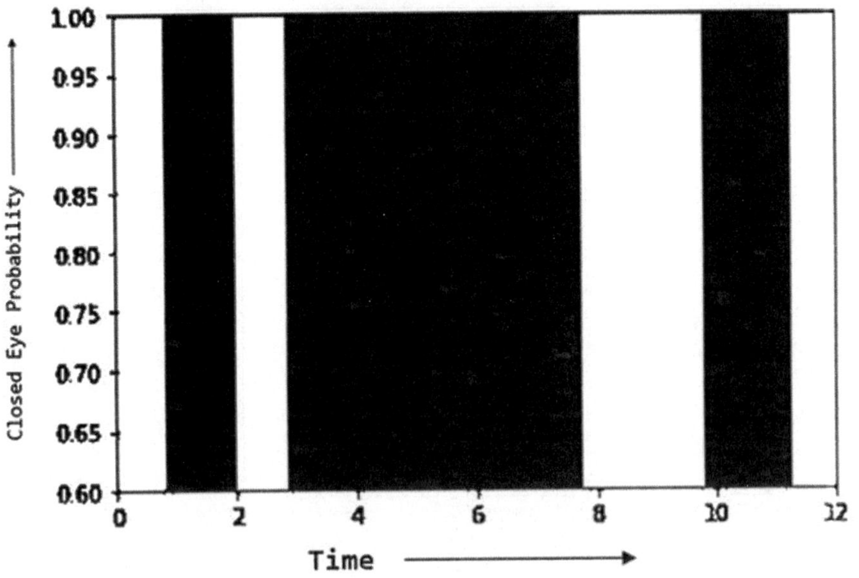

Fig. 22.14 Graphical representation of eye pattern of drowsy person

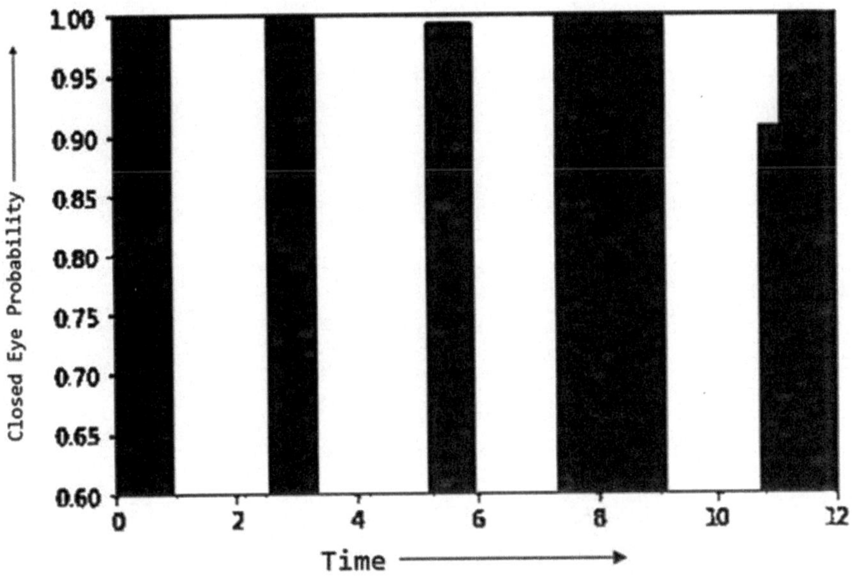

Fig. 22.15 Graphical representation of eye pattern of normal person

22.4 Conclusion

In previous approaches to implement a drowsiness detection system while driving, no notification system was integrated with the system. Our approach includes CNN along with a less dimensional image feed for faster prediction of the state of the eye whether closed or open. This system is supposed to be triggered when the probability of a closed state of the eye is high for a time interval and that time interval must pass the threshold value. Also, the model has been trained with the dataset which includes images in various lighting conditions and for every gender type which makes the detection more precise.

Implementation of this model will prevent accidents and will decrease fatalities caused by road accidents and will save the lives of many people. Moreover, if an accident occurs family members or guardians can act quickly so the life of the person in danger may be saved.

The limitations of our proposed work are that the system has to be internet-enabled and the person must not wear shady glasses. The system is capable to detect the state from naked eyes or transparent glasses. The system's region of interest is currently the eye which can be further expanded over facial expressions also.

The scope of this proposed work in the future can be implemented with a physical device that will be installed in the front dashboard of the vehicle and with any internet-enabled system, real-time images will be fed through which the state of the eyes is detected and the system works accordingly. Moreover, if any misshappening occurs, then the emergency contacts will get the info such as location and person's info of the incident.

Acknowledgements We would like to express our heartfelt gratitude to Prof. Debraj Chatterjee of the Department of Computer Science Engineering, whose significant role as mentor was invaluable for this research work. We are extremely thankful for his encouraging and supportive efforts in advising and for the guidance and reference materials provided for moral support. Last but not least, we convey our gratitude to all the teachers for providing the technical skill that will always remain an asset.

References

1. Margineantu, D., Wong, W.-K., Dash, D.: Machine learning algorithms for event detection. Mach. Learn. **79**(3), 257 (2010)
2. Wang, S.-C.: Artificial neural network. In: interdisciplinary computing in java programming, pp. 81–100. Springer, Boston, MA (2003)
3. Khalajzadeh, H., Mansouri, M., Teshnehlab, M.: Face recognition using convolutional neural network and simple logistic classifier. In: Soft computing in industrial applications, pp. 197–207. Springer, Cham (2014)
4. Chen, Z.X., Zhu, H.Y., Wang, Y.G.: A modified extreme learning machine with sigmoidal activation functions. Neural Comput. Appl. **22**(3), 541–550 (2013)
5. Tilley, D.H., Erwin, C.W., Gianturco, D.T.: Drowsiness and driving: preliminary report of a population. Survey. No. 730121. SAE Technical Paper, (1973)

6. Rosenfeld, A.: From image analysis to computer vision: an annotated bibliography, 1955–1979. Comput. Vis. Image Underst. **84**(2), 298–324 (2001)
7. Magamedova, D.M.: OpenCV-computer vision tool. Trends in the Development of Science and Education (2020)
8. Ali, S.M., Krishna, V.D.S., Srinivas, S.: Overview on open source machine learning platforms-tensorflow. J. Innov. Dev. Pharm. Tech. Sci. (JIDPTS) 3(11) (2020)
9. Hanif, S: Eyes open closed dataset. Dataset from Kaggle (2020)
10. Ranganathan, G.: A Study to Find Facts behind preprocessing on deep learning algorithms. J Innov Image Process (JIIP) **3**(01), 66–74 (2021)
11. Forcier, J., Bissex, P., Chun, W.J.: Python web development with Django. Addison-Wesley Professional (2008)
12. Loreto, S., Romano, S.P.: Real-time communication with WebRTC: peer-to-peer in the browser. O'Reilly Media, Inc. (2014)
13. Wessels, A., Purvis, M., Jackson, J. Rahman, S.: Remote data visualization through websockets. In 2011 Eighth international conference on information technology: new generations, pp. 1050–1051 (2011)
14. Handley, M., Jacobson, V., Perkins, C.: SDP: session description protocol (1998)
15. Vogelsang, D.C., Erickson, B.J.: Magician's corner: 6. TensorFlow and TensorBoard, p. e200012 (2020)
16. Sinha, S., Singh, T.N., Singh, V.K., Verma A.K.: Epoch determination for neural network by self-organized map (SOM). Comput. Geosci. **14**(1), 199–206 (2010)

Chapter 23
Parkinson's Disease Prediction Through Machine Learning Techniques

Angeline Lydia, K. Meena, R. Raja Sekar, and J. N. Swaminathan

Abstract Parkinson's disease is a progressive Neurodegenerative disorder that affects the quality of life. As per World Health Organization (WHO) survey, 95% affected people with Parkinson's disease have speech disorders. The symptoms occur slowly over time. The most important symptoms are shaking, rigidity, slowness of movement and difficulty with walking. In this paper, heterogeneous data in the form of numbers and text are collected from 245 patients. Machine learning techniques are used for early prediction of disease which helps to increase the lifespan of the elder person and improve lifestyle with Parkinson. This research work identifies the appropriate model for early prediction of this disease, by comparing K-nearest neighbor classifier and Adaboost classifier. After experiential analysis the most appropriate algorithm in terms of accuracy will be used for real-time data analytics. The performance of the algorithm also evaluated using various metrics such as True Positive (TP), True Negative (TN), False Positive (FP), False Negative (FN), Precision (P), Recall (R) and accuracy. Our experimental results depict that performance of Adaboost classifier is better than KNN classifier in terms of accuracy.

A. Lydia (✉) · K. Meena
Department of Computer Science and Egineering, Vel Tech Rangarajan Dr.Sagunthala R&D Institute of Science and Technology, Chennai, Tamil Nadu, India
e-mail: drangelinelydia@veltech.edu.in

K. Meena
e-mail: drkmeena@veltech.edu.in

R. Raja Sekar
Department of Computer Science and Engineering, Kalasalingam Academy of Research and Education, Krishnan kovil, Tamil Nadu, India
e-mail: rajasekar@kluniversity.in

J. N. Swaminathan
Department of Electronics and Communication Engineering, QIS College of Engineering and Technology, Ongole 523272, Andhra Pradesh, India
e-mail: swaminathan@qiscet.edu.in

J. I.-Z. Chen et al. (eds.), *Machine Learning and Autonomous Systems*, Smart Innovation, Systems and Technologies 269, https://doi.org/10.1007/978-981-16-7996-4_23

23.1 Introduction

Parkinson's disease is a disorder of the central nervous system which shows the symptoms as the moment of body and inducing tremors along with the stiffness. This is also known as neurodegenerative disorder that affect dopamine which is produced from neurons in the brain. Parkinson's disease is very difficult to find even for doctors. The Parkinson's disease was detected in the clinics in past years. After machine learning technology came into the limelight, they started implementing the machine learning technology in medical field also, as a result of that a new method called voice propagation using machine learning algorithm has been proposed to detect the Parkinson's disease [1, 2]. The deflection in the voice of the patient will be recorded and checked by the machine learning algorithm and those results are transferred to data sets to verify whether the patient is affected by Parkinson's disease or not. It requires heavy machines to detect the Parkinson's disease, which is not cost-effective [3, 4].

There is no proper medication to treat the patient if they reach the final stage of Parkinson's disease. In our research work, Parkinson's disease is detected using voice as input and we used machine learning technology for classification. A huge data was collected from previously affected people and healthy people. These data are been trained and tested with the new user input data to check whether he/she is affected by Parkinson's disease or not. To show whether the person is affected or not, there is a section called status value which represents the result in either as a "0" or "1". If the status value is displayed as "1" then the person is affected or else if it shows "0" then the person is not affected by Parkinson's disease [5, 6].

23.1.1 Literature Survey

In 2019 July, Nicholas Ho and Glenda M. Halliday found that the serum samples from a clinic are tested to find Parkinson's disease and the same samples are tested using ML algorithm to detect Parkinson's disease. The conformation of a person affected with Parkinson's will take time in a clinic, but using Machine Learning, it is too simple and fast [6, 7].

Blauwendraat, C. Bandres-Ciga, and S. Singleton found that Voice change is also a symptom of Parkinson's disease. Deflections in the voice from previous days also cause Parkinson's disease. Voice can be changed due to any reason but it will cure within a few days. If the change of voice is continued for some days consulting a doctor and knowing the results takes time. Machine learning algorithm is used for this purpose to detect voice changes in earlier stage [1, 8].

It is very difficult to predict Parkinson's symptoms in earlier days, but nowadays ML techniques play a vital role in the early prediction of Parkinson's [13, 14]. As per WHO survey, 90% of affected people with Parkinson's disease have speech disorders. The average age of onset is about 70 years and the symptoms of incidence rise significantly with advancing age [15, 16]

23.1.2 Organization of the Paper

This paper is structured as follows. Section 23.2 portrays the proposed methodology to predict the Parkinson's disease from the UCI data set. Section 23.3 narrates the extensive experimental analysis for real-time data collected on different timestamp. Section 23.4 concludes the paper.

23.2 Research Methodology

In this model, we are considering all the parameters that come under Parkinson's disease. The features considered for experimental purposes are tabulated in Table 23.1.

Machine Learning (ML) algorithms are used to verify whether the user is affected by Parkinson's disease or not. The collected data is separated into training data and

Table 23.1 Parameters used for experimental analysis

Sl. no	Parameters	Description
1	MDVP:Fo(Hz)	Average vocal fundamental frequency
2	MDVPFhi(Hz)	Maximum vocal fundamental frequency
3	MDVP:Flo(Hz)	Minimum vocal fundamental frequency
4	MDVP: Jitter(%), MDVP Jitter(A bs), MDVP:RAP.MD VP:PPQ, Jitter:DDP	Several measures of variation in fundamental frequency
5	MDVP: Shimmer, MDVP: Shimmer(dB), Shimmer:APQ3, Shimmer: APQ5, MDVP : APQ, Shimmer:DDA	Several measures of variation in amplitude NHR
6	NHR, HNR	Two measures of ratio of noise to tonal components in the voice
7	status	Health status of the subject (one)—Parkinson's, (zero)—healthy
8	RPDE, D2	Two nonlinear dynamical complexity measures
9	DFA	Signal fractal scaling exponent
10	spread1. spread2. PPE	Three nonlinear measures of fundamental frequency variation

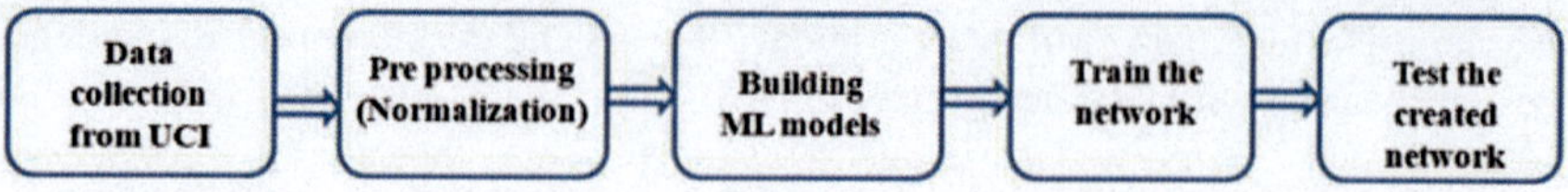

Fig. 23.1 Proposed Machine learning framework architecture

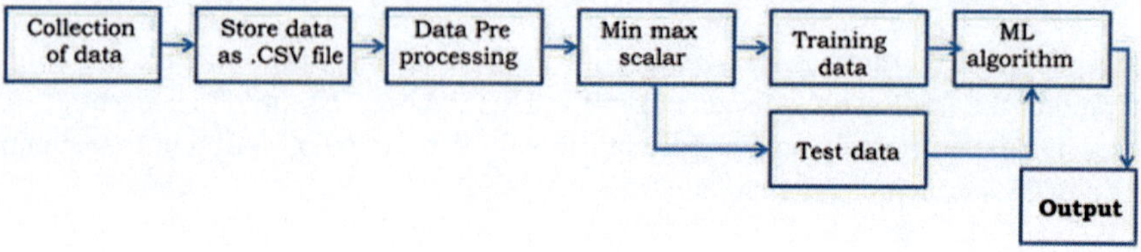

Fig. 23.2 Architecture diagram

testing data [9, 10]. These data will help the new user to verify the disease status. Figure 23.1 shows the proposed machine learning architecture.

The data is processed to train and test the data. A minmax scaler is adjusted to set the status values from -1 to $+1$ and then boosting algorithm is used to speedup the process. Finally, new user data is processed to classify whether the person is affected by Parkinson's disease or not (Fig. 23.2).

23.2.1 Data Collection from Various Resources

Collecting the data from various resources and storing them in the excel sheet and converting them into the CSV format. These data are used to access the data and visualize the data. A minmax scaler is used to check the status value in datasets to classify into affected or not; these data are collected from data flair with 195 columns and 24 rows. The 24 rows are the parameters.

23.2.2 Split of Training and Test Data

The visualized data is divided into training set and testing set. The training set has 60% of whole dataset and testing set has 40% of whole dataset. These data are used to classify the new user data. This will show the accuracy of the project and some data is also tested to check the accuracy [11, 12].

23.3 Machine Learning Algorithms

There are many machine learning algorithms to classify the data into datasets. This paper compares the performance of KNN and Adaboost algorithm for the above-mentioned dataset to predict whether the user is affected by Parkinson's disease or not (Fig. 23.3).

23.3.1 Experiment Using KNN

K-nearest neighbor algorithm is very popular and effective classification algorithm. In this experiment, K value is varied from 1 to N to find the optimal K value. The size of the training and test data is varied and the result is tabulated in Table 23.2.

From, the above table, it is observed that K = 3 gives maximum accuracy of 96% for tarin and test data size of 80 and 20%. Beyond the value of K = 3, the accuracy is decreased. It shows that noisy data is also included for classification.

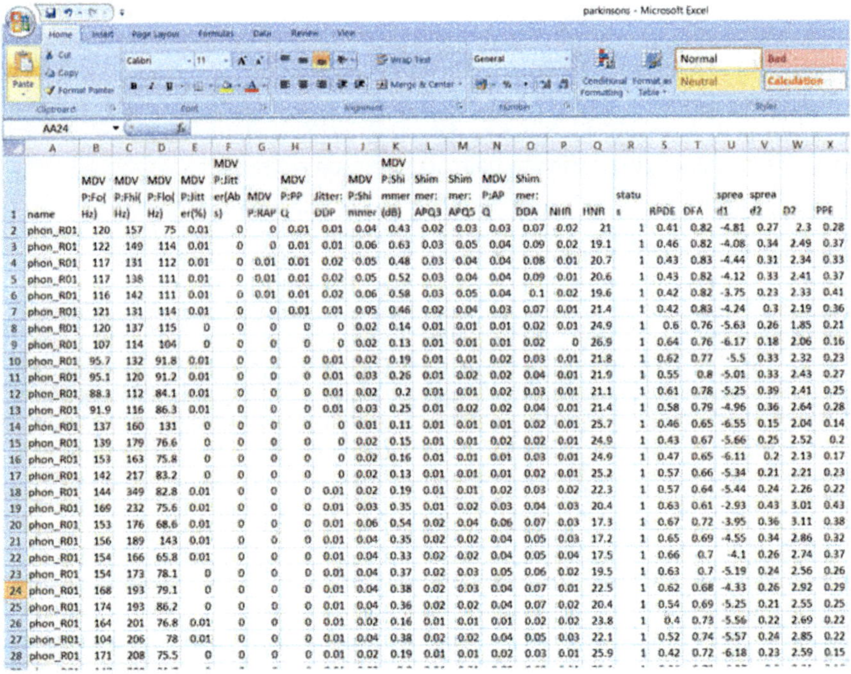

	A	B	C	D	E	F	G	H	I	J	K	L	M	N	O	P	Q	R	S	T	U	V	W	X
1	name	MDV P:Fo(Hz)	MDV P:Fhi(Hz)	MDV P:Flo(Hz)	MDV P:Jitt er(%)	MDV P:Jitt er(Ab s)	MDV P:RAP	MDV P:PP Q	Jitter: DDP	MDV P:Shi mmer	MDV P:Shi mmer (dB)	Shim mer: APQ3	Shim mer: APQ5	MDV P:AP Q	Shim mer: DDA	NHR	HNR	statu s	RPDE	DFA	sprea d1	sprea d2	D2	PPE
2	phon_R01	120	157	75	0.01	0	0	0.01	0.01	0.04	0.43	0.02	0.03	0.03	0.07	0.02	21	1	0.41	0.82	-4.81	0.27	2.3	0.28
3	phon_R01	122	149	114	0.01	0	0	0.01	0.01	0.06	0.63	0.03	0.05	0.04	0.09	0.02	19.1	1	0.46	0.82	-4.08	0.34	2.49	0.37
4	phon_R01	117	131	112	0.01	0	0.01	0.01	0.02	0.05	0.48	0.03	0.04	0.04	0.08	0.01	20.7	1	0.43	0.83	-4.44	0.31	2.34	0.33
5	phon_R01	117	138	111	0.01	0	0.01	0.01	0.02	0.05	0.52	0.03	0.04	0.04	0.09	0.01	20.6	1	0.43	0.82	-4.12	0.33	2.41	0.37
6	phon_R01	116	142	111	0.01	0	0.01	0.01	0.02	0.06	0.58	0.03	0.05	0.04	0.1	0.02	19.6	1	0.42	0.82	-3.75	0.23	2.33	0.41
7	phon_R01	121	131	114	0.01	0	0	0.01	0.01	0.05	0.46	0.02	0.04	0.03	0.07	0.01	21.4	1	0.42	0.83	-4.24	0.3	2.19	0.36
8	phon_R01	120	137	115	0	0	0	0	0	0.02	0.14	0.01	0.01	0.01	0.02	0.01	24.9	1	0.6	0.76	-5.63	0.26	1.85	0.21
9	phon_R01	107	114	104	0	0	0	0	0	0.02	0.13	0.01	0.01	0.01	0.02	0	26.9	1	0.64	0.76	-6.17	0.18	2.06	0.16
10	phon_R01	95.7	132	91.8	0.01	0	0	0	0.01	0.02	0.19	0.01	0.01	0.02	0.03	0.01	21.8	1	0.62	0.77	-5.5	0.33	2.32	0.23
11	phon_R01	95.1	120	91.2	0.01	0	0	0	0.01	0.03	0.26	0.01	0.02	0.02	0.04	0.01	21.9	1	0.55	0.8	-5.01	0.33	2.43	0.27
12	phon_R01	88.3	112	84.1	0.01	0	0	0	0.01	0.02	0.2	0.01	0.01	0.02	0.03	0.01	21.1	1	0.61	0.78	-5.25	0.39	2.41	0.25
13	phon_R01	91.9	116	86.3	0.01	0	0	0	0.01	0.03	0.25	0.01	0.02	0.02	0.04	0.01	21.4	1	0.58	0.79	-4.96	0.36	2.64	0.28
14	phon_R01	137	160	131	0	0	0	0	0	0.01	0.11	0.01	0.01	0.01	0.02	0.01	25.7	1	0.46	0.65	-6.55	0.15	2.04	0.14
15	phon_R01	139	179	76.6	0	0	0	0	0	0.02	0.15	0.01	0.01	0.02	0.02	0.01	24.9	1	0.43	0.67	-5.66	0.25	2.52	0.2
16	phon_R01	153	163	75.8	0	0	0	0	0	0.02	0.16	0.01	0.01	0.01	0.03	0.01	24.9	1	0.47	0.65	-6.11	0.2	2.13	0.17
17	phon_R01	142	217	83.2	0	0	0	0	0	0.02	0.13	0.01	0.01	0.01	0.02	0.01	25.2	1	0.57	0.66	-5.34	0.21	2.21	0.23
18	phon_R01	144	349	82.8	0.01	0	0	0	0.01	0.02	0.19	0.01	0.01	0.02	0.03	0.02	22.3	1	0.57	0.64	-5.44	0.24	2.26	0.22
19	phon_R01	169	232	75.6	0.01	0	0	0	0.01	0.03	0.35	0.01	0.02	0.03	0.04	0.03	20.4	1	0.63	0.61	-2.93	0.43	3.01	0.43
20	phon_R01	153	176	68.6	0.01	0	0	0	0.01	0.06	0.54	0.02	0.04	0.06	0.07	0.03	17.3	1	0.67	0.72	-3.95	0.36	3.11	0.38
21	phon_R01	156	189	143	0.01	0	0	0	0.01	0.04	0.35	0.02	0.02	0.04	0.05	0.03	17.2	1	0.65	0.69	-4.55	0.34	2.86	0.32
22	phon_R01	154	166	65.8	0.01	0	0	0	0.01	0.04	0.33	0.02	0.02	0.04	0.05	0.04	17.5	1	0.66	0.7	-4.1	0.26	2.74	0.37
23	phon_R01	154	173	78.1	0	0	0	0	0.01	0.04	0.37	0.02	0.03	0.05	0.06	0.02	19.5	1	0.63	0.7	-5.19	0.24	2.56	0.26
24	phon_R01	168	193	79.1	0	0	0	0	0.01	0.04	0.38	0.02	0.03	0.04	0.07	0.01	22.5	1	0.62	0.68	-4.33	0.26	2.92	0.29
25	phon_R01	174	193	86.2	0	0	0	0	0.01	0.04	0.36	0.02	0.02	0.04	0.07	0.02	20.4	1	0.54	0.69	-5.25	0.21	2.55	0.25
26	phon_R01	164	201	76.8	0.01	0	0	0	0.01	0.02	0.16	0.01	0.01	0.01	0.02	0.02	23.8	1	0.4	0.73	-5.56	0.22	2.69	0.22
27	phon_R01	104	206	78	0.01	0	0	0	0.01	0.04	0.38	0.02	0.02	0.04	0.05	0.03	22.1	1	0.52	0.74	-5.57	0.24	2.85	0.22
28	phon_R01	171	208	75.5	0	0	0	0	0.01	0.02	0.19	0.01	0.01	0.02	0.03	0.01	25.9	1	0.42	0.72	-6.18	0.23	2.59	0.15

Fig. 23.3 Input data

Table 23.2 Classification accuracy of KNN

Train data	Test data	Accuracy rate (%)				
		K = 1	K = 2	K = 3	K = 4	K = 5
50	50	40	45	56	56	52
60	40	52	59	62	61	56
70	30	56	67	82	75	68
80	20	84	91	96	88	76

Table 23.3 Classification accuracy of Adaboost classifier

Train data	Test data	Accuracy rate (%)
50	50	65
60	40	77
70	30	89
80	20	94

23.3.2 Experiment Using Adaboost Algorithm

Parkinson's dataset is experimented using Adaboost algorithm for different numbers of training and test data. The experimental results are tabulated in Table 23.3.

XG boost is used to speedup the process and performance of the dataset and even shows the high accuracy of processing data. It gives the maximum accuracy of 94%. The XG boost is a better option to find the Parkinson's disease.

23.3.3 Comparison of KNN and Adaboost Algorithm

This section shows the results of classification by KNN and Adaboost algorithm. From Tables 23.2 and 23.3 results, it is observed that the train and test data combination of 80 and 20% gives maximum accuracy. Hence, in this experiment, the same combination is considered and different performance metric is calculated and tabulated in Table 23.4. The performance metrics such as True positive, True negative, False positive, False negative, precision and recall are calculated.

From Table 23.2, it is observed that the performance of Adaboost algorithm is better than KNN algorithm. Precision and recall values are obtained as 0.92 and 0.96, respectively, for Adaboost classifier.

Table 23.4 Comparison of KNN and Adaboost classifier

Performance metric	Classifier	
	KNN	Adaboost
TP	20	24
TN	10	12
FP	5	2
FN	4	1
Precision	0.8	0.92
Recall	0.83	0.96

23.4 Conclusion

In this paper, Parkinson's disease is predicted in patient's body using machine learning algorithm and this method makes the process easy for the user. Adaboost model provides very accurate performance in detecting Parkinson's disease. In future work, Convolution Neural Network and Recurrent neural network classifier can also be used to predict the Parkinson's disease. After the successful implementation of our model, it may be extended in the future for analysis of finger moments of the user to know whether he/she is affected by Parkinson's disease or not with more accuracy.

References

1. Blauwendraa, C., Bandres-Ciga, S. Singleton, A.B.: Predicting progression in patients with Parkinsons disease using voice. Lancet Neurol. **16** (2017)
2. Chaudhuri, K.R., Healy, D.G., Schapira, A.H.: Non-motor symptoms of Parkinson's disease: diagnosis and management. Lancet Neurol. **5** (2006)
3. Das R.: A comparison of multiple classification methods for diagnosis of Parkinson disease. Expert Syst. Appl. (2010)
4. Hon, V., Lokaj, J. Rektor, I.: Interleukin-6 may contribute to mortality in Parkinson's disease patients: a 4-year prospective study. Park. Dis. (2015)
5. Leclair-Visonneau, L. et al.: REM sleep behavior disorder is related to enteric neuropathology in Parkinson disease. Neurology 89 (2017)
6. Ho N, Halliday G.M.: Parkinsons progression prediction using machine learning and serum cytokines. 25 July 2019
7. Pfister, F.M.J., Um, T.T., Pichler, D.C., et al.: High-resolution motor state detection in Parkinson's disease using convolutional neural networks. Sci Rep. **10**(1), 5860 (2020). https://doi.org/10.1038/s41598-020-61789-3
8. Ahmadi Rastegar, D., Ho, N., Halliday, G.M., Dzamko, N.: Parkinson's progression prediction using machine learning and serum cytokines. npj Park Dis. **5**(1), 14 (2019). https://doi.org/10.1038/s41531-019-0086-4
9. Angeles, P., Tai, Y., Pavese, N., Wilson, S., Vaidyanathan, R.: Automated assessment of symptom severity changes during deep brain stimulation (DBS) Therapy for Parkinson's Disease. In: International Conference On Rehabilitation Robotics (ICORR) (2017)
10. Nilashi, M., Ibrahim, O., Ahani, A.: Accuracy ımprovement for predicting Parkinson's disease progression. Sci. Rep. **6**, 34181 (2016)

11. Grover, S., Bhartia, S., Akshama, Yadav, A., Seeja, K.R.: Predicting severity of parkinson's disease using deep learning. In: International conference on computational ıntelligence and data science (ICCIDS 2018)
12. Benmalek, E., Elmhamdi, J., Jilbab, A.: UPDRS tracking using linear regression and neural network for Parkinson's disease prediction. Int. J. Emerg. Trends Technol. Comput. Sci. (IJETTCS) **4**, 189–193 (2015)
13. Mugunthan, S.R.: Decision tree based interference recognition for fog enabled IOT architecture. J. Trends Comput. Sci. Smart Technol (TCSST) **2**(01), 15–25 (2015)
14. Duraipandian, M.: Adaptive algorithms for signature wavelet recognition in the musical sounds. J. Soft Comput. Paradig. (JSCP) **2**(02), 120–129 (2020)
15. Jie Mei , Christian Desrosiers and Johannes Frasnelli.: Machine learning for the diagnosis of parkinson's disease: A review of literature, frontiers in ageing neuro science, vol.13, pp. 1–41 (2021)
16. Tiwari, H., Shridhar, S.K., Patil, P.V., Sinchana, K.R., Aishwarya, G.: Early prediction of parkinson disease using machine learning and deep learning approaches. EasyChair Preprint **4889**, 1–14 (2021)

Chapter 24
Goal-Oriented Obstacle Avoidance by Two-Wheeled Self Balancing Robot

Rajat Gurnani, Shreya Rastogi, Simrat Singh Chitkara, Surbhi Kumari, and Abhishek Gagneja

Abstract A solution to an age-old problem of keeping the inverted pendulum-based robot upright and its collision-free navigation through an unknown environment is presented in this paper. Two deep reinforcement learning models run simultaneously on a two-wheeled self-balancing robot. The robot acts as the agent in the learning environment created in Unity3D, a real-time development platform. Proximal Policy Optimization algorithm, an on-policy gradient method for reinforcement learning, has been used in both models. The navigation model feeds the intermediate position of the robot to the balancing model, which then outputs a set of actions to the robot. In addition, the navigation model uses curiosity-driven learning thus expanding the set of solvable tasks. The algorithm consequently generates an intrinsic reward function, which makes the robot generalize better with unexplored environments. Furthermore, the resultant intelligent controller is made to navigate both dynamic and static obstacle-filled courses. The simulations were run using Unity and the Nvidia PhysX engine.

24.1 Introduction

As ingenious as we humans are, we do have our limitations. In areas and domains wherein we need assistance, mobile robots are increasingly coming to our rescue. Mobile robotics is an enormously growing field of research. A two-wheeled self-balancing robot [1, 2] is one such wheeled mobile robot based on the principle of an inverted pendulum. The robot tends to fall as it is an inherently unstable system. The aim is to keep it upright during traversal as well as at a point. The robot will fall either forward or backward. Just like we balance a stick on our finger by moving our finger in the direction in which it is falling, similarly, we balance the TWSBR [1] by driving its wheels in the direction in which it is falling. The wheel velocity

R. Gurnani (✉) · S. Rastogi · S. S. Chitkara · S. Kumari · A. Gagneja
Bharati Vidyapeeth's College of Engineering, New Delhi, Delhi 110063, India
e-mail: coedelhi@bharatividyapeeth.edu
URL: http://coedelhi.bharatividyapeeth.edu

J. I.-Z. Chen et al. (eds.), *Machine Learning and Autonomous Systems*, Smart Innovation, Systems and Technologies 269, https://doi.org/10.1007/978-981-16-7996-4_24

is decided based upon the state of the robot such as the tilt angle (by how much angle the robot has tilted), by how much speed it is falling, and in which direction. Only when the robot is balanced can it perform other tasks like traversing, picking and placing things, and avoiding obstacles. Due to its design and the ability to rotate around a single point [3], it finds use in many applications like security, industry, personal transport systems, law enforcement, tourism, etc.

24.1.1 Related Works

First, we surveyed all the existing techniques for balancing the TWSBR [1]. Second, we surveyed the obstacle avoidance methods present for mobile robots. After which a survey of existing methods for obstacle avoidance by a TWSBR [1] was done.

JOE [4] is an early version of the robot and Segway [5], developed in 2000, is a popular commercial application. A lot of research exists about the control mechanisms for a TWSBR [1]. One of them being PID [6], a conventional closed-loop feedback controller. LQR [7] is a more modern model-based linear control mechanism. While comparing the two [8], LQR [7] turns out to be the better performer as it has a faster response time and can be used if the system includes motion. Another method depicted in [9] is Fuzzy logic, an intelligent non-model-based controller. When compared with the Pole Placement controller, it has better dynamic performance and higher robustness. In [10], out of the FLC [9], the LQR [7], and PID [6], FLC [9] performs better in terms of speed response but consumes higher energy. Other variants of Fuzzy logic include Fuzzy PD controller [11], T-S Adaptive Neural Network Fuzzy Control [12], and adaptive neuro-fuzzy inference (ANFIS) controller [13]. ANFIS controller is much better as compared to conventional controllers like PID [6] and Fuzzy logic [9] in terms of parameter variation, overshoot, and settling time. In addition to these control algorithms, there also exist non-model-based controllers based on Reinforcement Learning. Neural Network [14] has been a common controlling method because of its self-tuning ability. Q-Learning [15] offers optimal control of a TWSBR [1] in the absence of or lack of knowledge about system control parameters. A comparison amongst traditional controllers, Q-Learning [15], and Deep Q-Learning algorithm [16] shows that while DQN [16] is more stable than fuzzy and LQR [7], it is less so when compared to PID [6]. In this paper, PPO is used as the control algorithm to achieve balancing. PPO is an on-policy gradient method for reinforcement learning. PPO's core focus is on minimizing the step size so that there is little deviation from the previous step. Moreover, PPO shows a much better convergence rate as compared to other reinforcement learning algorithms.

Due to the significantly increased maneuverability of TWSBR [1], it can easily navigate in complex environments from a start position to a target position. Path planning and obstacle avoidance techniques are crucial for the robot to navigate intelligently following a collision-free path in both static and dynamic environments. The location of the obstacles does not change in a stationary environment whereas it keeps changing in a dynamic environment [17]. Navigation essentially can be categorized into two types: global navigation and local navigation [18]. In global navigation, prior knowledge of the environment is available. It consists of static obstacles. Several classical approaches exist for global navigation like Voronoi Diagram [19], Visibility graph [20], Cell decomposition method [21], and many more. It has been observed that classical approaches provide a collision-free path but offer many drawbacks such as trapping in local minima and a large amount of computation due to complex real-world environments [22]. In local navigation, the environment is unstructured, and the robot is unaware of its surroundings. The robot, in such a case, makes use of the sensor readings to reach a point. Local navigation requires path planning, where the robot learns to navigate by exploring the environment. Heuristic approaches such as Fuzzy logic [23], Neural networks [24], Genetic algorithms [25], Ant Colony Optimization [26], Particle swarm optimization algorithms [27] exist for local navigation. In short, all the methods have their strengths and drawbacks. Continuous research is going on for new and optimized algorithms.

While doing the literature survey, we found that very little research exists on the navigation of a TWSBR [1] through a cluttered environment. In [28], implementation of a back-propagation neural network-integrated modified DAYANI method is developed for path control of a TWSBR [1]. In this method, as soon as the sensor detects the obstacles, the algorithm forces them into action. It then generates the next feasible point in the direction of the goal. In [29], a combination of TLBO and ANFIS is implemented. TLBO is introduced to improve the performance of the ANFIS by training the parameters of the membership functions. In [30], it presents an obstacle avoidance control system based on ultrasonic sensors and fuzzy logic control. This paper has defined 24 fuzzy rules of three inputs and one output designed for the fuzzy controller.

24.1.2 Proposed Work

After performing an extensive literature survey regarding balancing and navigation of a TWSBR [1], we found the following things. There exists a lot of research related to the modeling and balancing algorithms of a TWSBR [1]. Navigation techniques for mobile robots exist but are not specific to two-wheeled robots. Lastly, work related to balancing and navigation of a TWSBR [1] in an obstacle cluttered environment is minimal. That is what we implement in this paper. The paper introduces a method to solve this problem using two PPO models, one for balancing the robot and one for obstacle avoidance. Both models work simultaneously for the successful navigation of the robot to reach its target while avoiding obstacles. Also, the models working together prove to be robust as the agent can navigate in both static and dynamic environment.

Fig. 24.1 shows the flow of the proposed work.

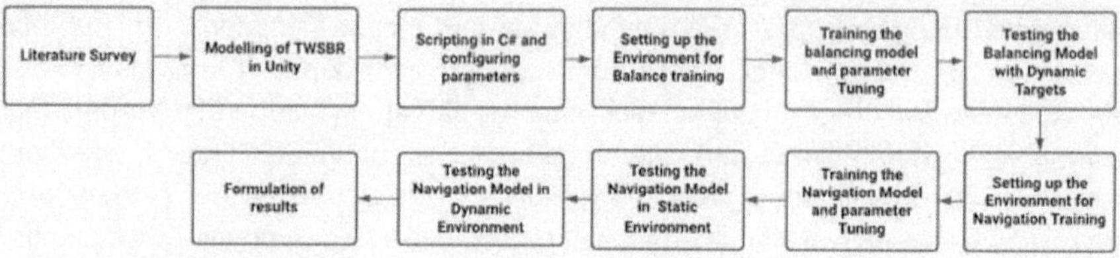

Fig. 24.1 Flow of the work

24.2 Reinforcement Learning

Reinforcement learning is a type of machine learning algorithm based on cause and effect experience. The two main elements [31] of a RL system are the agent and the environment. The agent performs a certain action A, based on the state s, the environment it is in. The agent is then rewarded or penalized based on his action. The four main subelements of a RL [31] system except for agent and environment are: a policy, a reward signal, a value function, and, optionally, a model of the environment. A policy is a mapping of states to actions. The reward signal helps the agent in deciding the right move from the wrong one. The value function of a state is the future rewards that can be expected or returned in terms of rewards, if that state is chosen at that time. A model is like a map of the environment with estimated actions for each state the environment can be in.

RL environments are formally represented using MDP [31]. A reinforcement learning task that satisfiẹs the Markov property is called a Markov decision process. If an RL system satisfies the Markov property, then the environments' next response at time $t + 1$ depends solely on the current state and action representations, that is, at time t. In order to keep the calculations simple, we consider a finite number of states and reward values. Hence, we talk in terms of probability. Given any state s, and action a, the probability of each possible pair of next state s' and reward r, according to [31] is denoted by

$$p(s', r|s, a) = P_r\{S_{t+1} = s', R_{t+1} = r|S_t = s, A_t = a\} \tag{24.1}$$

Reinforcement learning can be either model-based, here, the learning environment acts as the source for the policy, or Model-free, where the agent learns about the environment through trial-and-error and sets up the optimal policy. The final goal is to find a policy π, which maximizes the reward function R. The agent uses the probability distribution to sample its actions. A policy can be either deterministic, which maps states to action without uncertainty, or stochastic, which specifies probabilities, $\pi(a \mid s)$, for taking each action, a, in each state, s.

24.2.1 Policy Gradient Method

In policy-based methods, we directly learn the policy function that maps states to actions, that is, we select actions without using a value function. The policies are parameterized by neural networks and optimized with respect to the rewards. The quality of a policy π_θ is measured using an objective function J(θ). The local maximum value of J(θ) is found by gradient ascent of the policy with respect to parameters θ. The final objective is to maximize the cumulative reward starting with that state until the terminal time T. One of the gradient estimators frequently used to find the gradient estimate $\hat{g}$ according to [32] is:

$$\hat{g} = \hat{E}_t[\nabla\theta log\pi_\theta(a_t|s_t)\hat{A}_t] \tag{24.2}$$

where $\nabla\theta log$ is the gradient, π_θ is a parameterized stochastic policy and $\hat{A}$ is an estimator of the advantage function at time step t. The expectation $\hat{E}t$ [. . .] is an empirical average of a finite batch of samples, which are revised at repeated intervals of time. With every fresh set of samples, the policy π_θ is improved.

24.2.2 Proximal Policy Optimization (PPO)

Proximal policy optimization, just like TRPO, is an on-policy learning approach, which means that the experience samples collected update the current policy only once. PPO [32] is an improved version of TRPO. It is empirically less complex and significantly easier to implement. Let $r_t(\theta)$ denote the probability ratio

$$r_t(\theta) = \left[\frac{\pi_\theta(a_t|s_t)}{\pi_{\theta_{old}}(a_t|s_t)}\right] \tag{24.3}$$

Here, θ_{old} is the vector of policy parameters before the update. In TRPO and PPO both, the following "surrogate" objective function L^{CPI} (conservative policy iteration) is maximized:

$$L^{CPI}(\theta) = \hat{E}_t\left[\frac{\pi_\theta(a_t|s_t)}{\pi_{\theta_{old}}(a_t|s_t)}\hat{A}_t\right] \tag{24.4}$$

A very large policy update is inevitable without a constraint. Therefore, in the maximization of L^{CPI}, we penalize changes to the policy that move $r_t(\theta)$ away from 1. Hence, the new objective function is given by

$$L^{CLIP}(\theta) = \hat{E}_t[min(r_t(\theta)\hat{A}_t), clip(r_t(\theta), 1-\epsilon, 1+\epsilon)\hat{A}_t)] \tag{24.5}$$

where epsilon is a hyperparameter and which roughly says how far away the new policy is allowed to go from the old. Finally, a minimum of the unclipped, L^{CPI} and the clipped probability ratio is taken. This ensures that a lower value of the unclipped objective is considered. The main idea of Proximal Policy Optimization is to have medium policy updates so that there is less deviation in training. It also ensures that the agent does not go down an irreversible path of incomprehensible actions.

24.3 Modelling of TWSBR in Unity

Unity is a 3D game engine with the ability to simulate rigid body/soft-body physics in real-time. It is a powerful cross-platform IDE for developers, which offers various functions such as modeling, prototyping, physics simulation, etc. We used Unity to make the learning environment and the model of the robot/agent. In Unity, a Gameobject (Entity) comprises multiple modules known as Components, which are accessed via a Script attached to that object. For example, the position and scale of an agent/robot can be viewed and edited by taking a reference of the agent's Transform Component in a Script. Unity also supports hierarchical relationships among entities. These were used while training multiple agent instances, each in their separate environment for the model to train quickly. All the simulations were run with a physics update frequency set 100 Hz. The Unity Machine Learning Agents (ML-Agents) is an open-source project that enables simulations to serve as environments for training intelligent agents. It consists of implementations of state-of-the-art algorithms to train intelligent agents for 2D and 3D environments. In this paper, we used reinforcement learning (PPO). It is a part of the ML-Agents package. Furthermore, the ray perception sensors used for sensing the environment are present in the ML-Agents package itself.

The Two-Wheeled Self Balancing robot is modeled using primitive shapes such as Cylinder and Cuboid in Unity. The body of the balancing robot consists of a cuboid attached to a set of cylinders as can be seen in Fig. 24.2. The cylinders represent the wheels of the robot. The body of the balancing robot has a box collider attached to it, accurately representing its physical dimensions. The wheels also have capsule colliders of appropriate dimensions attached to them. Rigidbodies component is attached to each shape. Rigidbodies enable the GameObjects to act under the control of physics. The robot body has the highest mass followed by the wheels having 0.25 times the mass of the body. It was kept this way to depict the real-life scenario where the balancing robot has to carry a load on its body, which significantly increases its mass compared to the wheels. The entire structure of the balancing robot is held together with the use of Hinge joints. Hinge joints pass through the wheels and connect them to the lower body of the balancing robot. Unity software has a built-in implementation of Hinge joints. The hinge joint is then customized by adding a motor to it and assigning it a maximum force value of 5000. The velocities of the motors were controlled via scripts.

The TWSBR [1] has parameters that affect its stability. Such as, a high center of gravity improves the stability of the bot due to reduced angular acceleration while tilting. However, it will also reduce the maximum recoverable angle of the TWSBR [1]. Another factor that affects the stability of the bot is its mass. An increase in the

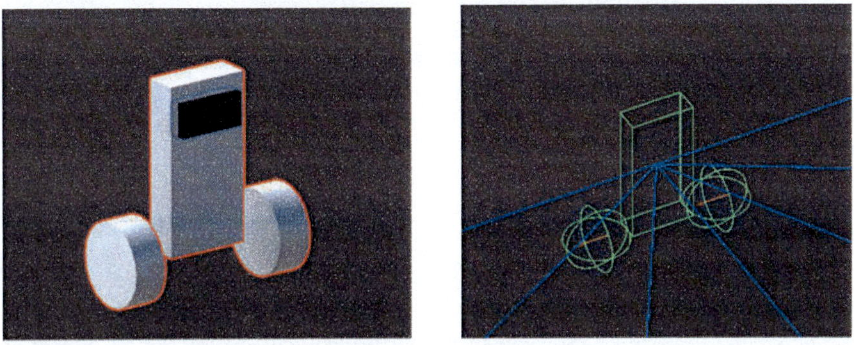

(a) (b)

Fig. 24.2 **a** Isometric view of robot. **b** Collider view of robot

mass value improves the stability of the bot. An offset in position center of mass from its mean position also affects the stability as now a tilt angle offset has to be added to the 0° upright position of the bot.

Ray casts: For the agent to sense its environment, we equipped it with distance measuring sensors on its sides. In real life, this corresponds to ultrasonic sensors or sharp sensors. In Unity, the same functionality is implemented using Ray Perception Sensor. They detect the obstacle and return various information to the system such as the normalized hit distance, hash id, Boolean value, indicating whether a certain kind of object was hit or not. The agent is equipped with Ray Perception Sensor 3D component. It shoots out seven rays 30° apart spanning a 180-degree cone in front of the Agent. Sensors were only used in front to reduce the size of the model. The perception sensor is configured to detect the obstacles and walls with appropriate tags assigned to them. Each ray of the sensor goes out to a maximum distance of 10 units. The output of the RPS 3D is automatically added to the Observation Buffer(Input) of the navigation model.

24.4 Methods

24.4.1 Balancing of TWSBR in Simulation Framework

For training the balancing model, the agent was initialized with a random position and the desired behavior expected of the agent was to move towards the goal while maintaining balance. Once it reached the goal, the goal was spawned at a different random location and the agent continued to chase the goal. We used a four-layered neural network consisting of an input layer, two hidden layers, and an output layer. The input to the neural network was distance, direction toward the target, the velocity of the bot, pitch angle, and angular velocity of the robot, all of which summed up to

12 neurons. The output had two neurons, which were the desired velocities for each left and right wheel. Both the hidden layers consisted of 512 neurons. The desired velocities could have been higher than the maximum specified velocity of wheels. To prevent this, we clipped them up to a specified maximum value to provide a finite input to the wheels.

24.4.2 Navigation of TWSBR in Simulation Framework

In our second model, we used PPO with an intrinsic curiosity module. This is an inverse model that takes the current state S_t and next state S_{t+1} as inputs and estimates the action A_t which was taken by the agent. This helps the agent in the exploration of the environment. The desired behavior of the model is to reach the target while avoiding obstacles that come in its path. For the navigation model as well, we used a four-layered neural network consisting of an input layer, two hidden layers, and an output layer. The inputs provided to this model were the direction to the final target, normalized endpoint distance, raycast(s) information, and agent's normalized velocity held in nine neurons. The output was an intermediate direction in the XZ plane held in two neurons. Both the hidden layers consisted of 512 neurons.

From Fig. 24.3, we can see that the navigation model provides an intermediate direction to the balancing model. Since our balancing model is trained for reaching a goal while maintaining balance. It takes the intermediate points as a target and keeps following them till it reaches the actual goal.

24.4.3 Rewards

In Reinforcement Learning, the reward signal tells the objectivity of a particular action in a state. A reward signal is a number that is computed in the environment and sent to the agent when the latter performs a certain action. By computing the reward outside the agent, we ensure that the agent has imperfect control over its ultimate goal. Selecting an appropriate reward function is an intricate and essential task. Due to the multi-tasking nature of our robot, the Reward function takes into account multiple facets of our system, to get the desired behavior.

24.4.3.1 Balancing Rewards

For the training of our first model, we rewarded the agent based on four parameters:

- maintaining an upright position
- minimizing the angular velocity
- aligning in the direction of the target

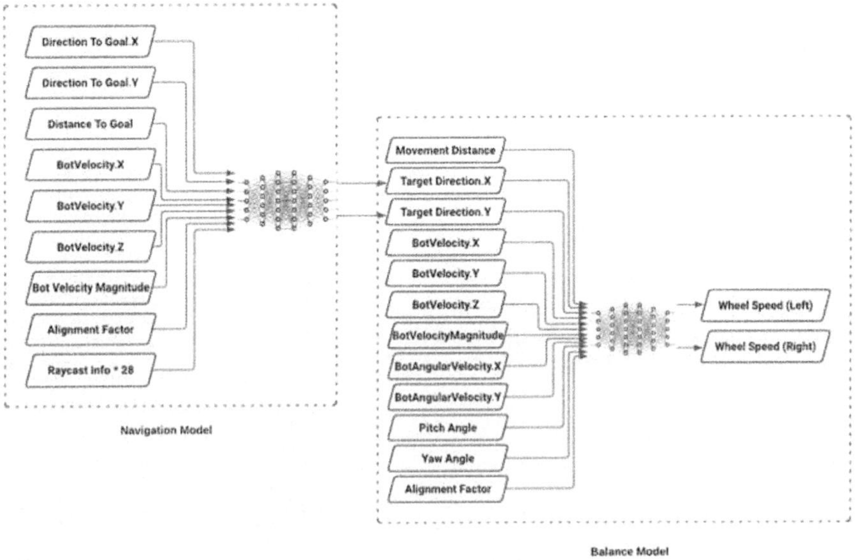

Fig. 24.3 Hybrid model system showing communication between navigation model having dimensions [36 ∗ 512 ∗ 512 ∗ 2] and balance model with dimensions [12 ∗ 512 ∗ 512 ∗ 2]

- reducing the distance toward the target

$$r_t = r_{ang} * r_{vel} * r_{alignment} + r_{dis} - 0.01 \tag{24.6}$$

Here,

- $\mathbf{r_{ang}}$ is the reward for maintaining an upright position and is given by:

$$r_{ang} = 1 - 0.1 * (|\Theta|) - 0.01 * (|\omega|) \tag{24.7}$$

  where θ is angle difference from stable/mean position and ω is the angular velocity along x-axis.
- $\mathbf{r_{alignment}}$ is the reward for aligning in the direction of the target and is defined as

$$r_{alignment} = \hat{v}_{botdirection} . \hat{v}_{targetdirection}, \tag{24.8}$$

  i.e. a vector dot product of the direction in which the agent is moving and the direction in which the goal/endpoint is located.
- $\mathbf{r_{vel}}$ maximizes our reward for matching the desired velocity.
- $\mathbf{r_{dis}}$ is the reward for reducing the distance between the agent and the goal. We have rewarded the agent with +1 reward for reaching the target and gave it a negative terminal penalty of −1 for when it lost its balance. Factor of **0.01** was subtracted in each step to minimize the episode length.

Table 24.1 Balancing hyperparameters

Hyperparameters	Value
Gamma	0.95
Lambda	0.95
Buffer size	20480
Batch size	2048
Number of epochs	5
Learning rate	0.0003
Beta	0.01
Epsilon	0.2.
Hidden units	512
Number of layers	3

24.4.3.2 Navigation Rewards

$$r_n = 2 * (distance_{t-1} - distance_t) + 0.3 * r_{alignment} - 0.01 \quad (24.9)$$

Here, the ($\mathbf{distance_{t-1}} - \mathbf{distance_t}$) term calculates the difference between the previous position and the current position of the agent. The aim is to keep reducing the distance between the agent and the endpoint. There's another reward called $\mathbf{r_{alignment}}$ for incentivizing the agent to stay aligned in the direction of the final goal. Here again, a factor of **0.01** was subtracted in each step to minimize the episode length.

24.4.4 Configuration of PPO Algorithm

The process of training a Reinforcement Learning model can often involve the need to tune the hyperparameters in order to achieve a level of performance that is desirable. We tuned the hyperparameters to certain values according to our requirements. Table 24.1 gives the balancing hyperparameters, and Table 24.2 gives the hyperparameters provided to the navigation model.

Table 24.2 Navigation hyperparameters

Hyperparameters	Value
Gamma	0.95
Lambda	0.95
Buffer size	20480
Batch size	2048
Number of epochs	5
Learning rate	0.0003
Beta	0.01
Epsilon	0.2.
Hidden units	512
Number of layers	3

24.5 Results and Simulations

24.5.1 Cumulative Rewards

We executed the policy training routine of the balance model with the hyperparameters given in Table 24.1. In Fig. 24.4a, we can see that till the 300k steps mark, the model struggled to balance. At around 5M steps, the agent was able to move in the direction of the target. We limited our model to a maximum of 5000 steps. The sudden dips observed in the graph in Fig. 24.4a, are because the training was stopped and then resumed at 10, 15, and 25 M steps. While training the navigation model (Fig. 24.4b), the agent started avoiding the obstacles at around 1 million steps. The model was able to generalize well, with different placements of obstacles. With a little more training, the model could traverse in a dynamic environment. Both models have trained on AMD Ryzen 5600x workbench with Nvidia RTX 3070 GPU. It took the Balancing model around 3 h 37 min to train while having 100 similar environments running in parallel. The Navigation model, however, took 5.5 h due to its longer episodes. Both graphs showed a continuous increase in reward as the training time increased.

24.5.2 Robot Navigation in Static and Dynamic Environment

In this paper, we attempt to prove the robustness of the designed method. To do so, we took the help of both **Static** and **Dynamic** environments. By definition, in a static environment, only the agent takes actions. Whereas, in a dynamic environment, apart from the actions of the agent, other processes operate on it as well. Thus, the agent is affected by both, its actions and that of its environment.

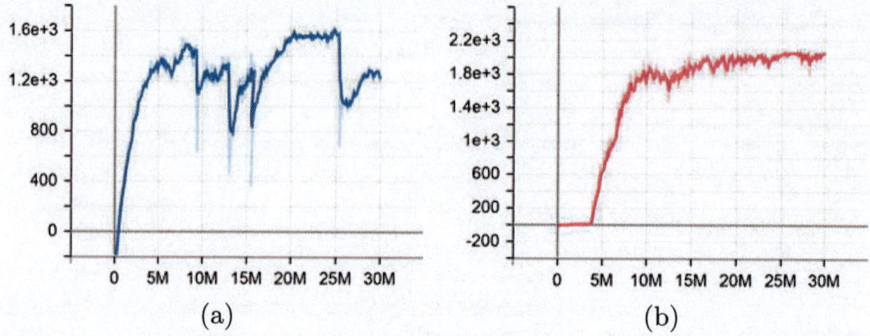

Fig. 24.4 **a** Rewards accumulated versus number of steps taken by Balancing Model **b** Rewards accumulated versus number of steps taken by Navigation Model

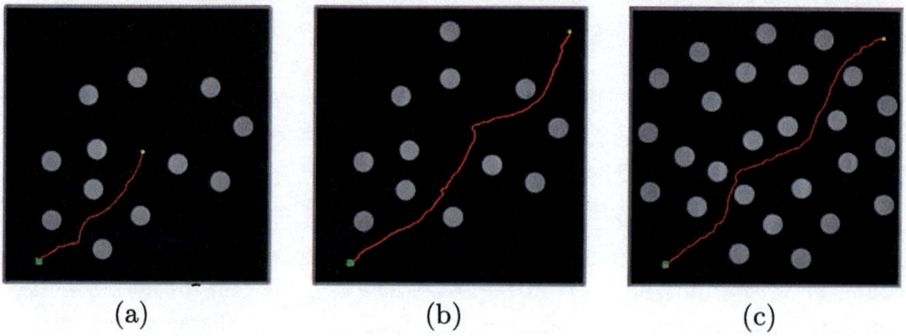

Fig. 24.5 Path traced by the robot from start to goal in static environment with varying level of obstacles

Figure 24.5 shows the three levels of a static environment having cylindrical obstacles and the path traced by the agent from the start point to the endpoint. We can see that the agent has successfully performed the desired task of reaching the goal in a varying difficulty level static environment. Figure 24.6 shows a four-step process in a dynamic environment. There are four moving cuboidal obstacles in the path of the agent. It is evident that the agent successfully navigates around mobile obstacles to reach the goal.

24.5.3 Robot Navigation Metrics

From the graphs in Fig. 24.7 we can make the following conclusions:

1. The distance–time graph in Fig. 24.7a is nearly linear with a constant slope. It intercepts the x-axis at distance = 0. Here, the distance signifies the difference in positions of the start point and the end goal. Therefore, we can conclude that

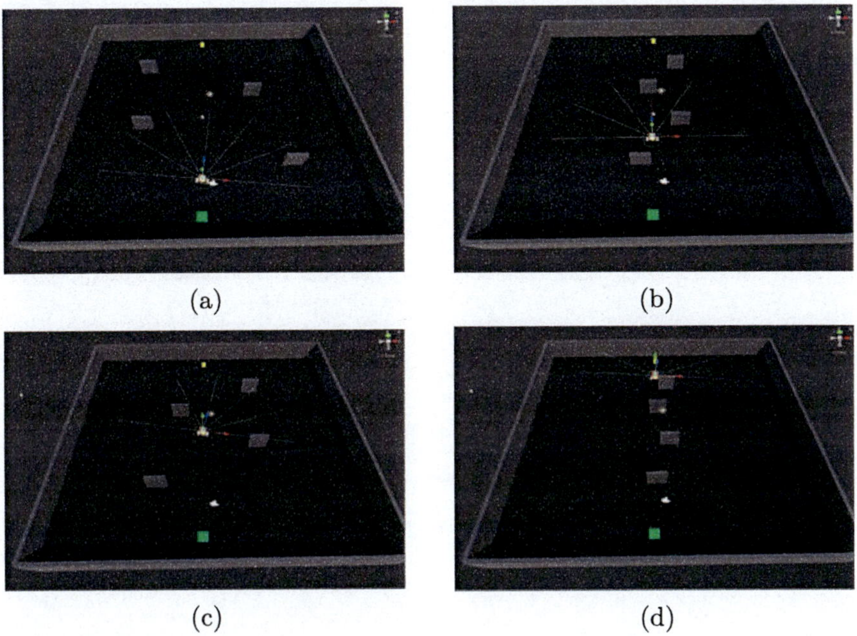

Fig. 24.6 Position of agent as it advances in time in a dynamic environment (obstacle having to and fro motion with constant velocity)

the robot continuously keeps closing the gap between itself and the target and ends up reaching it.

2. In the speed–time graph, in Fig. 24.7b, we can see that the speed is maintained at a mean value of 0.68 m/s. Since this is the desired velocity of the system, we are getting the desired response from the designed model.
3. The tilt angle–time graph, in Fig. 24.7c, oscillates across the x-axis with a small margin of error. This graph shows that the agent can stay upright or at 0° for the entirety of its run. There are slight variations from the 0° mark, but they are well within the margin of error. It's a testament to the fact that throughout the run, the robot can maintain its balance and traverse the environment towards the goal.
4. In the angular speed–time graph, in Fig. 24.7d, we can see that the angular speed is maintained at a mean value of 38.8°C/s.

24.6 Conclusion

In this paper, we have shown that the Proximal Policy Optimization algorithm can be used to simultaneously balance and navigate a two-wheeled self-balancing robot successfully. PPO is a model-free control technique that balances a TWSBR [1]

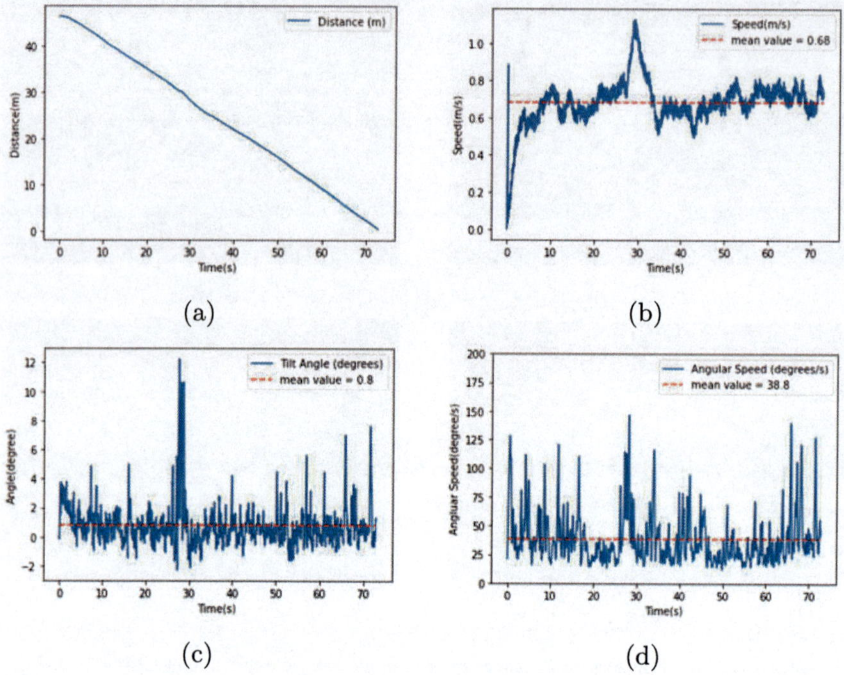

Fig. 24.7 **a** Distance from target w.r.t time, **b** Speed–time graph, **c** Tilt angle–time graph, **d** Angular speed–time graph

without prior knowledge about its system parameters. Furthermore, the designed method works in both static and dynamic environments. This method also helps in faster convergence towards the end goal as compared to other control algorithms.

In future work, we aim to implement this method on a robot in real environment and test its performance in comparison with other navigation algorithms. We also aim to make the system more robust and adaptable.

References

1. Ha, Y.-S., Yuta, S.: Trajectory tracking control for navigation of the inverse pendulum type self-contained mobile robot. Robot. Autonom. Syst. **17**(1–2), 65–80 (1996). ISSN 0921-8890. https://doi.org/10.1016/0921-8890(95)00062-3
2. Lee, D.-Y., Kim, Y., Kim, B., Kwak, Y.: Dynamics and Control of Non-holonomic Two Wheeled Inverted Pendulum Robots
3. Pathak, K., Franch, J., Agrawal, S.K.: Velocity and position control of a wheeled inverted pendulum by partial feedback linearization. IEEE Trans. Rob. **21**(3), 505–513 (2005). https://doi.org/10.1109/TRO.2004.840905
4. Grasser, F., D'Arrigo, A., Colombi, S., Rufer, A.C.: JOE: a mobile, inverted pendulum. IEEE Trans. Industr. Electron. **49**(1), 107–114 (2002). https://doi.org/10.1109/41.982254

5. https://www.segway.com/
6. Thao, N.G.M., Nghia, D.H., Phuc, N.H.: A PID backstepping controller for two-wheeled self-balancing robot,. Int. Forum Strat. Technol. **2010**, 76–81 (2010). https://doi.org/10.1109/IFOST.2010.5668001
7. Uddin, N., Nugroho, T.A., Pramudito, W.A.: Stabilizing Two-wheeled robot using linear quadratic regulator and states estimation. In: 2nd International Conferences on Information Technology. Information Systems and Electrical Engineering (ICITISEE), vol. 2017, pp. 229–234 (2017). https://doi.org/10.1109/ICITISEE.2017.8285501
8. Jamil, O., Jamil, M., Ayaz, Y., Ahmad, K.: Modeling, control of a two-wheeled self-balancing robot. In: International Conference on Robotics and Emerging Allied Technologies in Engineering (iCREATE), vol. 2014, pp. 191–199 (2014). https://doi.org/10.1109/iCREATE.2014.6828364
9. Wu, J., Zhang, W.: Design of fuzzy logic controller for two-wheeled self-balancing robot. In: Proceedings of 2011 6th International Forum on Strategic Technology, pp. 1266–1270 (2011). https://doi.org/10.1109/IFOST.2011.6021250
10. Bature, A., Buyamin, S., Ahmad, M., Muhammad, M.: A comparison of controllers for balancing two wheeled inverted pendulum robot. Int. J. Mech. Mech. Eng. **14**, 62–68 (2014)
11. Wu, J., Zhang, W., Wang, S.: A two-wheeled self-balancing robot with the fuzzy PD control method. Math. Probl. Eng. **2012**, Article ID 469491, 13 pages (2012). https://doi.org/10.1155/2012/469491
12. Wu, J., Jia, S.: T-S adaptive neural network fuzzy control applied in two-wheeled self-balancing robot. In: Proceedings of 2011 6th International Forum on Strategic Technology, 2011, pp. 1023–1026 (2011). https://doi.org/10.1109/IFOST.2011.6021194
13. Agrawal, R., Mitra, R.: Adaptive neuro fuzzy inference structure controller for rotary inverted pendulum. In: Kumar, M.A., R. S., Kumar, T., (eds.), Proceedings of International Conference on Advances in Computing. Advances in Intelligent Systems and Computing, vol. 174. Springer, New Delhi (2013). https://doi.org/10.1007/978-81-322-0740-5_141
14. Liang, S., Gan, F.: Balance control of two-wheeled robots based on reinforcement learning. In: Proceedings of 2011 International Conference on Electronic and Mechanical Engineering and Information Technology (2011). https://doi.org/10.1109/emeit.2011.6023706
15. Guo, L., Rizvi, S.A.A., Lin, Z.: Optimal control of a two-wheeled self-balancing robot by reinforcement Q-learning. In: 2020 IEEE 16th International Conference on Control and Automation (ICCA), pp. 955–960 (2020). https://doi.org/10.1109/ICCA51439.2020.9264485
16. Rahman, M.M., Rashid, S.M.H., Hossain, M.M.: Implementation of Q learning and deep Q network for controlling a self balancing robot model. Robot. Biomim. **5**, 8 (2018). https://doi.org/10.1186/s40638-018-0091-9
17. Gul, F., Rahiman, W., Alhady, S.S.N., Chen, K., (Reviewing, (eds.): A comprehensive study for robot navigation techniques. Cogent Eng. **6**, 1 (2019). https://doi.org/10.1080/23311916.2019.1632046
18. Pandey, A., Pandey, S., Parhi, D.R.: Mobile robot navigation and obstacle avoidance techniques: a review. Int. Rob. Auto. J. **2**(3), 96–105 (2017). https://doi.org/10.15406/iratj.2017.02.00023
19. Garrido, S., Moreno, L., Abderrahim, M., Martin, F.: Path planning for mobile robot navigation using voronoi diagram and fast marching. In: IEEE/RSJ International Conference on Intelligent Robots and Systems, Beijing, China, pp. 2376–2381 (2006)
20. Kaluder, H., Brezak, M., Petroviæ, I.: A visibility graph-based method for path planning in dynamic environments. In: MIPRO, Proceedings of the 34th International Convention, pp. 717–721. Opatija, Croatia (2011)
21. Park, K.H., Kim, Y.J., Kim, J.H.: Modular Q-learning-based multi-agent cooperation for robot soccer. ELSEVIER Robot. Autonom. Syst. **35**(2), 109–122 (2001)
22. Montiel, O., Sepúlveda, R., Orozco-Rosas, U.: Optimal path planning generation for mobile robots using parallel evolutionary artificial potential field. J. Intell. Robot. Syst. **79**, 237–257 (2015). https://doi.org/10.1007/s10846-014-0124-8
23. Montaner, M.B., Ramirez-Serrano, A.: Fuzzy knowledge-based controller design for autonomous robot navigation. ELSEVIER Expert Syst. Appl. **14**(1), 179–186 (1998)

24. Engedy, I., Horvath, G.: Artificial neural network based local motion planning of a wheeled mobile robot. In: IEEE International Symposium on Computational Intelligence and Informatics (CINTI), pp. 213–218 (2010)
25. Ghorbani, A., Shiry, S., Nodehi, A.: Using genetic algorithm for a mobile robot path planning. In: IEEE International Conference on Future Computer and Communication, pp. 164–166 (2009)
26. Dorigo, M., Birattari, M., Stutzle, T.: Ant colony optimization. IEEE Comput. Intell. Mag. **1**(4), 28–39 (2006). https://doi.org/10.1109/MCI.2006.329691. Nov
27. Ahmadzadeh, S., Ghanavati, M.: Navigation of mobile robot using the PSO particle swarm optimization. J. Acad. Appl. Stud. (JAAS) **2**(1), 32–38 (2012)
28. Chhotray, A., Parhi, D.: Navigational control analysis of two-wheeled self-balancing robot in an unknown terrain using back-propagation neural network integrated modified DAYANI approach. Robotica **37**(8), 1346–1362 (2019). https://doi.org/10.1017/S0263574718001558
29. Aouf, A., Boussaid, L., Sakly, A.: TLBO-based adaptive neurofuzzy controller for mobile robot navigation in a strange environment. Comput. Intell. Neurosci. Article ID 3145436, 8 pages (2018). https://doi.org/10.1155/2018/3145436
30. Ruan, X., Li, W.: Ultrasonic sensor based two-wheeled self-balancing robot obstacle avoidance control system. In: IEEE International Conference on Mechatronics and Automation, vol. 2014, pp. 896–900 (2014). https://doi.org/10.1109/ICMA.2014.6885816
31. Sutton, R.S., Barto, A.G.: Reinforcement Learning: An Introduction, vol. 1, 1st edn. MIT Press, Cambridge (1998)
32. Schulman, J., Wolski, F., Dhariwal, P., Radford, A., Klimov, O.: Proximal policy optimization algorithms (2017). arXiv preprint arXiv:1707.06347

Chapter 25
A Novel CNN Approach for Detecting Breast Cancer from Mammographic Image

Suneetha Chittineni and Sai Sandeep Edara

Abstract The most common cancer that affects women all around the world is the breast cancer. The death rate of the breast cancer could be reduced if it is detected and treated early. The factors like technological reasons related to image quality and human error may misdiagnosis the breast cancer, which can happen by the radiologists'. A convolutional neural network (CNN) system is developed that can solve these issues by detecting the breast cancer from mammographic images automatically. The system acquires the mammography images from the MIAS dataset. After pre-processing these images, gray-level co-occurrence matrix (GLCM) extracts the cancerous features from the pre-processed images. The GLCM extracts the statistical cancerous texture features from the mammography images. And then, from that images, breast cancer patients are classified by the Marine Predators Algorithm (MPA) that consists of Levy approach and Brownian motion. This system identifies the breast cancer patients accurately and its performance is superiorly high compared to other CNN approaches.

25.1 Introduction

Breast cancer is one of the most common cancers that demise people worldwide, and it mainly affects women [1]. The detection of breast cancer at its early stage and by providing efficient treatment, the death rate of the breast cancer can be reduced significantly. The human body has the ability to produce, expand and control the death cells. The abnormal growth of the cells leads the way to forming the cancerous lumps. Nowadays, there are various tools such as thermography, ultrasound, mammography and magnetic resonance imaging (MRI) [2] that helps the radiologists' to detect the breast cancer at its early.

Ultrasound imaging [3] does not produce correct results for the patients having dense breasts. The breast cancer image result relies on the expert taking the ultrasound for the patients. Therefore, most of the times it ultrasound images leads to unwanted

S. Chittineni (✉) · S. S. Edara
RVR and JC College of Engineering, Chowdavaram, Guntur, India

J. I.-Z. Chen et al. (eds.), *Machine Learning and Autonomous Systems*, Smart Innovation, Systems and Technologies 269, https://doi.org/10.1007/978-981-16-7996-4_25

biopsy due to high false-positive rate. For breast thermography [4], infrared cameras are utilized to capture the breast images. An infrared sensor is present inside the camera that records the temperature of the breast. The cancerous lumps in the breast are detected based on the variation in the breast temperature. In this screening method, the original breast temperature can be changed by applying cold pressed in the breast, which produces a false result. The magnetic resonance imaging (MRI) [5] 3D view breast cancer images are utilized to detect the breast cancer. The MRI produces non-ionizing radiation so it is safe for the patients. But in breast MRI, it is hard to distinguish the normal lumps and cancerous lumps, and also the cost of MRI is high. Hence, in this study to detect the breast cancer in early stage, mammographic images are utilized. Mammography is a best breast cancer screening method that decreases the death caused by breast cancer.

The existence of cancerous lumps in the mammography images is seen very tiny and also the image contrast is low. So, in recent years, the misdiagnosis of breast cancer by the radiologist raises because of technical problems, poor quality of image and by human mistakes. To avoid these limitations and to help the radiologist, a novel convolutional neural network (CNN) system is proposed that utilizes the mammography images to detect the cancerous lumps in the breast. The proposed convolutional neural network acquires the mammography images from the Mammographic Image Analysis Society (MIAS) image database. The convolutional neural network (CNN) [6] algorithm, gray level co-occurrence matrix (GLCM) [7] extract the features of breast cancer from pre-processed images. Then the extracted features are classified by using the CNN Marine Predators Algorithm (MPA) [8]. Figure 25.1 depicts the block diagram of the proposed system.

This article's remaining sections are organized as follows. A brief literature reviews on various breast cancer images, their feature extraction and classification approaches are discussed in Sect. 25.2. Section 25.3 explains the proposed method, the approaches used in the proposed method for image pre-processing, extracting the features of the image and classifying the image. In Sect. 25.4, experimental results with simulation are explained and discussed. The conclusion of the article is presented in Sect. 25.5.

25.2 Literature Review

Garg and Dhiman [9] introduce an image retrieval approach based on the content of the image. This approach undergoes the following stages, at first, the features are extracted from the image. Next, the similitude indicators are computed from the image, and at last, the features are indexed and downloaded. The image information was gathered from the local binary patterns. The gray level co-occurrence matrix (GLCM) algorithm was implemented in this approach to extract the statistical features of the image. On implementing this GLCM algorithm, the image retrieval approach produces good accuracy, and other performance metrics value is also high. Priyanka and Kumar [7] developed an artificial neural network that is used to diagnose kidney diseases in their early stage from ultrasound images. While diagnosing the ultrasound images, the performance was based on the quality of the image. The problem was to improve the image quality without losing any information from the image. For this, the image was pre-processed then the features of the image were extracted by GLCM algorithm, and lastly, principal component analysis helps to decrease the dataset. High accuracy was produced when GLCM and PCA were hybrid together.

Aslam et al. [10] introduced a deep convolutional neural network (DCNN) method that automatically spots the cancerous lumps in the breast. Two datasets were used in this method to gather the data. Convolutional neural network has multiple layers, these layers train the gathered data, and then it classifies the cancerous patients from the trained dataset. The result of this method was based on the training data. If large number of datasets were trained then its performance was good otherwise the performance reduces, which, in turn, cause wrong prediction. Chen et al. [5] develop a protocol called as an abbreviated protocol (AP), which spots the breast cancer from the magnetic resonance imaging (MRI). The abbreviated protocol (AP) was the hybridization of abbreviated protocol1 (AP1) and abbreviated protocol2 (AP2). Abbreviated protocol1 (AP1) was formed by grouping the Maximum Intensity Projection [MIP] and first post-contrast subtracted [FAST] images. Abbreviated protocol2 (AP2) was formed by grouping the AP1 protocol with diffusion-weighted imaging (DWI). The AP1and AP2 protocols are applied on the breast ultrasound images for finding the cancerous lumps in the breast. The disadvantage was that the previous patients records were not considered for examination, and also, the tiny cancerous lumps in the breast were not detected.

Singh et al. [11] modify and expand various already existing methods for segmentation, feature extraction, feature selection and classification. These modified methods were applied in the thermography images. And then the cancerous lumps were detected from the thermography images. The limitations in this method were it can only be applied to database having a less number of thermography images also the result may change based on the false-positive rate and the dimension of the lump. Cao et al. [12] implemented a convolutional neural network (CNN) framework that spots the cancerous lumps in the breast ultrasound images. Various object detection and classification methods were packed into this framework to spot the cancerous lumps. It first spots the presence of lumps in the breast and then classifies the kind of lumps. The limitation was underfitting problem occurs when a malignant lump was detected and also only less parameters were considered for identifying the lumps.

Mohammed et al. [13] designed an enhanced adaptive neuro-fuzzy inference system (ANFIS) to detect the coronavirus 2. The marine predator algorithm (MPA), which was a nature-inspired algorithm, was implemented in the enhanced ANFIS to detect the COVID-19. The performance of MPA-ANFIS was relatively high compared to approaches such as social-spider optimization, sine-cosine algorithm (SCA) and particle swarm intelligence (PSO). Ibrahim et al. [14] utilizes thermal infrared images to discover the cancerous lumps in the breast. From the DMR-IR dataset, the thermal images are collected. Then the collected thermal infrared images are pre-processed and segmented. Then from the segmented image, the features related to breast cancer are extracted. From the feature extracted images, the cancerous patients were classified. In this approach, three processes were carried out such as pre-processing, segmentation and classification. In each of this process, various techniques were applied that may cause certain problems like k-value setting and after combining the data, the image is fully modified from its original size and density, which unexpectedly produces a false prediction. To overcome these entire limitations, mammography images are proposed in this work. From the mammography image, to identify the breast cancer, radiologists are needed. All the time, the radiologists are not available, and there is a chance to occur human mistakes while examining the mammographic image. So, an automated convolutional neural network with marine predator algorithm is developed to identify the breast cancer.

25.3 Methodology Used

In the proposed method, the mammography images for chest are acquired from the Mammographic Image Analysis Society (MIAS) [15] image database. All the images acquired from the database undergo image pre-processing to make all the images of same size. Then from the pre-processed image, the features of breast cancer are extracted using a gray level co-occurrence matrix (GLCM) algorithm. Finally, the breast cancer patients are classified from the feature extracted image dataset by the convolutional neural network marine predator algorithm (CNN-MPA).

25.3.1 Image Pre-processing

Image pre-processing [16] is applied to the mammography image dataset for removing the unwanted noise from the image thus improving the quality of the image for further processing. There are various pre-processing approaches that help to improve the contrast of the image, rotating the image, remove noise from the image, scaling, smoothening, filtering, etc. [17]. Cropping the image, cubic interpolation, zero degree rotation are the pre-processing approaches implemented in the proposed approach.

Cropping the image denotes the way of eliminating the data that are not required from the image, and it also defines the region of interest (ROI). By cropping the image, only the infected area is selected for examining the cancerous lumps so that it improves the accuracy, decreases the error rate and increases the processing speed. After cropping, the resolution of the resultant image is 250*230 pixels.

Cubic Interpolation defines the image as a set of co-ordinates and converts it into another set of co-ordinates thus it removes the unwanted noise from the image and keeps the needed information.

Zero degree rotation is done next to interpolation to keep the major reference axis of the image to 0 °C.

25.3.2 Gray-Level Co-occurrence Matrix for Feature Extraction

The process of feature extraction extracts the specific features that are related to cancer from the pre-processed image. In the proposed approach, gray-level co-occurrence matrix (GLCM) algorithm helps to retrieve the statistical texture features from the pre-processed image. The texture feature defines the contents of the pre-processed images that retain the texture of the images. In GLCM, there are 22 texture features. Some of them are computed as follows. Energy in GLCM matrix is defined as the total amount of square elements. The homogeneous regions are related to non-homogeneous regions. If the frequency of the recursive pixel image increases then the energy value is high. The energy value is computed in Eq. (25.1).

$$Energy = \sum_{l,m=0}^{S-1} (G_{lm})^2 \tag{25.1}$$

The image randomness is computed by the entropy. The entropy value is less for the image, which is homogeneous. The entropy value is computed in Eq. (25.2).

$$Entropy = \sum_{l,m=0}^{S-1} -\ln(G_{lm})\, G_{lm} \tag{25.2}$$

The pixel resembles is defined as homogeneity. If the homogeneity value is one then the image is homogeneous otherwise the image texture should be improved because the value of homogeneity is extremely low. Homogeneity value is computed in Eq. (25.3).

$$Homogeneity = \sum_{l,m=0}^{S-1} \frac{G_{lm}}{1+(l-m)^2} \tag{25.3}$$

where, G represents the normalized matrix GLCM, the elements of the matrix is represented as l,m, total number of gray levels present in the image is represented as S, and the average value of the GLCM matrix is represented as $/mu$. The value of $/mu$ is computed in Eq. (25.4).

$$\mu = \sum_{l,m=0}^{S-1} l G_{lm} \tag{25.4}$$

For all the pixels, the variance in the intensity $/sigma$ is computed in Eq. (25.5).

$$\sigma^2 = \sum_{l,m=0}^{S-1} G_{lm}\,(l-\mu)^2 \tag{25.5}$$

25.3.3 *CNN Marine Predators Algorithm for Classification*

Marine Predators Algorithm [MPA] The MPA initially has random values that are allocated for a group of solutions based on the search space. This value is computed in Eq. (25.6).

$$P_0 = B_{min} + r * (B_{max} - B_{min}) \tag{25.6}$$

where, lower search space boundary is represented as B_{min}, higher search space boundary is represented as B_{max}, r refers the random number ranges from 0 to 1. In MPA, the search agents are both the hunter and the prey. After each search, process ends the elite, i.e., the matrix of the top hunter is modified. The elite and prey are computed in Eq. (25.7).

$$Elite = \begin{bmatrix} P_{11}^{1}\, P_{12}^{1} \cdots P_{1d}^{1} \\ P_{21}^{1}\, P_{22}^{1} \cdots P_{2d}^{1} \\ \cdots\cdots\cdots\cdots \\ P_{s1}^{1}\, P_{s2}^{1} \cdots P_{sd}^{1} \end{bmatrix} \tag{25.7}$$

where, number of search agents is referred as s, d refers the number of dimensions and $P_l m$ represents the $m^t h$ dimension of the $l^t h$ prey. After this, the location of the prey P should be upgraded. This can be done by imitating the whole hunting process of hunter and prey and also changing various velocity ratios such as high velocity ratio, unit velocity ratio and low velocity ratio.

Eddy Formation and FADs' Effect The behavior of the marine predators can change due to eddy formulation and fish aggregating devices (FADs). The FADs effect is computed in Eq. (25.8).

$$P_k = \begin{cases} P_k + SZ\left[R \otimes B_{max} - B_{min}\right] \otimes BS\ r_5 < FAD \\ P_k + [FAD\,(1 - r) + r]\,(P_{r1} - P_{r2})\ r_5 > FAD \end{cases} \tag{25.8}$$

where, BS refers the binary solution and FAD is set to 0.2. The prey indices are denoted as r1 and r2.

25.4 Results and Discussions

In MATLAB R2018a software, the proposed system is implemented. For the implementation of the proposed system, the Mammographic Image Analysis Society (MIAS) image database is chosen. In MIAS, there are totally 322 mammography breast images out of that 206 are normal images, and 113 are cancerous lumps in the breast images. Convolutional neural network (CNN), Decision Tree (DT), Naive Bayes (NB), K-Nearest Neighbor (KNN) and Random Forest (RF) approaches are

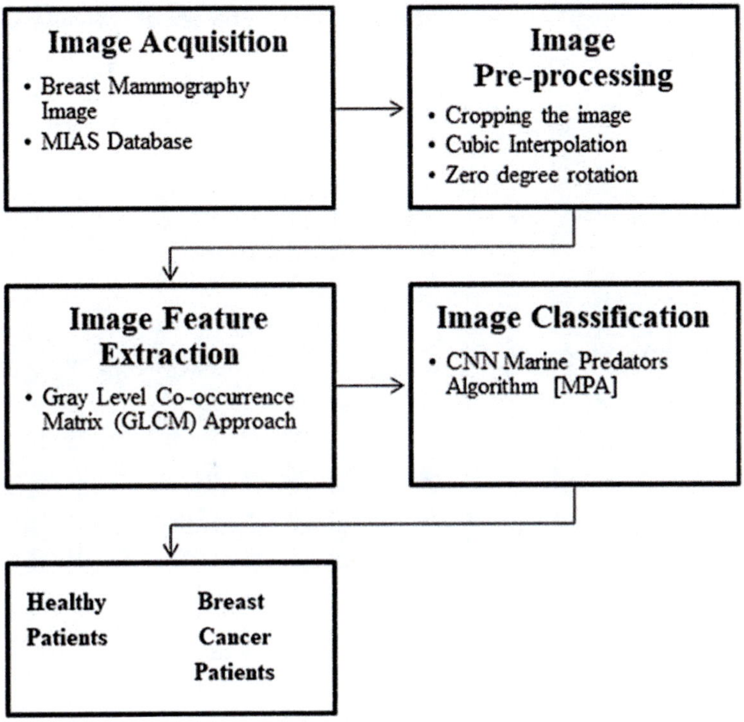

Fig. 25.1 Block diagram of the proposed system

compared with the proposed system. Recall, F1-Score, precision, specificity and Kappa statistic are the performance metrics considered for evaluating the algorithms. Figure 25.2 depicts the performance of various algorithms. From the figure, it is evident that the performance of the proposed convolutional neural network with marine predator algorithm (CNN-MPA) is superiorly high compared to other algorithms.

The accuracy for the proposed CNN-MPA and the various other algorithms are represented in Fig. 25.3. The accuracy for DT approach is 0.7857, KNN approach accuracy is 0.8214, NB approach accuracy is 0.9257, CNN approach accuracy is 0.9286 and the CNN-MPA accuracy is 0.9989. From this, it is clear that the accuracy of the proposed convolutional neural network with marine predator algorithm is relatively high.

Figure 25.4 shows the convergence curve of the proposed CNN-MPA. The number of iterations applied in the proposed CNN-MPA approach is 100. The proposed CNN-MPA produces fast convergence toward the near global optimum. So this proposed approach can detect the breast cancer patients faster. The error rate of the proposed CNN-MPA is very low and is nearly 0.0011. The Matthews Correlation Coefficient (MCC) value of the proposed approach is 0.9970. Since the MCC is relatively high, the system produces accurate result with less occurrence of error.

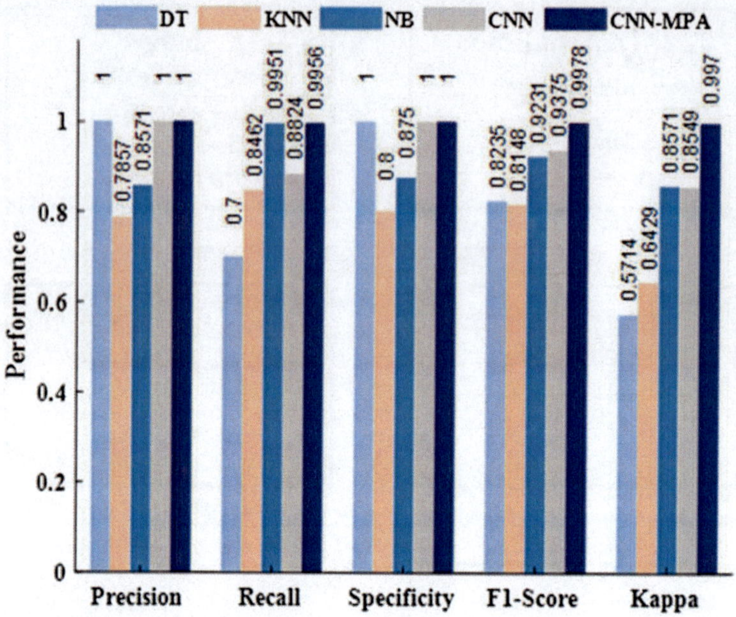

Fig. 25.2 Comparison of the proposed CNN-MPA with other approaches

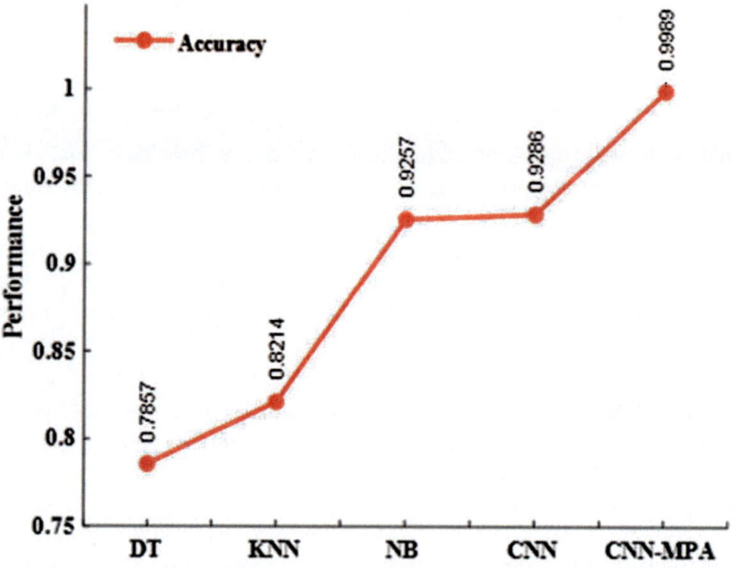

Fig. 25.3 Accuracy of the proposed CNN-MPA with existing DT, KNN, NB, CNN

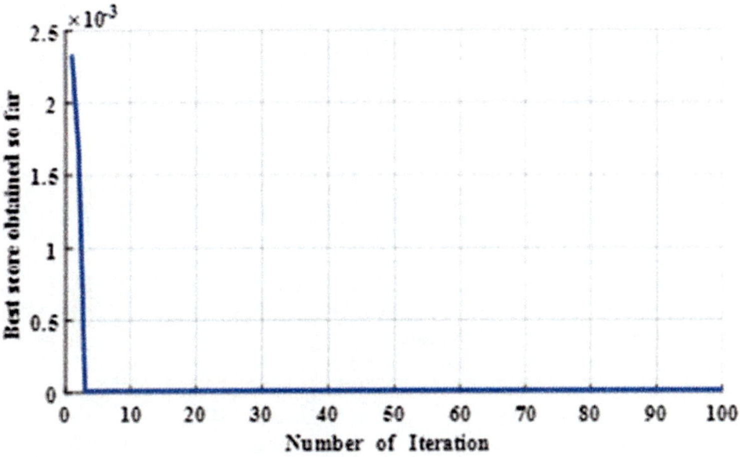

Fig. 25.4 Convergence curve of CNN-MPA

25.5 Conclusion

In this article, a novel CNN is designed to detect breast cancer from the chest mammographic images. The system acquires the data from the MIAS mammography image dataset. The dataset undergoes image pre-processing after that from the pre-processed image dataset the cancerous features are extracted by gray-level co-occurrence matrix. Then from the extracted feature dataset, breast cancer patients are classified by convolutional neural network with marine predator algorithm. This system helps the radiologist to diagnose the breast cancer at its early stage by reducing human error, bad image quality and error due to technical problems. The result of the proposed CNN-MPA performance is analyzed based on precision, F1-score, recall, specificity, error, MCC and kappa. The high MCC indicates that the proposed CNN-MPA produces better output with less error rate. While analyzing the performance of the CNN-based MPA classifier, the performance of the proposed CNN-MPA is superiorly high and also produces high accuracy in identifying the cancerous lumps from the mammography images. The proposed CNN-MPA provides effective design and efficient statistical outputs when compared to other approaches. For future work, the proposed system is for expanded other imaging modalities.

References

1. Jalalian, A., Mashohor, S., Mahmud, R., Karasfi, B., Iqbal, M., Saripan, B., Ramli, A.R.: Foundation and methodologies in computer-aided diagnosis systems for breast cancer detection. EXCLI J. **16**, 113—137 (2017)

2. He, Z., Chen, Z., Tan, M., Elingarami, S., Liu, Y., Li, T., Li, W.: A review on methods for diagnosis of breast cancer cells and tissues. Cell Prolifer. (2020). https://doi.org/10.1111/cpr.12822
3. Islam, M.S., Kaabouch, N., Hu, W.C.: A survey of medical imaging techniques used for breast cancer detection. In: IEEE International Conference on Electro-Information Technology (2013). https://doi.org/10.1109/eit.2013.6632694
4. Gonzalez-Hernandez, J.L., Recinella, A.N., Kandlikar, S.G., Dabydeen, D., Medeiros, L., Phatak, P.: Technology, application and potential of dynamic breast thermography for the detection of breast cancer. Int. J. Heat Mass Trans. **131**, 558–573 (2019). https://doi.org/10.1016/j.ijheatmasstransfer.2018.11.089
5. Chen, S.Q., Huang, M., Shen, Y.-Y., Liu, C.-L., Xu, C.-X.: Abbreviated MRI protocols for detecting breast cancer in women with dense breasts. Korean J. Radiol. **18**(3), 470 (2017). https://doi.org/10.3348/kjr.2017.18.3.470
6. Ramadan, S.Z.: Using convolutional neural network with cheat sheet and data augmentation to detect breast cancer in mammograms. In: Computational and Mathematical Methods in Medicine, pp. 1–9 (2020). https://doi.org/10.1155/2020/9523404
7. Priyanka, Kumar, D.: Feature extraction and selection of kidney ultrasound images using GLCM and PCA. Proc. Comput. Sci. **167**, 1722–1731 (2020). https://doi.org/10.1016/j.procs.2020.03.382
8. Faramarzi, A., Heidarinejad, M., Mirjalili, S., Gandomi, A.H.: Marine predators algorithm: a nature-inspired metaheuristic. Expert Syst. Appl. **113377** (2020). https://doi.org/10.1016/j.eswa.2020.113377
9. Garg, M., Dhiman, G.: A novel content-based image retrieval approach for classification using GLCM features and texture fused LBP variants. Neural Comput. Appl. (2020). https://doi.org/10.1007/s00521-020-05017-z
10. Aslam, M.A., Cui, D.: Breast cancer classification using deep convolutional neural network. J. Phys.: Conf. Ser. **1584**, 012005 (2020). https://doi.org/10.1088/1742-6596/1584/1/012005
11. Singh, D., Singh, A.K.: Role of image thermography in early breast cancer detection- past, present and future. Comput. Methods Programs Biomed. **183**, 105074 (2020). https://doi.org/10.1016/j.cmpb.2019.105074
12. Cao, Z., Duan, L., Yang, G., Yue, T., Chen, Q.: An experimental study on breast lesion detection and classification from ultrasound images using deep learning architectures. BMC Med. Imag. **19**(1) (2019). https://doi.org/10.1186/s12880-019-0349-x
13. Mohammed, A.A., Al-Ganess, M.A., Ewees, A.A., Abualigah, L., Abd Elaziz, M.: Marine predators algorithm for forecasting confirmed cases of COVID-19 in Italy, USA, Iran and Korea. Int. J. Environ. Res. Public Health **17**, 3520 (2020)
14. Ibrahim, A., Mohammed, S., Ali, H.A.: Breast cancer detection and classification using thermography: a review. Adv. Intell. Syst. Comput. 496–505 (2018). https://doi.org/10.1007/978-3-319-74690-6_49
15. Mustra, M., Stajduhar, A.: Segmentation masks for the mini-mammographic image analysis society (mini-MIAS) database. IEEE Cons. Electron. Mag. **9**(5), 28–33 (2020). https://doi.org/10.1109/mce.2020.2986799
16. Qin, B., Li, D.: Identifying facemask-wearing condition using image super-resolution with classification network to prevent COVID-19. Sensors **20**(18), 5236 (2020). https://doi.org/10.3390/s20185236
17. Vinoth, R., Anusha, C., SaiPriya, D., Sridhar, B.S., ManiKantaPavan, C., Aditya G.: A novel approach on ultrasound kidney images classification by using micro deep neural network algorithm. Int. J. Pure Appl. Math. **119**(12) (2018)

Chapter 26
Efficient Deep Learning Methods for Sarcasm Detection of News Headlines

Deepak Kumar Nayak and Bharath Kumar Bolla

Abstract Sarcasm is a combination of irony and mocking used to communicate contempt for someone or something. It can be found in textual form in various places, including reviews from customers about products or services, news stories, and most notably, social networking sites. Sarcastic utterances can conceal the negative connotation with a positive outlook. This deceptive nature of sarcasm has always been an exciting and crucial aspect of sentiment analysis. With the advent of false and manipulative news and the popularity of humorous articles, detecting sarcasm in news articles has turned out to be crucial. In this paper, we evaluated various vectorization and machine learning models to detect sarcastic headlines. Our experiments show that pre-trained transformer-based embeddings combined with Long Short-Term Memory (LSTM) networks provide better recall, accuracy, and F1 score.

26.1 Introduction

Sarcasm has recently gained popularity among social media users and the general public to criticize and make satirical comments about a product or service, or policy. Companies have recently begun to invest extensively in monitoring any negative remarks posted on social media platforms or in news articles that could affect their brand. As a result, identifying sarcasm has become an essential aspect of Sentiment Analysis in various Natural Language Processing (NLP) applications across the industry.

The misleading and metaphorical nature of sarcasm makes it challenging to detect. A sarcastic quote attempts to convey a negative message in an ostensibly favorable format. For example, let us consider a tweet that went viral a few years back "*Thank you for sending my baggage to Hyd and flying me to Calcutta at the same time.*

D. K. Nayak (✉)
Wells Fargo Center, Old Mumbai Hwy, Raidurgam, Madhura Nagar Colony, Hyderabad, Telangana 500032, India

B. K. Bolla
Verizon Data Services India-Torrey Pines, Koramangala-Indiranagar, Bangalore, Karnataka 560071, India

J. I.-Z. Chen et al. (eds.), *Machine Learning and Autonomous Systems*, Smart Innovation, Systems and Technologies 269, https://doi.org/10.1007/978-981-16-7996-4_26

Brilliant service," to which an automated bot from the company replied, "Glad to hear that." The preceding tweet is the best illustration of sarcasm's deceiving nature. If we remove the stop words from the previous tweet, we will only be left with words of a positive value, such as "*thank you*" and "*brilliant.*" But in reality, it's meant to communicate a negative message.

In recent years, automated journalism has advanced significantly, with Microsoft recently announcing that it will replace its MSN news production workforce with AI-powered bots. Each quarter, Bloomberg employs an automated new generator machine system known as Cyborg to assist reporters in producing thousands of stories on firm profit reports. These bots are programmed to scan the contents of a few websites to which they have access and share either the original post or a tale made up of other articles on the same subject. This was a significant motivation for our research, which aimed to assess the various conventional and current methodologies available for the task and determine the most effective method for detecting sarcasm in brief text.

In this work, the objective is to compare multiple combinations of word embedding techniques and supervised classification techniques to find which combination provides the best results in the most efficient way. This study contributes to the field of automated journalism for detecting sarcasm in short text format by comparing several strategies and providing results indicating which techniques have the most impact on performance. The findings of the research could also be applied in the field of sentiment analysis and automated journalism.

26.2 Related Work

Various linguists and psychologists have investigated the topic of sarcasm extensively [7]. Reference [3] investigated how sarcasm manifested itself in numerous dimensions and embedded in an ironic atmosphere. The ironic setting could be when the speaker's expectation was not met, resulting in an adverse event. They have also looked into how sarcasm breaches the pragmatic insecurity principle, believing that a speaker is truthful when conducting an act of speech.

Due to its ambiguous character, detecting sarcasm in textual format has been difficult in the Sentiment Analysis domain. The thin line which separates a non-sarcastic text from a sarcastic text is not very clear, unlike other examples in the Sentiment Analysis domain where the sentimentality of the text could be distinguished from each other. Furthermore, the distinguishing boundary frequently overlaps with other categories. As explored by [4], a few significant limitations faced while detecting sarcasm are difficulty detecting sarcasm from text, portraying negative sentiment using positive words, which is perhaps the biggest challenge, using short-length text and world knowledge, and use of hyperbole.

The majority of studies used Twitter datasets to detect sarcasm in brief text. The fact that Twitter data is readily available is one reason it is so popular among researchers. Using the Twitter API for developers, it is simple to retrieve a large volume of data with relevant hashtags. For example, most researchers utilize the hashtags *#sarcasm, #sarcastic,* etc.

The study conducted by [6, 11] has served as a tremendous source of inspiration for this study. Other studies using conversational data, such as [1], emphasize the relevance of context in detecting sarcasm. Extensive research on the application of LSTM in sarcasm detection [9, 13] also plays a significant role. Research has been conducted in similar or related domains, such as [2, 8], which investigated the use of similar techniques on Twitter and Reddit data, respectively.

26.3 Data Sourcing

In this study, we used the news headlines dataset from Kaggle. This dataset was submitted as part of the research [10]. The witty headlines come from websites such as The Onion, which is well-known for its satirical take on current events. HuffPost provided the non-sarcastic headlines. The dataset is provided as a JSON file. We used the most recent versions of the dataset available for our research, which was updated in 2019.

There are three attributes in the dataset:

1. *headline*: The article headlines.
2. *is_sarcastic*: A flag variable, where one represents a sarcastic record and zero represents a non-sarcastic record.
3. *article_link*: Contains link to the original news article. If necessary, this can be utilized to acquire more data.

A raw input dataset sample is given as below (see Table 26.1).

Table 26.1 Input data sample

is_sarcastic	headline	article_link
1	Study: use of phrase "don't skimp on the" linked to heart disease	https://www.theonion.com/study-use-of-phrase-dont-skimp-on-the-linked-to-heart-1819569623
0	Cat so scared in shelter won't even look at you	https://www.huffingtonpost.com, https://www.thedodo.com/backpack-cat-sad-rescue-1366450339.html
0	5 reasons trump's mika tweets are even worse than you think	https://www.huffingtonpost.com/entry/trumps-mika-tweets-are-even-worse-than-you-think-five_us_59559ccde4b0c85b96c66079
1	Ford develops new suv that runs purely on gasoline	https://www.theonion.com/ford-develops-new-suv-that-runs-purely-on-gasoline-1819575454

Table 26.2 Input data class distribution

Class	Count	Percentage (%)
Sarcastic	13,634	48
Non-sarcastic	14,985	52

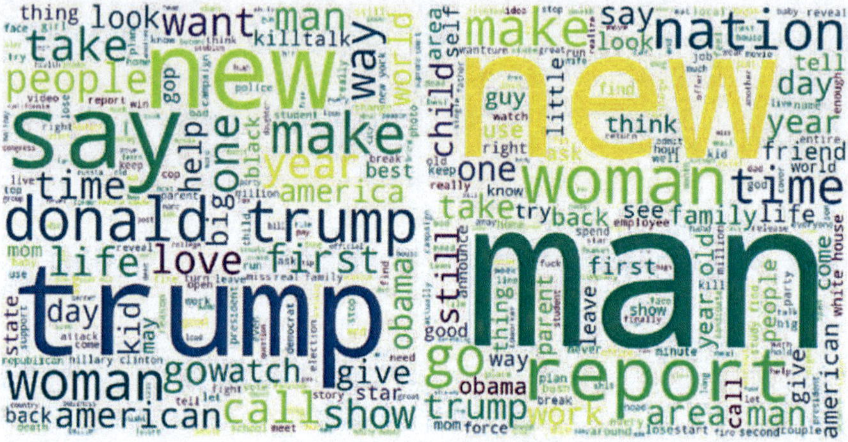

Fig. 26.1 Pictorial representation of the frequency of words through word cloud representation for sarcastic and non-sarcastic headlines, respectively

There are two classes in the available dataset: sarcastic and non-sarcastic. As a result, it is a binary classification task. The following is a list of classes and their distribution (see Table 26.2).

As shown in the table above, the dataset in concern is reasonably balanced, with 48% sarcastic and 52% non-sarcastic phrases. As a result, we do not need to take any action to address the class imbalance. Figure. 26.1 contains word clouds generated for sarcastic and non-sarcastic classes.

The character length distribution for the cleaned dataset is depicted in Fig. 26.2. Again, the length distribution for sarcastic texts is slightly more diffuse than for non-sarcastic texts.

The distribution of the number of words in each document is shown in Fig. 26.3. It is evident from Fig. 26.3 the distribution for the number of words is more scattered for sarcastic text than non-sarcastic text. It is interesting to note that a similar trend occurred in Fig. 26.1.

The average word length distribution in each text is shown in Fig. 26.4. The sarcastic text has an average length of 5.55 characters, while non-sarcastic writing has an average length of 5.25 characters. As a result, the groups' average length is virtually the same.

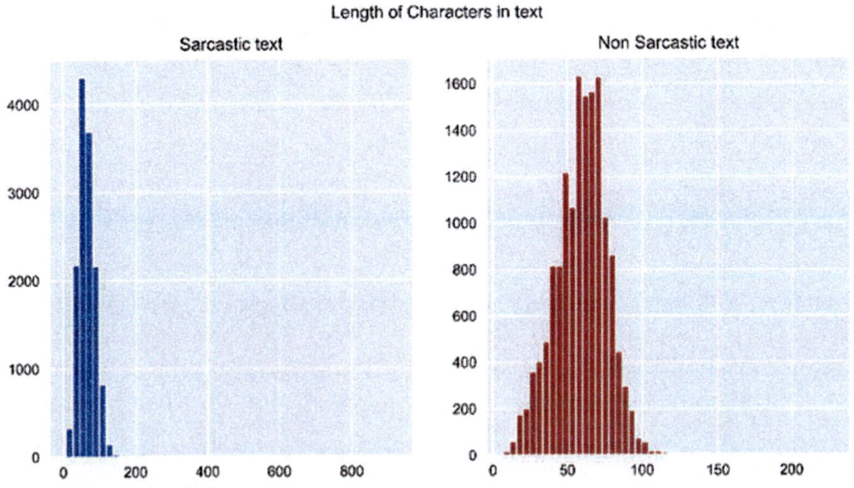

Fig. 26.2 Distribution of length of characters in each text

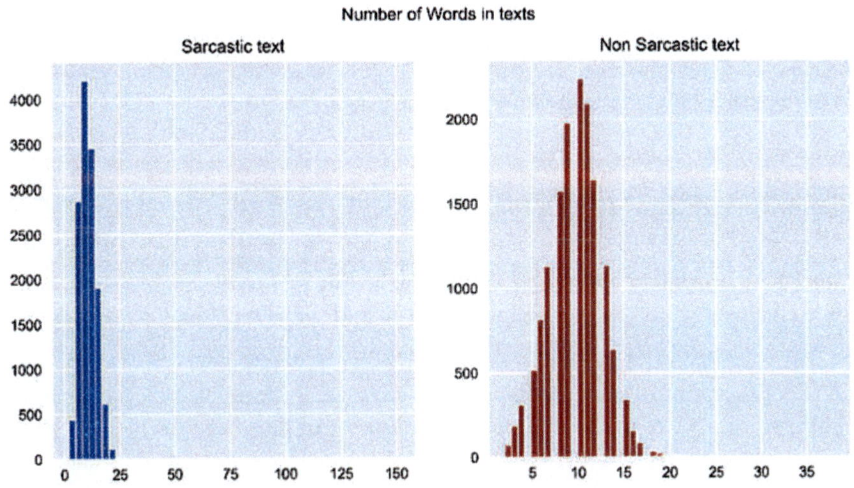

Fig. 26.3 The distribution of the word count in each text

26.4 Observations

- Compared to Twitter tweets, which are quite informal, news headlines are highly structured, professional, and formal. There are no misspelled words or emojis in the dataset.
- Non-sarcastic headlines consist of terms or phrases relating to current issues at the time of the collection. In contrast, sarcastic headlines contain more irony and are less likely to be relevant to current news trends.

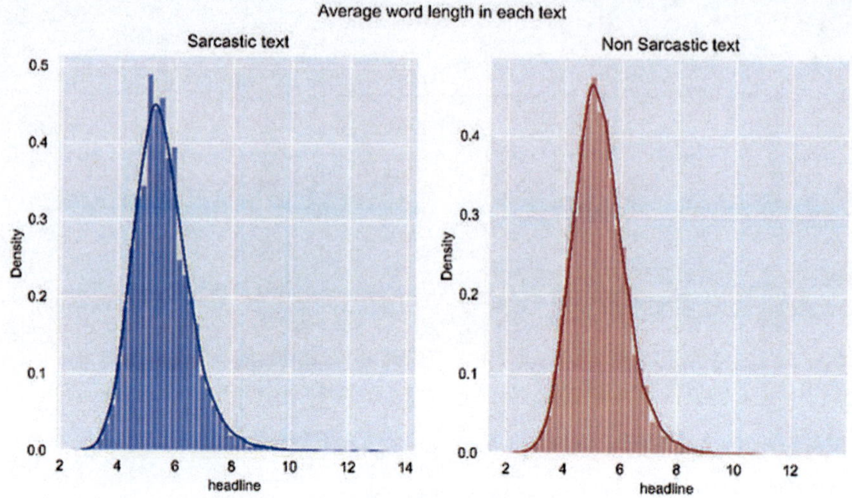

Fig. 26.4 The distribution of average word lengths in each text

- For sarcastic texts, the distribution of text length and word numbers is more scattered than non-sarcastic text.
- Both groups have approximately the same average length.

26.5 Research Methodology

There are four primary components of the methodology, i.e., data preparation, word embedding generation, classification, and evaluation. The methodology follows the following path shown in the flowchart (see Fig. 26.5).

Text Preprocessing. A critical component of any machine learning project is cleaning and preprocessing of the data. Every project or set of data requires its own set of

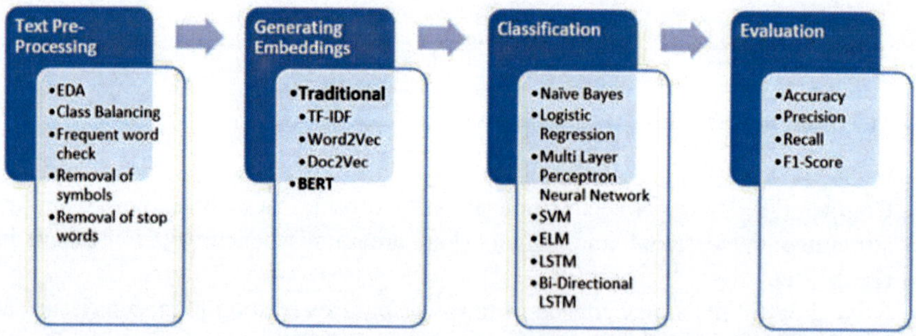

Fig. 26.5 Flowchart of the Methodology followed

data cleansing methods. For example, for jobs involving natural language processing, we sanitize the text headlines data as follows: symbols are removed, numerals are removed, word irregularities are removed, the text is converted to lowercase, stop-words and special words are removed, all single-letter words are removed, and empty rows are removed.

The stop words are removed only for data entered into the Term Frequency-Index Document Frequency (TF-IDF) [17] model, not for data entered into the other models. This is due to a decrease in total scores for Word2Vec [15, 16] and Dec2Vec when stop words were removed. After removing the stop words, the TF-IDF, on the other hand, performed better. Therefore, we did not perform any data cleansing on the data before sending it to Bidirectional Encoder Representations from Transformers (BERT) model. The model has been created to use the properties of the raw input text.

Generating Word Embeddings. The transformation of words into a numerical vector representation for an NLP project is vital. In order to apply any machine learning algorithms for text data, it must be translated into a form consisting of numerical vectors. We divided the word embedding process into two sections. The first portion comprises traditional techniques, including word vectorization using TF-IDF and word embedding techniques using Word2Vec, and Doc2Vec. The second section includes the most up-to-date transformer-based approach i.e., BERT. BERT [5] is a transformer, architecture-based generalized pre-trained model. The architecture of transformers was launched by Google's scientists in 2017 in paper [14]. For BERT embeddings, we used the DistilBERT transformer from Huggingface's transformer library. DistilBERT is a variant of Base BERT that is lighter, smaller, and faster. It retains 95% of BERT's performance while utilizing 40% fewer parameters than Bert-base-uncased and completing the GLUE language understanding benchmark 60% faster [12]. Due to the project's limited RAM and GPU resources, the fact that DistilBERT has 40% fewer parameters than BERT Base is critical in deciding between the two embedding models.

Classification. Classification is the task of assigning or predicting a class to an example of data using machine learning algorithms. It falls under the category of the supervised learning algorithms, which means that it requires a set of pre-labeled data called as the training dataset to be used in order for the algorithm to learn to predict a particular class to an input sample. We employed four different classification algorithms, starting with Naive Bayes and steadily rising in complexity with Logistic Regression, Multi-Layer Perceptron (MLP), Extreme Learning Machine (ELM), and lastly, the relatively complex LSTM approach.

Evaluation. The outcomes of the classification system should be evaluated using several evaluation metrics once the word vectors or embedding have been produced and passed through the classifier. These metrics are based on Confusion matrix (Table 26.3) as below.

The following metrics were utilized in this study:

Table 26.3 Confusion matrix

		Actual values	
		Positive (1)	Negative (0)
Predicted values	Positive (1)	True positive	False positive
	Negative (0)	False negative	True negative

- **Accuracy**: The most popular metric for calculating a model's performance is accuracy. It gives a good idea about performance in the case of balanced classes.

$$\text{Accuracy} = \frac{\text{True Positive} + \text{True Negative}}{\text{True Positive} + \text{False Positive} + \text{True Negative} + \text{False Negative}}$$

- **Precision**: Precision is the percentage of positive classes that were truly positive out of all the positive classes predicted by the model.

$$\text{Precision} = \frac{\text{True Positive}}{\text{True Positive} + \text{False Positive}}$$

- **Recall**: The recall percentage reflects how many of the actual positive classes are accurately predicted by the model.

$$\text{Recall} = \frac{\text{True Positive}}{\text{True Positive} + \text{False Negative}}$$

- **F1 score**: The Precision and Recall scores are combined into a single number to form F1 score. Instead of arithmetic, it uses a harmonic medium to make end values even more extreme.

$$F1\text{Score} = 2 * \frac{\text{Precision} * \text{Recall}}{\text{Precision} + \text{Recall}}$$

26.6 Results

We used TF-IDF, Word2Vec, Doc2Vec, and BERT for vectorization and evaluated seven different classification algorithms, including Naive Bayes, Logistic Regression, MLP-NN, SVM, ELM, LSTM, and Bi-LSTM. There are twenty-six different model combinations, with seven classification methods for each word embedding technique. The primary disadvantage of TF-IDF embeddings is that the size of the vector is too long for LSTM and Bi-LSTM which fall under deep learning category, as the embeddings are equal in length to the vocabulary size. Insufficient computer resources were available to manage such massive LSTM vectors, and the machine frequently crashed and rebooted. As a result, we chose not to combine TF-IDF and LSTM, resulting in the use of only five TF-IDF embedding combinations rather than seven. The comparison and assessment table of models (Table 26.4) shows that the BERT and LSTM or Bi-LSTM model combination provided the best outcome. For BERT embeddings, the outcomes from LSTM and Bi-LSTM were quite similar, with accuracy being nearly the same and only a slight difference in other evaluation metrics such as precision, recall, and F1 scores.

The graph (Fig. 26.6) compares the average metrics from vectorization (TF-IDF) and three embeddings (Word2Vec, Doc2Vec, and BERT) across the seven classification algorithms. We can deduce that in the assessment measures employed, the best results were observed with the BERT embeddings, with Word2Vec being the second-best, followed by TF-IDF and Doc2Vec. Furthermore, we can see that for every assessment parameter utilized, BERT embeddings exhibited an average performance boost of 7–10% over traditional embedding techniques, which is a considerable advance in the field of NLP classification tasks. BERT-based embeddings, for example, had an average accuracy of 87% throughout all classification algorithms. This demonstrates the importance of pre-trained models in Natural Language Understanding tasks.

The graph (Fig. 26.7) shows the comparison of the average of the metrics resulting from the different classification techniques used across the vectorization and embedding techniques. Here, we can infer that the LSTM and the Bi-LSTM have delivered the best overall accuracy; however, Logistic regression and SVM exhibited consistent and superior outcomes in all evaluation measurements.

26.7 Conclusion

According to the report's model results section, on every evaluation parameter evaluated, BERT alone provided for an average performance boost of 7–10% above standard embedding techniques. The overall superior performance was due to a mix of high-quality embedding from BERT and learning abilities from LSTM. The results from our work reinforce the importance of pre-trained models in NLP, as obtaining sufficient data for training models is one of the most challenging tasks. In the model results and evaluation section, the LSTM classifier did not considerably improve the metrics of other vectorization methods. We can deduce that high-quality pre-trained embeddings are alone sufficient for significant improvement in model metrics. Optimizing the model architecture may not yield substantial improvement in model metrics. For applications that require faster inference, Logistic Regression can be used with pre-trained BERT embeddings.

Table 26.4 Model results comparison and evaluation

Word embedding	Classifier	Accuracy (in %)	Precision (in %)	Recall (in %)	F1-score (in %)	Model training time
TF-IDF	Naïve Bayes	79.2	79.2	79.2	79.2	16 mil s
	Logistic Regression	79.0	79.0	79.0	79.0	273 mil s
	MLP-NN	75.5	75.6	75.5	75.5	1 min 44 s
	SVM	79.5	79.5	79.5	79.5	57 s
	ELM	70.9	70.2	66.8	68.8	44 s
Word2Vec	Naïve Bayes	76.6	77.1	76.6	76.6	142 mil sec
	Logistic Regression	80.1	80.3	80.1	80.1	22 s
	MLP-NN	79.7	79.7	79.7	79.6	16 s
	SVM	79.7	79.9	79.7	79.7	1 min 25 s
	ELM	80.1	77.9	80.8	79.3	6 s
	LSTM	79.4	77.6	77.7	77.7	1 min
	Bi-LSTM	79.2	77.3	77.7	77.5	1 min
Doc2Vec	Naïve Bayes	61.9	61.8	61.9	61.7	143 mil s
	Logistic Regression	77.2	77.3	77.2	77.3	1 min 5 s
	MLP-NN	76.0	76.2	76.0	76.0	8 s
	SVM	76.8	76.8	76.8	76.8	1 min 41 s
	ELM	77.2	76.3	75.1	75.7	5 s
	LSTM	76.0	74.0	71.5	72.5	1 min
	Bi-LSTM	76.1	73.7	72.5	73.0	1 min
BERT	Naïve Bayes	78.7	78.7	78.7	78.7	444 mil sec
	Logistic Regression	87.5	87.5	87.5	87.5	22 s
	MLP-NN	87.7	87.8	87.7	87.7	44 s
	SVM	87.7	87.7	87.7	87.7	7 min 22 s
	ELM	83.1	82.1	81.7	81.9	5 s
	LSTM	89.5	89.2	90.2	89.7	1 min
	Bi-LSTM	89.5	89.5	89.5	89.5	1 min

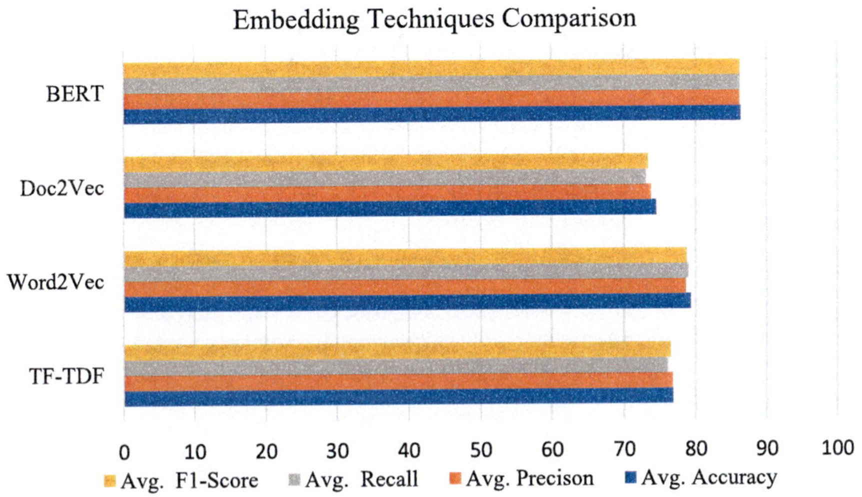

Fig. 26.6 Embedding techniques comparison

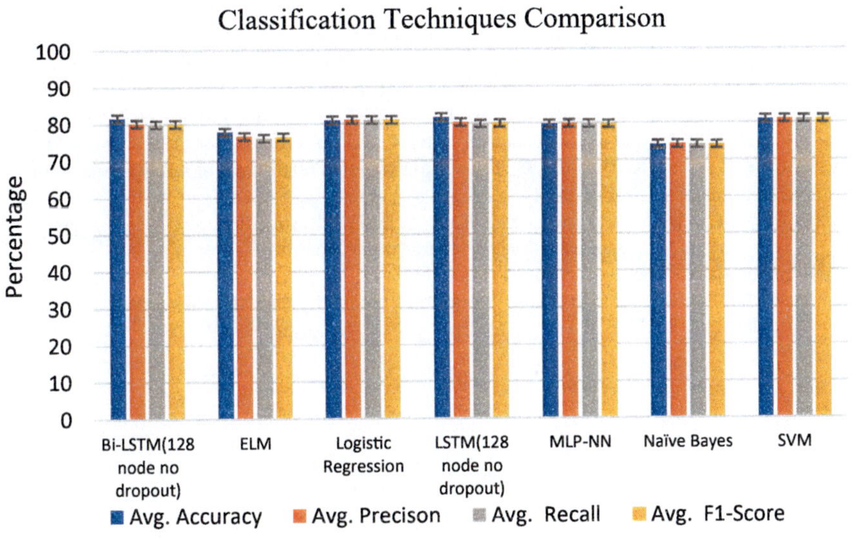

Fig. 26.7 Classification techniques comparison

References

1. Abercrombie, G., Hovy, D.: Putting sarcasm detection into context: the effects of class imbalance and manual labelling on supervised machine classification of twitter conversations (2016)
2. Baruah, A., Das, K., Barbhuiya, F., Dey, K.: Context-aware sarcasm detection using BERT, pp. 83–87 (2020). https://doi.org/10.18653/v1/2020.figlang-1.12

3. Campbell, J.D., Katz, A.N.: are there necessary conditions for inducing a sense of sarcastic irony? Discourse Process **49**, 459–480 (2012). https://doi.org/10.1080/0163853X.2012.687863
4. Chaudhari, P., Chandankhede, C.: Literature survey of sarcasm detection. In: Proceedings of the 2017 International Conference on Communications and Signal Process Networking, WiSPNET 2017, 2018 January, pp. 2041–2046 (2018). https://doi.org/10.1109/WiSPNET.2017.8300120
5. Devlin, J., Chang, M.W., Lee, K., Toutanova, K.: BERT: Pre-training of deep bidirectional transformers for language understanding. In: NAACL HLT 2019-Conference of the North American Chapter of the Association for Computational Linguistics: Human Language Technologies, Proceedings Conference, vol. 1, pp. 4171–4186 (2019)
6. Ghosh, D., Fabbri, A.R., Muresan, S.: Sarcasm analysis using conversation. Context (2018). https://doi.org/10.1162/coli
7. Joshi, A., Bhattacharyya, P., Carman, M.J.: Automatic sarcasm detection: a survey (2016)
8. Kalaivani, A., Thenmozhi, D.: Sarcasm identification and detection in conversion context using BERT, pp. 72–76 (2020). https://doi.org/10.18653/v1/2020.figlang-1.10
9. Kumar, A., Narapareddy, V.T., Srikanth, V.A., Malapati, A., Neti, L.B.M.: Sarcasm detection using multi-head attention based bidirectional LSTM. Ieee Access 8 (2020). https://doi.org/10.1109/ACCESS.2019.2963630
10. Misra, R., Arora, P.: Sarcasm detection using hybrid neural network (2019). https://doi.org/10.13140/RG.2.2.32427.39204
11. Porwal, S., Ostwal, G., Phadtare, A., Pandey, M., Marathe, M.V.: Sarcasm detection using recurrent neural network (2018)
12. Sanh, V., Debut, L., Chaumond, J., Wolf, T.: DistilBERT, a distilled version of BERT: smaller, faster, cheaper and lighter, pp. 2–6 (2019)
13. Son, L., Kumar, A., Sangwan, S., Arora, A., Nayyar, A., Abdel-Basset, M.: Sarcasm detection using soft attention-based bidirectional long short-term memory model with convolution network. IEEE Access 7 (2019). https://doi.org/10.1109/ACCESS.2019.2899260
14. Vaswani, A., Shazeer, N., Parmar, N., Uszkoreit, J., Jones, L., Gomez, A.N., Kaiser, Ł., Polosukhin, I.: Attention is all you need. Adv. Neural Inf. Process Syst. 5999–6009s (2017)
15. Mikolov, T., Chen, K., Corrado, G., Dean, J.: Distributed representations of words and phrases and their compositionality. pp. 1–9 (2013)
16. Mikolov, T., Chen, K., Corrado, G., Dean, J.: Efficient estimation of word representations in vector space. In: 1st International Conference on Learning Representations, ICLR 2013-Workshop Track Proceedings, pp. 1–12 (2013)
17. Jones, K.S.: A statistical interpretation of term specificity and its application in retrieval. J. Doc. **281**, 11–21 (1972)

Chapter 27
A Fish Biomass Prediction Model for Aquaponics System Using Machine Learning Algorithms

Pragnaleena Debroy and Lalu Seban

Abstract Most of the fish biomass estimation process till now, is based on human intervention, making the process time-consuming, expensive, imprecise, and laborious. Also, no studies have been done on fish biomass estimation for aquaponics systems. This paper proposes two prediction models for fish weight estimation of the aquaponics system using the Artificial neural network and its hybrid with fuzzy logic i.e., Adaptive neuro-fuzzy inference system (ANFIS). The Feed-forward back-propagation network is used in the ANN model, and for the fuzzification process in ANFIS model, the Gaussian membership function is employed with Sugeno structure as the fuzzy inference system (FIS). The performance evaluation indicated that ANFIS model had attained the best prediction accuracy in terms of MAE of 0.3141 and RMSE of 0.6379, and R^2 of 0.9808 in comparison with the conventional ANN model. This prediction model can help predict the fish growth of the aquaponics system simply and cost-effectively, supporting farmers in avoiding imbalances in market supply, and demand and economic management.

27.1 Introduction

Fish farming has acquired massive recognition in current years and this makes fish farming one of the fastest-growing ventures in the food production industry [1]. In intensive fish farming, fish biomass measurement is essential for stock estimation, increasing production rate, maximizing profit, and economic stability. The knowledge of fish biomass has a significant role in calculating oxygen consumption, managing feeding rate [2], estimating antibiotic dose, and determining the optimal harvesting time [3]. Water quality directly affects fish health and growth. Reliable fish biomass estimation is crucial for maintaining water quality parameters like temperature, pH, ammonia level, dissolved oxygen by avoiding the adverse effect of overfeeding [4]. These increase the need to accurately forecast the fish biomass to support the scientific fish production system's improvisation and development.

P. Debroy (✉) · L. Seban
NIT Silchar, Silchar, Assam, India
e-mail: pragnaleena_rs@ei.nits.ac.in

J. I.-Z. Chen et al. (eds.), *Machine Learning and Autonomous Systems*, Smart Innovation, Systems and Technologies 269, https://doi.org/10.1007/978-981-16-7996-4_27

The aquaponics system integrates aquaculture and hydroponics systems and involves three live components: fish, plant, and microbes [5]. This integration often causes unfavourable environmental conditions, which can directly affect fish growth. For ensuring the proper working of the aquaponics system, an adequate environment must be maintained continuously. Any irregular changes or unfavourable parametric conditions can damage the continuity of the whole process [6]. The success of the aquaponics business heavily depends on the time of yield, size of the product, cost of the product, market demand, etc. [7]. Being a promising field, the aquaponics system must assure economic stability; otherwise, people will not adopt this. Therefore, the accurate prediction of fish mass has become an urgent need to intensify the aquaponics system's fish production rate.

The conventional approach of fish biomass estimation is sampling and weighting, and for accurate measurement, the examination of the representative portion of the population must be there [3]. But manual sampling can cause stress to the fish, and it is labour-intensive too.

To avoid stress and injury to fish, machine vision systems (MVS), a non-invasive approach for predicting the size and mass of fish, have come into the picture that attracted the researchers and aquacultural industries over three decades [8]. The stereo-vision system reported by [9–11] and many more have become the most desired fish weight estimation method. Despite being a non-invasive and effective approach, the stereo-vision system involves multiple hardware and software collaboration to synchronize the cameras, which makes the estimation process more complicated. Some researchers mentioned other methods like laser scanner [12] or CCD camera [13–15], but for implementing these approaches, the fishes need to be in the same depth or fishes that pass through a tube. Indoor light is also a limitation for high-quality images [8]. So, near infrared-based computer vision system is used in some literature [16, 17] to get high-quality images in dim light. But these methods involve complexity as fishes move in the water freely, and taking good oriented pictures is not very easy, and only a limited number of fish can be sampled.

Some literature have mentioned machine learning methods for fish biomass estimation [18], but there is still a huge research gap in implementing machine learning models in fish estimation. Until now, no studies have been done on fish biomass estimation for aquaponics systems using machine learning algorithms to the best of our knowledge. Therefore, developing a non-invasive, efficient, and cost-effective fish biomass estimation is highly needed for the aquaponics system using machine learning algorithms.

This study's main purpose is to develop a non-invasive, cost-effective fish biomass prediction model of an aquaponics system using machine learning algorithms. The introduced methodology is based on the mathematical model for data collection and neural network models to predict fish biomass. Here, the artificial neural network with feed-forward back propagation structure and its hybrid with a fuzzy neural network model with a Gaussian membership function is used to estimate the fish biomass of the aquaponics system accurately. This approach for estimating fish biomass information during the rearing cycle would be valuable for aquaponics system management and fish farming industries.

27.2 Methodology

The proposed model's architecture is divided into two sections: Data pre-processing and designing the prediction model. The first module is collecting data from mathematical model and pre-processing the data by defining input–output variables in the dataset, correlation analysis, and formation of learning and testing dataset. Here fish growth model of the aquaponics system is used for data collection. The second part includes the development of the prediction models based on machine learning algorithms. Here, ANN and ANFIS models are used for predicting fish growth in terms of weights (g). The basic architectural view of the prediction model is summarized in Fig. 27.1.

The basic architectural view of the prediction model is summarized in Fig. 27.1.

27.2.1 Data Pre-processing

The dataset includes input and output data. Input data includes the information of recirculating water temperature (°C), dissolved oxygen (mg/l), ammonia (mg/l), and pH (ppm), and output data is the weight of the fish (g). Data have been collected through mathematical models.

Fish Growth Model. The base of the fish growth model is bioenergetics that includes the anabolism and catabolism process of fish established by [16] and reformed by [17]. The formula for the rate of change of fish weight is given as,

$$\frac{dW}{dt} = HW^m - kW^n \tag{27.1}$$

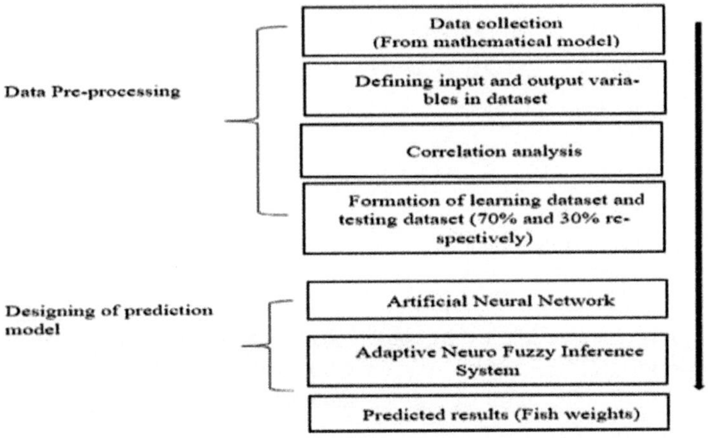

Fig. 27.1 Architectural view of the proposed fish biomass prediction model of aquaponics system

where, W represent fish weight (g), H is the net anabolism coefficient (g^{1-m} day^{-1}), k is the fasting catabolism coefficient (g^{1-m} day^{-1}), m denotes the order of body weight for net anabolism, and n is the order of body weight for fasting catabolism.

Anabolism increases the fish growth rate by food assimilation in terms of feeding rate, whereas catabolism decreases the fish growth rate by feeding and fasting catabolism [16]. Then the reformation of the Eq. (27.1) in respect of feeding rate $\frac{dR}{dt}$ (g day^{-1}) can be,

$$\frac{dW}{dt} = \left(b\frac{dR}{dt} - ab\frac{dR}{dt}\right) - kW^n \tag{27.2}$$

$$\frac{dW}{dt} = b\frac{dR}{dt} - \left(ab\frac{dR}{dt} + kW^n\right) \tag{27.3}$$

where, a denotes the fraction of food consumed which is used for feeding catabolism (dimensionless), b denotes the efficiency of food consumption (dimensionless). Equation (27.3) can be reformed as,

$$\frac{dW}{dt} = (1-a)b\frac{dR}{dt} - kW^n \tag{27.4}$$

The parameter b represents the proportion of the food intake available for the metabolizable energy [19], which is not a constant value. The modified equation for b given in [20] as,

$$b = d - \left[c/\left(1 + \left(e^{-\varepsilon(W-T_c)}\right)\right)\right] \tag{27.5}$$

The parameter b is in the range of $d - c \leq b \leq d$, where d-c represents minimum food assimilation efficiency and d represents maximum food assimilation efficiency. ε is the rate of change of b from its maximum to minimum value and T_c is the time span to shift the rate of change of b.

The literature [21] has given the modified equation for fasting catabolism coefficient as,

$$k = k_{\min} e^{j(T-T_{\min})} \tag{27.6}$$

where, $k_{\min}$ is the fasting catabolism coefficient (g^{1-n} day^{-1}) in $T_{\min}$ (°C) which is the minimum water temperature, T is the water temperature (°C), and j is the constant for describing the catabolism temperature (1 $°C^{-1}$).

The feeding rate depends on the fish body weight and several environmental factors like, temperature (T), dissolved oxygen (DO), ammonia (A), biological oxygen demand (BOD), feeding level (f). So, feeding rate can be written as,

$$\frac{dR}{dt} = \tau\delta\upsilon p f h W^m \tag{27.7}$$

where, τ denotes temperature factor ($0 < \tau < 1$, dimensionless), δ denotes the dissolved oxygen (DO) factor ($0 < \delta < 1$, dimensionless), v denotes unionized ammonia factor ($0 < v < 1$, dimensionless), p is the pH factor ($0 < p < 1$, dimensionless), f denotes relative feeding level ($0 < f < 1$, dimensionless), and h is the food consumption coefficient (g^{1-m} day^{-1}).

The equations for all these factors are given in the literatures [16, 17, 22–24].

$$\tau = e^{\left\{-4.6[(T_{opt}-T)/(T_{opt}-T_{min})]^4\right\}}, \text{ if } T < T_{opt} \quad (27.8)$$

$$\tau = e^{\left\{-4.6[(T-T_{opt})/(T_{max}-T_{opt})]^4\right\}}, \text{ if } T > T_{opt} \quad (27.9)$$

where, T is the water temperature of the rearing tank, T_{min} is the minimum water temperature (°C), T_{max} is the maximum water temperature (°C), and T_{opt} is the optimal water temperature (°C).

$$\delta = 1, \text{ if } DO > DO_{crit} \quad (27.10)$$

$$\delta = (DO - DO_{min})/(DO_{crit} - DO_{min}), \text{ if } DO_{min} \leq DO \leq DO_{crit} \quad (27.11)$$

$$\delta = 0, \text{ if } DO < DO_{crit} \quad (27.12)$$

where, DO is the dissolved oxygen of the rearing tank water (mg/l), DO_{min} is the minimum dissolved oxygen (mg/l), and DO_{crit} is the critical dissolved oxygen (mg/l).

$$v = 1, \text{ if } A < A_{crit} \quad (27.13)$$

$$v = (A_{max} - A)/(A_{max} - A_{crit}), \text{ if } A_{crit} \leq A \leq A_{max} \quad (27.14)$$

$$v = 0, \text{ if } A > A_{max} \quad (27.15)$$

where, A is the ammonia level in the water (mg/l), A_{crit} is the critical ammonia level (mg/l), and A_{max} is the maximum ammonia level (mg/l).

$$p = 1, \text{ if } p > p_{crit} \quad (27.16)$$

$$p = (p_{max} - p)/(p_{max} - p_{crit}), \text{ if } p_{crit} \leq p \leq p_{max} \quad (27.17)$$

$$p = 0, \text{ if } p > p_{max} \quad (27.18)$$

where, BOD is the biological oxygen demand, BOD_{crit} is the critical limit of biological oxygen demand, and BOD_{max} is the maximum level of biological oxygen demand.

$$f = \frac{r}{R} \tag{27.19}$$

where, r denotes actual feeding rate (g day^{-1}) and R denotes maximum feeding rate (g day^{-1}).

The final growth rate equation (modified by [20]) is given as,

$$\frac{dW}{dt} = \{(1 - a)b\tau\delta vpfhW^m\} - kW^n \tag{27.20}$$

Varying T, DO, A, and p in the Eq. (27.20), the weight values have enlisted in Microsoft excel, keeping other parameters constant. Here, relative feeding level (f) is kept at a constant value to investigate the other parameter effects in the fish weight, and an average of 150 days is taken as time (t). Parameter evaluation is done by referring to some existing literature, as discussed in [20].

The input–output dataset is formed in Microsoft Excel consisting of the information of temperature (°C), dissolved oxygen (mg/l), ammonia (mg/l), pH, and weight (g) of fish. A group of 100 samples were collected from the mathematical model to use in the prediction model. Here only 10 samples are presented in the Table 27.1. to give a idea about the dataset.

Correlation Analysis. Correlation analysis is a statistical method employed to assess the degree of relationship between two quantitative variables. The equation which is used for correlation analysis is given below.

Table 27.1 A group of dataset collected from the mathemetical model

Sl. no.	T (°C)	DO (mg/l)	A (mg/l)	pH (ppm)	Actual fish weight (g)
1	20	9.185	0.077	7.08	77.94
2	20.5	9.09755	0.081	7.0705	107.8
3	21	9.0101	0.082	7.061	138.3
4	21.5	8.92265	0.085	7.0515	154
5	22	8.8352	0.34	7.042	171.1
6	22.5	8.74775	0.595	7.0325	186.2
7	23	8.6603	0.85	7.023	197.7
8	23.5	8.57285	1.105	7.0135	208.5
9	24	8.4854	1.36	7.004	211.9
10	24.5	8.39795	1.615	6.9945	223.6

$$r = \frac{n(\sum V_i E) - (\sum V_i) * (\sum E)}{\sqrt{\left[n \sum V_i^2 - (\sum V_i)^2\right] * \left[n \sum E^2 - (\sum E)^2\right]}} \quad (27.21)$$

where, r represents the Pearson correlation coefficient (PCC), Vi represents the parameters that affect output, E represents output parameter, n represents total data points. The coefficient r is in the range of -1 to 1, in which 1 denotes strong positive relation between input and output, 0 shows no relationship between input and output, and -1 denotes strong negative relation between input and output. The relationship strength depends upon the absolute value of r. The greater value of r means the relationship is stronger [19].

70% of data is selected as a training dataset for training the model, and the remaining 30% of data is used for testing the prediction model.

27.2.2 *Artificial Neural Network (ANN)*

ANN is the most familiar type of soft computing techniques. ANN works on the human brain concept, an extremely complicated structure made up of neurons that are mainly processing units that have the capabilities to store knowledge and implement whenever required, mostly for future prediction and estimation. ANN has a highly distributed parallel computing structure comprised of a plurality of neurons trained to perform like humans. ANN stimulates messaging between neurons [25].

The traditional ANN modelling methodology is given in Fig. 27.2. First inputs are given to the model. Then Initialization of the ANN parameters (weights and biases) is done. Using the transfer function, the net input is created from all the inputs. As transfer functions, hyperbolic tangent sigmoid function (***tansig***), logarithmic sigmoid function (***logsig***), and pure linear function (***purelin***) can be used. ANNs can be used as a universal function approximator for enough hidden nodes, using tansig and logsig

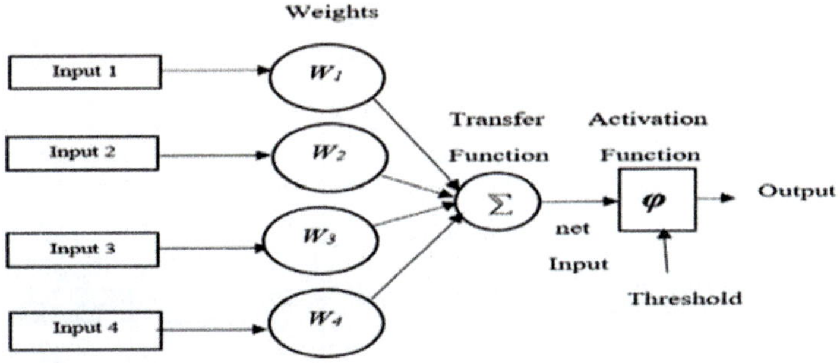

Fig. 27.2 ANN modeling methodology

transfer functions. Different types of training algorithms are there, like, Levenberg–Marquardt (LM), Broyden–Fletcher–GoldfarbShanno quasi-Newton (BFGS), scaled conjugate gradient (SCG), gradient descent with momentum and adaptive learning rate (GDX), and many more [26]. Among those, LM performs better for its robustness and higher convergence speed. Then the activation function will create the output based on the error between the net input and threshold [27].

In this research, the ANN model utilizes feedforward neural network structure with a backpropagation algorithm because, together these two creates a simple network with robustness and quick convergence of errors.

As, back propagation algorithm is based on the negetive gradient descent, it is having some flaws like, slow computational time, poor learning stability, and slow convergence speed. These disadvantages can be improved by emplying fuzzy rules in ANN.

27.2.3 Adaptive Neuro Fuzzy Inference System (ANFIS)

The hybridization of artificial neural networks and fuzzy logic is done in ANFIS. This system creates a relation between input–output datasets. For the learning process, ANFIS uses backpropagation gradient descent which depends on the backward calculations of the parameters from output nodes to the input nodes [25].

The basic working of the ANFIS model is described in Fig. 27.3. The first step of the ANFIS method is fuzzification, which refers to mapping input–output variables and their rank in the fuzzy space. The input to the fuzzy inference system (FIS) is a non-fuzzy value delimited within a range. Fuzzification of the inputs is executed by converting these crisp inputs into linguistic sets depending upon the fuzzy rules of the FIS. The fuzzification of the inputs are done using the Gaussian membership function, which gives efficient results with truncated computational time, smooth curves, and non-zero points. Some more membership functions are there like triangular and trapezoidal functions that are simple but sometimes not maintaining 0–1 criteria in modelling [19].

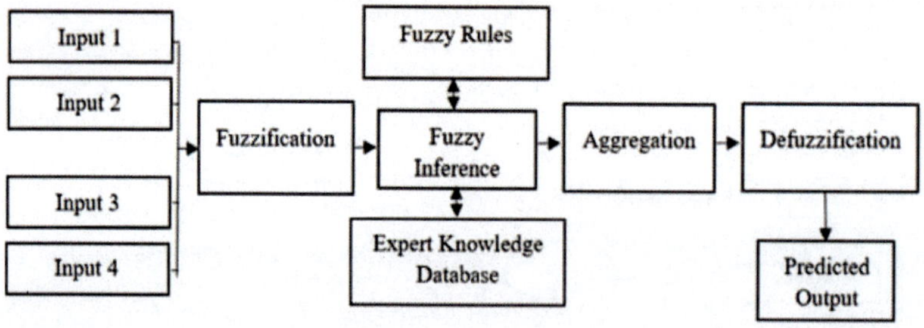

Fig. 27.3 The basic working of ANFIS based prediction model

After the fuzzification, the decision-making process is carried out by the Fuzzy inference system (FIS) of the ANFIS model based on the parallel inference of fuzzy rule and expert knowledge database. For this model, the Sugeno structure is adapted as the FIS for its ease of rule formation, defuzzification technique-based center of gravity, and generation of output with effective interpretation. Other types are also there like, Tsukamoto and Mamdani, the only difference in the defuzzification process. They give the output as a constant or linear equation.

Next step is the Aggregation which is the combination of fuzzy set outputs of all rules to generate the output fuzzy set. Aggregation stage receives the inputs as all the shortened outputs for each rule from the FIS. The rule firing in aggregation stage is done in parallel as it is always commutation, but not associative. This process supports following methods:

a. Maximum (max)
b. Probabilistic OR (probOr)
c. Sum of the output sets of each rule (sum).

Defuzzification converts the set of fuzzy values which is the output from aggregation process to a crisp value as the predicted output [19]. The main function of the defuzzifier allows the identification of an object that best fits the fuzzy set created by aggregation process. This process is essential to get the crisp output from the fuzzy set that includes a range of fuzzified values. Centroid defuzzification method used in this ANFIS model that gives the center point of the area under the curve [19].

The fuzzy network that used in this work improves the fuzzy rules and membership function's parameter values as central value, and width of the membership functions to get a improved result.

27.3 Results

27.3.1 Correlation Analysis

The correlation analysis is executed in the Microsoft Excel using the Eq. (27.10). The Pearson's correlation coefficient results for each feature are shown below.

- Temperature (T) = 0.81101
- Dissolved oxygen (DO) = −0.63112
- Ammonia (A) = 0.57759
- pH = −0.72291.

27.3.2 *Modelling of Prediction Models*

ANN Modelling. Modelling parameters of the ANN model are given as,

a. Input variable count = 4
b. Output variable count = 1
c. Network type = Feed-forward back propagation.
d. Training function = TRAINLM
e. Adaption learning function = LEARNGDM
f. Number of layers = 2
g. Training algorithm = Levenberg–Marquardt
h. Transfer function = TANSIG
i. Epoch = 1000.

ANFIS Modelling. Modelling parameters of ANFIS model are given as,

a. Input variable count = 4
b. Output variable count = 1
c. FIS type = Sugeno
d. Number of rules = 81
e. Membership function = Gaussian MF
f. Optimization method = Hybrid
g. Defuzzification method = *wtaver*
h. Epoch = 1000.

27.3.3 *Prediction Model Evaluation Results*

The graphical plots of both the prediction model evaluation results with minimized magnitude relative error (MRE) are presented in Fig. 27.4 and Fig. 27.6, respectively.

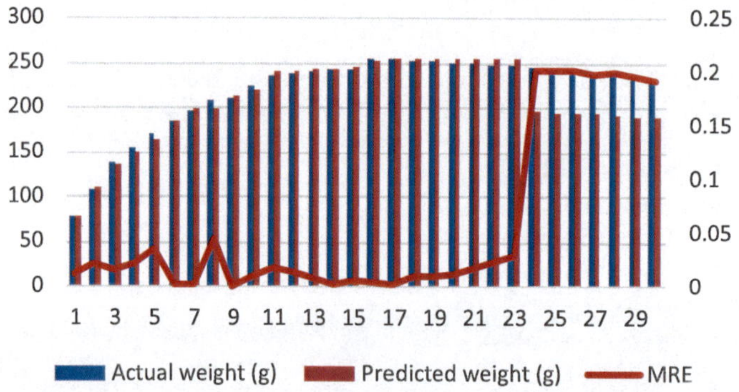

Fig. 27.4 Graphical plots of actual weight, predicted weight, and magnitude relative error (MRE) of ANN model

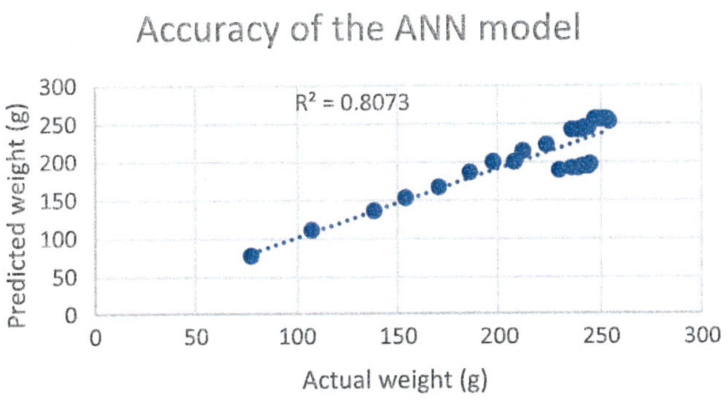

Fig. 27.5 Graphical plot of accuracy of ANN model

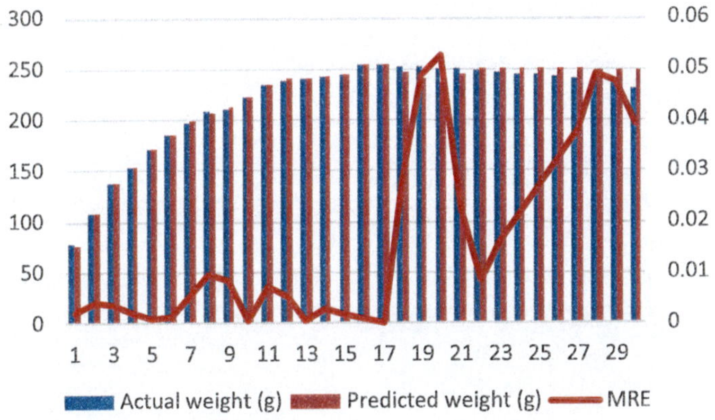

Fig. 27.6 Graphical plots of actual weight, predicted weight and magnitude relative error (MRE) of ANFIS model

The accuracy of the prediction models is also shown in Fig. 27.5 and Fig. 27.7, respectively.

27.3.4 Performance Evaluation of the Prediction Results

For performance evaluation of the prediction models, three performance matrices are considered. They are,

a. Mean Absolute Error (MAE)
b. Root Mean Square Error (RMSE)
c. Coefficient of determination (R^2)
d. Computational time.

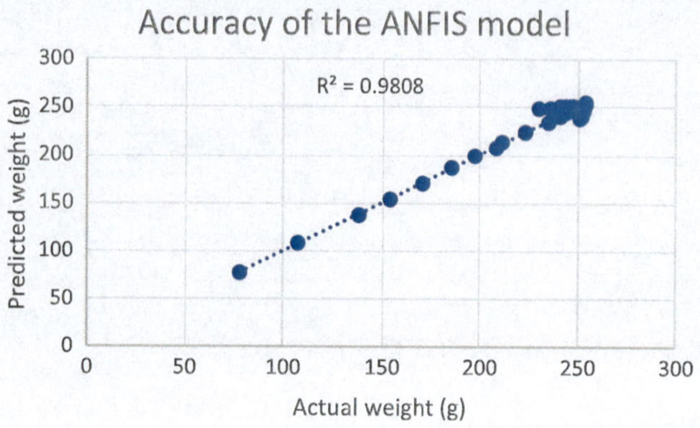

Fig. 27.7 Graphical plot of accuracy of ANFIS model

Table 27.2 Performance comparison of prediction models

Prediction model	MAE	RMSE	R^2	Computational time
ANN	0.6314	0.8671	0.8073	20.003421 s
ANFIS	0.1414	0.3079	0.9808	16.762934 s

The performance comparisons of the ANN and FNN models in terms of MAE, RMSE, R^2, and computational time are shown in Table 27.2.

27.3.5 Comperative Analysis with Other Models

Table 27.3 shows proposed model eveluation result with other 2 Fuzzy neural network model reviewed in related literature. Most of the reviewed literature evaluated their models using RMSE and R^2.

Table 27.3 Comparative analysis with other prediction models

Prediction model	RMSE	R^2
Petkovic et al. [25]	2.1305	0.9771
Ren et al. [19]	0.4807	–
Proposed model	0.3079	0.9808

27.4 Discussion of Results

The analysis of the graphical plots of MRE of the neural network models (Figs. 27.4, and 27.6), depicts that the ANFIS model has the smaller values of MRE ranging from 0.000317 to 0.05289, while the ANN model has higher values than ANFIS model of 0.000772 to 0.201731 range. The accuracy analysis of the models (Figs. 27.5 and 27.7) showed that the ANFIS model has the highest coefficient of determination value (R2) of 0.9808 than the ANN model with R2 value of 0.8073. The fish biomass prediction model's performance analysis that showed in Table 27.2, illustrates that compared to ANN, the ANFIS model has performed best with lower MAE values as 3.1414 and RMSE of 0.6379. Also, computational time of ANFIS model is 16.762934 s which is less compared to traditional ANN model. Table 27.3 shows that proposed Fuzzy based neural network model is performing better compared to other fuzzy models reported in literature. However, these errors can be minimized further if test instances are increased, and an enhanced range of input values are used for the prediction model rather than the clumped range of input values. Also, the accuracy and error rate can be improved by minute tuning of the membership functions of the ANFIS model. The prediction accuracy of the ANFIS model shows an encouraging possibility for neuro-fuzzy application in fish weight prediction. The correlation analysis shows that temperature, dissolved oxygen, and pH play a more determinant effect in fish weight prediction than ammonia. Furthermore, other parameters that influence fish weight can be explored and applied in fuzzy based neural network models for fish weight prediction. Other machine learning algorithms can also be designed to show the best performing model for predicting fish biomass of the aquaponics system.

27.5 Conclusion

This work explored various factors that magnify the highly improved performance of prediction models for fish weight in the aquaponics system. A limited sample data of fish weight and the parameters affecting the fish weight were utilized from the mathematical models of aquaponics fish growth and temperature-pH correlation to establish the prediction models. Temperature, dissolved oxygen, and pH showed a strong correlation effect on fish weight. In the ANN model, the Feed-forward backpropagation network is used, and for the ANFIS model, the Gaussian membership function is used with Sugeno structure as the FIS. The performance evaluation analysis of the prediction models clearly depicted that the proposed models are efficient in forecasting the fish biomass of the aquaponics system. ANFIS model has performed best compared to ANN with the best values of MAE, RMSE, and R^2.Computational efficiency of ANFIS model is better compared to ANN model. Also, compared to other fuzzy based NN models reported in literature, this proposed model is performed better in terms of RMSE and R^2.This work confirmed that fish biomass of aquaponics system could be predicted with acceptable accuracy from a small set of variables using neural network models, especially the fuzzy-based neural network model. Future work for this research can be carried out by collecting a large number of time-series data with other aquaponics factors that affect the fish biomass and analysing those data using semi-supervised machine learning algorithms towards effective fish biomass prediction of the aquaponics system.

References

1. Olsen, R.L.: A limited supply of fishmeal: impact on future increases in global aquaculture production. Trends Food Sci. Technol. **27**(2), 120–128 (2012)
2. Kubitza, F.: Formulated diets, feeding strategies, and cannibalism control during intensive culture of juvenile carnivorous fishes. Rev. Fish. Sci. **7**, 1–22 (1999)
3. Lines, J.: An automatic image-based system for estimating the mass of free-swimming fish. Comput. Electron. Agric. **31**, 151 (2001)
4. Lopes, F.: Fish farming autonomous calibration system. In: OCEANS 2017 –Anchorage, pp. 1–6. IEEE, Aberdeen, UK (2017)
5. Simon, G.: Challenges of sustainable and commercial aquaponics. Sustainability **7**, 4199–4224 (2005)
6. Shortis, M.: Progress in the automated identification, measurement, and counting of fish in underwater image sequences. Mar. Technol. Soc. J. **50**, 4 (2016)
7. Dios, J.: Computer vision and robotics techniques in fish farms. Robotica **21**, 233 (2003)
8. Hufschmied, P.: Automatic stress-free sorting of sturgeons inside culture tanks using image processing. J. Appl. Ichthyol. **27**, 622 (2011)
9. Miranda, J.: A prototype to measure rainbow trout's length using image processing. Aquacult. Eng. **76**, 41–49 (2017)
10. Trobbiani, G.: A novel method to obtain accurate length estimates of carnivorous reef fishes from a single video camera. Neotropical Ichthyol. **13**, 93–102 (2015)
11. Zhou, C.: Near-infrared imaging to quantify the feeding behavior of fish in aquaculture. Comput. Electron. Agric. **135**, 233–241 (2017)
12. Hung, C.: A highly sensitive underwater video system for use in turbid aquaculture ponds. Sci. Rep. **6**, 1–7 (2016)
13. Saberioon, M.: Automated within tank fish mass estimation using infrared reflection system. Comput. Electron. Agric. **150**, 484–492 (2018)
14. Ursin, E.: A mathematical model of some aspects of fish growth, respiration, and mortality. J. Fish. Res. Board Can. **24**, 2355–2453 (1967)
15. Bolte, J.P., Nath, S.S., Ernst, D.H.: POND: A decision support system for pond aquaculture. In: Twelfth Annual Administrative Report. PD/A CRSP, Corvallis, OR, pp. 48–67 (1995)
16. Karimanzira, D.: Dynamic modeling of the INAPRO aquaponic system. Aquacult. Eng. **75**, 29–45 (2016)
17. Light, T.S.: Temperature dependence and measurement of resistivity of pure water. Anal. Chem. **56**(7), 1138–1142 (1984)
18. Sidda, S.: A Review on Artificial Intelligence Techniques in Electrical Drives. In: International Conference on Smart Technologies for Smart Nation, pp. 55–60. IEEE. REVA University, Bangalore, India (2017)
19. Ren, Q., Zhang, L., Wei, Y., Li, D.: A method for predicting dissolved oxygen in aquaculture water in an aquaponics system. Comput. Electron. Agric. **151**, 384–391 (2018)
20. Statistics How To, https://www.statisticshowto.com/probability-and-statistics/regression-analysis/rmse-root-mean-square-error/

21. Omomule, T.G.: Fuzzy prediction and pattern analysis of poultry egg production. Comput. Electron. Agric **171**, 105301 (2020)
22. Brett, J.R.: Environmental factors and growth. In: Hoar, W.S., Randall, D.J., Brett, J.R. (eds.) Fish Physiology. Academic Press, vol. 8, pp. 599–675. New York, NY (1979)
23. Chen, C.Y.: Fish nutrition, feeds, and feeding with special emphasis onsalmonid aquaculture. Food Rev. Int. **6**, 333–357 (1990)
24. Ivlev, V.S.: Experimental Ecology of the Feeding of Fishes. Yale University Press, New Haven, CT (1961)
25. Petkovic, B.: Neuro-fuzzy estimation of reference crop evapotranspiration by neuro fuzzy logic based on weather conditions. Comput. Electron. Agric. **173**, 105358 (2021)
26. Plumb, A.P.: Optimisation of the predictive ability of artificial neural network (ANN) models: a comparison of three ANN programs and four classes of training algorithm. Eur. J. Pharm. Sci. **25**, 395–405 (2021)
27. Mean Absolute Error, https://en.wikipedia.org/wiki/Mean_absolute_error. Accessed 2021/01/13

Chapter 28
Feature Selection Technique for Microarray Data Using Multi-objective Jaya Algorithm Based on Chaos Theory

Abhilasha Chaudhuri and Tirath Prasad Sahu

Abstract Microarray datasets are high-dimensional data that contain the gene expression profiling information used for cancer classification. This causes a curse of dimensionality and poses multiple challenges in gene classification. Moreover, the microarray datasets are imbalanced as they have a very small sample size, which deteriorates the performance of machine learning algorithms. In order to address these problems, dimensionality reduction technique is required. Meta-heuristic optimization algorithm-based feature selection is the efficient dimensionality reduction technique. This paper introduces a feature selection technique for microarray data classification using multi-objective Jaya algorithm based on chaos theory. The two objectives: minimization of selected gene subset and maximization of classification accuracy are considered. Chaos theory is used to improve the convergence of Jaya algorithm. Five benchmark microarray datasets were used to evaluate the performance of the proposed feature selection technique. Furthermore, a comparative study is carried out where the performance of the proposed technique is compared with four other competitors.

28.1 Introduction

In the realm of bioinformatics, dealing with the high dimensionality of microarray data is a difficult problem. This challenge becomes even more difficult with the fact that microarray data have a very small sample size, which makes the dataset imbalanced. The high dimensionality, low sample size and imbalance in microarray datasets deteriorate the performance of machine learning models. Microarray datasets contains the gene expression profiling of thousands of genes, which have the potential to diagnose and classify different cancer [1], therefore, it is important to study the

A. Chaudhuri (✉) · T. P. Sahu
National Institute of Technology, Raipur 492001, CG, India
e-mail: achaudhuri.phd2018.it@nitrr.ac.in

T. P. Sahu
e-mail: tpsahu.it@nitrr.ac.in

J. I.-Z. Chen et al. (eds.), *Machine Learning and Autonomous Systems*, Smart Innovation, Systems and Technologies 269, https://doi.org/10.1007/978-981-16-7996-4_28

microarray datasets. In a microarray dataset, the number of features is in the range 2000–20,000, and due to its expensiveness, the number of samples is limited in the range 50–250. Thousands of features and only hundreds of samples make the microarray dataset very imbalanced. However, all the features are not meaningful, and the huge number of features causes the curse of dimensionality. Dimensionality reduction is necessary to improve the performance of machine learning models on high-dimensional, imbalanced datasets such as microarray datasets.

Feature selection is the well-known dimensionality reduction technique that I used for microarray data. In high-dimensional data, all features (here genes) are not helpful in training the machine learning models. Feature selection (FS) technique selects the most significant subset of genes that improve the performance of the machine learning model and remove the noisy genes that deteriorate the performance. For an N dimensional problem, there exist $2N$ possible combinations of feature subset that can be chosen. This is a huge number for higher values of N, it means that it is not feasible to explore each of the possible combination. Meta-heuristic optimization-based feature selection algorithms are best suitable for this type of combinatorial optimization problems [2]. A large number of meta-heuristic optimization algorithms are available in the literature. A meta-heuristic optimization algorithm gives the optimum result only when its parameters are tuned properly according to the problem under consideration. Parameter tuning is not a trivial task, it has to be done very carefully, which takes huge effort and time. However, Jaya algorithm does not have any algorithm-specific parameter, therefore, parameter tuning is not required hence it saves a considerable amount of time and effort. Moreover, it also has fast convergence compared to other meta-heuristic algorithms. Therefore, we have utilized Jaya algorithm in this work for developing an efficient feature selection technique.

28.2 Related Works

Feature selection using meta-heuristic optimization is the recent area where researchers have focused. Meta-heuristic optimization algorithm-based feature selection has been used for dimensionality reduction in the work proposed by [2–5] on various high-dimensional benchmark datasets. Researchers have proposed a complete binary version of Jaya algorithm based on Jaccard similarity measure [6]. Some researchers have used feature selection to address a specific problem, like Ref.[7] has used a hybrid FS technique for Parkinson's disease classification, Ref.[8] has used it for global optimization problem, Refs[9–11] have used feature extraction for image processing. Some authors have used feature engineering in another set of applications like depression, decision-making on the basis of false information and recommender systems [12–14]. Gene classification is done by Refs[15–18] using meta-heuristic optimization-based FS technique. Diagnosis of various diseases like Cancer, Alzheimer can also be done by gene classification with the help of microarray data [19].

The majority of the algorithms mentioned above have considered FS as a single objective optimization problem. However, FS is a multi-objective optimization problem because there are two objectives that are equally important, to reduce the number of features and to improve the classification accuracy. The multi-objective approach for feature selection on microarray datasets and other benchmark datasets are introduced by Refs[20–24].

28.3 Preliminaries

This section articulates the pre-requisites for the proposed feature selection technique.

28.3.1 Feature Selection

Feature selection is an important preprocessing step for high-dimensional data. In the literature, there are three types of feature selection approaches: filter, wrapper and hybrid [25]. Filter feature selection takes into account all of the dataset's features for studying their underlying statistical and information-theoretical properties and choses the feature subset based on the result of this study. The selected subset is then applied to the classifier. In case of wrapper-based feature selection approach, classifier is involved in each and every iteration of the algorithm. A search technique is applied to get a feature subset, thereafter; the chosen subset is assessed based on the classifier. This operation is carried out again and again until the termination criteria are reached. The collection of features that produces the best classification result is chosen. Hybrid feature selection approach combines the best properties of filter and wrapper feature selection methods.

Feature selection (FS) means selecting a subset of optimum number of features that not only reduces the computational cost of the machine learning model but also improves its performance. Basically, there are two objectives for FS: to identify the smallest subset of features and to obtain the highest level of accuracy in data classification task. The two objectives stated above are mutually contradictive, i.e., when we try to achieve the first objective it adversely affects the second objective and vice versa. This type of problem can be solved efficiently with multi-objective optimization algorithms.

28.3.2 Multi-objective Optimization

The simultaneous solution of multiple objective functions is referred to as multi-objective optimization (MOO) [26]. The nature of objective functions is such that they contradict each other. In other words, an optimal solution according to one objective may deteriorate the performance of another objective. This means that one solution cannot be designated as the best solution for all the objectives of the problem. Therefore, it is necessary to find a compromise solution that meets the need of every objective, which is referred to as a pareto-optimum solution. A pareto-optimal solution is a solution that balances the trade-off between the objective functions.

28.3.3 Jaya Algorithm

Jaya is a parameter-less meta-heuristic optimization algorithm. It is simple yet effective and fast converging population-based algorithm. Parameter tuning takes huge effort in other meta-heuristic algorithms but this is not the case with Jaya [27] because it does not have any algorithm-specific parameter. Population update mechanism of this algorithm is elicited from two parameters: (i) best solution and (ii) worst solution. Jaya algorithm follows the greedy approach and tries to move closer to the best solution and far from the worst solution, this property of the solution update mechanism is depicted in (28.1).

$$X_{p,d}^{I+1} = X_{p,d}^{I} + r1 \times \left(\text{Best} - \left|X_{p,d}^{I}\right|\right) - r2 \times \left(\text{Worst} - \left|X_{p,d}^{I}\right|\right) \quad (28.1)$$

where $p = (1, 2,...N)$; $d = (1, 2,...D)$ for a problem which has population of size N and dimensionality D. The variable $X_{p,d}^{I}$ represents the current solution (i.e., solution at Ith iteration), $X_{p,d}^{I+1}$ denotes the updated solution, the variable named Best implies the population's best solution and the variable Worst exhibits the worst solution so far, the random numbers $r_1 r_2 \in [0, 1]$.

28.3.4 Chaotic Maps

Chaotic maps represent a pseudo-random, deterministic process that is bounded and sensitive to the initial parameter [28]. These maps are nonlinear and dynamic in nature. The use of chaotic sequence is proven to be helpful in avoiding the local optima and leads to better convergence. The random numbers in a meta-heuristic algorithm are replaced with the chaotic maps in order to improve the convergence. Many chaotic maps are available in literature [29], tent chaotic map is used in this work, which is defined in (28.2).

$$R_{(k+1)} = \begin{cases} \frac{R_k}{0.7} & R_k < 0.7 \\ \left(\frac{10}{3}\right) \times (1 - R_k) & R_k \geq 0.7 \end{cases} \quad (28.2)$$

here R_k is the chaotic number that is calculated previously whereas $R_{(k+1)}$ is newly quantified chaotic numeral. The random numbers used in the Jaya algorithm will be replaced by the chaotic number sequence generated by (28.2). The use of chaotic numbers will result in better exploration capability of search space and chances of being trapped in local optima will be less.

28.4 Proposed Technique

28.4.1 Multi-objective Chaotic Jaya Algorithm (MO-CJaya)

Multi-objective Chaotic Jaya algorithm works in a similar manner as the Multi-Objective Jaya algorithm [30]. First of all, the population is randomly initialized, and termination criteria are decided. All the N solutions of population are evaluated on the basis of fitness functions (here we have two fitness functions). The concept of non-dominated sorting and crowding distance introduced by Deb et al. [31] is used to rank the agents of the population. The solution vector having the highest non-dominance rank and maximum crowding is designated as best. Whereas, the solution vector with least non-dominance rank and minimum crowding distance is designated as worst. Identification of best and worst solutions is done in each iteration of the algorithm. All of the population's solutions are updated depending on the best and worst solutions determined in (28.1) and (28.2). The two random numbers $r1$ and $r2$ of (28.1) are replaced with the chaotic sequence obtained by (28.2). Modified solutions are pooled with the preliminary solution of the population, and the combined group of solutions (i.e., $2N$ solutions) is again ranked with the help of non-dominated sorting and crowding distance.

Top N solutions out of $2N$ combined solutions are chosen on the basis of two criteria: 1. non-dominance rank and 2. crowding distance. This procedure is repeated until the termination criteria are met. This whole procedure is clearly shown in Fig. 28.1.

28.4.2 Objective Function Formulation for MO-CJaya

Multi-objective feature selection considers two objectives: (i) maximize the classification accuracy (which is equivalent to minimizing classification error rate) and (ii) minimize the selected number features. The formulation of the above two objectives as fitness functions for MO-CJaya is depicted in (28.3a–28.3c).

$$\text{Minimize } F(x) = \begin{cases} f_1(x) \\ f_2(x) \end{cases} \tag{28.3a}$$

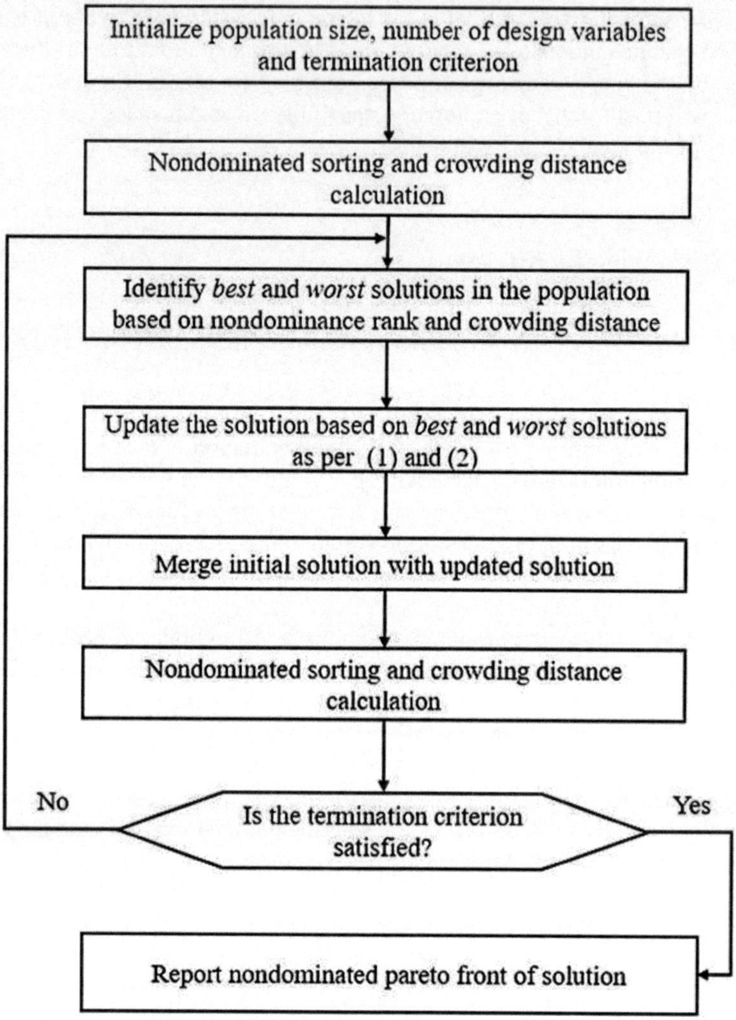

Fig. 28.1 Flowchart of Multi-objective Chaotic Jaya algorithm

$$f_1(x) = \frac{\text{Number of features selected}}{\text{Total number of features}} \tag{28.3b}$$

$$f_2(x) = \frac{\text{Number of incorrectly classified samples}}{\text{Size of original dataset}} \tag{28.3c}$$

Here, $f_1(x)$ represents the feature selection ratio and $f_2(x)$ represents the classification error rate, both the functions are to be minimized simultaneously.

28.4.3 Feature Selection Using MO-CJaya

Feature selection is a combinatorial optimization problem, which is discrete in nature. The MO-CJaya algorithm discussed here is designed for continuous problems, this calls for the remodeling of the continuous workspace into a discrete workspace. This conversion task can be performed with the help of a transfer function [2].

In feature selection, a solution in the population is represented as a binary row vector, which can have only two values 0 and 1. Consider a D dimensional problem, where solution vector is represented by $S = \{s_1, s_2, s_3, \ldots s_D\}$, A 1 in the ith position of solution vector denotes that the ith feature will be included in solution, whereas 0 in that position indicates that the ith feature is excluded.

This work uses the sigmoid transfer function for converting the continuous solution variables to binary solutions. The transfer function is described in (28.4), and the procedure for continuous to binary conversion is described in (28.5).

$$T(x) = \frac{1}{1 + e^{-2x}} \tag{28.4}$$

$$X(p, d) = \begin{cases} 1 & \text{if } r\text{and} < T\ (X(p, d)) \\ 0 & \text{otherwise} \end{cases} \tag{28.5}$$

The variable $X(p, d)$ represents the value at dth dimension of pth solution where p ranges from 1 to N, and d ranges from 1 to D. Here, D and N denote the dimensionality and population size of the problem, respectively. This conversion takes place at each iteration after execution of (28.1).

28.5 Result and Discussion

The effectiveness of the proposed technique is assessed on five benchmark microarray datasets, the description about these datasets is shown in Table 28.1. The results reported in this work are average of 20 independent runs of all the algorithms. The upper limit for number of iteration is 100, and population size is kept 30. Maximum

Table 28.1 Dataset description

Dataset name	Number of genes	Number of classes	Number of samples
Colon	2000	2	62
SRBCT	2308	4	83
CNS	7129	2	60
Lung	12,533	2	181
MLL	12,582	3	72

iteration number is the termination criteria in all algorithms. Implementation is done in Python 3.0, and the code is executed on an Intel Core i5-8265U-1.6 GHz CPU with 8 GB of RAM. The fitness functions are evaluated using Gaussian Naïve Bayes classifier. Results are validated using tenfold cross-validation on each dataset. Parameters of all the algorithms are set as suggested by their respective authors. The performance is evaluated based on average accuracy, average feature selection ratio and execution time of the algorithm.

The performance of the proposed multi-objective feature selection technique MO-CJaya is compared with four other feature selection algorithms. The FS technique based on MOJaya [30] and NSGA II [31] is a multi-objective feature selection algorithm, whereas TOPSIS Jaya [18] and MBEGA [32] employ single objective feature selection.

Inspecting the result shown in Table 28.2, it is clear that MO CJaya and TOPSIS Jaya algorithms are giving the best performance in two out of five datasets in terms of average classification accuracy. Whereas, MOJaya and MBEGA are able to outperform other algorithms only once. However, it cannot be concluded from the result of Table 28.2 alone that MO CJaya and TOPSIS Jaya algorithms have equivalent performance because MO CJaya is a multi-objective FS algorithm and TOPSIS Jaya is a single-objective algorithm. MO CJaya has a second objective function related to number of features selected. The related result is shown in Table 28.3.

It is evident from Table 28.3 that MO CJaya outperforms other algorithms in three out of five datasets, and TOPSIS Jaya gives better results in two out of five datasets. The combined analysis of Tables 28.2 and 28.3 gives the clear picture about the performance of MO CJaya algorithm because it is a multi-objective feature selection technique. Its one objective is to minimize classification error rate (i.e., improve

Table 28.2 Average accuracy comparison

Dataset name	MO CJaya	MO Jaya	NSGA II	TOPSIS Jaya	MBEGA
Colon	96	84.62	85.64	**97.76**	86.66
SRBCT	97.06	**100**	99.615	**100**	99.23
CNS	**98.32**	75	73.605	96.22	72.21
Lung	97.8	87.8	93.38	94.24	**98.96**
MLL	**100**	86.67	90.5	99.62	94.33

Table 28.3 Average feature selection ratio comparison

Dataset name	MO CJaya	MO Jaya	NSGA II	TOPSIS Jaya	MBEGA
Colon	**12.3**	19.3	21.9	18.9	24.5
SRBCT	**7.9**	10.9	35.8	15.8	60.7
CNS	10.4	34.2	27.35	**8.7**	20.5
Lung	18.3	16.7	15.4	**9.9**	14.1
MLL	**11.1**	18.5	25.3	12.9	32.1

classification accuracy), and another goal is to keep the number of features chosen to a minimum. It should be noted that classification accuracy of MO CJaya on Colon and SRBCT dataset is slightly lower than TOPSIS Jaya. However, the feature selection ratio of MO CJaya is better than TOPSIS Jaya on the very same datasets. This implies that MO CJaya has compromised a little on classification accuracy to reduce the number of chosen features (this behavior is exactly as expected as per the concept of pareto-optimal set of solutions). Moreover, it is worth noticing that MO CJaya outperforms MO Jaya in terms of classification accuracy and feature selection ratio. Therefore, it can be inferred that the use of chaotic map instead of random numbers is helpful in improving the performance of the meta-heuristic optimization algorithm.

The average running time of the algorithms is reported in Table 28.4. The running time is mentioned in seconds. The TOPSIS Jaya algorithm is the fastest algorithm among all the algorithms under consideration. The last row of the Table 28.4 shows the average running time of the algorithms across all the datasets. We can see that the proposed MO CJaya algorithm is the second fastest algorithm, despite the fact that it has two fitness functions to evaluate. MO CJaya outperforms all the multi-objective FS algorithms under consideration (MO Jaya and NSGA II). In other words, MO CJaya is the fastest algorithm in the multi-objective category of FS.

Figure 28.2 shows the mean accuracy of all the algorithms over all the datasets. The proposed MO CJaya has the mean accuracy of 97.84%, which is the highest,

Table 28.4 Average running time (in seconds) comparison

Dataset name	MO CJaya	MO Jaya	NSGA II	TOPSIS Jaya	MBEGA
Colon	18.87	20.51	45.555	**12.71**	70.6
SRBCT	22.72	34.64	140.42	**16.38**	246.2
CNS	21.97	80.5	80.8	**14.79**	81.1
Lung	45.1	540.1	790.9	**35.03**	1041.7
MLL	28.42	120.55	151.325	**18.33**	182.1
Average	27.42	159.26	241.8	**19.45**	324.34

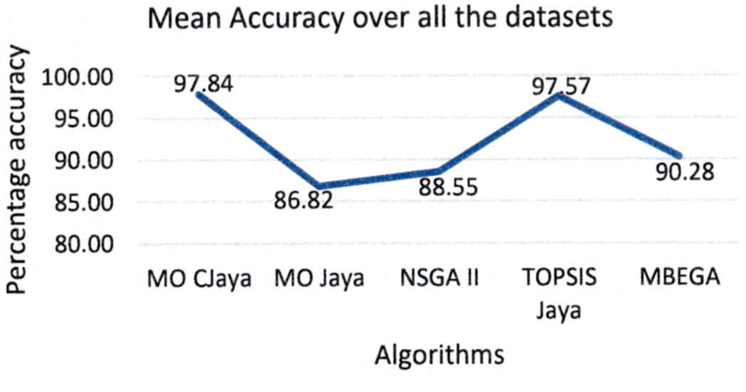

Fig. 28.2 Mean accuracy comparison of MO CJaya algorithm with other algorithms on all datasets

as can be seen in Fig. 28.2. This indicates that multi-objective feature selection is a better way to reduce the dimensionality of the dataset.

28.6 Conclusion

A new multi-objective feature selection algorithm based on chaotic map has been proposed for microarray dataset. It has been made evident that the use of chaotic map and multi-objective approach for feature selection outperforms the single objective feature selection approaches. On the basis of five benchmark microarray datasets, the performance of the proposed multi-objective feature selection technique is compared with four other popular techniques. Results of experiments indicated that the proposed technique outperforms other aspirants. It has the highest mean accuracy of 97.84%. The average running time is 27.42 s, which is also satisfactorily small. The feature selection ratio is also considerably small for all the datasets. Hence, it can be concluded that the proposed technique is able to achieve the highest classification accuracy in a short amount of time with a minimal feature subset. Therefore, the proposed technique is well suited for solving the feature selection problem for microarray data classification.

References

1. Baliarsingh, S.K., Vipsita, S., Dash, B.: A new optimal gene selection approach for cancer classification using enhanced Jaya-based forest optimization algorithm. Neural Comput. Appl. 0123456789 (2019). https://doi.org/10.1007/s00521-019-04355-x
2. Chaudhuri, A., Sahu, T.P.: Feature selection using Binary Crow Search Algorithm with time varying flight length. Expert Syst. Appl. **168**, 114288 (2021). https://doi.org/10.1016/j.eswa.2020.114288
3. Abdel-Basset, M., El-Shahat, D., El-henawy, I., de Albuquerque, V.H.C., Mirjalili, S.: A new fusion of grey wolf optimizer algorithm with a two-phase mutation for feature selection. Expert Syst. Appl. **139**, 112824 (2020). https://doi.org/10.1016/j.eswa.2019.112824
4. Nguyen, B.H., Xue, B., Andreae, P.: A new binary particle swarm optimization approach: momentum and dynamic balance between exploration and exploitation. IEEE Trans. Cybern., 1–15 (2019). https://doi.org/10.1109/TCYB.2019.2944141
5. Tran, B., Xue, B., Zhang, M.: Variable-length particle swarm optimization for feature selection on high-dimensional classification. IEEE Trans. Evol. Comput. **23**, 473–487 (2019). https://doi.org/10.1109/TEVC.2018.2869405
6. Chaudhuri, A., Sahu, T.P.: Binary Jaya algorithm based on binary similarity measure for feature selection. J. Ambient Intell. Humaniz. Comput. (2021). https://doi.org/10.1007/s12652-021-03226-5
7. Chaudhuri, A., Sahu, T.P.: PROMETHEE-based hybrid feature selection technique for high-dimensional biomedical data: application to Parkinson's disease classification. Electron. Lett. (2020)
8. Chen, H., Li, W., Yang, X.: A whale optimization algorithm with chaos mechanism based on quasi-opposition for global optimization problems. Expert Syst. Appl. **158**, 113612 (2020). https://doi.org/10.1016/j.eswa.2020.113612

9. Dewangan, D.K., Sahu, S.P.: Deep learning-based speed bump detection model for intelligent vehicle system using raspberry Pi. IEEE Sens. J. (2021). https://doi.org/10.1109/JSEN.2020.3027097
10. Dewangan, D.K., Sahu, S.P.: Driving behaviour analysis of intelligent vehicle system for lane detection using vision-sensor. IEEE Sens. J. **21**, 6367–6375 (2020). https://doi.org/10.1109/JSEN.2020.3037340
11. Dewangan, D.K., Sahu, S.P.: RCNet: road classification convolutional neural networks for intelligent vehicle system. Intell. Serv. Robot. (2021). https://doi.org/10.1007/s11370-020-00343-6
12. Joe, M.C.V., Raj, J.S.: Location-based orientation context dependent recommender system for users. J. Trends Comput. Sci. Smart Technol. **3**, 14–23 (2021)
13. Hamdan, Y.B.: Faultless decision making for false information in online: a systematic approach. J. Soft Comput. Paradig. **2**, 226–235 (2020)
14. Smys, S., Raj, J.S.: analysis of deep learning techniques for early detection of depression on social media network-a comparative study. J. Trends Comput. Sci. Smart Technol. **3**, 24–39 (2021)
15. Cleofas-Sánchez, L., Sánchez, J.S., García, V.: Gene selection and disease prediction from gene expression data using a two-stage hetero-associative memory. Prog. Artif. Intell. **8**, 63–71 (2019). https://doi.org/10.1007/s13748-018-0148-6
16. Baliarsingh, S.K., Vipsita, S., Muhammad, K., Dash, B., Bakshi, S.: Analysis of high-dimensional genomic data employing a novel bio-inspired algorithm. Appl. Soft Comput. J. **77**, 520–532 (2019). https://doi.org/10.1016/j.asoc.2019.01.007
17. Coleto-Alcudia, V., Vega-Rodríguez, M.A.: Artificial Bee Colony algorithm based on Dominance (ABCD) for a hybrid gene selection method. Knowl. Based Syst. **205**, 106323 (2020). https://doi.org/10.1016/j.knosys.2020.106323
18. Chaudhuri, A., Sahu, T.P.: A hybrid feature selection method based on Binary Jaya algorithm for micro-array data classification. Comput. Electr. Eng. **90**, 106963 (2021). https://doi.org/10.1016/j.compeleceng.2020.106963
19. Chaudhuri, A., Sahu, T.P.: a case study on disease diagnosis using gene expression data classification with feature selection: application of data science techniques in health care. In: Data Science and Its Applications. pp. 239–254. Chapman and Hall/CRC (2021)
20. Zhang, Y., Gong, D.W., Gao, X.Z., Tian, T., Sun, X.Y.: Binary differential evolution with self-learning for multi-objective feature selection. Inf. Sci. (Ny) **507**:67–85 (2020). https://doi.org/10.1016/j.ins.2019.08.040
21. Chen, C.W., Tsai, Y.H., Chang, F.R., Lin, W.C.: Ensemble feature selection in medical datasets: Combining filter, wrapper, and embedded feature selection results. Expert Syst. **37**, 1–10 (2020). https://doi.org/10.1111/exsy.12553
22. Sharma, A., Rani, R.: C-HMOSHSSA: gene selection for cancer classification using multi-objective meta-heuristic and machine learning methods. Comput. Methods Programs Biomed. **178**, 219–235 (2019). https://doi.org/10.1016/j.cmpb.2019.06.029
23. Lai, C.M.: Multi-objective simplified swarm optimization with weighting scheme for gene selection. Appl. Soft Comput. J. **65**, 58–68 (2018). https://doi.org/10.1016/j.asoc.2017.12.049
24. Chaudhuri, A., Sahu, T.P.: Feature weighting for naïve Bayes using multi objective artificial bee colony algorithm. Int. J. Comput. Sci. Eng. **24**, 74–88 (2021). https://doi.org/10.1504/IJCSE.2021.113655
25. Ang, J.C., Mirzal, A., Haron, H., Hamed, H.N.A.: Supervised, unsupervised, and semi-supervised feature selection: a review on gene selection. IEEE/ACM Trans. Comput. Biol. Bioinforma. **13**, 971–989 (2016). https://doi.org/10.1109/TCBB.2015.2478454
26. Hancer, E., Xue, B., Zhang, M., Karaboga, D., Akay, B.: Pareto front feature selection base d on artificial bee colony optimization. Inf. Sci. (Ny) **422**, 462–479 (2018). https://doi.org/10.1016/j.ins.2017.09.028
27. Rao, R.V.: Jaya: An Advanced Optimization Algorithm and its Engineering Applications
28. dos Santos Coelho, L., Mariani, V.C.: Use of chaotic sequences in a biologically inspired algorithm for engineering design optimization. Expert Syst. Appl. (2008). https://doi.org/10.1016/j.eswa.2007.02.002

29. Sayed, G.I., Hassanien, A.E., Azar, A.T.: Feature selection via a novel chaotic crow search algorithm. Neural Comput. Appl. **31**, 171–188 (2019). https://doi.org/10.1007/s00521-017-2988-6
30. Venkata Rao, R., Rai, D.P., Balic, J.: Optimization of abrasive waterjet machining process using multi-objective Jaya algorithm. Mater. Today Proc. **5**, 4930–4938 (2018). https://doi.org/10.1016/j.matpr.2017.12.070
31. Deb, K., Agrawal, S., Pratap, A., Meyarivan, T.: A Fast elitist non-dominated sorting genetic algorithm for multi-objective optimization: NSGA-II. Presented at the 2007 (2007)https://doi.org/10.1007/3-540-45356-3_83
32. Zhu, Z., Ong, Y.S., Dash, M.: Markov blanket-embedded genetic algorithm for gene selection. Pattern Recognit. **40**, 3236–3248 (2007)

Chapter 29
Audio-Based Staircase Navigation System for Visually Impaired

Jay S. Bhatia, Nimit K. Vasavat, Manali U. Maniyar, Neha N. Doshi, and Ruhina Karani

Abstract Visual impairments are life-altering and often make an individual dependent on others for primary navigation. A population of 62 million people in India alone has compromised vision. Accidental falls and collisions adversely affect the health of these individuals. To avoid such mishaps and get some assistance, visually impaired individuals look into the support of other humans that might be around them. Technology has had an immense impact on easing the day to day struggles of the differently-abled. This paper aims to help individuals with navigation of staircases and look into available assistance nearby. Focus remains on staircases and humans around that might be present in the frame captured by the user's mobile camera. The system discussed in the paper uses YOLO (You Only Look Once), which is a deep learning-based algorithm, for the staircase and human detection. Then the concepts of Hough Line transform and morphological closing are used to count the number of stairs. The system also estimates the distance of the approaching object using the area covered by the detected object on a particular frame. The closer the object, the greater the area covered. Finally, the individual receives audio-based alerts regarding the outputs of the steps discussed earlier. Thus, the system provides an efficient navigation system.

29.1 Introduction

Applications of deep learning have helped to automate many systems in recent times. Thus, leading to greater ease in the day-to-day tasks, an individual might want to carry. Technology has been constantly being put to use to help differently-abled people as well. Mobility and detecting obstacles are some of the few problems faced by blind or visually impaired people. Navigating through unknown areas, several possible causes of danger can be met, which include locating stairs and people to help

J. S. Bhatia (✉) · N. K. Vasavat · M. U. Maniyar · N. N. Doshi · R. Karani
Computer Department, DJSCE, Mumbai, India

R. Karani
e-mail: ruhina.karani@djsce.ac.in

J. I.-Z. Chen et al. (eds.), *Machine Learning and Autonomous Systems*, Smart Innovation, Systems and Technologies 269, https://doi.org/10.1007/978-981-16-7996-4_29

them overcome problems. Navigation systems can be put together for the visually impaired to provide them with the independence of movement. Such a system would be expected to provide all necessary alerts to an individual to avoid fall accidents. The most existing solutions require both hardware and software setup, which makes use of either specialized sensors or dual cameras, or both. These increase the cost of the system, and the setup can be overwhelming for a common man. This paper proposes a system for staircase navigation that eliminates the use of expensive hardware. The key area of focus is to use just smartphone cameras for detection in real time. The system uses Image Processing and Deep Learning. Object detection techniques based on deep learning and feature extraction by convolutional neural networks have been used.

For stair detection, the system uses YOLOv4 (You Only Look Once—v4) Darknet pre-trained model (AlexeyAB/darknet) and is retrained with a custom dataset of stairs (TransferLearning). The custom-made dataset consists of images of staircases of various steps and sizes. YOLOv4-tiny is a compressed version of YOLOv4, which takes less computing power [9], and hence can be used for faster implementation in real time. For human detection, a pre-trained YOLOv3-tiny model is used. Then the concepts of Hough Line transform and morphological closing are used to count the number of stairs. The system also estimates the closeness of the approaching object using the area covered by the detected object on a particular frame. The closer the object, the greater the area covered. Finally, the individual receives audio-based alerts regarding the outputs of the steps discussed earlier. Thus, the system provides an efficient navigation system.

29.2 Review of Literature

Patil [1] proposed a system that uses a custom dataset trained on Darknet YOLOv3 using Transfer Learning. The lines obtained from standard edge detector and Standard Hough Transform are used to construct a line that passes through the center of the previous lines. Then the angle is measured, which gives the desired heading of the tracked robot.

In the system proposed by Romić et al. [2], the frame initially undergoes through Canny edge detector and morphological operations, and after three iterations of closing operation, rectangular segments of the image are obtained. The interesting region is selected with horizontal rectangular shapes and a highlighted stairs counter. Then the vertical analysis is performed, and the length of arrays obtained is counted based on an increase in size.

Ciobanu et al. [3] proposed a system that uses the camera's absolute orientation information, provided by an IMU sensor. The system uses RGB-D cameras for staircase detection. Sensors are used for ground-level calibration. It carries out detection and clustering of patches that have the surface normal vectors oriented towards the "up" direction. This system requires a significant amount of setup and only works on Apple devices.

The system proposed by Habib et al. [4] uses an RGB-D camera, buzzer, an ultrasonic sensor, and a Raspberry pi to aid the visually impaired in navigation. An efficient model is developed that classifies between upward and downward preceding staircases. The drawback of the prototype lies in the requirement of multiple hardware devices.

Munoz et al. [5] proposed a system that makes use of RGB as well as RGB-D cameras. Sobel edge detector is used to acquire edge maps, which are fed to Hough transform to obtain a set of concurrent parallel lines, i.e. staircase candidates. These are then validated via the depth channel images acquired from the RGB-D camera.

Sultana et al. [6] proposed a system that uses the mathematical morphology method for image feature extractions and also performs binary morphology operations on extracted images for various applications. Operations such as dilation, erosion, opening, and closing have been used to extract information of particular shapes and structures of objects within an image.

The most existing solutions require both hardware and software setup, which makes use of either specialized sensors or dual cameras, or both. These increase the cost of the system, and the setup can be overwhelming for a common man. In this paper, an attempt to overcome commonly observed drawbacks is made. Elimination of hardware and camera requirements, provision of the count of stairs, nearby assistance, closeness to the staircase are the focal points. The proposed system achieves these goals to provide the user with information sufficient for navigation in such spaces. The models are trained on custom datasets, and the system is implemented in real time.

29.3 System Methodology

29.3.1 System Workflow

The workflow is segmented into various stages, namely, Staircase detection, Closeness Estimation (by calculating area fraction-AF), and Stair Count Estimation. The real-time input frames are fed to the Staircase Detection Module. If the staircase is detected, the coordinates of the bounding boxes obtained in the previous detection model are fed to the closeness estimation model to calculate the area fraction, and, thus, estimate the closeness of the stairs. If the area fraction is greater than a certain threshold then the count of the stairs and the humans ahead on the stairs is estimated. The output of the above model is notified during the navigation when necessary. Figure 29.1 describes the overall system workflow.

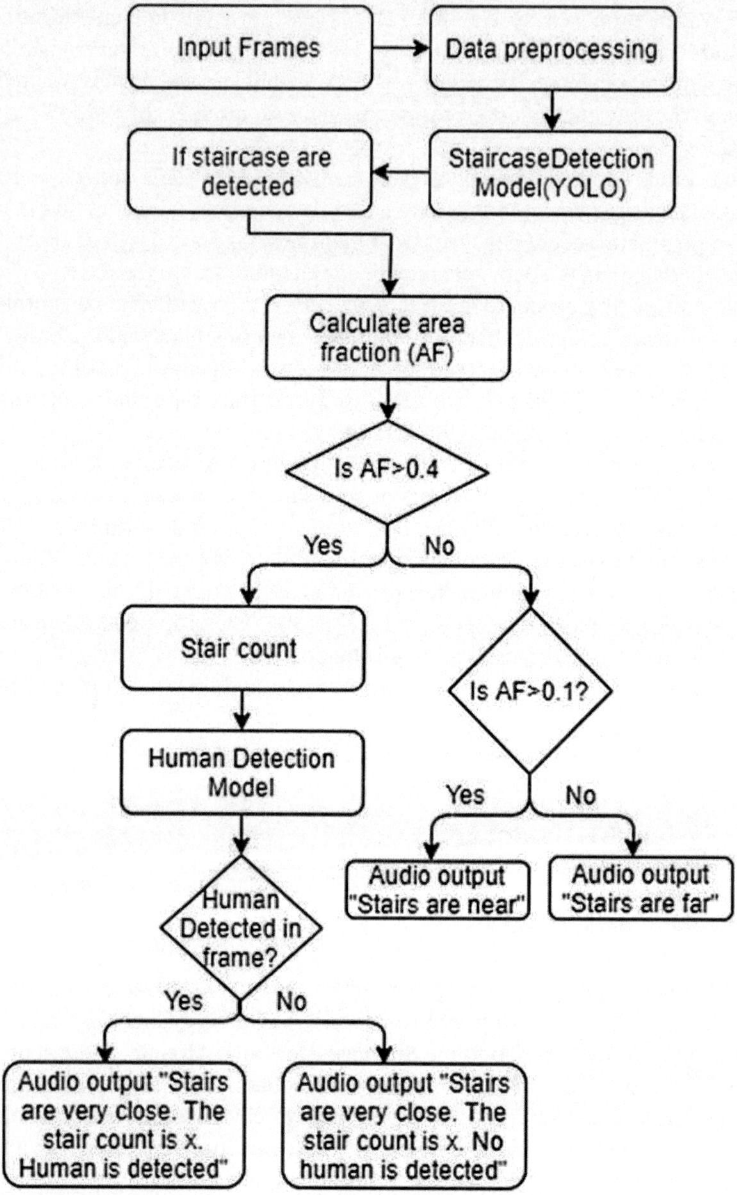

Fig. 29.1 System workflow

29.3.2 Dataset

The model has been trained on a custom dataset. Dataset consists of various images of stairs, mainly from the campus of Visvesvaraya National Institute of Technology, Nagpur, and the remaining from the other Internet sources. The images are taken from left as well as right angles to train the model better. Thus the system is trained on a total of 848 collected images. The LabelImg tool is used to annotate the images and get YOLO format labels. To augment the size of the dataset, a data augmentation technique has been applied to augment the size of our dataset. All 848 images are horizontally flipped, so the total images in the dataset are 1696 then. The dataset was split into an 80:20 ratio of training, and testing samples consisted of 1357 training set images and 339 testing set images.

29.3.3 Staircase Detection

In [7], a comprehensive description is provided about the recent innovations in staircase detection for the visually impaired. Most of the methods mentioned in this paper use RGBD Cameras and sensors, which is not a feasible solution for the majority of the population due to its high cost and complexity. Moreover, some of the methods are suitable only for either indoor or outdoor applications. Our proposed method is a cheap and feasible option for a wider range of the population and is easy to use. It worked well in indoor as well as outdoor environments.

The system proposed in this paper uses YOLOv4-tiny model for staircase detection. It is a compressed version of YOLOv4, which is an object detection system in real time that detects and localizes the object present in the input image in a single enclosure. It can estimate up to 9000+ seen unseen classes of objects. YOLO is based on the Convolutional Neural Network (CNN). YOLO algorithm uses a neural network on the image and divides it into different regions [8]. It then predicts the different bounding boxes and probabilities of those classes. YOLOv4-tiny has two YOLO layers that are trained on 29 pre-trained convolutional layers. It is 8 times the FPS of the YOLOv4 dataset. Thus, YOLOv4-tiny is used for faster training and detection.

According to Bochkovskiy et al. [9], Yolov4 has AP (accuracy) and FPS (frame rate per second) improved by 10% and 12% as compared to Yolov3. In [10], they also prove that the YOLOv4 model has achieved 55.5% AP for the MS COCO dataset at a speed of 16 FPS on Tesla V100 and with augmentation, it has achieved 56.0% AP. Whereas, YOLOv4-tiny model has achieved 22.0% AP at speed of 443 FPS on RTX 2080Ti.

Therefore, for real-time object detection, YOLOv4-tiny is chosen over YOLOv4 and YOLOv3 since for real-time object detection environments, faster inference time is more important than accuracy.

29.3.4 Stair Count Estimation

In [1, 4, 5], the proposed prototype does not provide any implementation of stair count. Our method aims to provide a fair count estimation for smoother navigation.

The proposed system uses Hough transform (HT) algorithm to detect the edges of the stairs. It is a method to segment geometrical objects from the image. Once all edges are detected, Hough transform returns the coordinates of the endpoints of edges. Horizontal edges are filtered out using morphological operation on the edges detected with the help of structuring elements. The input images are preprocessed based on kernels using morphological operations. The value of each pixel in the output image is based on a comparison of the corresponding pixel in the input image with its neighbors.

A morphological operation that extracts only horizontal lines from all the edges that the Hough Transform returns using the shape and size of the kernel was the requirement of the system. Hence, the system uses dilation and erosion morphological operations. These two operations are useful for binary image feature extraction. Dilation adds pixels to the inferred bounding box whereas erosion removes the pixels from the inferred bounding box. On the basis of the shape and size of the structuring element that was used to process the image, the above two morphological operations decide the number of pixels that are to be added or removed. A structuring element can have any arbitrary shape and size and is a matrix consisting of 0's and 1's. Typically, the neighborhood is defined by the pixels with values of 1. Origin, the center pixel of the structuring element helps to identify the pixel that is being processed.

Structuring elements can consist of common shapes, such as lines, circles, disks, and squares and sizes. Choice of the structuring element is made on the basis of the size and shape of the objects that are to be extracted in the input. The input frames are preprocessed, and the binary image is generated. Figure 29.2 shows the workflow of the stair count module.

With the help of the structuring element created and the morphological operations such as erosion and dilation, the binary image is structured and the required feature is extracted, i.e. horizontal edges of the staircase. Filtering operation is performed further by considering only those lines, which are at a distance of 30px apart. Since this can result in high inaccuracy, as it can detect any horizontal line in the image, the steps are counted only when they are too close to the camera. The count of horizontal lines represents the edges of the stair thus helping the model to evaluate the stair count value.

29.3.5 Human Detection

COCO dataset, i.e. "Common Objects in Context (COCO)" is large-scale object detection, segmentation, and captioning dataset. It has features like object segmentation, recognition in context, etc. Since YOLO is a model pre-trained on the COCO

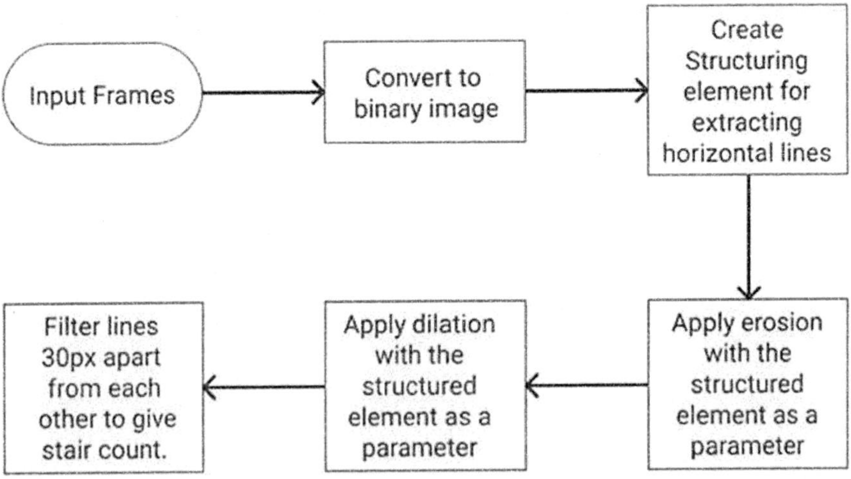

Fig. 29.2 Workflow of stair count module

dataset and humans are already defined class in the dataset, the pre-trained weight file for the network is imported. That's how a human detection model using YOLOv3 is generated.

29.3.6 Closeness Estimation

The comparative closeness of the staircase from the user is done using the area fraction method. If the staircase is detected then the fraction of screen occupied by the detected staircase is calculated. On the basis of the AF (area fraction) calculated, the system decides if the stairs are far, near, or very close. After multiple trials, optimal values of area fraction are taken for the result of closeness estimation. When the stairs are detected, the bounding box coordinates are passed to the closeness estimation module by the Staircase Detection module, and the area fraction of the box with respect to the input frame is calculated.

29.3.7 Audio

In the model, the audio feature is included, which assists the visually impaired person in navigation and notifies the person about the incoming staircase. Once the staircase is detected by the model, the person is apprised of the closeness of the staircase at that moment, and the same is repeated after every 20 s depending on their closeness from the staircase. According to the closeness module, once the person is very close to the staircase, the count of the staircase and the number of people ahead who are climbing the staircase are also conveyed to the person. All the textual output that is printed on the images in the **Results** below is communicated to the person during their navigation with the help of audio.

29.4 Results and Discussions

The mAP and IoU values at multiple iterations are noted down as the custom dataset is used to train the model. The above-mentioned values are used as a parameter to judge the accuracy of the model. Throughout the training process, the validation accuracy is examined, which helps to prevent overfitting. Overfitting is prevented with the help of early stopping, which is a form of regularization. Early stopping is a method that stops the training of the model when the validation score and model performance keep declining through further iterations. The training process is terminated when the performance metric declines over the previous iteration. mAP@0.50 is used as the performance metric in the proposed system. YOLOv4-tiny and YOLOv3-tiny are the compressed versions of YOLOv4 and YOLOv3, respectively. YOLO tiny versions are used to get a simpler network structure and lessen the parameters such that it becomes viable for mobile and embedded devices to work with. The accuracy of YOLOV3 and YOLOV4 is better than their respective tiny versions. Objects should be detected at a quicker pace when the proposed system is in real time. Hence, when working in a real-time detection environment, faster inference time is given more priority than precision.

In Table 29.1, the results obtained while training the YOLOv3-tiny, YOLOv4-tiny, YOLOv3, and YOLOv4-based models are displayed. It can be inferred from the table that the best results were recorded at 4000 iterations for YOLOv3-tiny and 5000 iterations for YOLOv4-tiny. The highest mAP and IoU values achieved are 0.9895 and 0.7578 for YOLOv3-tiny at 4000 iterations and 0.9811 and 0.7462 for YOLOv4-tiny at 5000 iterations. Figure 29.4 plots mAP at 0.50 score v/s the number of iterations.

The results of the performance metrics obtained for the YOLOv3, YOLOv4, YOLOv3 tiny, and YOLOv4 tiny models are represented graphically in Figs. 29.3 and 29.4. For the performance evaluation, mAP and IoU are used as the metrics. IoU, also known as Intersection Over Union, specifies the amount of overlap between the ground truth and the predicted bounding boxes after object detection.

The mAP metric can be defined as a combined metric that takes precision as well as recall into account. Observations made a state that YOLOv4 tiny gives better results as compared with YOLOv3 tiny. The Staircase detection model forms a bounding box around the staircase that is detected. After calculating the area fraction of the bounding box with respect to the image, the model estimates the closeness of the user to the stairs and prints a textual output. The value of the area fraction is also printed on the bounding box figure showing the output of the closeness estimation module.

Table 29.1. mAP and IoU values over different iterations

Model	Iterations	mAP	IoU
Yolov4 tiny	3000	0.9758	0.7498
	4000	**0.9895**	**0.7578**
	5000	0.9812	0.7549
	6000	0.9793	0.7635
Yolov3 tiny	3000	0.9626	0.7007
	4000	0.9771	0.7080
	5000	**0.9811**	**0.7462**
	6000	0.9782	0.7569
Yolov3	3000	0.9891	0.8168
	4000	0.9926	0.8525
	5000	**0.9977**	**0.8842**
	6000	0.9845	0.8875
Yolov4	3000	0.9921	0.8098
	4000	**0.9976**	**0.8258**
	5000	0.9906	0.8294
	6000	0.9893	0.8262

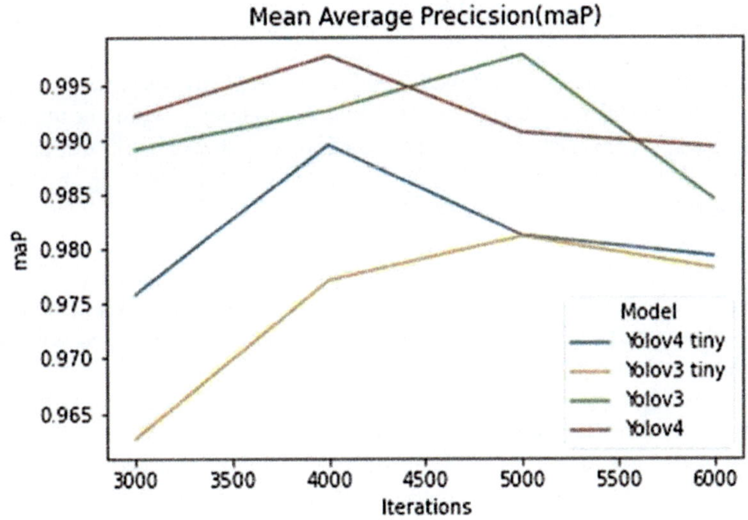

Fig. 29.3 mAP analysis for all YOLO model

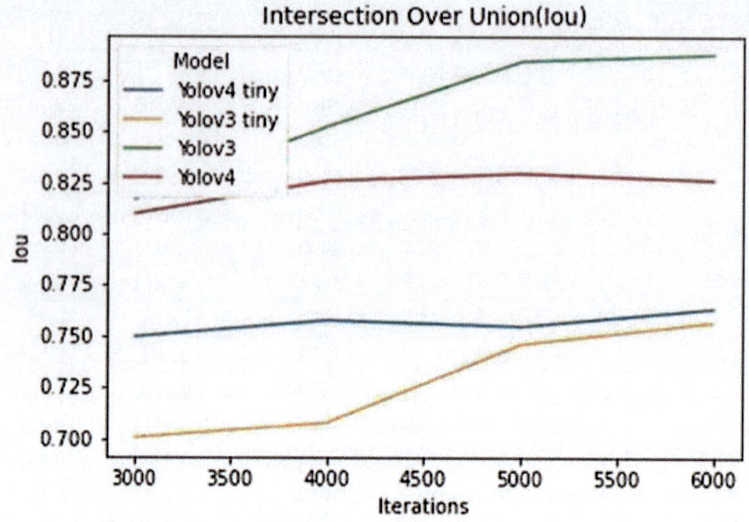

Fig. 29.4 IoU analysis for all YOLO model

As shown in Fig. 29.5, the area fraction of the bounding box is **0.04,** which is less than 0.1, thus, it is estimated that the staircase is far from the user. As shown in Fig. 29.6, the area fraction of the bounding box is **0.22,** which is greater than 0.1, thus, it is estimated that the staircase is near to the user. Only when the user reaches very close to the stairs, the Stair Count Estimation module will estimate the number of stairs the staircase ahead has, and the detection module will notify the user of the number of people which are on the stairs. As shown in Fig. 29.7, the area fraction bounding box is **0.44,** which is greater than 0.4, thus, it is estimated that the staircase is very close to the user.

The accuracy obtained in the stair count is around 86% with an error of ±2 stairs. This is performed on the staircase dataset used earlier to train the YOLO model. The stair count depends on the distance between the stairs and the person holding the smartphone and works best when the area fraction is more than 0.4.

When the stairs are detected, the audio output would be "Staircase detected. Stairs are far/near" on the basis of the closeness estimation method. When the estimated distance of the staircase is very close, the models would also detect humans and audio output would be, "Staircase detected. Stairs are very close. X stairs ahead. Y humans ahead" where X denotes the number of stairs detected and Y denotes the number of humans detected by the proposed model.

Fig. 29.5 Output when stairs are far

29.5 Conclusions and Future Work

This paper proposes a solution that would reduce the reliance of visually impaired people on others. The model would help the user detect the staircase, its counts, its distance from the user and let them know if they have any help nearby. Using the methods described, a result of the maximum mAP value of 0.9895 and IoU value of 0.7578 for the YOLOv4-tiny-based model are obtained. The values are 0.9811 and 0.7462 for mAP and IoU, respectively, for the YOLOv3-tiny-based model.

Since most of the dataset only contains images of upstairs, the model can be improved further by detecting both upstairs as well as downstairs efficiently. Future implementation involves extending the system by detecting other indoor navigation elements like lifts, doors, lift buttons, balcony windows, etc. Also, face recognition for known people can be carried out. The models of the staircase and human detection can be used to inherit these features. Future implementation involves extending the system by detecting other indoor navigation elements like lifts, doors, lift buttons, balcony windows, etc. Also, the method used for stair count is uncertain and rudimentary and gives accurate results only when the user is very close to the stairs. The

Fig. 29.6 Output when stairs are near

models of the staircase and human detection can be used to inherit these features. This would make indoor navigation in everyday places like offices and homes very smooth.

Fig. 29.7 Output when stairs are very close

References

1. Patil, U., et al.: Deep learning based stair detection and statistical image filtering for autonomous stair climbing. In: 2019 Third IEEE International Conference on Robotic Computing (IRC), pp. 159–166 (2019).https://doi.org/10.1109/IRC.2019.00031
2. Romić, K., Galić, I., Galba, T.: Technology assisting the blind—video processing based staircase detection. In: 2015 57th International Symposium ELMAR (ELMAR), pp. 221–224 (2015). https://doi.org/10.1109/ELMAR.2015.7334533
3. Ciobanu, A., Morar, A., Moldoveanu, F., Petrescu, L., Ferche, O., Moldoveanu, A. (2017) Real-time ındoor staircase detection on mobile devices. In: 2017 21st International Conference on Control Systems and Computer Science (CSCS), pp. 287–293. https://doi.org/10.1109/CSCS.2017.46
4. Habib, A., Islam, M.M., Kabir, M.N., Mredul, M.B., Hasan, M.: Staircase detection to guide

visually impaired people: a hybrid approach. Revue d'Intell. Artif. **33**(5), 327–334 (2019)
5. Munoz, R., Rong, X., Tian, Y.: Depth-aware indoor staircase detection and recognition for the visually impaired. In: 2016 IEEE International Conference on Multimedia & Expo Workshops (ICMEW), pp. 1–6 (2016). https://doi.org/10.1109/ICMEW.2016.7574706
6. Sultana, S., Ganesh, R.: FPGA ımplementation of binary morphological processing for ımage feature extraction. Int. J. Eng. Res. V4 (2015). https://doi.org/10.17577/IJERTV4IS100506.
7. Islam, M.M., Sheikh Sadi, M., Zamli, K.Z., Ahmed, M.M.: Developing walking assistants for visually ımpaired people: a review. IEEE Sens. J. **19**(8), 2814–2828 (2019). https://doi.org/10.1109/JSEN.2018.2890423
8. Redmon, J., Divvala, S., Girshick, R., Farhadi, A.: You only look once: unified, real-time object detection. In: 2016 IEEE Conference on Computer Vision and Pattern Recognition (CVPR), pp. 779–788 (2016)https://doi.org/10.1109/CVPR.2016.91
9. Bochkovskiy, A., Wang, C.-Y., Liao, H.: YOLOv4: optimal speed and accuracy of object detection (2020). abs/2004.10934
10. Wang, C.-Y., Bochkovskiy, A., Liao, H.-Y.: Scaled-YOLOv4: scaling cross stage partial network (2020)
11. Ren, S., He, K., Girshick, R., Sun, J.: Faster R-CNN: towards real-time object detection with region proposal networks. IEEE Trans. Pattern Anal. Mach. Intell., 39(2015). https://doi.org/10.1109/TPAMI.2016.2577031
12. Molchanov, V.V., Vishnyakov, B.V., Vizilter, Y.V., Vishnyakova, O.V., Knyaz, V.A.: Pedestrian detection in video surveillance using fully convolutional YOLO neural network. In: Proceedings of the SPIE 10334, Automated Visual Inspection and Machine Vision II, 103340Q (26 June 2017)
13. Liu, Z., Chen, Z., Li, Z., Hu, W.: An efficient pedestrian detection method based on YOLOv2. Math. Probl. Eng., 10 (2018). Article ID 3518959
14. Bauer, Z., Dominguez, A., Cruz, E., Gomez-Donoso, F., Orts-Escolano, S., Cazorla, M.: Enhancing perception for the visually impaired with deep learning techniques and low-cost wearable sensors. Pattern Recognit. Lett. (2019)
15. Ganesan, P., Sajiv, G.: A comprehensive study of edge detection for image processing applications. In: 2017 International Conference on Innovations in Information, Embedded and Communication Systems (ICIIECS), pp. 1–6 (2017). https://doi.org/10.1109/ICIIECS.2017.8275968.
16. Rani, S.: A novel mathematical morphology based edge detection method for medical images. CSIT **4**, 217–225 (2016)

Chapter 30
Deep Learning-Based Automated Classification of Epileptic and Non-epileptic Scalp-EEG Signals

Pooja Prabhu, Karunakar A. Kotegar, N. Mariyappa, H. Anitha, G. K. Bhargava, Jitender Saini, and Sanjib Sinha

Abstract Epilepsy is the most common neurological disorder that has affected 50 million worldwide population. Epilepsy is a chronic condition that is characterized as recurrent and unprovoked seizures. A seizure is defined as abnormal, excessive paroxysmal discharge of the cerebral neurons. In general, epilepsy evaluation, the high time resolution (~1–2 ms) electrophysiological data recorded using electroencephalography (EEG) is commonly used. The EEG data are visually inspected for epileptic seizure instances. However, the manual interpretation may lead to subjective error-causing misdiagnosis, and identifying the seizure instances in high temporal resolution data may be time-consuming. To mitigate these problems, the proposed study uses a hand-crafted deep learning model to classify the epileptic and non-epileptic EEG data. This study used EEG data acquired from two databases—University of Bonn and Physiobank Children's Hospital Boston—Massachusetts Institute of Technology (CHB-MIT). We split the EEG data in 80:20 ratio for training and

P. Prabhu · K. A. Kotegar (✉)
Department of Computer Applications, Manipal Institute of Technology, Manipal Academy of Higher Education, Manipal, Karnataka 576104, India
e-mail: karunakar.ak@manipal.edu

N. Mariyappa · S. Sinha
Department of Neurology, National Institute of Mental Health and Neurosciences, Bengaluru, Karnataka 560029, India

N. Mariyappa · G. K. Bhargava · S. Sinha
MEG Research Centre, National Institute of Mental Health and Neurosciences, Bengaluru, Karnataka 560029, India

H. Anitha
Department of Electronics and Communication, Manipal Institute of Technology, Manipal Academy of Higher Education, Manipal, Karnataka 576104, India
e-mail: anitha.h@manipal.edu

G. K. Bhargava
Department of Clinical Neurosciences, National Institute of Mental Health and Neurosciences, Bengaluru, Karnataka 560029, India

J. Saini
Department of Neuroimaging and Interventional Radiology, National Institute of Mental Health and Neurosciences, Bengaluru, Karnataka 560029, India

J. I.-Z. Chen et al. (eds.), *Machine Learning and Autonomous Systems*, Smart Innovation, Systems and Technologies 269, https://doi.org/10.1007/978-981-16-7996-4_30

testing the model. The model was trained under three network specifications, which were designed based on the number of high-level features and low-level features. The result showed that the proposed study successfully classified epileptic and non-epileptic signals with the accuracy of 67% and 92% for University of Bonn data and CHB-MIT EEG data, respectively, for the network specification which had a high number of low-level features. We tested the model for optimal value of number of epochs (20) and learning rate (0.001). The study shows that the MIT-CHB data are suitable for classification as they have a good number of samples and balanced epileptic and non-epileptic signals.

30.1 Introduction

The neurological disorder affects the central nervous system (brain and spinal cord) and nerves in the human body. The symptoms observed in neurological disorders are structural abnormalities, biochemical and electrical abnormalities in the nervous system. One of the most common neurological disorders is epilepsy. Epilepsy is a neurological disorder in which the brain is believed to be in a hyperexcitable state [1]. Recurrent and unprovoked seizures characterize it. Seizures are abnormal, excessive, paroxysmal discharge of the cerebral neurons causing the electrical disruption in brain activity [2]. In clinical practice, if the patient is observed to have two or more seizures, the patient is diagnosed as a victim of epilepsy [3].

During the epileptic evaluation, a neurologist examines the patient's electrophysiological data to identify the occurrence of seizures. The electrophysiological data are examined from electroencephalography (EEG) [4]. The EEG recording of the epileptic patient is captured for 24–30 h. EEG records the electrical activity of the brain by placing the electrodes on the surface of the head. The electrode placement follows the standard 10–20 montage, covering the entire head with good precision. EEG data have a poor spatial resolution of 19–256 channels but have a high temporal resolution in the range of milliseconds. The clinicians examine the occurrence of seizures in each channel manually by visual inspection. During visual inspection, the clinicians observe the morphology of brain signals like spikes (<70 ms) and sharp (>70 ms) waves in the entire EEG recording to identify the seizures. Manually classifying the channel instances as epileptic and non-epileptic is time-consuming and is subjective, which may cause misdiagnosis. Thus, there is a need for an automated system that can identify the channel instance to be epileptic or non-epileptic.

Several studies have attempted to automate the classification of epileptic and non-epileptic in scalp-EEG and invasive EEG. The study by Tsiouris et al. [5] used Long Short-Term Memory (LSTM) deep learning network to classify preictal and interictal EEG signals. Preictal is the state before the seizure, and interictal is the state between seizures. They used the CHB-MIT EEG dataset [6], and each EEG signal is segmented to 5 s, which is the input to LSTM. The study by Aristizabal et al. [7] used Recurrent Neural Network (RNN) via LSTM to classify ictal, healthy, and inter-ictal EEG data. They used University of Bonn EEG data [8]. The study

Yao et al. [9] used the independently RNN architecture to classify the seizure and non-seizure data. They extracted features from 17 common channels of the CHB-MIT EEG dataset [6], and each time series is segmented for 23 s. Medvedev et al. [10] used the invasive EEG data to classify the ripples, spikes, and ripple-on-spike using the LSTM network. The study by Ullah et al. [11] used a one-dimensional Convolutional Neural Network (CNN) to classify normal versus epileptic, normal versus interictal versus ictal, and seizure versus non-seizure. They used University of Bonn EEG data [8], and they used the data in two ways—EEG data segmented with 512 samples and unsegmented EEG data. From all these studies, we can sum up that they had tested the model on a single dataset to classify either state of epilepsy or types of seizures or normal versus epileptic using LSTM or CNN. Also, these studies used pre-processed data or extracted handcrafted features before training the deep learning model.

The proposed study aims to classify the given EEG signal as epileptic or non-epileptic. In the proposed study, we are not performing any pre-processing or segmentation of EEG data to not lose any crucial information. Also, we are handcrafting the deep learning model, and then we are testing it on scalp-EEG obtained from two datasets—University of Bonn EEG data [8] and CHB-MIT EEG data [6]. We designed the Convolutional Neural Network (CNN) to predict the epileptic and non-epileptic time series with three different network specifications. Then, we tested which specification work classifies better for two datasets. Alongside, we also determined the optimal model parameter (number of epochs and learning rate) values for each specification over two different datasets. This study was able to predict the time series in each EEG channel as epileptic or non-epileptic, which can help the clinician to detect the occurrence of seizures by looking into only those channels, which are being predicted as epileptic.

30.2 Materials and Methods

In this section, we describe the scalp-EEG data obtained from two databases. Then, we emphasize the different layers used in the handcrafted deep learning model, followed by the specification of the deep learning model.

30.2.1 Dataset

In this study, we are using two datasets to test the handcrafted deep learning model. The reason for choosing the two standard datasets is due to the nature of recording, and both the datasets have a good number of epileptic and non-epileptic EEG data.

University of Bonn EEG data [8]. This is the publicly available dataset from the University of Bonn. This dataset comprises EEG data under five sets named from

Table 30.1 Dataset summary of the proposed study

Database name	Sampling frequency (Hz)	Number of samples	Number of time series	
			Epileptic	Non-epileptic
University of Bonn	173.6	4096	300	200
CHB-MIT	256	5000–10,000	3000	2000

A to E. Each set is composed of 100 time series. Each time series is single-channel EEG data containing 4096 samples recorded for the sampling frequency of 173.6 Hz. The first two sets (A and B) are the EEG data recorded from five healthy subjects. At the same time, the latter three sets (C, D, E) are EEG data recorded from Epileptic subjects. Thus, for this study, we are using 300 EEG time series under the Epileptic category and 200 EEG time series under the non-Epileptic category. Table 30.1 gives the summary of the dataset used in this study.

CHB-MIT EEG dataset [6]. This dataset consists of EEG data recorded for subjects whose ages range from 1.5 to 22 years. This dataset has EEG data recorded for 23 subjects at the sampling frequency of 256 Hz. Overall, this dataset has 686 EEG recordings, out of which 198 EEG data have seizures. All these recordings are taken using a 10–20 electrode system with 18–23 channel EEG. Figure 30.1 demonstrates the epileptic and non-epileptic signals. In the proposed study, we are using 3000 and 2000 single-channel EEG data under the epileptic and non-epileptic category. Table 30.1 shows the summary of the dataset used in the proposed study.

30.2.2 Convolution Neural Network (CNN)

In the proposed study, we are using the deep learning model, specifically CNN, for classifying epileptic and non-epileptic EEG data.

Architecture. Figure 30.2 shows the CNN architecture imbibed in the proposed study. The one-dimensional EEG signals are given as input to the handcrafted CNN model to extract the features and to classify the epileptic and non-epileptic EEG signals. The three convolutional blocks do the feature extraction, and the SoftMax layer does classification. The features of the input EEG signal are extracted in three convolutional blocks, and then these high-level features are sent to three dense layers to flatten the feature map, which can then be passed on to the softmax layer.

The first convolutional block consists of four layers: the convolution layer, batch normalization layer, ReLU (non-linear activation function), and max pooling. The convolution layer is the main component of the CNN model. In the convolution layer, the one-dimensional signal is convolved with an input filter to generate the feature map of a given signal. The user specifies the filter size and the number of filters. The feature map from the convolution layer is passed on to the batch normalization layer, which normalizes the network parameters (activation and gradients) propagating

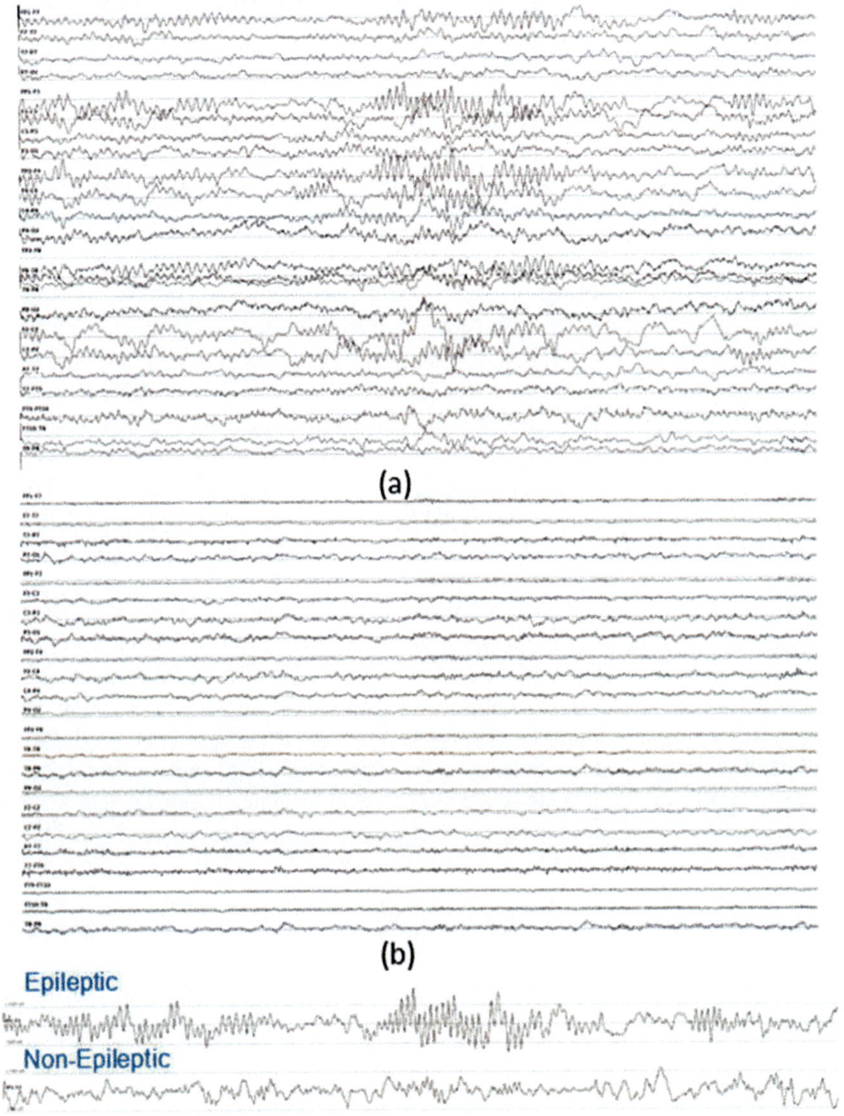

(c)

Fig. 30.1 Illustration of **a** Epileptic and **b** non-epileptic in 23 channels and **c** single channel (Fp1-F3) Epileptic and non-epileptic EEG signal

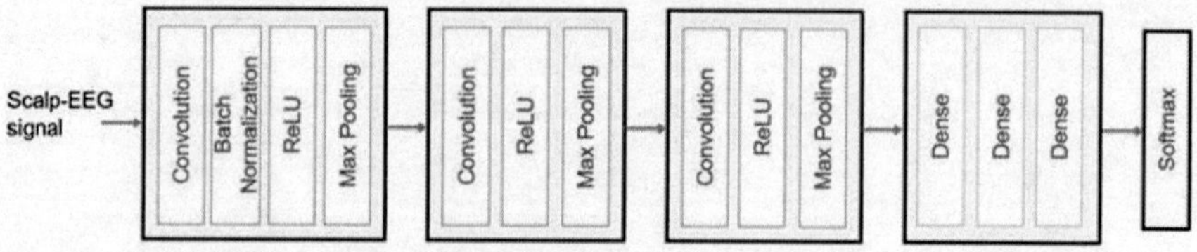

Fig. 30.2 CNN architecture used in the proposed study

through the network. This layer helps in easing the optimization problem during the training of the network. To overcome the overfitting during training, the non-linearity to features is added by non-linear activation functions like ReLU (Rectified Linear Unit). In this layer, the negative feature values are mapped to zero, and the positive feature values remain intact. Then the features are passed on to the max-pooling layer. This reduces the dimension of features and removes redundant spatial information. The main parameters of the pooling layer are pool size and stride. This layer helps in increasing the number of filters in subsequent convolution layers without increasing the computational load.

The output of the max-pooling layer of the first block is passed on to the second convolution block, which has three layers such as convolution, ReLU, and max pooling. The output from the second block is given as input to the third convolution block, which has a convolution layer, ReLU, and max-pooling layer. As we move from the first convolution block to the third convolution block, the number of filters used in the convolution layer is increased to get high-level features. Thus, the third block extracts the high-level features. These high-level features are flattened in three dense or fully connected layers. This layer has interconnected neurons, and the dimension of the final feature map is decided by the number of neurons used in the last dense layer. The feature map from the last dense layer is sent to the softmax layer. The feature map of each input, along with the labels ("1" for epileptic and "0" for non-epileptic), is sent to the softmax layer. The softmax layer approximates the probability of expected output between 0 and 1.

Table 30.2 shows the three sets of specifications of the CNN model, which is partially adopted from [12]. The parameter values for each specification are determined based on the number of high-level features and low-level features. We are training the CNN model separately for these three sets of specifications. Then these three models are tested on two datasets to see the optimal model for the dataset.

Details of Learning. The training options that have to be specified during the model training are the optimization method, number of epochs, and learning rate. In the proposed study, we used minibatch and stochastic gradient descent method for optimization. We used 100 as the minibatch size. We set 20 as the number of the epoch, and the initial learning rate is set to 0.001, which falls by the factor of five after every ten epochs.

Table 30.2 Specifications of CNN model

Layers	Parameters	Specification 1	Specification 2	Specification 3
Conv_1	Number of filters	32	256	64
	Filter size	12	12	12
Maxpool_1	Pool size	4	4	4
	Stride	2	2	2
Conv_2	Number of filters	64	128	64
	Filter size	8	8	8
Maxpool_2	Pool size	4	4	4
	Stride	2	2	2
Conv_3	Number of filters	128	64	32
	Filter size	8	8	8
Maxpool_3	Pool size	4	4	4
	Stride	2	2	2
Dense_1	Number of neurons	128	128	64
Dense_2	Number of neurons	64	64	32
Dense_3	Number of neurons	2	2	2

30.3 Results and Discussions

In this study, we used Dell Precision 7530 system with an intel core i9 processor and clock speed of 2.9 GHz. The training and testing of the model are performed on this system, which has a 1 TB hard drive and 125 GB RAM. We implemented the model in Google Colab. From both datasets, we used an 80:20 split for training and testing the model. This makes training data size of 240 and 2400 epileptic time series of the University of Bonn and CHB-MIT data, respectively. The training dataset consists of 240 epileptic time series and 160 non-epileptic time series, and the testing dataset consists of 60 epileptic time series and 40 non-epileptic time series from the University of Bonn. Whereas the training dataset of CHB-MIT data consists of 2400 epileptic time series and 1600 non-epileptic time series, and the testing dataset of CHB-MIT data consists of 600 epileptic time series and 400 non-epileptic time series.

30.3.1 Evaluation of CNN Model Using University of Bonn Data

We used a total of 400 EEG time series of University of Bonn data as a training dataset. This training dataset had 240 Epileptic time series and 160 non-epileptic time series. Using this training dataset, we trained the CNN model under three specifications for

Table 30.3 Model evaluation for University of Bonn data and CHB-MIT data over three specifications

Specification	Accuracy (%)	
	University of Bonn data	CHB-MIT EEG data
1	60	90
2	67	89
3	58	92

20 epochs. We obtained training accuracy of 68%, 72%, and 65% for specification 1, specification 2, and specification 3, respectively.

Using these three trained CNN models, we tested the model using 100 EEG time series. These time series consisted of 60 epileptic time series and 40 non-epileptic time series. We obtained testing accuracy of 60%, 67%, and 58% for specification 1, specification 2, and specification 3, respectively. Table 30.3 shows the testing accuracy obtained for University of Bonn data over three model specifications. From the testing result, we can say that model performed better for specification 2 when compared to the model with specification 1 and specification 3. Specification 2 has more number filters in convolution layers, which helps in extracting more features from low sampled EEG data.

30.3.2 Evaluation of CNN Model Using CHB-MIT EEG Data

We used a total of 4000 EEG time series of CHB-MIT EEG data as a training dataset. This training dataset had 2400 Epileptic time series and 1600 non-epileptic time series. Using this training dataset, we trained the CNN model under three specifications for 20 epochs. We obtained training accuracy of 95%, 92%, and 95% for specification 1, specification 2, and specification 3, respectively.

Using these three trained CNN models, we tested the model using 1000 EEG time series. These time series consisted of 600 epileptic time series and 400 non-epileptic time series. We obtained testing accuracy of 90%, 89%, and 92% for specification 1, specification 2, and specification 3, respectively. Table 30.3 shows the testing accuracy obtained for CHB-MIT EEG data over three model specifications. From the results, we can see that model with specification 3 performed better when compared to other models. The reason may be specification 3 has the same number of filters in all the convolution layers, making the model extract the same number of features and reduce the redundancy. All the three model performance was consistent even though the specification varied. This may be because the data have a high number of samples per second, and training data had a good number of time series.

30.3.3 Comparative Analysis

The proposed study used two different datasets for classifying epileptic and non-epileptic EEG signals using a handcrafted CNN model designed under three different network specifications. Other studies in the literature [11, 12] used only one dataset to evaluate the deep learning model. Their work not only involved pre-processing of raw signals but also to increase the size of training and testing data they fragmented each EEG signal. Fragmenting the data without prior knowledge on epileptic instance could cause the loss of crucial information from EEG signals.

There are studies like [13, 14] that used CNN model to classify the epileptic and non-epileptic signals but on depth EEG or intracranial EEG. The studies like [15] used the machine learning algorithm not only to classify epileptic instances but also to detect them in each channel using high-frequency oscillation characteristics of EEG signals.

30.3.4 Ablation Study

We trained the model with a different combination of a number of epochs and initial learning rate using training data obtained from both databases. We used the number of epochs to be 20, 50, and the learning rate to be 0.01, 0.05, 0.001. Using these values, we obtained six combinations for which the CNN model is trained. Table 30.4 shows the training accuracy of a model with specification 1, specification 2, and specification 3, which is trained using two EEG databases for six combinations of epochs and learning rates. For each combination, the first three bars correspond to University of Bonn data, and the next three bars correspond to CHB-MIT data. From Table 30.4, we can say that when the learning rate decreases below 0.001, the training accuracy drops for both the training dataset. When the number of epochs is increased to 50, the training accuracy saturates, which means the model is overfitting the data. Thus, the optimal value of the number of epochs and learning rate is 20 and 0.001, respectively.

30.4 Conclusion

The proposed study was successful in classifying the epileptic and non-epileptic time series obtained from two different databases. Both the datasets did not show high accuracy for the same model. In other words, the model designed depends on the nature of the data like the number of samples, the number of time series for each category, and so on. The study deduced that if the dataset has a low number of samples, and then used the model with a high number of filters to get good classification accuracy.

Table 30.4 Comparison of training accuracy of a model with specification 1, specification 2, and specification 3, which is trained using two EEG databases for six combinations of the number of epochs and learning rate

Number of epochs	Learning rate	Dataset	Specification 1 (accuracy %)	Specification 2 (accuracy %)	Specification 3 (accuracy %)
20	0.01	University of Bonn	60	71	60
		CHB-MIT EEG data	89	89	90
20	0.05	University of Bonn	55	69	62
		CHB-MIT EEG data	80	85	90
20	0.001	University of Bonn	68	72	65
		CHB-MIT EEG data	90	89	92
50	0.01	University of Bonn	60	69	60
		CHB-MIT EEG data	85	85	83
50	0.05	University of Bonn	63	71	58
		CHB-MIT EEG data	89	89	90
50	0.001	University of Bonn	95	99	93
		CHB-MIT EEG data	97	96	95

This study can be extended to detect the seizures within the epileptic time series. Using the detected seizures, one can derive the nature of the seizure and the duration of the seizure.

Acknowledgements This project was supported by CSIR-SRF under the file number 08/0602(11528)/2020-EMR-I. We thank the Manipal Institute of Technology for providing the computational facility.

References

1. Goetz, C.G.: Textbook of Clinical Neurology. Elsevier Health Sciences (2007)
2. Steinlein, O.K.: Genetic mechanisms that underlie epilepsy. Nat. Rev. Neurosci. **5**, 400–408 (2004)

3. Ahmed, S.N., Spencer, S.S.: An approach to the evaluation of a patient for seizures and epilepsy. WMJ Off. Publ. State Med. Soc. Wis. **103**, 49–55 (2004)
4. Staba, R.J., Stead, M., Worrell, G.A.: Electrophysiological biomarkers of epilepsy. Neurotherapeutics **11**, 334–346 (2014)
5. Tsiouris, K.M., Pezoulas, V.C., Zervakis, M., Konitsiotis, S., Koutsouris, D.D., Fotiadis, D.I.: A long short-term memory deep learning network for the prediction of epileptic seizures using EEG signals. Comput. Biol. Med. **99**, 24–37 (2018)
6. Goldberger, A.L., Amaral, L.A., Glass, L., Hausdorff, J.M., Ivanov, P.C., Mark, R.G., et al.: PhysioBank, PhysioToolkit, and PhysioNet: components of a new research resource for complex physiologic signals. Circulation **101**, e215–e220 (2000)
7. Ahmedt-Aristizabal, D., Fookes, C., Nguyen, K., Sridharan, S.: Deep classification of epileptic signals. In: 2018 40th Annual International Conference of the IEEE Engineering in Medicine and Biology Society (EMBC), pp. 332–335. IEEE (2018)
8. Andrzejak, R.G., Lehnertz, K., Mormann, F., Rieke, C., David, P., Elger, C.E.: Indications of nonlinear deterministic and finite-dimensional structures in time series of brain electrical activity: dependence on recording region and brain state. Phys. Rev. E **64**, 061907 (2001)
9. Yao, X., Cheng, Q., Zhang, G.-Q.: A novel independent RNN approach to classification of seizures against non-seizures (2019). arXiv:190309326
10. Medvedev, A.V., Agoureeva, G.I., Murro, A.M.: A Long Short-Term Memory neural network for the detection of epileptiform spikes and high frequency oscillations. Sci. Rep. **9**, 19374 (2019)
11. Ullah, I., Hussain, M., Aboalsamh, H.: An automated system for epilepsy detection using EEG brain signals based on deep learning approach. Expert Syst. Appl. **107**, 61–71 (2018)
12. Zuo, R., Wei, J., Li, X., Li, C., Zhao, C., Ren, Z., et al.: Automated detection of high-frequency oscillations in epilepsy based on a convolutional neural network. Front. Comput. Neurosci. **13**, 6 (2019)
13. Jrad, N., Kachenoura, A., Merlet, I., Nica, A., Benar, C.G., Wendling, F.: Classification of high frequency oscillations in epileptic intracerebral EEG. In: Proceedings of the Annual International Conference of the IEEE Engineering in Medicine and Biology Society, pp. 574–577 (2015)
14. Jrad, N., Kachenoura, A., Merlet, I., Bartolomei, F., Nica, A., Biraben, A., et al.: Automatic detection and classification of high-frequency oscillations in depth-EEG signals. IEEE Trans. Bio-Med. Eng. **64**, 2230–2240 (2017)
15. Sciaraffa, N., Klados, M.A., Borghini, G., Di Flumeri, G., Babiloni, F., Aricò, P.: Double-step machine learning based procedure for HFOs detection and classification. Brain Sci. **10** (2020). https://res.mdpi.com/d_attachment/brainsci/brainsci-10-00220/article_deploy/brainsci-10-00220.pdf

Chapter 31
Clustering of MRI in Brain Images Using Fuzzy C Means Algorithm

Md. Rawshan Habib, Ahmed Yousuf Suhan, Abhishek Vadher, Md. Ashiqur Rahman Swapno, Md. Rashedul Arefin, Saiful Islam, Khan Anik Rahman, and Md Shahnewaz Tanvir

Abstract This paper concentrates on the analyses of medical images of patients with brain tumor obtained via magnetic resonance imaging (MRI). Clustering algorithms, statistical techniques and various distribution functions are used to find patterns in the computed parameters to help finding the desired clustering technique. In this paper, Fuzzy C Means clustering algorithm is applied for each patient, and the negative functions and the iteration counts are noted. All simulations are carried out in MATLAB. The results reveal the objective functions and the iteration count of pre-operation and post-operation stages of MRI image frames. The thresholding method and multiple histograms are used to observe the images. An overview of how imaging techniques and MRI functions work is also reviewed along with the different medical image processing established methods.

31.1 Introduction

Medical imaging technologies were first introduced in the form of radiography in the early 1900s after the discovery of X-rays. It was soon put into diagnostic use after discovering that the radiation from the X-rays gets absorbed by the bones or solid structures and can penetrate the tissues. The importance of medical imaging was soon realized. Locating internal injuries, diagnosing and treatment of diseases, study of the internal parts, etc. were some of the many benefits appreciated by the medical world. During the 1950s, more imaging technologies began to emerge. The development of ultrasound imaging, magnetic resonance imaging (MRI), computed tomography (CT) showed the world different and enhanced the forms of images with

Md. R. Habib · A. Vadher
Murdoch University, Murdoch, Australia

A. Y. Suhan
Curtin University, Bentley, Australia

Md. A. R. Swapno · Md. R. Arefin (✉) · S. Islam · K. A. Rahman · Md S. Tanvir
Ahsanullah University of Science and Technology, Dhaka, Bangladesh

J. I.-Z. Chen et al. (eds.), *Machine Learning and Autonomous Systems*, Smart Innovation, Systems and Technologies 269, https://doi.org/10.1007/978-981-16-7996-4_31

good visual clarity. Currently, these are the leading techniques to attain images of the internals of a human body.

One of the major areas of interest in medical imaging is the brain and discovering information about brain tumors. The cell by which brain tumors arise is given its name. These might be intrinsic (beginning inside the brain) or tertiary (beginning elsewhere in the body) (spreading to the brain from another area). Medical therapy changes based on the kind, shape, and condition of the patient, if it has progressed, and the patient's maturity and health condition. Medical therapy might be remedial or focused on pain relief. Magnetic resonance imaging (MRI) and ultrasound scans are noninvasive tests, which utilize RF impulses and magnetization, the latter one uses ultrasound signals, to offer a comprehensive picture of the brain's mucous membranes. It sees the cortex in three dimensions, in cross-sections, which may be obtained from its sides or through the above. It may also be used to assess neurological symptoms including its consequences mostly on element contains.

Due to the large number of brain tumor images that are currently being generated in medical center, that are not possible for clinicians to selectively annotate and segment such visuals in an acceptable period. Hence, the development of automatic image processing techniques has become inevitable. A Fuzzy C Means (FCM) clustering method is used in [1] to highlight the fundamental cluster from the raw data of individual segmentation of brain MRI. The proposed framework is capable of displaying the Image segmentation and classification used for future analysis and clustering. A Modified Fuzzy C Means algorithm is developed in [2], which is capable of showing the change of bias field accommodated with the intensity variations across patients, scans and equipment changes. An improved Fuzzy C Means algorithm is proposed in [3], which reaches the maximum value of dice similarity coefficient (DSC) to make sure the generated result is pre-eminent.

Greedy Snake Model and Fuzzy C Means are optimized in [4] to develop a successful automated brain tumor classification. FCM clustering method and preliminary draft are used in [5] to minimize the inaccuracy rate and attain excellent segmentation findings. The proposed prototype has enough strength than FCM algorithm with a very strong clinical defense ability in the utility of AGI digitally created visuals, MRI images and CT images to reduce the error rate of the blurred boundary areas in compared with CGMM method. Gaussian filtering and improved FCM are adopted in [6] to reduce noise by gray histograms. To minimize analysis time, K Mean Clustering and FCM are combined in [7] for segmentation of brain tumor, which shows accuracy and reproducibility when compared to manual segmentation. With this proposed technique, it is possible to extract tumor from MR image with exact position and shape. An enhanced K-means and kernelized FCM-based algorithm is developed in [8] for determining the starting foci of the suggested grouping techniques. Research observations validate the supremacy of the suggested algorithm. FCM grouping and watershed algorithm are improved in [9] where histogram calculation-based initial centroid selection method is proposed for FCM, and atlas-based marker detection method is used for watershed algorithm to avoid the issue of over-segmentation. Rigorous Fuzzy C Means clustering method utilizing non-parametric Bayesian estimate in wavelet transform region is proposed in [10] to reduce the major limitation of FCM from the noise sensitivity by retrieving good image with a replacement of noisy defective image. The proposed method illustrates a new gateway of tentative results of both simulated and real MRI having the closed form of expression to get better execution.

The basic purpose of this study is to build digital image clustering algorithms and apply them to medical images obtained from MRI. To understand and comprehend the images formed by MRI, a firm knowledge of the basics and the physics behind how these imaging technologies work and functions is necessary. Using various clustering algorithms over the MRI, also observations were made by identifying the patterns and distributions. Patterns are expected in different images of pre-operation and post-operation stages due to different distributions of pixel densities because of the existence of tumors, it can help us to identify the frame containing the tumor. Fuzzy C Means clustering algorithm and thresholding were utilized for identifying the tumor frame and segmentation. Analysis is done using both numerical and graphical methods for clustering the images using the patterns in the computed data.

31.2 Fuzzy C Means Algorithm

The Fuzzy C Means method is an optimization grouping approach, which minimizes the weighting among cluster mean squared fault optimization technique to create an ideal amount of divisions. Joe Dunn published the initial paper on the Fuzzy C Means grouping method in 1973 [11]. Jim Bezdek of Cornell University refined this afterward in 1981 [12]. Figure 31.1 depicts the findings generated by FCM clustering. Here is a formula for determining WGSS.

$$Q(A, B) = \sum_{i=1}^{n}\sum_{j=1}^{c} N_{ij}^{m} \left\| x_i - r_j \right\|^2 \tag{31.1}$$

Here $\|x_i - r_j\|^2$ is considered as objective function among ith information and jth data point, m is the factor of Fuzziness that is a true value larger than 1. Assume $X = \{x_1, x_2, x_3 \ldots, x_n\}$ be the set of n dimensional data points and $R = \{r_1, r_2, r_3 \ldots, r_c\}$ stands for the group of radii. The overall purpose of such Fuzzy C Means method seems to be to reduce the optimization problem to its smallest value. The following is the technique.

1. Reset pattern of membership function $N = [N_{ij}]$ matrix, $N(0)$
2. Obtain individual group center variables $R(k) = [Rj]$ with $N(k)$ at each k-step.
3. Find the fuzzy edges Rj by using formula beneath:

$$\frac{\sum_{i=1}^{n} x_i M_{ij}^{m}}{\sum_{i=1}^{n} M_{ij}^{m}} \tag{31.2}$$

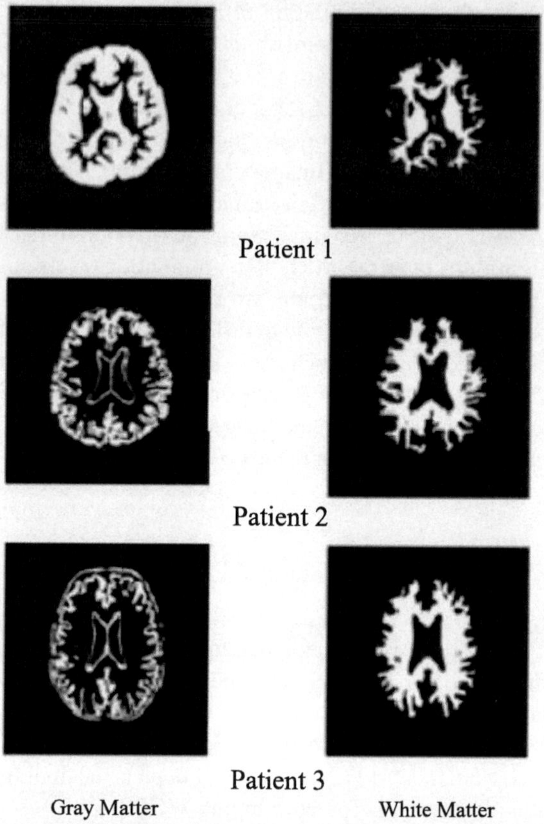

Fig. 31.1 Findings obtained by FCM method

4. $Rj = \forall j = 1, 2 \ldots e$, e is the group edges.
5. Upgrade $A(k), A(k+1)$. Here $A = (N_{ij})n + e$ is the fuzzy membership function.
6. Pattern of x_i's membership in such group j is as follows,

$$N_{ij} = \frac{1}{k = \sum_{k=1}^{c} \left(\frac{\|x_i - e_j\|}{\|x_i - e_k\|} \right)^{\frac{2}{m-1}}} \tag{31.3}$$

7. The action comes to a halt if, $\max_{ij} \| N^{(k+1)} - N^{(k)} \| < \partial$ or least Q is acquired, here ∂ is a factor for deciding whether or not to continue among [0, 1].
8. If not, proceed to stage 3.

31.3 K-Means Clustering Algorithm

K-means segmentation is indeed an edge detection approach that first originated within data analysis and has been widely used in information gathering for supervised classification. The goal of K-means clusters is to divide n sightings among k groups, upon every view belonging toward the grouping with the closest average, which serves as such cluster's pattern. As a result, the space of data is partitioned in Feature vectors. Due to the k within title, such method is similar to such k-nearest neighboring predictor, a prominent deep learning approach of categorization that is sometimes mistaken as K-means. To categorize updated information in the following groups, the 1-nearest neighboring predictor may be used to the clusters generated using K-means. It is recognized as the Rocchio technique or closest central filter [13]. Figure 31.2 indicates the results obtained by K-means clustering. The method is given below.

1. Input: K, set of points $x_1 \ldots x_n$.
2. Put clusters $c_1 \ldots c_k$ at a place randomly.
3. Continue till results are found:

   - for each point x_i:
     - obtain closest centroid c_j
     - decide on a spot x_j to group j
   - For each cluster $j = 1 \ldots K$: Input: K, set of points $x_1 \ldots x_n$

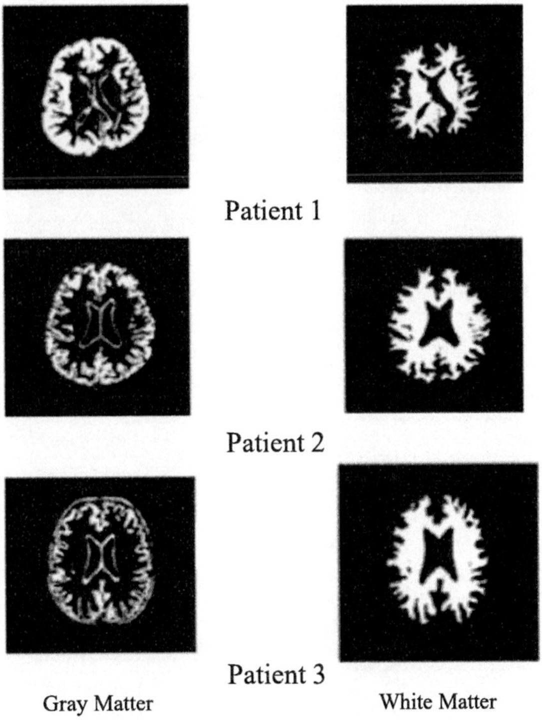

Fig. 31.2 Findings obtained by K-means method

$$C_j(a) = \frac{1}{n_{j_{x_i \to c_j}} \sum x_i(a)} \text{for} a = 1 \ldots d \tag{31.4}$$

– latest cluster c_j = average of every spot x_i allocated to group j in prior level

4. If no cluster allocations alter, come to a halt.

31.4 Proposed Image Processing Techniques

The basic focus of this study is in analyzing biological images. Biological or medical images can have a plethora of items and structures that might reveal details regarding basic biological mechanisms. The free online database [14] used for this study is released for the purpose of validating the latest image processing algorithms as finding real clinical images for research can be challenging for computer scientists and engineers. This database is created with clinical images of 14 patients, pre- to post-operation T1-weighted MR images containing gadolinium enrichment are performed on all the patients involved in the research. Several postoperative B-mode ultrasonic data are indeed recorded for every individual, allowing for easier analysis of diverse brain regions and diseases with MR data. In certain picture pairings, identical features are explicitly picked for verification. The instances of brain tumors covered under this investigation are summarized in Table 31.1. Gliomas have been the unique form of initial tumor considered in the research since they are the most frequent. The tumors studied are supratentorial, only with bulk of them in a forebrain.

Table 31.1 A summary of the brain tumor samples involved in this research

Patient number	Sex	Age	Tumor location	Tumor vol (cm^2)
1	F	42	Right frontal	2.3
2	M	31	Right frontal	79.2
3	M	70	Left parieto-temporal	53.7
4	M	72	Right parietal	31.6
5	M	39	Left frontal	0.2
6	M	40	Left frontal	32.3
7	M	62	Left temporal	13.9
8	M	23	Left frontal	66.0
9	F	42	Left frontal	4.8
10	F	49	Left frontal	10.4
11	F	40	Right frontal	39.7
12	F	72	Left frontal	49.1
13	M	76	Right frontal	31.9
14	M	70	Right Temporal	37.7

The easy way to divide a visual is by utilizing threshold value [15]. When $f(a, b) > T$, $f(a, b)$ is zero otherwise it will be 255.

1. Automated segmentation

   - To ensure separation very reliable, the limit must be typically set mostly by scheme.
   - To automatically determine the threshold, information of the assets, its service, and the surroundings must be employed:

     – The item's luminosity properties
     – The item's dimensions
     – The items take up portions of a frame
     – The quantity of distinct sorts of things that emerge in a photo.

2. Have used picture entropy to determine the threshold

   - Large maxima in the spectrum result from zones of homogeneous luminosity.
   - It's also feasible to use multilayer threshold method even if it is very challenging in reality
     When $f(a, b) < T1$
     $f(a, b)$ is 255
     else if $T1 \leq f(a, b) < T2$
     $f(a, b)$ is 128
     otherwise $f(a, b)$ is 0.
   - The graphic maxima should be high, thin, symmetrical, as well as divided by deep valleys in order to pick an appropriate threshold.

3. Hysteresis thresholding

   - When an object's entropy has no obvious valley, this implies there are numerous backdrop units having equal gray values to item units, and likewise.
   - Throughout this scenario, hysteresis thresholding could be employed.
   - Pixel values being beyond the higher bar being identified as objects, whereas pixels even under the lower limit are classed as foreground.
   - Those pixels that are close to several item pixels get identified as item pixels among the higher and lower criteria.

As an insert picture, a gray or disperse tumor scan from an MRI Image is generated. Additionally, the picture is resized, improved, and quantitative extracted features such as aperture are done before the tumor is separated utilizing removal of the linked component method.

31.5 Results Analysis

In the first stage, we have applied FCM algorithm on different images obtained by MRI, and then iteration counts and objective functions are observed (see Table 31.2 and Figs. 31.3, 31.4, 31.5 and 31.6).

In the second stage, we applied FCM only on tumors and noted iteration count, initial and final objective function (see Table 31.3 and Fig. 31.7).

Table 31.2 Iteration continues until convergence is reached

Iteration	Objective
İnitial objective function = 561,075,688.814849	
1	457,644,487.981064
2	457,660,245.944319
3	457,863,074.772509
4	458,904,774.769113
5	468,232,805.975744
6	586,638,764.934994
7	1,754,586,646.499107
8	2,874,878,601.507988
9	3,137,512,466.090066
10	3,187,378,746.521156
11	3,196,861,275.986648
12	3,198,669,377.271853
13	3,199,014,373.761649
14	3,199,080,210.562194
15	3,199,092,774.774880
16	3,199,095,172.527126
17	3,199,095,630.114251
18	3,199,095,717.440238
19	3,199,095,734.105503
20	3,199,095,737.285910
21	3,199,095,737.892904
22	3,199,095,738.008695
23	3,199,095,738.030809
24	3,199,095,738.034976
25	3,199,095,738.035829

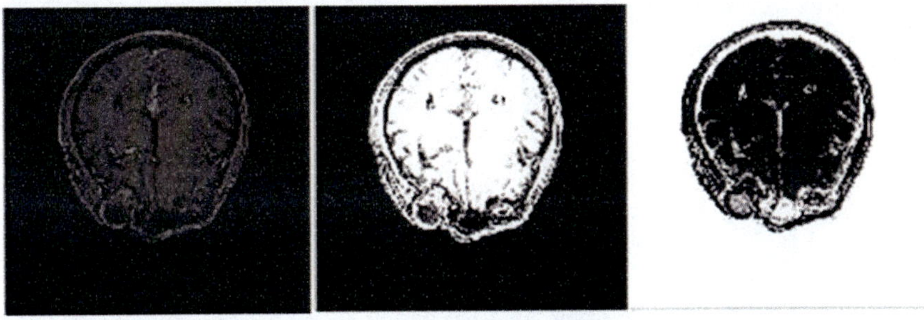

Fig. 31.3 Application of FCM algorithm on pre-operation MRI frames

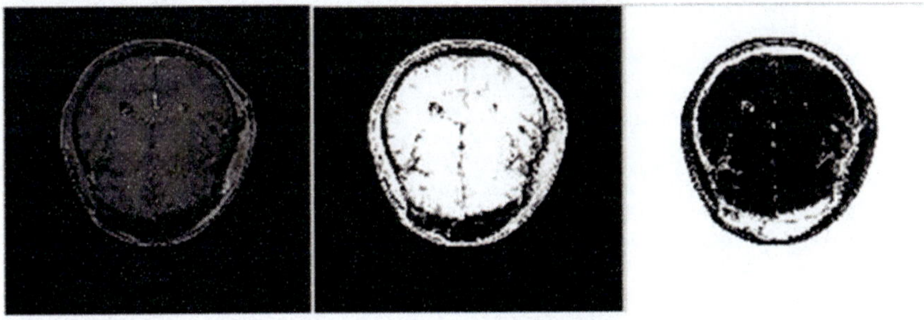

Fig. 31.4 Application of FCM algorithm on post-operation MRI frames

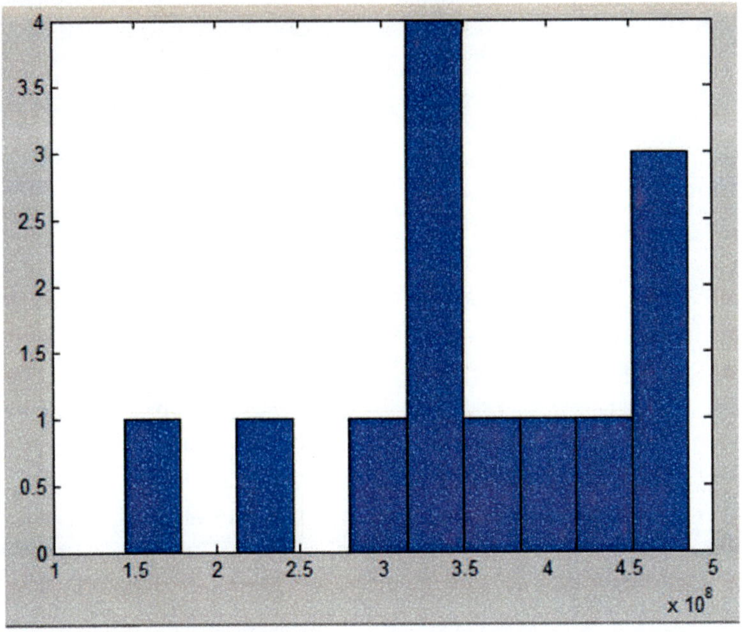

Fig. 31.5 Histogram of pre-operation data

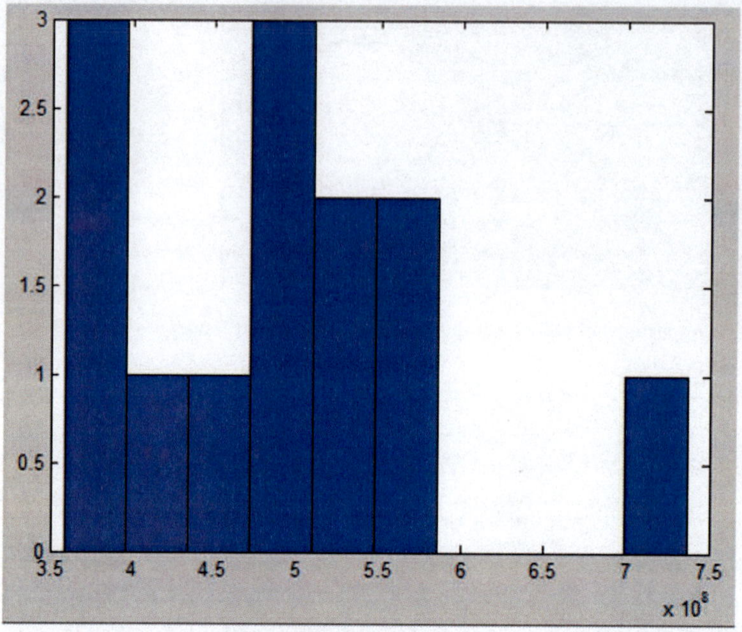

Fig. 31.6 Histogram of post-operation data

Table 31.3 Obtained data from applying FCM algorithm on tumors

Patient number	Initial objective function	Final objective function	Iteration count
1	18,355,744.796894	37,584,449.992009	6
2	356,220,763.647836	950,730,525.000000	7
3	61,963,905.057932	548,615,782.819662	7
4	87,978,456.765014	204,438,600.000000	7
5	5,043,432.109127	17,037,395,324.981459	6
6	145,627,714.551021	247,745,245.718788	7
7	76,027,734.923151	191,498,576.167858	7
8	89,930,435.542279	143,900,310.714015	7
9	42,128,216.752127	30,301,650.000000	7
10	33,427,629.831381	116,656,386.126817	31
11	174,326,843.688452	16,762,274,549.975018	7
12	191,766,702.359413	425,393,548.679964	7
13	167,265,084.426502	227,262,374.997069	7
14	216,021,617.690291	313,160,350.594556	7

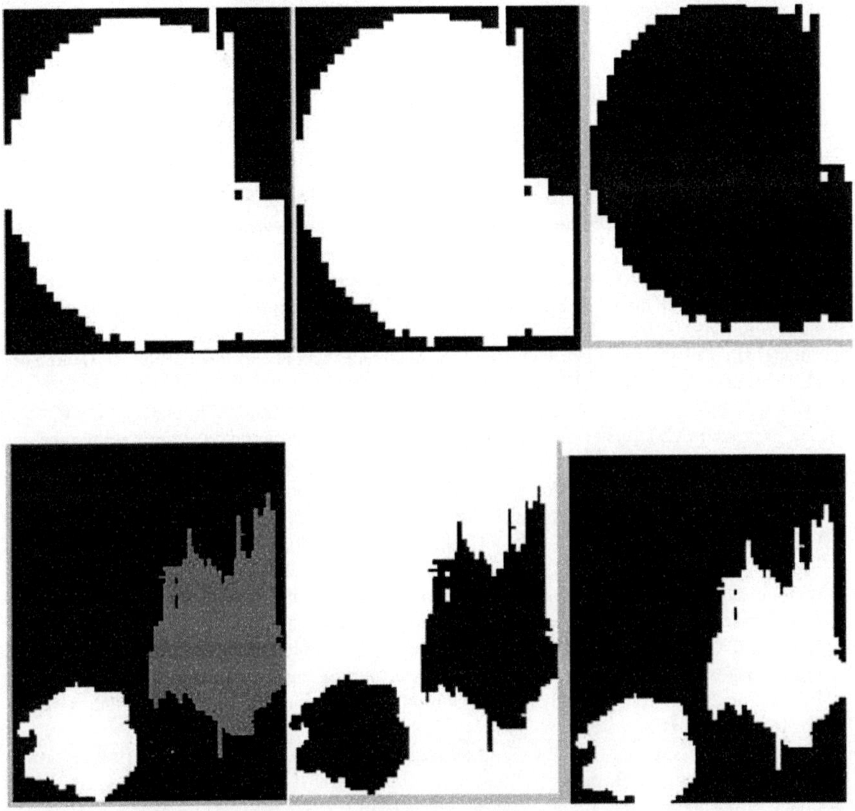

Fig. 31.7 Results obtained when FCM is applied to tumors

31.6 Conclusion

FCM algorithm is being used extensively all over the world. It's been widely used in the fields of artificial intelligence, neural mapping, even detecting tumor and other abnormalities. In this paper, the emphasis is on different clustering methods and briefly observed the Fuzzy C Means clustering method for MRI image segmentation. The images are observed and extracted by using ImageJ. FCM algorithm is applied on several MRI frames of pre-operation and post-operation images. The iteration counts and objective functions are noted. Later, the algorithm is only applied on only tumors.

References

1. Paul, G., Varghese, T., Purushothaman, K.V., Singh, N.A.: A Fuzzy C Mean clustering algorithm for automated segmentation of brain MRI. Adv. Intell. Syst. Comput. **247**, 59–65 (2013)
2. Ahmed, M.N., et al.: A modified Fuzzy C-Means algorithm for bias field estimation and segmentation of MRI data. IEEE Trans. Med. Imaging **21**, 193–199 (2002)
3. Li, M., et al.: An improved Fuzzy C-Means algorithm for brain MRI image segmentation. In: 2016 IEEE International Conference on Progress in Informatics and Computing, pp. 336–339, Shanghai (2016)
4. Sheela, C.J.J., Suganthi, G.: Automatic brain tumor segmentation from MRI using Greedy Snake Model and Fuzzy C-Means optimization, J. King Saud Univ. Comput. Inform. Sci. https://doi.org/10.1016/j.jksuci.2019.04.006
5. Huang, H., et al.: Brain image segmentation based on FCM clustering algorithm and rough set. IEEE Access **7**, 12386–12396 (2019)
6. Wan, C., Ye, M., Yao, C., Wu, C.: Brain MR image segmentation based on Gaussian filtering and improved FCM clustering algorithm. In: 2017 10th International Congress on Image and Signal Processing, BioMedical Engineering and Informatics, Shanghai, pp. 1–5 (2017)
7. Selvakumar, J., Lakshmi, A., Arivoli, T.: Brain tumor segmentation and its area calculation in brain MR images using K-Mean clustering and Fuzzy C-Mean algorithm. In: IEEE International Conference on Advanced Management Science, Tamil Nadu, pp. 186–190 (2012)
8. Rajini, N.H., Bhavani, R.: Enhancing K-means and Kernelized Fuzzy C-Means clustering with cluster center initialization in segmenting MRI brain images. In: 2011 3rd International Conference on Electronics Computer Technology, Kanyakumari, pp. 259–263 (2011)
9. Benson, C.C., Deepa, V., Lajish, V.L., Rajamani, K.: Brain tumor segmentation from MR brain images using improved Fuzzy C-Means clustering and watershed algorithm. In: 2016 International Conference on Advances in Computing, Communications and Informatics, Jaipur, pp. 187–192 (2016)
10. Chetih, N., Messali, Z., Serir, A., Ramou, N.: Robust Fuzzy C-Means clustering algorithm using non-parametric Bayesian estimation in wavelet transform domain for noisy MR brain image segmentation. IET Image Process. **12**, 652–660 (2018)
11. Dunn, J.C.: A Fuzzy relative of the ISODATA process and its use in detecting compact well-separated clusters. J. Cybernetics **3**, 32–57 (2008)
12. Bezdek, J.C.: Pattern Recognition with Fuzzy Objective Function Algorithms. Plenum, New York (1981)
13. MacQueen, J.: Some methods for classification and analysis of multivariate observations. In: Proceeding 5th Berkeley Symposium on Mathematical Statistics and Probability, vol. 1, pp. 281–297 (1967)
14. Mercier, L., et al.: Online database of clinical MR and ultrasound images of brain tumour. Am. Assoc. Phys. Med. **39**, 3253–3261 (2012)
15. Thresholding, https://www.cse.unr.edu/~bebis/CS791E/Notes/Thresholding.pdf

Chapter 32
Conversational Artificial Intelligence in Healthcare

Jatin Gupta, Nupur Raychaudhuri, and Min Lee

Abstract Innovative conversational artificial intelligence (AI) powered systems have been gaining momentum in the healthcare industry in recent years. Automated artificial intelligence programs are built with the purpose of allowing effective communication by providing an interface between the computer and the user. Conversational AI are making a significant impact on the healthcare industry for both medical health providers and patients. Several natural language processing (NLP) platforms, in particular using natural language understanding (NLU), such as Google Dialogflow, IBM Watson and Rasa are used in conversational AI. This paper intends to present an architecture adopted to deploy a successful conversational AI agent, named Ainume, using Google Dialogflow on the Google Cloud Platform (GCP). Ainume identifies symptoms of common and chronic diseases, accordingly, suggesting nutraceutical solutions to reduce the symptoms of these diseases. The focus of this paper is on one aspect that Ainume is equipped to deal with, that is, cardiovascular diseases.

32.1 Introduction

Conversational artificial intelligence (AI) is an emerging technology and is generating interest in research institutions, businesses and government organisations, as this technology is proving to be useful in areas of education, health, entertainment, retail and journalism [1]. Conversational AI involves conversations between human and computer systems and refers to technologies such as voice assistants and chatbots. Conversational AI technologies make use of machine learning and natural language processing (NLP) to build human like conversations between humans and computers [2].

J. Gupta (✉) · N. Raychaudhuri · M. Lee
IPMD Inc., San Mateo, CA, USA
e-mail: jatin@ieee.org

M. Lee
e-mail: minlee@berkeley.edu

J. I.-Z. Chen et al. (eds.), *Machine Learning and Autonomous Systems*, Smart Innovation, Systems and Technologies 269, https://doi.org/10.1007/978-981-16-7996-4_32

According to researchers [3], conversational agents such as dialogue systems, chatbots, personal assistants and chatterboxes, are using similar fundamental principles of AI. However, researchers [4] grouped conversational systems into question and answer bots, social bots and task-focused dialogue agents.

Conversational assistants, for example Siri and Alexa, are convenient to users as they can answer users' queries and questions in a matter of seconds. In addition to their convenience and ease of use, they can be used to entertain and interact with users. There has been a rise in the application of voice and text-based search engines and voice assistants such as Google Chrome, Siri and Cortana [5].

An important tool to create an efficient conversational AI is Google Dialogflow. Google Dialogflow is a Natural Language platform which streamlines the process of designing, developing and building a conversational AI. It provides a console where developers can build and train both text-based and voice-based conversation interfaces to create AI powered conversational assistant. Through the Dialogflow platform, these conversational interfaces can be integrated with different systems such as websites, Facebook Messenger, Slack and Twilio [6].

32.2 Literature Review

Conversational AI agents are computer programs which are created, designed and developed to have effective communication between a computer machine and a human. These communications are made possible through natural language processing technologies on mobile applications, dynamic websites, virtual chat rooms and smartphones. Further, these are known to be intelligent systems as they are developed mainly by using AI and NLP algorithms [7].

Conversational AI is known to be extensively used across various disciplines such as psychology, computer science, communication science and marketing and is greatly associated with machine learning, natural language processing and deep learning [1].

Researchers in their paper [8], stated that conversational AI can held automated text and verbal communications due to the advancements in machine learning models, deep learning techniques and natural language interfaces and their applications in healthcare.

Conversational AI assistants can be trained to discuss the general health of a user, and through this, they can diagnose what illness the user is likely to have. In the cases of common illnesses this can prove to reduce the valuable time of hospitals and doctors and they can provide further information to the user on how to treat it. This also helps in monitoring users' health conditions by providing the relevant information to the users [9]. For example, the mobile health service chatbot proposed for suggesting quick treatment in answering to questions related to everyday life accidents [10] and a virtual assistant to talk with users to monitor health symptoms and treatments [11]. There are also other chatbot systems designed for chronic health conditions such as cancer [12], acute patient support [13, 14] and paediatric asthmatic patients [15]. However, the use and impact of most of these chatbots in the long term are still to be evaluated.

According to researchers [16], literature on conversational AI agents in regard to chronic conditions is rare and mainly consists of quasi-experimental research studies which use natural language processing for prototype chatbots. Further, authors in their study concluded that future research could include a more organised development of AI conversational agents on specific chronic diseases to improve the quality of AI agents and their impact on the users.

Researchers [17] in their paper indicated that although there exists a number of chatbots which are providing the services for the healthcare domain, most are in the form of providing answers to common healthcare related frequently asked questions. Further, the authors stated that these chatbots are not able to develop a natural conversation with the user in the same manner as medical staff. To create the experience of doctor and patient communication in a natural manner, the authors suggested smart communication could be created by including technologies such as machine learning, natural language processing and natural language understanding.

Further, descriptive literature is available on conversational assistants in the healthcare domain and is mainly intended for the monitoring and providing of support regarding healthcare services. However, further work is required on the formats of conversational assistants in the healthcare domain [18].

Based upon the comprehensive literature review conducted, it has been found that there are limited research studies on the design and deployment of conversational AI using Google Dialogflow in the healthcare domain, which focus on cardiovascular disease. Thus, the present research paper provides an architecture to create, design and deploy a conversational AI to identify cardiovascular disease related symptoms, based upon symptoms shown by the user, and then accordingly, the conversational AI can suggest nutraceutical solutions in the form of food menus. Cardiovascular disease has been chosen in this study as it is one of the leading causes of death in the world [19].

32.3 Methodology

In the development of Ainume, a conversational AI to identify disease-related symptoms and suggest nutraceutical solutions, NLP and natural language understanding (NLU) were two key concepts used. NLP was used in the processing of text or speech in a similar way to how humans would and NLU was used in understanding the text or speech provided by a user in the conversation. In the following section, key terminologies and components used in the development of conversational AI are discussed.

32.3.1 Key Terminologies and Components

a. Natural Language Processing

Natural Language Processing (NLP) is a subset of Computer Science and Artificial Intelligence which involves the interaction between natural human language and computers. It focuses on programming computers which can analyse, understand and process natural human language in a similar way to how humans would do so, and it focuses on converting text to structured data.

b. Natural Language Understanding

Natural Language Understanding (NLU) is a subset of Natural Language Processing that aims to understand text or speech. Two concepts of NLU include intent and entity classification. As discussed, intent classification determines the user's intent in the conversation, and entity classification recognises the entities in a conversation and extracts these. The objective of NLU is simply to understand the text or speech given by a user in the conversation.

c. Components of Conversational AI

There are several components that form the conversational AI. These include intents, entities, training phrases, contexts and fulfilment. The parts of the conversational AI work together in order to achieve a functioning conversational AI which effectively converses with the user regarding cardiovascular disease. These components are discussed below further.

i. Agent

In Dialogflow, an agent describes the conversational agent that the user developed, which can converse with end-users and carry out data processing on the input it obtains from the user. There are several components of a conversational agent which complement each other to form the agent. When one or more of the several components are modified or changed, the agent is retrained for these changes to be reflected in the agent.

In this research, the agent is the conversational AI, named Ainume.

ii. Intent

An intent is the user's end goal for each conversation when talking to an agent, and an intent is used by the agent to recognise the end goal of a conversation. In other words, it involves what the user hopes to achieve from the conversation with the user. There would be several intents that come together as part of an agent. Intents can be linked together using contexts.

An intent in this agent may be 'What is cardiovascular disease?', which the conversational AI would be able to provide a response to.

iii. Entity

An entity is the method that Dialogflow allows specific data to be extracted from user input. For example, an entity for days of the week. A user may include a day of the week in their input to the agent, which may be important to the conversation. Some entities are predefined by Dialogflow and custom entities can also be created.

An example of an entity could be a list of the symptoms of cardiovascular disease, and if a user were to mention a symptom, the agent would understand the symptom to be from the entity of symptoms of cardiovascular disease.

iv. Training phrase

Training phrases for each intent allow the agent to understand and recognise intents from users. The greater the number of training phrases for an intent, the higher the accuracy of the agent to correctly identify user intents.

v. Contexts

The flow of conversation can be controlled by contexts. Input and output contexts can be used to match the upcoming user intents.

vi. Fulfilment

This allows an agent to give a more specialised response to intents. This may involve calling a service to execute an action for the user.

32.3.2 Architecture and Deployment

As this conversational AI agent was designed, developed and implemented in Dialogflow, a new agent was created in the Google Dialogflow Console, as well as a new Google Cloud Project. To do this, the agent was given the appropriate name, and the other basic details regarding the agent were entered. The agent could then be created, and after this, the Dialogflow Console for the agent was visible. Here, the basic intents and entities, alongside other key components of the conversational AI agent, could be created and developed. To the right of the Console, the agent could also be tested and conversed with.

The first steps in making the conversational AI agent be able to converse with users would be to add intents. These intents could classify user input (text or speech) and give an appropriate response to the user, based on the literal intent of the user, whether that is to ask a question or have a conversation with the conversational AI agent.

Usually, the conversational AI agent starts the conversation. Therefore, in this, the first intent added was a basic welcome intent, which would by default welcome the user when they open the conversational AI agent. An intent was created and named as welcome intent. The context of an intent is the scenario or circumstance in which it would be used, so the context was entered as the welcome intent. There were no

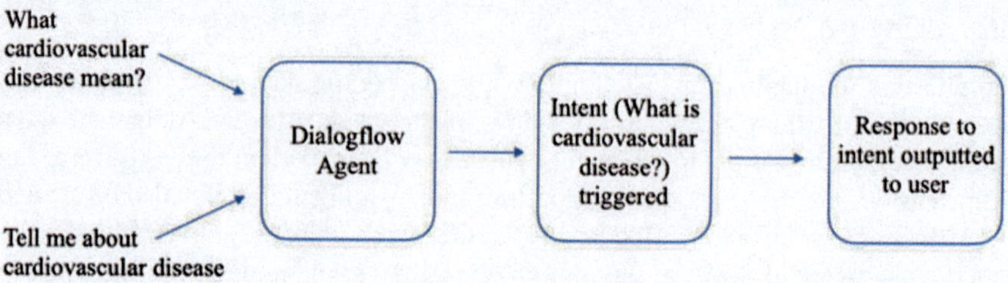

Fig. 32.1 Architecture of single intent

training phrases for this intent, as this intent would automatically and by default be the first output to the user from the conversational AI agent. To complete this intent, the conversational AI's output was added in the response section and the intent was then saved. For this, an appropriate welcome message to the user could be something similar to 'Hello. How may I assist you with any questions or concerns regarding cardiovascular disease?'

At this point, the conversational AI can only welcome the user. If the user replies to the conversational AI's welcome, by asking a question, the conversational AI wouldn't be able to give an appropriate reply. The conversational AI only has a welcome intent, and so wouldn't be able to classify the user input to an intent, as any other intents do not exist at that time. As a result, another intent was created. It was given an appropriate name and based on the intent; the relevant training phrases could be added to the intent. For example, the intent name may be 'What is cardiovascular disease?' The training phrases would be several different conversational lines in which the user asks what cardiovascular disease is. This intent would be for when a user asks what cardiovascular disease is, or something similar to that. Therefore, if the user asks what cardiovascular disease is, the conversational AI would be able to give an appropriate reply to this. An overview of this is presented in Fig. 32.1.

In a similar way, multiple more intents can be added from the Google Dialogflow Console. However, if the conversation a user is having with the conversational AI depends on responses to previous parts of the conversation, contexts can be added to each intent. Contexts allow a previous line of conversation to be passed along different intents to assist the conversational AI in answering questions or having a conversation where responses depend on previous responses in the conversation.

The following figure shows the architecture of the conversational AI. Once the user receives a response, they have the choice of continuing the conversation or starting a new one (Fig. 32.2).

Entities were used in order to extract specific information from the user as shown in Fig. 32.3. For example, an entity was created, which contained a list of symptoms of cardiovascular disease. If a user mentioned that they have any of the symptoms to the conversational AI, the conversational AI would extract the part of the user input which matched to a symptom, and thus, the conversational AI would recognise this, and converse with the user about their other symptoms.

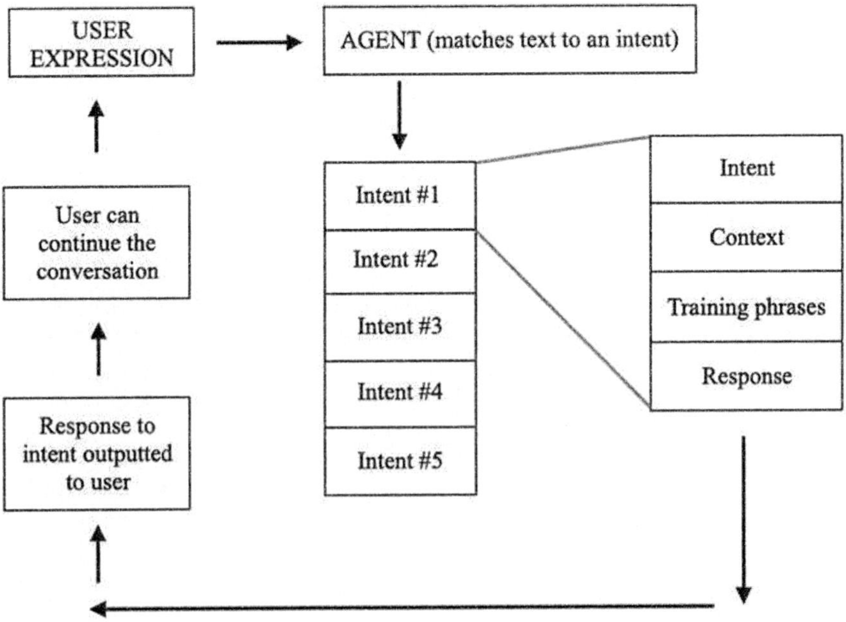

Fig. 32.2 Architecture of conversational AI

Through this method, a conversational AI to answer basic information and converse with users about symptoms of cardiovascular disease, suggesting foods which can lessen the impact of cardiovascular disease, was created. Each intent was extensively trained before the final testing of the conversational AI, through adding several training phrases. Then, the conversational AI was heavily tested in the Console, and the appropriate modifications made to the conversational AI.

The intents range from basic information regarding cardiovascular disease, to discussing symptoms of cardiovascular disease with the user, in order to make a judgement on whether the user may have cardiovascular disease, using NLP. The user also has the option to request nutraceutical food menus, to improve or maintain their health, and links are provided to these if requested by the user.

32.4 Conclusion

Conversational AI agents are machine learning and natural language processing based computer programs which are able to interpret human text and human voice, and provide appropriate responses to the human, in the conversation. Google Dialogflow in Google Cloud Platform has been leveraged in this research to design and build an NLP-based conversational AI assistant named Ainume which converses with users about cardiovascular disease in a simplified manner. An effective agent, which

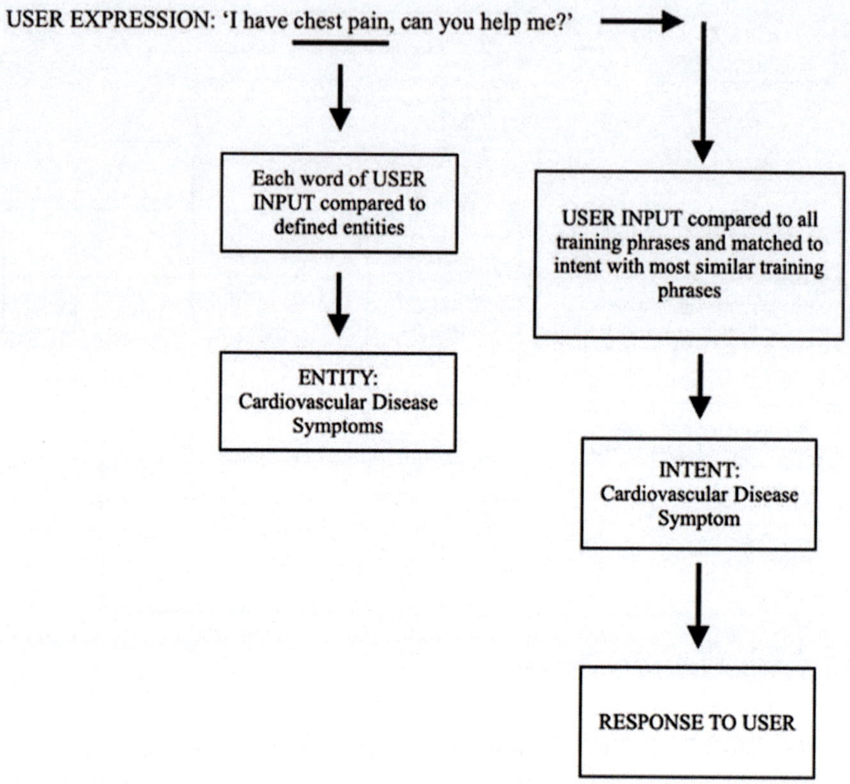

Fig. 32.3 The classification of entities and intents from user conversation

includes intents, entities, training phrases, fulfilment and contexts, was deployed. This assisted Ainume in conversing with users regarding both cardiovascular disease information and symptoms, and accordingly provide suitable nutraceutical solutions in the form of food menus to the user.

References

1. Song, X., Xiong, T.: A survey of published literature on conversational artificial intelligence. In: 7th International Conference on Information Management (ICIM). IEEE, pp. 113–117 (2021)
2. IBM. https://www.ibm.com/cloud/learn/conversational-ai/ (2021)
3. Basnyat, B., Singh, N., Roy, N., Gangopadhyay, A.: Vision powered conversational AI for easy human dialogue systems. In: IEEE 17th International Conference on Mobile Ad Hoc and Sensor Systems (MASS). IEEE, pp. 684–692 (2020)
4. Gao, J., Galley, M., Li, L.: Neural approaches to conversational AI. Found. Trends® Inf. Retr. **13**(2–3), 127–298 (2019)
5. Abdul-Kader, S.A., Woods, J.C.: Survey on chatbot design techniques in speech conversation systems. Int. J. Adv. Comput. Sci. Appl. **6**(7) (2015)

6. Google Dialogflow. https://cloud.google.com/dialogflow/docs/ (2021)
7. Nithuna, S., Laseena, C.A.: Review on implementation techniques of chatbot. In: International Conference on Communication and Signal Processing (ICCSP). IEEE, pp. 157–161 (2020)
8. Kulkarni, P., Mahabaleshwarkar, A., Kulkarni, M., Sirsikar, N., Gadgil, K.: Conversational AI: an overview of methodologies, applications and future scope. In: 5th International Conference On Computing, Communication, Control And Automation (ICCUBEA). IEEE, pp. 1–7 (2019)
9. Huang, C.Y., Yang, M.C., Huang, C.Y., Chen, Y.J., Wu, M.L., Chen, K.W.: A chatbot-supported smart wireless interactive healthcare system for weight control and health promotion. In: IEEE International Conference on Industrial Engineering and Engineering Management (IEEM) (2018)
10. Chung, K., Park, R.C.: Chatbot-based healthcare service with a knowledge base for cloud computing. Clust. Comput. (2018)
11. Polignano, M., Narducci, F., Iovine, A., Musto, C., De Gemmis, M., Semeraro, G.: HealthAssistantBot: a personal health assistant for the Italian language. IEEE Access **8**, 107479–107497 (2020)
12. Piau, A., Crissey, R., Brechemier, D., Balardy, L., Nourhashemi, F.: A smartphone chatbot application to optimize monitoring of older patients with cancer. Int. J. Med. Inform. **128**, 18–23 (2019)
13. Roca, S., Sancho, J., Garcia J., Alesanco, A.: Microservice chatbot architecture for chronic patient support. J. Biomed. Inform. **102** (2020)
14. Hauser-Ulrich, S., Kunzli, H., Meier-Peterhans, D., Kowatsch, T.: A smartphone-based health care chatbot to promote self-management of chronic pain (SELMA): pilot randomized controlled trial. JMIR mHealth uHealth **8**(4) (2020)
15. Kadariya, D., Venkataramanan, R., Yip, H.Y., Kalra, M., Thirunarayanan, K., Sheth, A.: kBot: knowledge-enabled personalized chatbot for asthma self-management. In: IEEE International Conference on Smart Computing (SMARTCOMP) (2019)
16. Schachner, T., Keller, R., von Wangenheim, F.: Artificial intelligence-based conversational agents for chronic conditions: systematic literature review. J. Med. Internet Res. **22**(9) (2020)
17. Bhirud, N.S., Tataale, S., Randive, S., Nahar, S.: A literature review on chatbots in healthcare domain. Int. J. Sci. Technol. Res. **8**, 225–231 (2019)
18. Tudor Car, L., Dhinagaran, D.A., Kyaw, B.M., Kowatsch, T., Joty, S., Theng, Y.L., Atun, R.: Conversational agents in health care: scoping review and conceptual analysis. J. Med. Internet Res. **22**(8) (2020)
19. Gupta, J.: The accuracy of supervised machine learning algorithms in predicting cardiovascular disease. In: IEEE International Conference on Artificial Intelligence and Computer Science Technology (ICAICST). IEEE Press (2021)

Chapter 33
Deep-CNN for Plant Disease Diagnosis Using Low Resolution Leaf Images

Ashiqur Rahman, Md. Hafiz Al Foisal, Md. Hafijur Rahman, Md. Ranju Miah, and M. F. Mridha

Abstract Plant disease diagnosis is critical in agriculture since diseases frequently restrict plant production receptivity. Physical strategies to identify plant diseases are mostly timely, objecting, and lengthy. As a consequence, agricultural automation with automated identification of plant diseases is widely preferred. Most of the implemented models could only identify diseases of a particular plant using high-resolution images, which is quite expensive from a farmer's position. Because of the variation in leaf colors, aspect ratios, and congested backgrounds, detecting plant disease by low-quality images is difficult. This paper explores an efficient plant disease identification model that combines multiple plant diagnoses for low-resolution images using deep convolutional neural networks (DCNNs). This system acquires a multilabel classification to classify both the plant type and the specific disorder simultaneously.

33.1 Introduction

A disease occurring in food crops is a significant threat that has a significant economic effect, including its productivity. Plant diseases are responsible for 10–16 percentage of crop damage worldwide every year. Plant disorders affect the growth of individual plants, and early detection is crucial. So, if disease outbreaks are not identified in advance, food disruption will grow. Plant disease study is associated with the examination of visually noticeable patterns on plants and especially on leaves. It takes quite a lot of fieldwork, such as looking at any leaf or plant to find symptoms to detect diseases at an early stage. This method could be sped up using advanced computational methods to analyze real-time images of plants or leaves and recognize the disease. Numerous Machine Learning (ML) frameworks were used to detect and

A. Rahman · Md. H. Al Foisal · Md. H. Rahman (✉) · Md. R. Miah · M. F. Mridha
Bangladesh University of Business and Technology, Dhaka, Bangladesh

M. F. Mridha
e-mail: firoz@bubt.edu.bd

J. I.-Z. Chen et al. (eds.), *Machine Learning and Autonomous Systems*, Smart Innovation, Systems and Technologies 269, https://doi.org/10.1007/978-981-16-7996-4_33

identify plant diseases, such as SVM classifiers, KNN classifiers, ANN classifiers, etc.

The advancement makes far toward deep learning approaches, which is also proven to be a novel approach for further research. Deep learning methods have been developed to detect any objects or similar patterns and have seen great success due to their reliability. Researchers tried many Deep Learning approaches to build a system that can detect plant diseases more frequently with real-time input. Convolutional neural networks (CNNs) were already introduced by DL researchers that overcame the image-related plant disease recognition challenges. The development of the CNN approach has become more reliable and shows improvement after getting in touch with the transfer learning (TF) strategy. In TF, the CNN is primarily evaluated on a pretrained dataset. The pretrained architecture identifies related images out of the similar or dissimilar dataset ranges. This strategy helps to prevent overfitting issues with the small datasets.

Several works have been done on high-resolution image datasets when analyzing some literature on detecting or diagnosing plant disease. Most farmers have no access to high-resolution cameras and cannot afford to detect diseases of plants. Consequently, a system for diagnosing and recognizing plant diseases on low-resolution images that would be both effective and highly accurate must be developed. And so, we are determined to develop a framework that can effectively detect the patterns of low-resolution images. We have chosen an already experimented dataset for this work, and we will work on three different plants as Apples, tomatoes, and corns. The overall contribution can be summed up as:

- We have investigated previously done researchers to determine the general knowledge behind the architectures and their inabilities in deep learning.
- We gathered images of 3 plants (Tomato, Apple, and Corn) 14 diseases to construct our study and evaluate the deep-CNN architecture.
- We tested six deep convolutional neural network architectures, including Densenet, Inception, Mobilenet, ResNet, VGG, and Xception, and explored MobileNet compelling low-resolution plant leaf images.

The continuation of this research is designed as the previous literature is exhibited in Sect. 2. The dataset is introduced in Sect. 3. Section 4 describes the basic structure of deep-CNN. Section 5 evaluates the model and measures the outcome. Finally, Sect. 6 brings the chapter to a conclusion.

33.2 Related Work

Numerous works on automated plant diagnostic techniques and detection using deep learning techniques have developed in recent years. Chen [1] introduced transfer learning of the CNN as a method for identifying plant diseases using pretrained models learned from large datasets and transferring them into the particular task in their very own dataset. This method advances in this area by not using random weights; alternatively, they utilized ImageNet, a pretrained large-labeled dataset. Shruthi [2] showed a general phase of the plant disease diagnosis system and a comparison with some of the most commonly used machine learning classification methods. They compared SVM classifiers, ANN classifiers, KNN classifiers, Fuzzy classifiers, and Deep Learning approaches, and CNN outperformed them all. Ferentinos [3] presented an established deep learning model for detecting diseased plants from an openly available dataset that a particular CNN develops. The VGG Convolutional Neural Network was the most efficient.

Barbedo [4] investigated different diseases which affect the same leaf but in different parts. The researchers divided the leaf criteria into isolated symptoms taken individually and defects that were a portion of classes observed as a cluster to detect disease. This study demonstrated how dataset limitations could influence any techniques used to classify plant disease. And this solution leads them to get more reliable results. Kawasaki [5] presented an approach to achieve high classification efficiency using a Convolutional Neural Network and the K-fold Cross-Validation technique and obtained an average accuracy. Durmuş [6] shared research on plant infection classification using modern deep learning, which intended to detect diseases in real-time data collected by the robot. They worked hard to investigate cases such as physical changes in leaves that were observed by RGB cameras. They tested their network on AlexNet and SqueezeNet. Francis [7] Presented a CNN model constructed to identify plant diseases using leaf images of apple and tomato of normal and abnormal trees. They noticed the overfitting problem, eliminated that dropout value, and evaluated their output on the GPU. They attempted to provide the researchers with creative insight to create an integrated plant disease identification method. This study explained the architectures utilized to identify plant infection utilizing images of leaves and a few segmentation and feature extraction algorithms applied during tree infection diagnosis [8].

Ramesh [9] proposed to identify anomalies that arise on plants in their natural environment. To remove occlusion, they shot photographs with a plain backdrop. In terms of precision, the method was applied to other machine learning models. The classifier was tested using 160 images of papaya leaves and the Random forest classifier. Mohanty [10] proposed a deep-CNN trained on a public dataset of 54,306 images to recognize 14 plant species and 26 types of infections. The proposed method achieved an efficiency of 99.35 percentage on the validation set. Venkataramanan [11] presented a Deep Learning technique for recognizing tree infections by analyzing the leaf of a particular crop. The classifier is trained in multiple stages to remove possibilities at each point. An object detector, YOLO v3, extracted the leaf and analyzed it through ResNet18 models. Many strategies for recognizing disease symptoms were examined, and how researchers were using the PlantVillage dataset on DL models to assess results. A systematic research is required to illustrate plant disease detection factors, including dataset classes and scale, learning rate, visibility, etc. Also, DL techniques must be enhanced to identify pathogens during their entire life cycle [12]. Sharma [13] proposed a work investigating a possible solution for automatic detection in plants by training convolutional neural network (CNN) models with segmented image data. They compared the F-CNN model and S-CNN model and tested more than double the performance in this work. Sladojevic [14] proposed a novel approach for crop infection identification algorithm depending on leaf picture classification, distinguishing 13 different plant diseases among healthy leaves and distinguishing plant leaves from their surroundings. This study [15] employs a minor change of the CNN model known as LeNet to recognize infections in tomato leaves to achieve comparatively better outcomes for detecting tomato plant diseases.

33.3 Dataset

Three openly prepared plant disorder datasets are applied throughout the investigations. The images were accumulated from Kaggle. It carries pictures of 3 separate plants where every plant has a different number of similar diseases. Figure 33.1 shows a few sample images from the dataset.

Fig. 33.1 The figure shows the disorder images of apple (top left), corn (bottom left), and tomato (right)

33.4 Methodology

To diagnose plant diseases, various CNN architectures are introduced and compared. The basic concept and layers of different CNN architectures are instructed in the following sections.

33.4.1 Image Preprocessing

Data normalization is a method in image analysis that alters the area of pixel intensities [16]. It assures that each input image has an equal distribution. Since the CNN architectures need the input images to be of the identical form, each picture is restructured into 128 by 128 pixels, resulting in faster CNN convergence. As a consequence, every channel of the reconfigured leaf images is normalized as follows [17]:

$$Normalize\,(D) = \begin{bmatrix} d_{1,1} & \cdots & d_{1,m} \\ \vdots & \ddots & \vdots \\ d_{n,1} & \cdots & d_{n,m} \end{bmatrix} /255 \tag{33.1}$$

Here D is the image matrix, n refers to the number of rows, and m means the number of columns.

33.4.2 Baseline Architecture

Convolutional Neural Networks are renowned for their intelligence to identify patterns in images. This field allows and enables machines to understand the world in the same way individuals do and apply that knowledge to several devices and methods such as image identification, image analysis, classification, etc. The first step is to feed the image's pixels in the formation of arrays to the neural network's input layer (used to classify images). The hidden layers extract features by conducting various measurements. Convolutional neural networks are feedforward neural networks commonly used to interpret visual images by analyzing information and are referred to as a ConvNet [18].

To recognize and identify objects in an image, CNN is used. CNN has many hidden layers that enable the extraction of information from images. The essential layers are given below with a brief discussion:

We emphasize analyzing reviewed CNN architecture for detecting plant disease on low-resolution images in this article. The CNN structure is depicted in Fig. 33.2, with the input layer (images of the leaves from the dataset), convolutional layers, a dense layer, and an output layer.

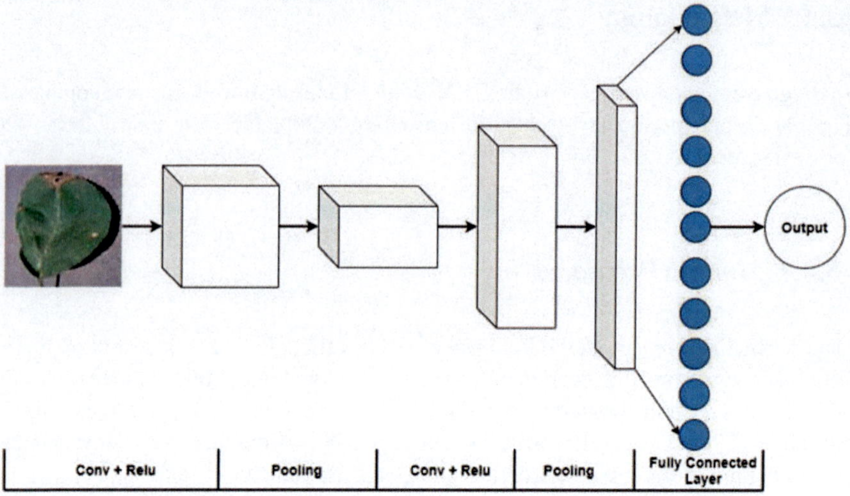

Fig. 33.2 The CNN structure of plant diagnosis from leaf images

Input layer: In the input layer, the original images of different plants from the dataset are input. But the authentic images are reconstructed with different widths, heights in the shape of $120 * 120 * 3$ before given to the architecture as inputs.

Convolutional layers: This is the first step in extracting useful information from images. The convolution process is performed by plenty of filters (matrix of values known as weights that have been trained to detect unique features) in a convolution layer. CNN's utilize filters, also identified as kernels, to determine what features, such as edges, are present in images. The filter iteratively checks the images to see if the function it is designed to detect is available. The mathematical procedure can be found in [19]. The Batch Normalization process converts the images into a standard shape based on the mean and variance of a particular batch. It also improves the reliability of the network and leads to faster convergence.

ReLU Layer: The rectified linear unit (ReLU) is an abbreviation. After the function maps have been removed, they must be moved to a ReLU layer. This is a non-linear activation function, and its sole aim is to incorporate non-linearity into the network. The ReLU function is straightforward; values less than or equal to zero become zero, while all positive values remain unchanged.

Pooling layer: Pooling is a type of downsampling that reduces the dimensions of a feature map. Reduce the redundancy in the input function to boost up the training process and decrease the amount of storage used by the network. To generate a pooled feature map, the rectified feature map is now passed through a max-pooling pooling layer. Max pooling dramatically decreases the size of the image, lowering the amount of system memory and the number of operations conducted further in the network.

Fully Connected Layer: The following step in the process is known as flattening (convert all the resultant 2-Dimensional arrays into a single long continuous linear vector). With the fully connected neural network operation, the input image is flattened into a feature vector and transferred through a network that predicts the output possibilities. This fully connected layer is followed by several dense layers of neurons, which eventually produce accurate observations.

33.4.3 Transfer Learning

In general, for transfer learning the initial step is to obtain the pretrained model that will be used to solve the specified task. To go for one of the frameworks, such as ResNet or Xception, to initialize the base model. The base model would generally have much more units in the final output layer than desired and would need to remove the final output layer. Then, add a final output layer that is task-compatible. Freezing the layers from the pretrained model is crucial because all previous learning would be lost if not. The following move introduces additional trainable layers, converting existing features into predictions on the new dataset. After training the new layers on the dataset, we will have a model to make predictions on the dataset. And we can improve the performance by fine-tuning. Fine-tuning is accomplished by unfreezing the base model or a part of it and training the whole model on the data set at a shallow learning rate. The low learning rate would improve the model's efficiency on the new dataset while avoiding overfitting.

33.5 Evaluation

This section is designed with some subsections, including evaluation metrics, experimental setup, experiments, and comparison that we have done for this specific task. The datasets on which the tests were performed.

33.5.1 Evaluation Metric

The following equations evaluated various baseline architectures by using the confusion matrix's accuracy, precision, and recall evaluation metrics.

Accuracy: By following the equation, we can find the accuracies of the model.

$$Accuracy = \frac{TP + TN}{TP + TN + FP + FN} \tag{33.2}$$

Recall: The equation which helps to gain an efficient recall can be stated as

$$Recall = \frac{TP}{TP + FN} \tag{33.3}$$

Precision: This equation will give the precision of the model, and it can be stated as

$$Precision = \frac{TP}{TP + FP} \tag{33.4}$$

33.5.2 Experimental Setup

We used Python for data collection, pre-processing, tests, and model evaluation [20]. And Keras [21] is used to implement the neural network architecture also for deep learning models. We used Numpy [22], which is a Python library for performing simple mathematical operations. Then TensorFlow [23] is used to produce neural network GPU execution.

33.5.3 Experiments and Comparisons

Experimented the proposed system using various pretrained Keras-implemented transfer learning models like Mobilenet, MobileNetV2, Xception, ResNet50, ResNet50V2, ResNet101, DenseNet121, DenseNet169, DenseNet201, InceptionV3, VGG16, VGG19, and so on in this section. As the pretrained learned features were applied in the new diagnosis task, so it certainly identifies the diseases from the infected plants. Thus, we present the effectiveness of each model to determine which had the best accuracy, recall, and precision. Table 33.2 shows the dimensions of the developed system. We could see that MobileNet performs admirably, with 99.79% accuracy, 99.08% precision, and 98.87% recall. This is the present state of the performance, as far as we can learn. We applied different architectures such as MobileNet, ResNet, DenseNet, Inception, VGG, Xception, and their latest version, producing better performance. However, MobileNet provides the best performance for disease diagnosis in various plants. The concept of transfer learning was greatly used to run the experiment using these CNN architectures.

However, in the investigation, the MobileNet [19] architecture provides the highest performance. Apart from the first layer, this method is based on depthwise separable convolutions. The first layer is a dense layer. Behind every sheet, batch normalization and ReLU non-linearity are applied. MobileNet especially gives better accuracy for low-resolution images because of its lightweight, streamlined structure that utilizes depthwise separable convolutions.

Table 33.1 This table represents the accuracy, precision, and recall of different architectures

Model	Accuracy	Precision	Recall
Xception [24]	98.7	98.6	98.56
ResNet50 [19]	96.8	93.41	94.13
ResNet50V2 [25]	96.90	97.42	96.67
ResNet101 [26]	97.28	98.82	98.65
DenseNet121 [27]	97.87	94.1	95.54
DenseNet169 [27]	97.36	97.86	96.63
DenseNet201 [27]	97.08	97.88	96.67
InceptionV3 [26]	95.55	95.9	94.53
VGG16 [28]	94.56	92.53	93.45
VGG19 [28]	91.05	93.54	92.13
MobileNet [29]	**99.79**	**99.08**	**98.87**

33.6 Conclusion

This section incorporates and justifies a multi-plant diagnosis approach based on various image classifier baselines for low-resolution images. We used a transfer learning strategy to train and evaluate our method systematically. We collected a publicly introduced dataset that consisted of different images, and we evaluated three different plants, apples, corns, and tomatoes, with several different diseased leaf images. We find that separate and distinct convolution and skipped connections greatly enhance plant disease detection efficiency. The paper focused on image processing using CNN for its excellent feature extraction capability and superiority over conventional systems in terms of real-time, precision, and resilience. We evaluate and observe different architectures such as Xception, ResNet, DenseNet, and MobileNet, and in comparison, MobileNet performs better for low-resolution plant leaf images. This work does have a lot of potential for expansion. Changing the architecture of CNN will help to reduce feature extraction losses. We are confident that perhaps the objectives of this research work will lead the road for intensive studies on image processing methods in the future and improve the knowledge and usefulness of object classification based on CNN.

References

1. Chen, J., Chen, J., Zhang, D., Sun, Y., Nanehkaran, Y.A.: Using deep transfer learning for image-based plant disease identification. Comput. Electron. Agric. **173**, 105393 (2020)
2. Shruthi, U., Nagaveni, V., Raghavendra, B.K.: A review on machine learning classification techniques for plant disease detection. In: 2019 5th International Conference on Advanced Computing & Communication Systems (ICACCS). IEEE, pp. 281–284 (2019)
3. Ferentinos, K.P.: Deep learning models for plant disease detection and diagnosis. Comput. Electron. Agric. **145**, 311–318 (2018)

4. Barbedo, J.G.A.: Plant disease identification from individual lesions and spots using deep learning. Biosys. Eng. **180**, 96–107 (2019)
5. Kawasaki, Y., Uga, H., Kagiwada, S., Iyatomi, H.: Basic study of automated diagnosis of viral plant diseases using convolutional neural networks. In: International Symposium on Visual Computing. Springer, Cham, pp. 638–645 (2015)
6. Durmuş, H., Güneş, E.O., Kırcı, M.: Disease detection on the leaves of the tomato plants by using deep learning. In: 2017 6th International Conference on Agro-Geoinformatics. IEEE, pp. 1–5 (2017)
7. Francis, M., Deisy, C.: Disease detection and classification in agricultural plants using convolutional neural networks—a visual understanding. In 2019 6th International Conference on Signal Processing and Integrated Networks (SPIN). IEEE, pp. 1063–1068 (2019)
8. Khirade, S.D., Patil, A.B.: Plant disease detection using image processing. In: 2015 International Conference on Computing Communication Control and Automation. IEEE, pp. 768–771 (2015)
9. Ramesh, S., Hebbar, R., Niveditha, M., Pooja, R., Shashank, N., Vinod, P.V.: Plant disease detection using machine learning. In: 2018 International Conference on Design Innovations for 3Cs Compute Communicate Control (ICDI3C). IEEE, pp. 41–45 (2018)
10. Mohanty, S.P., Hughes, D.P., Salathé, M.: Using deep learning for image-based plant disease detection. Front. Plant Sci. **7**, 1419 (2016)
11. Venkataramanan, A., Honakeri, D.K.P., Agarwal, P.: Plant disease detection and classification using deep neural networks. Int. J. Comput. Sci. Eng. (IJCSE) **11**(08), 40–46 (2019)
12. Saleem, M.H., Potgieter, J., Arif, K.M.: Plant disease detection and classification by deep learning. Plants **8**(11), 468 (2019)
13. Sharma, P., Berwal, Y.P.S., Ghai, W.: Performance analysis of deep learning CNN models for disease detection in plants using image segmentation. Inf. Process. Agric. **7**(4), 566–574 (2020)
14. Sladojevic, S., Arsenovic, M., Anderla, A., Culibrk, D., Stefanovic, D.: Deep neural networks based recognition of plant diseases by leaf image classification. Comput. Intell. Neurosci. (2016)
15. Tm, P., Pranathi, A., SaiAshritha, K., Chittaragi, N.B., Koolagudi, S.G.: Tomato leaf disease detection using convolutional neural networks. In: 2018 Eleventh International Conference on Contemporary Computing (IC3). IEEE, pp. 1–5 (2018)
16. Kabir, M.M., Ohi, A.Q., Mridha, M.F.: A Multi-Plant Disease Diagnosis Method using Convolutional Neural Network. arXiv:2011.05151 (2020)
17. Krizhevsky, A., Sutskever, I., Hinton, G.E.: Imagenet classification with deep convolutional neural networks. Adv. Neural. Inf. Process. Syst. **25**, 1097–1105 (2012)
18. Kabir, M.M., Ohi, A.Q., Rahman, M.S., Mridha, M.F.: An evolution of CNN object classifiers on low-resolution images. In: 2020 IEEE 17th International Conference on Smart Communities: Improving Quality of Life Using ICT, IoT and AI (HONET). IEEE, pp. 209–213 (2020)
19. Howard, A.G., Zhu, M., Chen, B., Kalenichenko, D., Wang, W., Weyand, T., Adam, H.: Mobilenets: Efficient Convolutional Neural Networks for Mobile Vision Applications. arXiv: 1704.04861 (2017)
20. Van Rossum, G., Drake, F.L.: Python 3 Reference Manual CreateSpace. Scotts Valley, CA (2009)
21. Chollet, F.: Keras: Deep learning library for theano and tensorflow. https://keras.io/k, **7**(8), T1 (2015)
22. Van Der Walt, S., Colbert, S.C., Varoquaux, G.: The NumPy array: a structure for efficient numerical computation. Computing in science & engineering **13**(2), 22–30 (2011)
23. Abadi, M., Barham, P., Chen, J., Chen, Z., Davis, A., Dean, J., Zheng, X.: Tensorflow: a system for large-scale machine learning. In: 12th {USENIX} symposium on Operating Systems Design and Implementation ({OSDI}'16). IEEE, pp. 265–283 (2016)
24. Chollet, F.: Xception: deep learning with depthwise separable convolutions. In: Proceedings of the IEEE Conference on Computer Vision and Pattern Recognition. IEEE, pp. 1251–1258 (2017)

25. He, K., Zhang, X., Ren, S., Sun, J.: Identity mappings in deep residual networks. In: European Conference on Computer Vision. Springer, Cham, pp. 630–645 (2016)
26. Huang, G., Liu, Z., Van Der Maaten, L., Weinberger, K.Q.: Densely connected convolutional networks. In: Proceedings of the IEEE Conference on Computer Vision and Pattern Recognition. IEEE, pp. 4700–4708 (2017)
27. He, K., Zhang, X., Ren, S., Sun, J.: Deep residual learning for image recognition. In: Proceedings of the IEEE Conference on Computer Vision and Pattern Recognition. IEEE, pp. 770–778 (2016)
28. Simonyan, K., Zisserman, A.: Very Deep Convolutional Networks for Large-Scale Image Recognition. arXiv:1409.1556 (2014)
29. Szegedy, C., Ioffe, S., Vanhoucke, V., Alemi, A.: Inception-v4, inception-resnet and the impact of residual connections on learning. In: Proceedings of the AAAI Conference on Artificial Intelligence, vol. 31 (No. 1). IEEE, (2017)

Chapter 34
Analysis and Evaluation of Machine Learning Classifiers for IoT Attack Dataset

H. Jagruthi and C. Kavitha

Abstract IoT connects millions of smart devices through internet. IoT is vulnerable to cyber-attacks because of its constrained/controlled environment. To provide security for IoT devices, in this paper experiments are conducted and evaluated to check various machine learning classifiers like SVM with linear, radial basis function, polynomial function, decision tree, logistic regression on the UNSWNB-15 dataset to solve security issues. The dataset was computed for various metrics to analyze the various classifiers. The conducted experiments gave an insight that the decision tree classifier and random forest classifier achieved a good accuracy rate.

34.1 Introduction

Internet of Things (IoT) is a connection of smart devices for actuating, analyzing, controlling, and monitoring physical devices (environments). Due to the growth in IoT, many devices are connected through sensors, actuators, and various software to fulfill functionality. Some of the examples of IoT are smart homes, smart agriculture, and smart cities to automate daily tasks with the goal of easy life and becoming more efficient and effective.

IoT is growing rapidly, and the number of smart devices connected via internet is increasing day by day. The major concern is IoT security and protecting the data from various attacks. Here the attacker can broadcast messages and inject false data into the network. IoT networks compromise the node attacks; compromising may result in false information, and the whole may lead to failure in data protection. There are many different types of attacks that are intimidating the confidentiality, availability,

H. Jagruthi (✉)
Department of Information Science and Engineering, BNM Institute of Technology, Bengaluru, India

C. Kavitha
Department of Computer Science and Engineering, Dayanand Sagar Academy of Technology & Management, Udayapura, India

J. I.-Z. Chen et al. (eds.), *Machine Learning and Autonomous Systems*, Smart Innovation, Systems and Technologies 269, https://doi.org/10.1007/978-981-16-7996-4_34

and integrity of the network environment. One of the common dangerous attacks can be a denial of service attack. The main aim of DoS attack is to deny multiple services of users.

34.2 Literature Review

Intrusion detection is a tool used for network attack detection. An attack is a major threat to the abrupt information security over internet. In this regard the intrusion detection system plays an important role in identifying attacks and improving confidentiality, integrity, and security by notifying attacks to users and customers. For a long time, many machine learning techniques are used for the intrusion detection system. There are many widely and publicly available datasets like NSL-KDD, KDD 99, and DARPA [1]. The first dataset was KDD 99 and it was later improved as DARPA-98 in the year 1999. The NSL-KDD was an improved version of KDD 99 which removes many duplicate and redundant records.

Several studies have been carried out on intrusion detection systems and many anomaly detection techniques are proposed. Bhavsar et al. [2] used the NSL-KDD dataset and support vector machine classifier with tenfold cross-validation and achieved 98.57% accuracy. Mohiuddin et al. [3] reviewed various datasets and intrusion detection methods from 2009 to 2014 and compared the performances of various attack detection with clustering and statistical methods. Karan Bajaj et al. [4] used naïve Bayes, NB tree, support vector machine, and information gain model relying on j48 which has improved the binary class classification time complexity using feature selection method. Ingre et al. [5] used the NSL-KDD dataset and artificial neural network method LM and BFG and tested for both binary class and multi-classification that has achieved 81.2% and 79.9%. Basheer et al. [6] used KDD 99 and NSL-KDD with the SOM method for binary class classification and achieved 75.5% accuracy. Tsai et al. [7] have done a survey on support vector machine classifier, Bayesian classifier, genetic algorithms, decision tree, neural networks for intrusion detection system based on the survey and it was known that support vector machine and k-nearest neighbor was used more commonly. Elbasiony et al. [8] have given random forest and k-means hybrid algorithms and they found to have a good detection rate of 98.3%. Kim et al. [9] have proposed support vector machine and decision trees helpful for attack classification of the IDS model. Tang et al. [10] have proposed SVM with nonlinear scaling for binary class classification with 82.2% and multiclass classification with 79.6% using the NSL-KDD dataset.

In this paper different classifiers like SVM with linear, polynomial, RBF, decision tree classifier is proposed and compared with all proposed classifiers to better know about the method.

34.3 Motivational Objectives

Although many works are carried out on different datasets like KDD, NSL-KDD, KDD CUP-99, few works are carried on the UNSWNB-15 dataset. This paper focuses on the UNSWNB-15 dataset by applying linear SVM, RBF, polynomial, logistic regression, random forest, and decision tree classifier to analyze which machine learning classifier suits best the UNSWNB-15 dataset. To analyze these classifiers with the UNSWNB-15 dataset, there are very few or no research works available. This paper gives a complete analysis and evaluation of all linear SVM, RBF, polynomial, logistic regression, random forest, and decision tree classifiers at once.

34.4 Methods

34.4.1 UNSWNB-15 Dataset Description

The UNSWNB-15 dataset contains modern networks and various attacks behavior. The attacks are classified into different classes such as normal record class and nine different attack records such as Exploits, Dos, Backdoor, Fuzzes, Analysis, Generic, shellcode, Worms, and Reconnaissance. Different types of attacks and their details are shown in Table 34.1.

This UNSWNB-15 dataset has been created by IXIA perfect storm tool in the cyber range lab of ACCS Australian Centre of cyber security to generate modern and normal attack behaviors. The dataset contains in total 2,27,673 data, which are randomly divided into 1,75,341 training data and 82,332 test data. The data in training and testing does not contain any repetitions to ensure reliability in evaluating

Table 34.1 Different categories of UNSWNB-15 dataset

S. no.	Attack_type	Description of the attack
1	Normal	Normal transaction of data without errors
2	Analysis	An attack that occupies network through emails, ports, etc.
3	Backdoor	Stealing of personal and critical data by using unauthorized identity to hijack device security
4	DoS	Making system to shutdown abruptly and not giving access to intended users
5	Exploits	Attack on computer system to make network vulnerable
6	Fuzzers	Attack that exploits the system and make to cause unexpected behavior
7	Generic	An attack that works against cryptography
8	Reconnaissance	An attack to penetrate the network to get authorized data
9	Shellcode	A subset of code used to exploit software
10	Worm	A virus which will add itself to programs

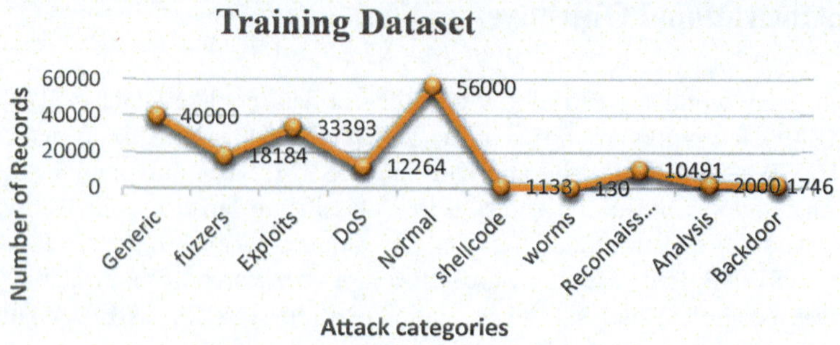

Fig. 34.1 Training dataset samples

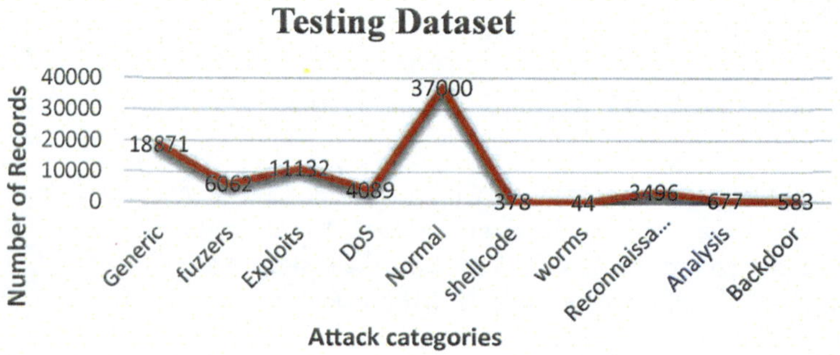

Fig. 34.2 Testing dataset samples

network intrusion detection system. The training and testing record contains the same attacks. Figure 34.1 shows training dataset samples which include 1,75,341 records divided among ten different attack categories, including normal records and Fig. 34.2 shows the testing dataset samples which include 82,332 records divided among nine different attack categories including normal records.

The dataset contains 44 features as listed in Table 34.2. The entire dataset is divided as (1–5) flow features, (19–26) content features, (27–35) time features, (6–18) basic features; 36–40 are considered as general-purpose features and 41–47 are mentioned as connection features, and (48–49) are called as labeled features.

34.4.2 An Overview of Machine Learning Classifiers

34.4.2.1 Support Vector Machine Model with Linear Scaling

For high-dimensional data, SVM is used as one of the classifiers for the intrusion detection system. In this work, SVM is proposed for intrusion detection. Here we use

Table 34.2 UNSWNB-15 dataset features

No.	Name of features	No.	Name of features
I	Id	XXIII	Dtcpb
II	Duration	XXIV	D-win
III	Protocol	XXV	Tcp-rtt
IV	Rate	XXVI	Sy-nack
V	State	XXVII	Acknowlede data
VI	Source pkts	XXVIII	S-mean
VII	Destination pkts	XXIX	D-mean
VIII	Source bytes	XXX	Trsna-dept
IX	Destination bytes	XXXI	Response-body
X	Service	XXXII	Ct-srv-src
XI	S-ttl	XXXIII	Ct-state-ttl
XII	D-ttl	XXXIV	Ct-dst-ltm
XIII	S-load	XXXV	Ct-src-dport-ltm
XIV	D-load	XXXVI	Ct-dst-sport-ltm
XV	S-loss	XXXVII	Ct-dst-src-ltm
XVI	D-loss	XXXVIII	İs-ftp-login
XVII	Sink pkt	XXXIX	Ct-ftp-cmd
XVIII	Din pkt	XXXX	Ct-flw-http-mthd
XIX	S-jit	XXXXI	Ct-src-ltm
XX	D-jit	XXXXII	Ct_srv_dst
XXI	S win	XXXXIII	İs-sm-ips-ports
XXII	S tcpb	XXXXIX	Attack_category

a linear and nonlinear scaling method for data pre-processing and the UNSWNB-15 dataset. Figure 34.3 shows the IDS using a support vector machine. There are several parameters available for evaluating different dimensions of a dataset, which in turn

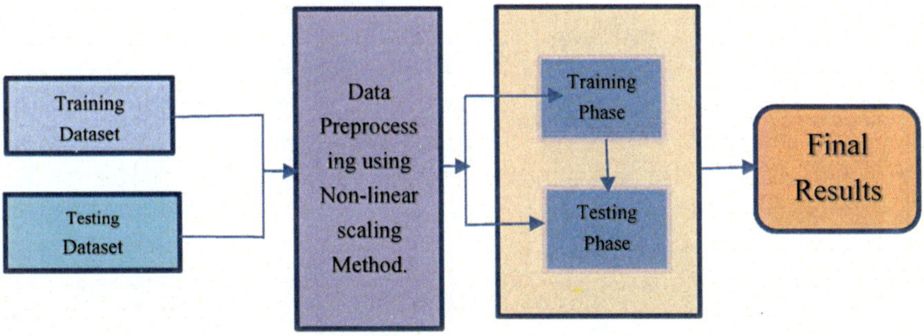

Fig. 34.3 IDS using proposed support vector machine method

affect the data classification, so the pre-processing state is very much necessary to eradicate this effect to data classification results.

There are different kernel functions available for SVM, like linear, RBF, sigmoid, polynomial, Gaussian, etc.

1. Linear Kernel

Linear kernel can be called dot product, and it can be written as

$$\text{K (X, Xi)} = \text{sum (X} * \text{Xi)} \tag{34.1}$$

It is the distance between the linear combination of inputs. Linear kernel tells the distance between data and support vectors.

2. Polynomial Kernel

It is the transformation of input to high dimension, and it can be written as

$$\text{K (X, Xi)} = 1 + \text{sum (X} * \text{Xi)} \tag{34.2}$$

If the degree is set to 1 then it is considered the same as a linear kernel. The polynomial kernel allows curve lines in input space.

3. Radial Kernel SVM

This is also the transformation of input to high dimension, and it can be written as

$$\text{K(X, Xi)} = \exp\left(-\text{gamma} * \text{sum}\left(\text{X, XI}^2\right)\right) \tag{34.3}$$

Here gamma is a parameter that must be used for learning the algorithm. The default value of gamma is 0.1 and gamma is often between $0 < \text{gamma} < 1$.

34.4.2.2 Decision Tree Classifier

Decision is the classifier used to develop a decision tree, which helps to recognize the prediction output class. Many different search algorithms are available, for example, genetic algorithm, cross-validation, and breadth-first search can be used to bring the decision table efficiency [11]. The generated decision tree consists of given conditions and the expected outputs with respect to predefined conditions. In the other way decision tree classifier gives the set of rules that helps in identifying the new set of inputs [12]. The lookup table of the decision tree classifier can be used to give the rules for the system which is complex.

34.4.2.3 Random Forest Classifier

Random forest classifier is a type of classification tree algorithm. The main objective of the algorithm is to intensify the trees classifiers based on the forest. Random forest classifier's results given by research [13] have a welcoming accuracy rate and RF also lifts up the loss in data values. İn order to implement this the classifier should first decide the trees within the forest since each tree inside the forest foresees the expected output for which voting will be calculated, and this technique is used to choose the expected outputs which have the highest votes [13].

34.4.2.4 Logistic Regression

Logistic regression is a machine learning classifier that uses a statistical model and the sigmoid function. It is just similar to linear regression. Logistic regression uses Eq. 34.4. The equation consists of y predicted value, b an error value, and b1 is the coefficient value or a real value of x.

$$y \frac{\hat{e}(b0 + b1 * X)}{(1 + \hat{e}(b0 + b1 * X)} \tag{34.4}$$

Here the input values are merged with beta values or coefficient values to give an output value y. The major dissimilarity of linear regression is the output class is predicted as binary values 0 and 1 instead of numerical values.

34.5 Experimental Setup

The dataset used for research is UNSWNB-15 dataset which has new categories of normal and attack classes. The dataset consists of nine different attack classes, which also include normal attack classes. İn this work, a confusion matrix is generated for all machine learning methods. The confusion matrix includes information about existing and predicted classes. The below metrics are used for evaluation.

34.5.1 Performance Metrics

There are performance metrics available to evaluate the performance of intrusion detection like accuracy, F1 score, false alarm rate (FAR), false positive rate (FPR), false negative (FN), precision, and recall.

True positive (TP): It gives the right classification of attack classes as attacks.

True negative (TN): It gives the right classification of normal classes as normal.

False negative rate (FN): Attacks will be categorized as normal packets if more values of false negative repeats. It represents a problem of availability and confidentiality of network data.

$$False\ negative\ rate = \frac{FN}{FN + TP} \tag{34.5}$$

False positive rate (FP): It gives the negative classification in which a normal dataset will be classified as an attack and if the FP value gets higher then the evaluation time also increases.

$$False\ positive\ rate = \frac{FP}{FP + TN} \tag{34.6}$$

Precision: The final number of values are commonly classified as attack divided by a total number of values classified as an attack.

$$Precision = \frac{TP}{TP + FP} \tag{34.7}$$

Accuracy: It gives the near measurement value of the correct value.

$$Accuracy = \frac{TP + TN}{TP + TN + FP + FN} \tag{34.8}$$

In addition to the above standard metrics, correct prediction and incorrect prediction records are also noted along with the time required for the training dataset. The SVM with linear kernel function is checked for binary class classification and the testing accuracy is promising when compared to other classifiers. The SVM with polynomial kernel function was tested for various degrees, like 1, 2, and 3, and the testing accuracy was improving accordingly with changing degrees. The SVM with RBF kernel function was tested with varying gamma values from 0.5 to 0.8. Various readings of testing accuracy were noted. RBF kernel function works for both binary and multiclass classification and the dataset was also checked for multiclass and the attack classification report was also generated for nine different attack classes and the accuracy was 78%. The random tree classifier was tested for 100 trees. A decision table classifier was tested based on minimum leaf node 1 and the split node is 2. The logistic regression classifier was tested for c = 1 and the scaling value is 1. The random forest classifier, logistic regression, and decision tree classifier are experimented with fivefold cross-validation techniques to achieve better accuracy rates. Table 34.3 shows the test precision and cross-validation fit time for different machine learning classifiers.

From Table 34.3, it can be observed that the decision tree classifier has achieved the highest accuracy of 100%, while the logistic regression has got the least precision of 74.93%. Confusion matrix was generated for all machine learning classifiers proposed for the evaluation model. Figure 34.4 shows the confusion matrix gener-

Table 34.3 Test precision and CV fit time

Algorithms	Test accuracy	CV fit time
SVM with linear kernel function	97.58	29.48
SVM with polynomial kernel function	76.54	45.67
SVM with RBF kernel function	77.90	44.29
Decision tree	**100**	**2.34**
Random forest	**99.86**	28.30
Logistic regression	74.93	43.38

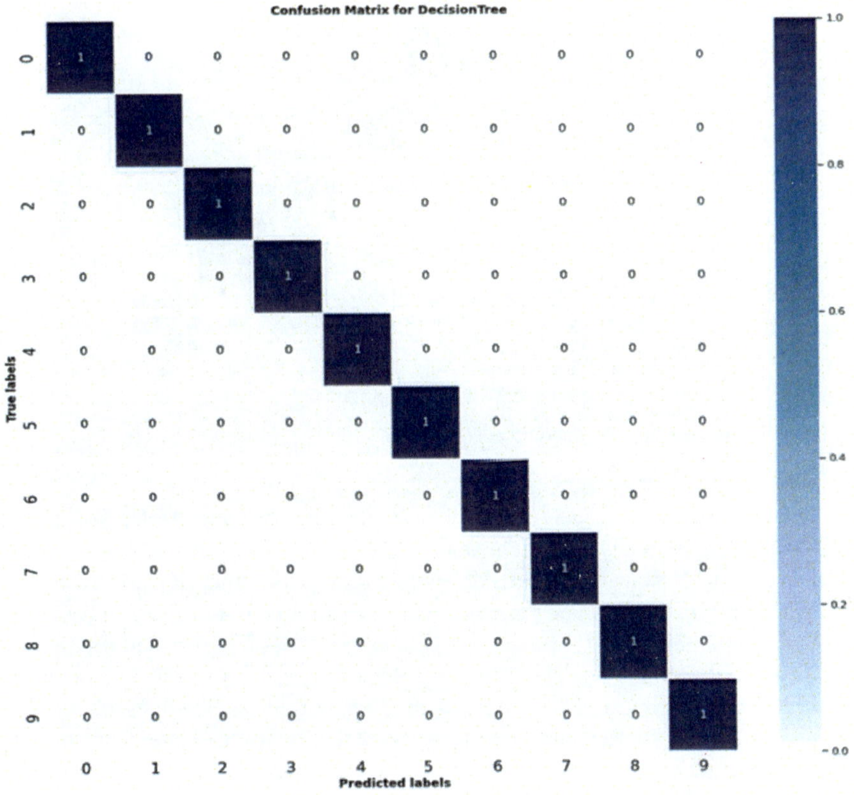

Fig. 34.4 Confusion matrix for decision tree

ated for the decision tree, which shows the predicted labels and true positive and true negative rates.

Figure 34.5 shows the confusion matrix generated for the random forest classifier which shows the predicted labels and true positive and true negative rates.

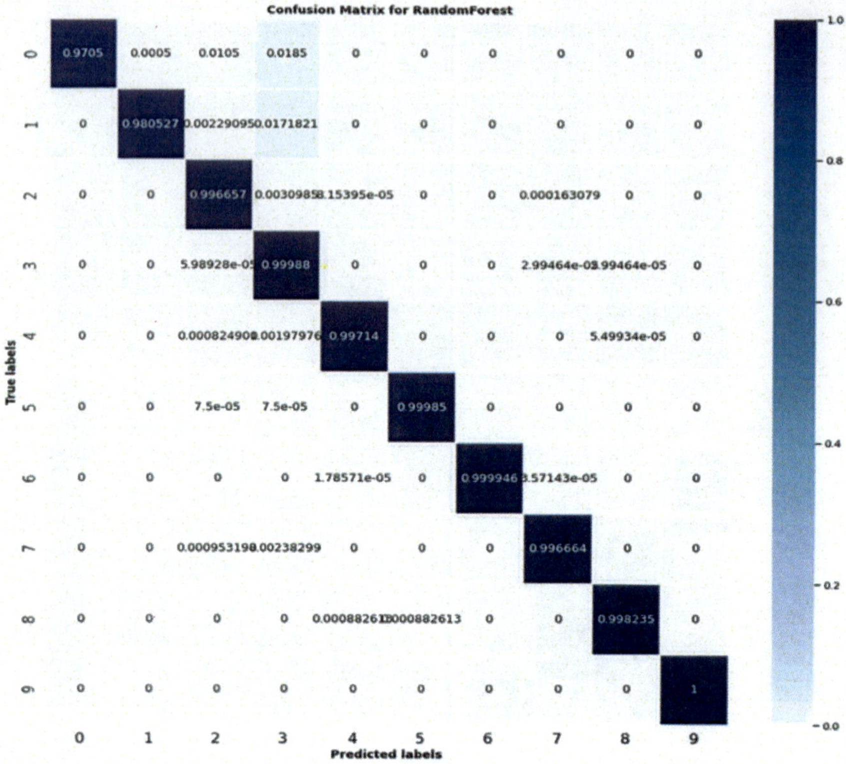

Fig. 34.5 Confusion matrix of random forest classifier

The SVM with linear kernel function gives the testing accuracy of 99.98% and the cross-validation fit time is 29.48. The major drawback here is it can be applied only for binary class classification. The SVM with radial basis kernel function gives the testing accuracy of 77.90% and the cross-validation fit time is 44.29. The advantage here is that the radial basis function can be applied for both binary and multiclass classification. Random forest classifier gives the testing accuracy of 99.86% and the cross-validation fit time is 28.30. All machine learning classifiers present acceptable precision rates for detecting normal packets. Decision tree classifier gained a maximum accuracy rate of 100%, testing accuracy, and the cross-validation fit time is 2.34. It can be assumed that a group of classifiers like the decision tree can present an acceptable accuracy rate with the lowest FN rate, which also increases the confidentiality and the availability of the network resources. Table 34.4 shows the comparison accuracy rate of different classifiers with respect to a different dataset.

Table 34.4 Comparison accuracy of different machine learning methods with different datasets

Author	Dataset	Method	Accuracy
Almseidin et al. [14]	KDD	J48	93.1
		Random forest	93.8
		Random tree	90.6
		Decision table	92.4
		MLP	91.9
		Naïve bayes	91.2
		Bayes network	90.7
Bhavani et al. [15]	NSL-KDD	Random forest	95
		Decision tree	81
Maniriho et al. [16]	NSL-KDD	KNN	98.7
		RC	98.6
	UNSW-NB15	KNN	97.3
		RC	98.9
Ponthapalli et al. [17]	KDDNSL	Random forest	73
		Decision tree	72
		Logistic regression	71
		Support vector machine-	68
Proposed methodology	**UNSW-NB15**	**Linear SVM**	**97.58**
		RBF	**76.54**
		Polynomial	**77.90**
		Random forest	**99.86**
		Decision tree	**100**
		Logistic regression	**74.93**

From the above table, it can be observed that very few works are carried out on the UNSWNB-15 dataset and in the proposed method UNSWNB-15 dataset has been analyzed and evaluated for different classifiers to get to know which machine learning classifier suits best for the UNSWNB-15 dataset and the results have proved that better intrusion detection accuracy has been seen with decision tree and random forest classifier.

34.6 Conclusion

Security is the major concern in today's environments since data are precious and should not be given access to unauthorized users which leads to the effect of data confidentiality, integrity, and availability. So here several experiments are carried out on various machine learning classifiers like SVM with linear, polynomial, RBF function, random forest classifier, logistic regression, multi-layer perceptron, and decision tree classifier. It is found that less works are carried with the UNSWNB-15 dataset, so all the tests are done for the UNSWNB-15 dataset. The experiments are conducted to monitor all the intrusions and to get to know no false activity and deviating from normal behaviors. Finally, the experiments show that decision tree classifier and random forest classifier are giving the best detection accuracy in finding the attack categories and best fit time compared to other classifiers.

References

1. Tavallaee, M., Bagheri, E., Lu, W., Ghorbani, A.A.: A detailed analysis of the KDD CUP 99 data set. In: IEEE International Conference on Computational Intelligence for Security and Defense Applications. IEEE, pp. 53–58 (2009)
2. Bhavsar, Y.B., Waghmare, K.C.: Intrusion detection system using data mining technique: support vector machine. Int. J. Emerg. Technol. Adv. Eng. **3** (2013)
3. Ahmed, M., Mahmood, A.N., Hu, J.: A survey of network anomaly detection techniques. J. Netw. Comput. Appl. **60**, 19–31 (2016)
4. Bajaj, K., Arora, A.: Improving the intrusion detection using discriminative machine learning approach and improve the time complexity by data mining feature selection methods. Int. J. Comput. Appl. **76**(1), 5–11 (2013)
5. Ingre, B., Yadav, A.: Performance analysis of NSL-KDD dataset using ANN. In: 2015 International Conference on Signal Processing and Communication Engineering Systems. IEEE, pp. 92–96 (2015)
6. Basheer, D.T., Ibrahim, L.M., Mahamod, M.S.: A comparison study for intrusion database (KDD99, NSL-KDD) based on self-organization map (SOM) artificial neural network. J. Eng. Sci. Technol. **8**(1), 107–119 (2013)
7. Tsai, C.-F., Hsu, Y.-F., Lin, C.Y., Lin, W.-Y.: Intrusion detection by machine learning: a review. Expert Syst. Appl. **36**(10), 11994–12000 (2009)
8. Elbasiony, R.M., Sallam, E.A., Eltobely, T.E.: A hybrid network intrusion detection framework based on random forests and weighted k-means. Ain Shams Eng. J. **4**(4), 753–762 (2013)
9. Kim, G., Lee, S., Kim, S.: A novel hybrid intrusion detection method integrating anomaly detection with misuse detection. Expert Syst. Appl. **41**(4), 1690–1700 (2014)
10. Tang, X., Tan, S.X.D., Chen, H.-B.: SVM Based Intrusion Detection Using Nonlinear Scaling Scheme (No. 978-1-5386-4441-6/18/$31.00). IEEE (2018)
11. Kohavi, R., Sommerfield, D.: Targeting business users with decision table classifiers. In: KDD, pp. 249–253 (1998)
12. Aditi, P., Hitesh, G.: A New Approach of Intrusion Detection System Using Clustering, Classification and Decision Table (2013)
13. Breiman, L.: Random forests. Mach. Learn. **45**(1), 5–32 (2001)
14. Almseidin, M., Alzubi, M., Kovacs, S., Alkasassbeh, M.: Evaluation of machine learning algorithms for intrusion detection system. In: Abnormality in Computer Networks. IEEE, (2018)
15. Bhavani, T.T, Kameswara, M.R., Manohar A.R.: Network intrusion detection system using random forest and decision tree machine learning techniques. In: International Conference on Sustainable Technologies for Computational Intelligence (ICSTCI). Springer, Berlin, pp. 637–643 (2020)
16. Maniriho, et al.: Detecting intrusions in computer network traffic with machine learning approaches. Int. J. Intell. Eng. Syst. (INASS), 433–445 (2020)
17. Ponthapalli, R., et al.: Implementation of machine learning algorithms for detection of network intrusion. Int. J. Comput. Sci. Trends Technol. (IJCST), 163–169 (2020)

Chapter 35
Analysis of Unsupervised Machine Learning Techniques for Customer Segmentation

Anant Katyayan, Anuja Bokhare, Rajat Gupta, Sushmita Kumari, and Twinkle Pardeshi

Abstract Customer relationship management (CRM) has played an important part in marketing tools for giving companies the model business intelligence which helps in raising, developing, and managing valuable long-term customer relationships with companies. Most organizations have understood the acceptance of customer relationship management and the application of computational capability to achieve cut-throat market aid. The current study expands the role of customer segmentation as a function of customer relationship management as well as the different framework for customer segmentation with the help of machine learning unsupervised techniques like clustering. The existing clustering methods of customer segmentation, and the vital techniques of K-means and hierarchical clustering, are considered in the current study. The righteousness and wickedness of the methods are pointed out during the study. Furthermore, as per the data collected a comparative analysis is done between hierarchical and K-means, and the results obtained from both the models were discussed. The difference arises when discussing the accuracy in resultant outcomes from the unsupervised learning technique.

A. Katyayan · A. Bokhare (✉) · R. Gupta · S. Kumari · T. Pardeshi
Symbiosis Institute of Computer Studies and Research (SICSR), Symbiosis International (Deemed University), Pune 411016, Maharashtra, India

A. Katyayan
e-mail: ank2041005@sicsr.ac.in

R. Gupta
e-mail: rrg2041039@sicsr.ac.in

S. Kumari
e-mail: spk2041057@sicsr.ac.in

T. Pardeshi
e-mail: tsp2041059@sicsr.ac.in

J. I.-Z. Chen et al. (eds.), *Machine Learning and Autonomous Systems*, Smart Innovation, Systems and Technologies 269, https://doi.org/10.1007/978-981-16-7996-4_35

35.1 Introduction

Customer segmentation is basically the technique of dividing a large consumer or market, containing an existing and capable consumer, into a group of consumers based on some type of common behavior. Many organizations retaining customer segmentation work under the impression that all customers are different and promotional effort would be well served. Consumers would find it relevant to buy what they need if organizations target particular and smaller groups. Organizations also gain a deep knowledge of the customers' choices and needs with the idea of discovering what each group finds more valuable to tail end or marketing materials toward that segment. Malls or shopping complexes are often moderate in the competition to increase their customers and also in making huge profits from it. To carry off this task machine learning (ML) techniques are being applied by many organizations already. The organization or shopping stores make use of their customers' data and develop ML models to target the right customers. This not only increases the sales but also makes the complex competitive.

Clustering is an unsupervised method in machine learning, which aids to analyze customers based on their key behavior. This study illustrates the use of two unsupervised machine learning techniques, i.e., the clustering methods: K-means clustering and hierarchical clustering for the identification and segmentation of customers. The results obtained from both techniques are analyzed for their accuracy of implementation in customer segmentation.

35.2 Previous Study

This unit defines the review in relation to the context available in the literature. Sanjana [1] has introduced various bunching calculations considering the highlights of big data like size, commotion, dimensionality, calculation computations, group shape and introduced a short outline of the few bunching calculations assembled under apportioning, progressive, thickness, network, and framework-based calculations. Azarnoush [2] has combined clustering of unclear c-means and genetic method so as to cluster steel organization consumers by implementing the LRFM variable (monetary value, recency, regularity, and length) system. Consumers are thus distributed into two clusters. Pedro [3] another researcher took a different method to DM (data mining) for customer segmentation clustering and innovation. The framework has acquired six segments of market and follows 49 rules in a highly modified way. The fashion producer gives a well understanding of customer likings, thus the difficulty of clustering method and their restrictions facilitate procedural refinement. Shreya [4] another researcher discovered what requirements are required for segmentation of the consumer using clustering method for working CRM. The frequently used KM & HCA were studied and the merit and demerit of these methods were mentioned. Coming to an end, the idea of generating a hybrid method was addressed by mixing the above two methods with a possibility to exceed the separate designs. According to Dr. Manju Kaushik [5] clustering is a procedure by which we can keep similar data into collections. As with every other problem of this kind, clustering too is an unsupervised learning method that deals with uncovering a structure in a company of untagged data. This includes numerous kinds of clustering approaches such as hierarchical, partitioning, density created, model-based, grid-based, and soft computing approaches. In this paper, a comparative study is done which focuses not only on strength but also on the weakness of both the clustering techniques, i.e., K-means clustering and hierarchical clustering and their methodology and processes too. According to the author [6], by stating the count of business enterprises, we can come into view of CRM and the competitive advantage achieved from the expert with a technical background. It identifies the significance of the customers based on different parameters, i.e., demographics, income, age, etc., which needs to be targeted and thus is known as customer segmentation. In one of the business canvases models play a main role of CRM along with many other models that are used in the clustering method which are applied on the sampling method and it is heterogeneous in nature. In the end, the opportunity to develop a hybrid solution that has a mixture of two methods and has the capability to outperform finite models is concluded. Data clustering is meant to be the process of initializing groups of homogeneous. A clustering algorithm divides the dataset into different groups in a structured manner and is also having similarities within a group. The study analysis of six types of clustering techniques, i.e., KM clustering, HCA, DBS cluster, density-based clustering, optics, and EM technique is performed with the help of WEKA, a tool that we use for clustering where we can compare and present all the six techniques, which helps us in implementation and analysis of the groups [7].

In this study researchers [8] define that clustering division can also be taken as a dataset, thus the observations are placed together in similar groups but different from other groups. K-means and hierarchical clustering are two ways having distinct advantages and disadvantages. Suppose HCA recognizes groups in an exceeding dendrogram but differs from computational difficulty in large data, KM clustering is well organized but designed to remember spherically shaped homogeneous clusters. It confers a non-parametric hybrid clustering method that merges the two techniques to spot normal-shaped clusters and may be implemented to larger data. Therefore, in the first part of the data K-means spherical groups are used. After that, merging of these groups using HCA technique is done with information obsessed to measure distance as a stopping criterion. The study proposes the prospect of disclosing segments with normal shapes and framework in a very set of data. While going through this case the author of [9] has identified a way to accelerate unsupervised learning method with hierarchical and KM showing HCA and KM speed up mean-based learning technique in both the training and query phases at less cost to check accuracy. Because of this more number of centroids can be used, which resulted in more better learning. For the microarray data analysis using clustering methods such as K-means and hierarchical clustering are advantageous today [10]. However, each of these clustering techniques has its own restrictions. In this thesis, the authors have introduced a new technique called hierarchical KM regulating divisive. The hierarchical KM initially implements the KM method in every cluster to regulate it while working and then it implements a hierarchical clustering technique to truncate merge clusters while creating a tree-like structure (dendrogram). The technique is applied in two array datasets. The output shows disruptive hierarchical KM is better than hierarchical clustering on cluster quality and K-means clustering on speed.

The study focuses on the clustering of citation contexts in scientific papers [11]. Two methods, KM and HCA, are accustomed to better understand the phenomenon and kinds of citations and to explore the multidimensional nature of the weather composing the contexts of citations in several sections of the papers. An information set of seven peer-reviewed academic journals published has been analyzed by PLOS. The obtained clusters show that the Methods section is restricted in nature, no matter the journal. A proximity between several of the journals may be observed [12].

In clustering, items of the dataset are grouped into clusters, such a way that each collection is very dissimilar from each other and the objects in the same group are very similar to each other. Analysis of K-means and hierarchical is achieved. The presentation of these two clustering algorithms is associated with the clustering toolkit WEKA, which is a platform-independent open-source toolkit. A presentation has been planned for customer segmentation. Data on many customers were analyzed and later handled in such a way that model can be trained which would help to allocate customers into three dissimilar customer groups. Two classification models are used during the study, i.e., the logistic regression model and a support vector classifier which have been qualified to put users into dissimilar categories as per user personas. Later all the testing of the outcomes show that the logistic regression model had an accuracy of 71% [13]. The method of ant colony was used to accomplish customer segmentation. In the ant colony algorithm, clustering is done by the Euclidean distance procedure. The contracts customer segmentation is more related to providing the compensation factor to the customer [14]. Based on the analysis of purchase behaviors Recency Frequency Monetary (RFM) is the most general method for customer segmentation. All customers are scored and ranked according to each dimension individually and the similar purchase pattern of customer groups was recognized. Recency is the duration since the date of last purchase, Frequency is the regularity of the total count of purchases, and Monetary is the average amount purchased. In data mining, the customer segmentation parallels the clustering task where the objective is nothing but to define the groups of customers with related purchase patterns according to purchase behaviors. The planned methodology helps to measure different RFM variations with individual scoring and splitting strategies and highlights enhanced outcomes [15].

An eCommerce website is established for farmers where they can sign up and make a buying for seeds or equipment as per their wants. Based on data segmentation was achieved on the same. The type of learning used here is unsupervised learning. The algorithm used is K-means which helps us to run the eCommerce website in a very efficient manner, providing a 94.44% accuracy rate [16]. Direct showcasing is finished by the bank to build deals of its items and administrations. Authors associated have compared several data mining approaches like NB, K-NN, RF, SVM, J48, AB J48 which before classified the SMOTE prepuce method is also done. The result describes that the SMOTE method with Random Forest acquired almost the same result practical to the information provided by the bank. The SMOTE and RF methods in this research give the maximum correctness level of 92.61% [17]. The telco organization thinks that bringing new clients is no longer a good method; instead, retaining the existing clients is more cost-effective. The study proposes a combined client analytics architecture for churn management that has been involved in the industry. It gives an integrated customer analytics architecture. As per the data collected from Dataset 1 and using AdaBoost, the framework achieves a higher F1-score at 63.11% and the AUC value at 84.52%. For the second data, we have used the Random Forest method and the defined framework gives F1-score and AUC value as 77.20% and 91.40, respectively [18]. A passive computing system is a heterogeneous communication technology DECD design that enhances the robustness of the system. The RRL process computes the request and enhances the flow of request; the experiment observation demonstrates design against the existing system [19]. Computer detection systems identify diagnoses of cancer cells and the proposed algorithm is implemented for segmentation of early detection of cancer modules. It focuses on improvement in the state of the process and accuracy level [20]. The wastewater from processing liquid cannot be used without proper treatment. This paper talks about fuzzy logic Acrator for the wastewater. The results show there is improvement in the performance of FLACS in terms of electrical energy uses and duration of working hours [21]. In this paper K-means technique is used for customer segmentation and provides key strategies based on associative customer behavior as cluster obtained [22].

Table 35.1 describes the comparison between K-means and hierarchical clustering.

35.3 Problem Statement and Methodology

An organization or a store, which maintains the details of its customers through a campaign membership, is considered during the study. It has information about customers, along with customer age, customer annual income, customer gender, and their spending score. The spending score is based on past purchasing or habits of buying from the stores or organization. Now, the store or organization is introducing a deluxe product and wants to reach out to capable consumers who can purchase it. Going through each customer will be a time-consuming and costly process. Hence, by using customer information, the store can differentiate consumers who have the capability to purchase a deluxe product. This issue can be solved using clustering, in which customers can be placed into distinct segments and further recognized capable customers. So, on two dimensions the customer will represent the X-axis which shows the annual income of customers, and on Y-axis it represents the spending score of customers. Once plotting all customers on the graph, then the clustering technique is used to form the segments of customers based on their income and spending score, and analyze the group of customers who have the capability to buy a deluxe product.

Table 35.1 Comparison between K-means and hierarchical clustering

	K-means clustering	Hierarchical clustering	References
Definition	K-means clustering gives a specific number of disjoint, flat, i.e., non-hierarchical clusters	Hierarchical clustering methods build a hierarchy of clustering, not just a single division of objects	[1–18]
Clustering criteria	It is used to generate globular clusters	Use a distance matrix as clustering criteria can be used for termination condition	[5]
Sensitive noise	K-means is very sensitive to noise in the dataset	It is less sensitive to noise in the dataset	[5, 12, 17]
Execution time	Increases its time of execution	Its performance is better	[5, 10, 12, 14]
Quality	In K-means algorithms it shows less quality	In hierarchical algorithms it shows more quality	[5]
Data set	It is good for large datasets	It is good for small datasets	[5, 7, 8, 11]
Categories	Partition and centroid based	Agglomerative and hierarchical based	[1, 6, 9, 18]
Method to find optimal cluster number	The elbow technique using WCSS	Dendrogram	[6, 7, 10, 11]
Directional approach	Only centroid	Top-down and bottom-up	[6, 15]

In the current study initially, K-means clustering is used and then hierarchical clustering is analyzed. Finally, the differentiation between these two methods, K-means, and hierarchical clustering is identified.

Clustering: It is basically done for dividing the dataset and converting it into several groups such that in a group data points are the same as other data points in other groups.

K-means: It is an iterative calculation that attempts to segment the dataset or population into non-covering subgroups that is clusters where every information point belongs to one group. K-means attempts to make the data points of intra-group focuses the same.

Hierarchical clustering: Also known as hierarchical cluster analysis, this is an algorithm that groups similar objects of the population into groups called clusters. The result is a set of clusters, where each cluster is distinct from each other cluster, and the objects in inter-cluster are broadly like each other. Figure 35.1 explains the customer segmentation model of the system.

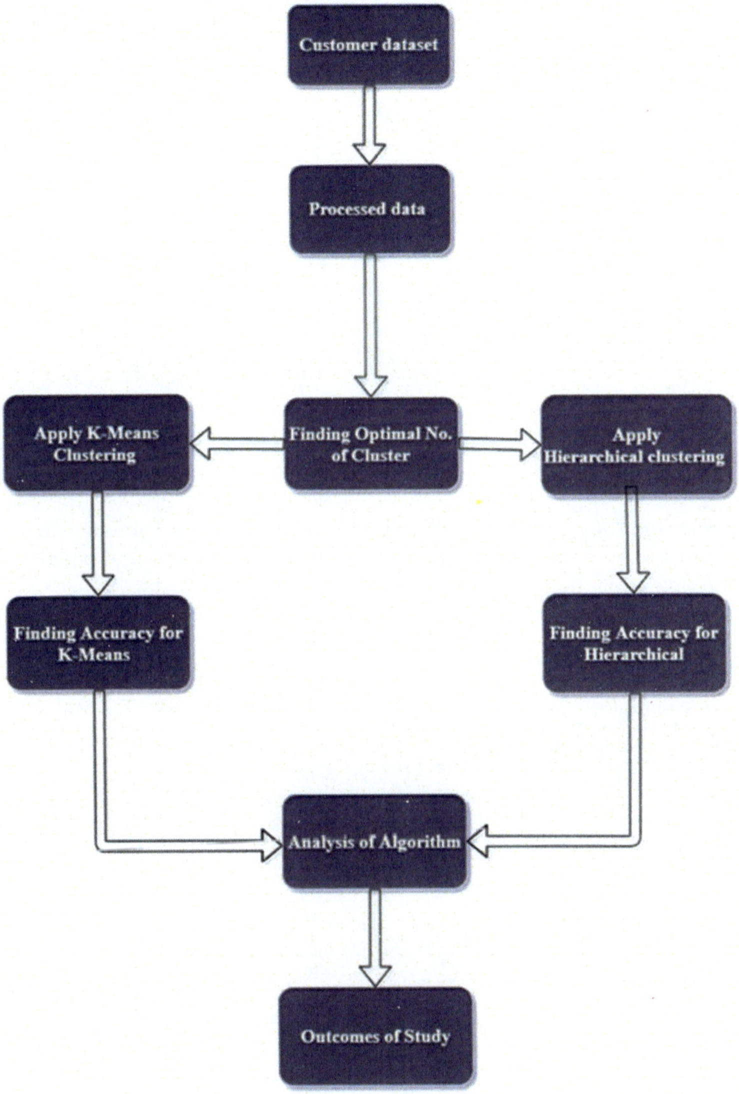

Fig. 35.1 Customer segmentation model using unsupervised ML techniques

35.4 Dataset and Experiment Discussion

Dataset considered for the study is based on the Store Customer dataset. This dataset contains basic information about customers such as id, age, gender, income, and spending score.

Attribute Description:

Customer id: It is the unique ID given to a customer for identifying them uniquely.

Gender: It shows gender of the customer.

Age: Age of the customer.

Annual income (k$): It is basically the annual income of the customer.

Spending score: It is the score, out of 100 given to a customer by the store authorities, based on the purchase done and the behavior of the customer.

35.4.1 Experiment Steps

1. Data preprocessing
   For data preprocessing, preprocessed data from data marts are considered for the experiment. Preprocessed data means all entities are already shortlisted and analyzed. On the basis of visualization, analysis has been performed.
2. Apply K-means

K-means clustering algorithm implementation steps:-
Step 1: Import the basic library to read the CSV file and visualize the data.
Step 2: Read the CSV file.
Step 3: To implement the K-means clustering, an optimal number of clusters is to be found. The Elbow method is used based on with-cluster-sum-square (WCSS) keeping the customer as a placed variable After implementation it will be visualized as a hand-like structure and from that we will identify the elbow location on X-axis by looking, so we will be able to find some approximate number on the X-axis. That approximate number will be the optimal number for the K-means clustering.
Step 4: Once the optimal number of clusters is known, then the prediction of cluster for each data element will be identified.
Step 5: After the algorithm predicts a cluster for each of the data items, we need to visualize the result. For better understanding, each of the clusters is given a unique color and name.

3. Apply Hierarchical Clustering

Hierarchical clustering algorithm implementation steps:-
Step 1: Import the basic library to read the CSV file and visualize the data.
Step 2: Read the CSV file.

(continued)

(continued)

Hierarchical clustering algorithm implementation steps:-
Step 3: To identify the optimal number of clusters in hierarchical clustering. Dendrograms are used for finding optimal clusters After a successful visualization of dendrogram, next we identify the combination of lines that are not joined on the Y-axis. That number of lines will be identified as the optimal number of clusters.
Step 4: After knowing the optimal number of clusters, next our aim is at the hierarchical clustering algorithm and predict the cluster for each data item.
Step 5: Once the predictions of the cluster are done, it can be visualized.

4. Analyze the results.
5. Result outcomes.

35.5 Result Analysis

This section focuses on the result analysis and related explanation of the visualizations, i.e., Figure 35.2 which describes that annual income is less than spending score. And in Fig. 35.3 a comparison is done between males and females on the basis of the number, which depicts that females doing shopping is more than male customers.

Figure 35.4 indicates that the age group which is doing more shopping is between 26–35 years and less are above 55+ years.

Figure 35.5 indicates that most people have spent a score between 41 and 60.

Figure 35.6 indicates that most of the customers have incomes between $60,001 and 90,000.

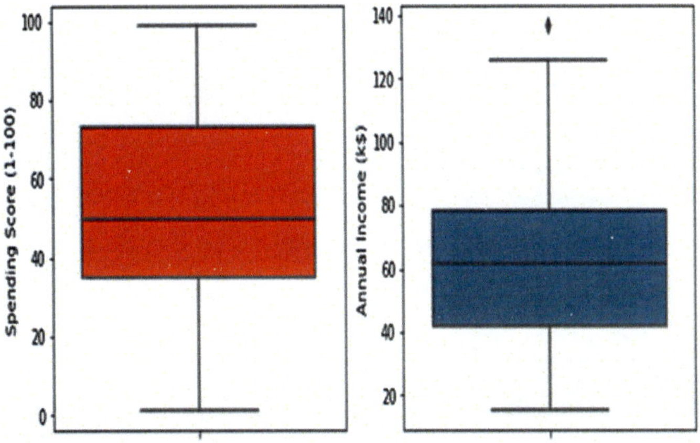

Fig. 35.2 Box plot for spending score and annual income of customers

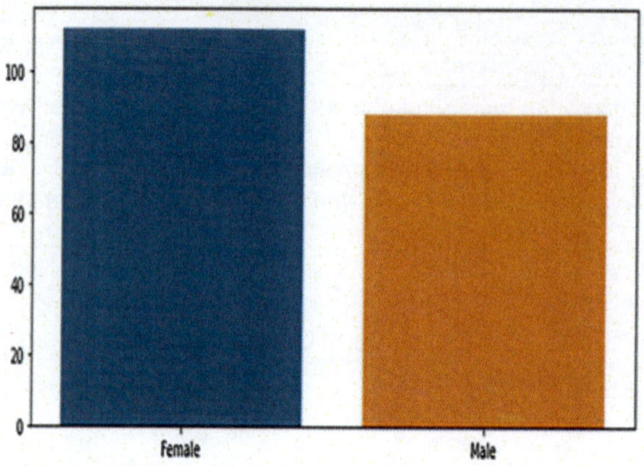

Fig. 35.3 Number of female (0) and male (1) customer

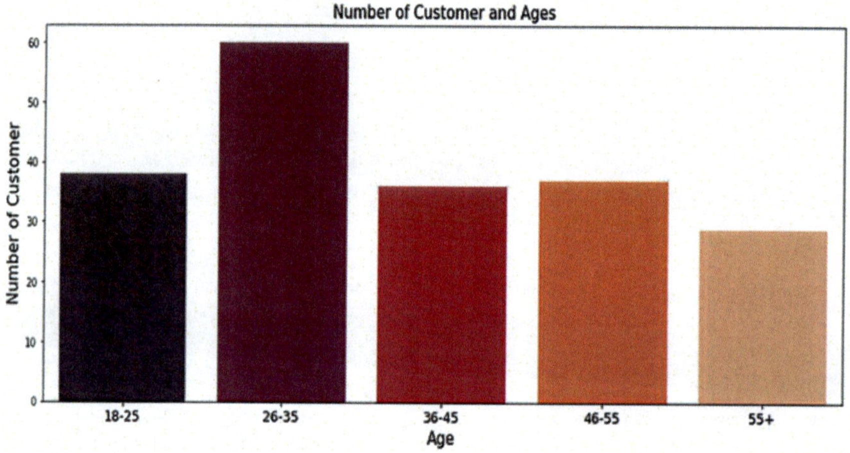

Fig. 35.4 Distribution of the number of customers according to age group

35.5.1 K-Means Clustering Analysis

This section explains the result analysis performed and the interpretations of the same. The initial task in K-means is to decide the value for the cluster. The subsection explains the elbow technique utilized to do the same.

Finding value for cluster

Here first the cluster sum of square error is calculated for all values of k and then selected the k for which within-cluster sum of square error, i.e., WSS starts to decrease. In Fig. 35.7 WCSS versus k is shown. This is visible as an elbow method which shows the value for (k) cluster can be taken as 5.

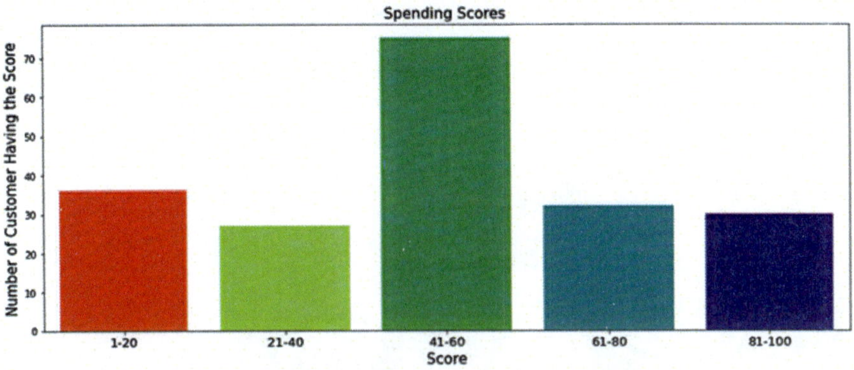

Fig. 35.5 Distribution of number of customers according to their spending score

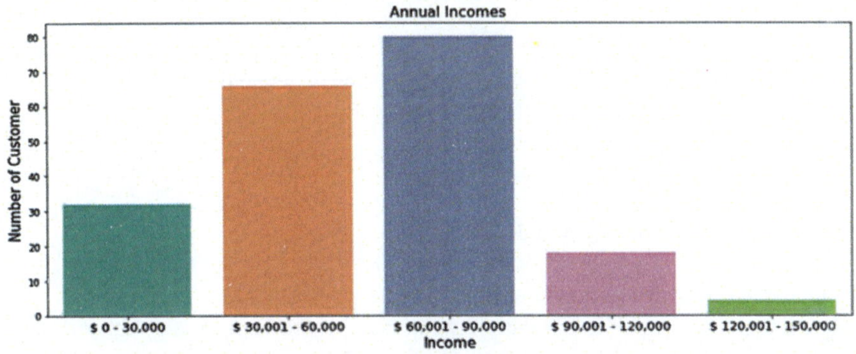

Fig. 35.6 Distribution of annual income of a customer

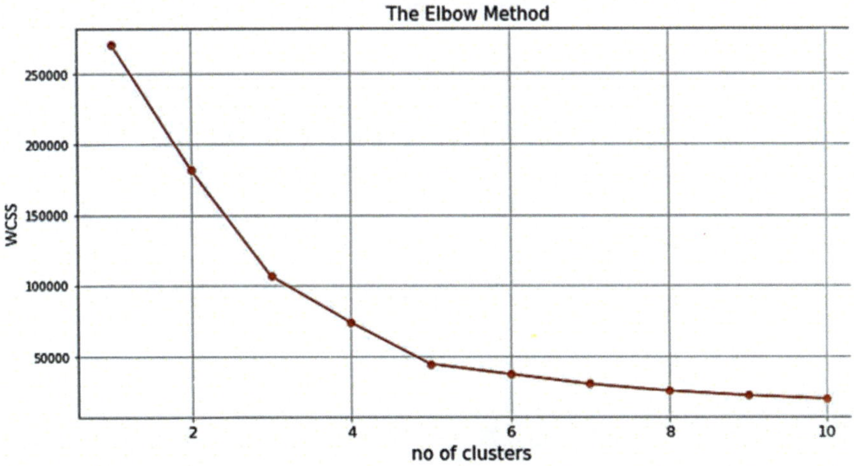

Fig. 35.7 Elbow technique to find the value of cluster K

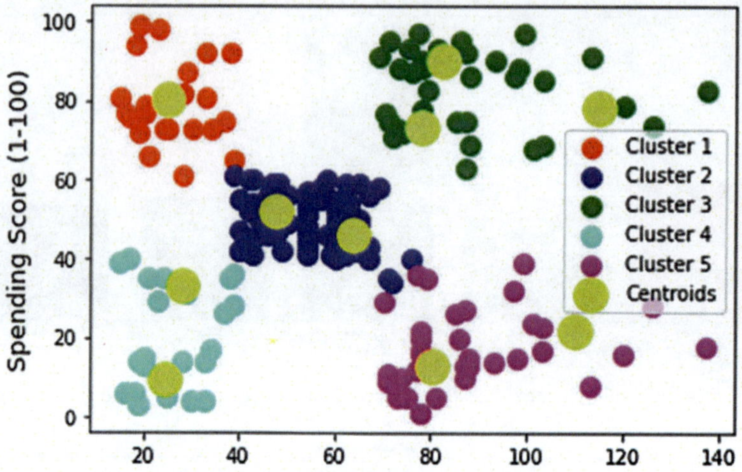

Fig. 35.8 K-means clustering of customer segmentation

In the end, the K-means implementation for customer segmentation can be summarized as:

- First, calculate clustering for all values of K by changing K from 1 to10 clusters.
- Compute the all inter-cluster WSS for every K.
- Plot the graph by using the number of clusters k and WCSS.
- The knee in the graph is basically an indicator of the appropriate number of k clusters.

K-means clustering visualization

Fig. 35.8 indicates the following:

- Red color represents cluster 1:

  It shows clusters of customers, who are earning more but purchasing less.
- Blue color represents cluster 2:

  It shows clusters of average customers in terms of earning and purchasing.
- Green color represents cluster 3:

  It shows a cluster of customers, who are earning more and purchasing more which is also a target group.
- Cyan color represents cluster 4:

  It shows clusters of customers, who are earning less but purchasing more.
- Magenta color represents cluster 5:

  It shows clusters of customers, who are earning less and purchasing less.

35.5.2 Hierarchical Clustering Analysis

This section explains hierarchical clustering is implemented for customer segmentation problem. To decide the value for the cluster, the subsection explains the dendrogram method utilized for the same.

Finding value for cluster

Figure 35.9 shows the dendrogram for finding the optimal number of clusters. Thus, the combination of five lines is not situated on the Y-axis from 100 to 240, for about 140 units. Therefore, hierarchical clustering is having five clusters.

Hierarchical clustering visualization

From Fig. 35.10, it has been observed that the clustering of customers is almost like what was obtained by K-means clustering. Further analysis is performed to check the accuracy of K-means and hierarchical clustering.

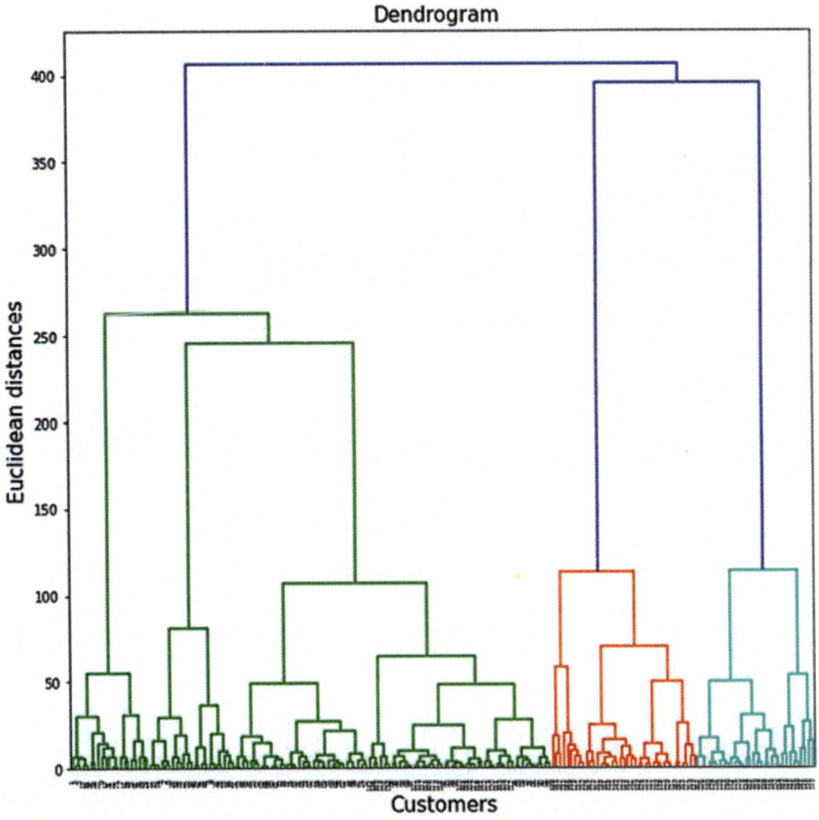

Fig. 35.9 Dendrogram to find the value for (k) cluster

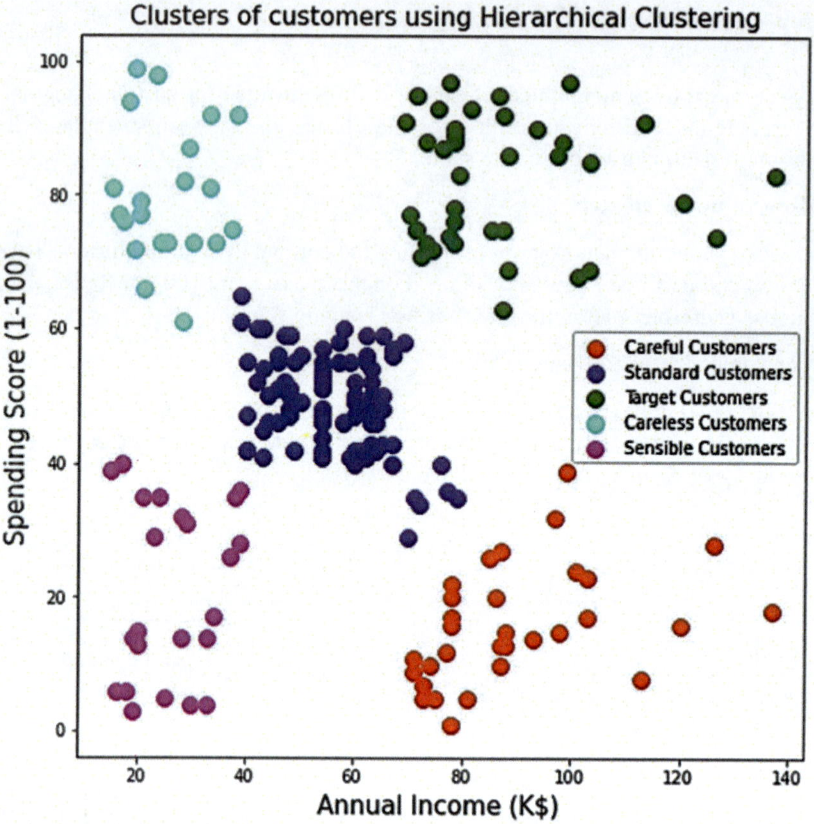

Fig. 35.10 Customer segmentation using hierarchical clustering

Table 35.2 Comparison between K-means and hierarchical accuracy score

Performance measure	K-means clustering	Hierarchical clustering
Accuracy	48%	60%

Hierarchical is more accurate than K-means. It has 120 out of 200 samples correctly labeled, whereas in K-means only 97 out of 200 samples were correctly labeled. Table 35.2 describes the accuracy analysis from both techniques.

35.6 Discussion

The result of K-means and hierarchical clustering is the same which has similar results for the same dataset. It is possible that for a big dataset, the cluster shape may differ. Moreover, there are many similarities between K-means and hierarchical but there are some differences.

35.7 Conclusion

The current study focuses on the analysis of unsupervised machine learning techniques. Due to the rise in profit-oriented organizations, customer information is increasing drastically. Organizations should use structured clustering techniques for segmenting the customer while managing this massive value of data. The study discusses clustering techniques that include upsides and downsides. The outcome of the study indicates that both techniques give the same result for an identical dataset. It may be possible that clusters differ as per different datasets. However, both clusters have many similarities, but they have some dissimilarities too. In this research, while comparing accuracy it has been observed that hierarchical clustering is more accurate customer segmentation that is 60%. The output of K-means is presented through flat-structured clusters, whereas hierarchical clustering is represented through dendrograms. Moreover, the choice process of those suitable techniques requires a decent time investment in studying and processing the information together with an adequate understanding of the goals and requirements.

References

1. Ngai, E., Xiu, L., Chau, D.: Application of data mining techniques in customer relationship management: a literature review and classification. Expert Syst. Appl. **36**(2), 2592–2602 (2009)
2. Peppard, J.: Customer relationship management (CRM) in financial services. Eur. Manag. J. 18(3), 312–327 (2000)
3. Ansari, A. and Riasi, A.: Taxonomy of marketing strategies using bank customers clustering. Int. J. Bus. Manag. **11**(7), 106–119 (2016)
4. Ghzanfari, M., et al.: Customer segmentation in clothing exports based on clustering algorithms. Iran. J. Trade Stud. **14**(56), 59–86 (2010)
5. Rygielski, C., Wang, J.C. and Yen, D.C.: Data mining techniques for customer relationship management. Technol. Soc. **24**(4), 483–502 (2002)
6. Arthur, D., Andvassılvıtskıı, S.: K-means++: the advantages of careful seeding. In: Proceedings of the Eighteenth Annual lACM-SIAM Symposium on Discrete Algorithms (Philadelphia, PA, USA, 2007), SODA '07. Society for Industrial and Applied Mathematics, pp 1027–1035 (2007)
7. Coates, A., Andng, A.Y.: Learning feature representations with k- means. In: Neural networks: tricks of the trade, 2nd edn., pp. 561–580 (2012)
8. Coates, A., Ng, A.Y., Andlee, H.: An analysis of single-layer networks in unsupervised feature learning. In: AISTATS, pp. 215–223 (2011)

9. Hartıgan, J.A., Andwong, M.A.: A K-means clustering algorithm. Appl. Stat. **28**, 100–108 (1979)
10. Hınton, G.E., Osındero, S., Andteh, Y.-W.: A fast-learning algorithm for deep belief nets. Neural Comput. **18**, 1527–1554 (2006)
11. Bertin, M., Atanassova, I.: K-means and hierarchical clustering method to improve our understanding of citation contexts. BIRNDL@SIGIR (1) (2017)
12. Singh, P., Surya, A.: Performance analysis of clustering algorithms in data mining in weka. Int. J. Adv. Eng. Technol. **7**, 1866–1873 (2015)
13. Johansson, A., Wikström, J.: Customer Segmentation Using Machine Learning (2021)
14. Gulluoglu, B., et al.: Customer segmentation for churn management by using ant colony algorithm. Glob. J. Comput. Sci. Theory Res. **10**(1), 18–26 (2020)
15. Kabasakal, İ: Customer segmentation based on recency frequency monetary model: a case study in E-retailing. Bilişim Teknolojileri Dergisi. **13**, 47–56 (2020)
16. Bepary, A.C., Ferdous, Z.: Customer Segmentation by Using Machine Learning and E-Commerce Solution (2020)
17. Fitriani, M.A., Febrianto, D.C.: Data mining for potential customer segmentation in the marketing bank dataset. JUİTA: Jurnal İnformatika **9**, 25–32 (2021)
18. Wu, S., et al.: Integrated churn prediction and customer segmentation framework for telco business. IEEE Access **9**, 62118–62136 (2021)
19. Shakya, S., Nepal, L.: Computational enhancements of wearable healthcare deviceson pervasive computing system. J. Ubiquitous Comput. Commun. Technol. (UCCT) **2**(02), 98–108 (2020)
20. Manoharan, S.: Early diagnosis of lung cancer with probability of malignancy calculation and automatic segmentation of lung CT scan images. J. Innov. Image Process. (JIIP) **2**(04), 175–186 (2020)
21. Vijayakumar, T., Vinothkanna, R., Duraipandian, M.: Fuzzy logic based aeration control system for contaminated water. J. Electron. 2(01), 10–17 (2020)
22. Gamarra, M., Manjarres, Y., Torres, M.T., Escorcia-Gutierrez, J., Zurek, E.: MC-Kmeans: an approach to cell image segmentation using clustering algorithms. Int. J. Artif. Intell. (2021)

Chapter 36
Application of Failure Prediction via Ensemble Techniques

A. Harshitha, D. S. L. Pravallika, D. Chandana, K. Rakesh Krishna, and Sreebha Bhaskaran

Abstract Fault is considered one of the most common issues in the software life cycle. Some of these errors are not identified during the testing phase, causing problems throughout the production process and execution time. The technology and complexity in the field of software development are increasing day by day, increasing the difficulties in fault prediction prior to the production process. Earlier, the prediction of failure in software is pushed into production by enhancing the system's reliability and quality. The wastage of time and cost due to failure in runtime can be avoided by implementing a better prediction system. Some traditional methods are available for predicting failures in different aspects and cannot provide proper output. This paper proposes an advanced failure prediction system by using a machine learning (ML) algorithm. It is capable to detect software faults in a very fast and accurate manner. This system is developed by combining different ML techniques for comparing the performance of each technique and it also determines how they influence results. Ensemble learners are used for combining ML algorithms by soft voting and plurality voting for performing early prediction in software module failure. A graphical user interface (GUI) and the dashboard have been developed for obtaining better output representation.

36.1 Introduction

Recent years have been witnessing significant growth in the field of software development. Every day lot of new technologies are implemented in the development of modern software. The complexity of coding is also increased. In this situation, the chances of failure increase which also affects the reliability of software. According to the complexity of the code, fault identification from the code is also a challenging

A. Harshitha (✉) · D. S. L. Pravallika · D. Chandana · K. R. Krishna · S. Bhaskaran
Department of Computer Science and Engineering, Amrita School of Engineering, Amrita Vishwa Vidyapeetham, Bengaluru, India

S. Bhaskaran
e-mail: b_sreebha@blr.amrita.edu

J. I.-Z. Chen et al. (eds.), *Machine Learning and Autonomous Systems*, Smart Innovation, Systems and Technologies 269, https://doi.org/10.1007/978-981-16-7996-4_36

process [1]. One of the primary reasons for failures may be attributed to this coding fault. Several organizations have a testing team that is responsible for debugging and testing new code [2]. In the case of large projects, manual debugging and testing are impracticable and time-consuming. The budgets of the development will be limited and clients are not ready to pay separately for error correction and testing. By utilizing limited resources, an automatic debugging and testing system is required. Several software models are used by different companies for performing testing and bug correction. It plays a vital role in cost reduction and in delivering software without failure [3]. The software runtime error is the most common one which can only be identified after the production and implementation phase. In some situations, it will turn into a system outage. In 2018, a 63-min outage of the Amazon website was reported. The company reported that it costs around 100 million dollars in loss [4]. Another critical incident happens in 2016, where the global positioning system (GPS) failed for several hours and caused several issues worldwide [5]. The above two failure scenarios explain the seriousness of the issues cause due to failure and how it will affect worldwide. These are only some samples, and failure in banking software cannot be imaginable. Because of these issues, the demand for quality software has increased. To develop reliable and quality software an automated failure detection system is required.

The majority of studies say that the faults are only shown in particular modules and some of these faults are common [6, 7]. These fault modules cause software failure and it will increase the development time and cost. Also, it decreases customer satisfaction. To improve the software quality an effective failure-prone module prediction can be implemented. The prediction of failure-prone modules at an early stage of development helps in planning activities. The fault-proneness cannot be measured directly. Unfortunately, there are no direct methods. There are several studies conducted based on failure prediction at an early stage. There are several techniques implemented in the past decade for predicting failure. Online failure prediction (OFP) is one of them. It records the past data of the system and compares it with current system data to predict failure in the future [8]. According to the prediction the measures taken are restarting the system before failure and saving data before losing. But these methods are not applicable in all scenarios and it is not reliable.

The development of technologies is increasing day by day, and one of the most promising technologies widespread in recent years is the machine learning algorithm. The implementation of machine learning in different complex projects delivers successful results. This algorithm can handle complex patterns of data and learn from the previous dataset. It can predict new input data with a previously learned model. Several well-known machine learning algorithms are support vector machine (SVM), decision tree, gradient boosting, nearest neighbor, etc. Using these classifier algorithms, the machine learning model is trained individually for prediction. One method introduced recently which is called ensemble uses multiple algorithm combinations for better accuracy and precision.

Why should we consider an ensemble classifier?

The predictive model developed by combining two or more other models is known as the ensembles method. When the project performance is most important then the ensemble technique is the best suite method for predictive modeling. There are two major reasons to consider an ensemble model over a single model; they are performance and robustness. When compared to a single contributing model, the ensemble can achieve better predictive performance. Also, the ensemble can reduce the dispersion of prediction and performance of the model. In this paper, we discuss the development and utilization of ensemble learning techniques in the field of software failure prediction. Also, we compare the performance of the ensemble method with other machine learning algorithms individually. The data recorded in the time of fault that occurred in software is used as the input for prediction. In our system initially, the machine learning model is trained with four ML algorithms separately: random forest (RF), decision tree (DT), bagging (BG), gradient boosting (GB). The trained model is evaluated for comparison. Then combining different machine learning algorithms using ensemble classifiers improves model performance. Two sets of ensemble learners are used. The first set is the combination of logistic regression, decision tree, random forest, bagging, neural network, and AdaBoost algorithms and the second set is the combination of XGB classifier, gradient boosting, and result of ensemble set one algorithms. A graphical user interface (GUI) using Tkinter and a dashboard using Plotly are developed for displaying the predicted output.

36.2 Literature Survey

There is a lot of research and development conducted for fault detection in recent decades. Among them, some are efficient and still used by several organizations and developers. In this section, we are going to discuss different methods used for software failure prediction.

Rathore et al. [9] reported a genetic programming and neural network-based method for detecting the number of fault events occurring in the runtime. They used camel 1.4 and camel 1.2 datasets for training. The system was tested with ten different software projects and the system achieved sufficient completeness for fault prediction.

The project proposed by Coelho et al. [10] used clustering techniques for detecting failure. Ensembling different clustering algorithms will generate a combined solution to predict fault. PSO, which is called a particle swarm optimization algorithm, is used in this system for delivering a better quality output. They trained five different models of Euclidean and Manhattan function. While comparing the results PSO had more accuracy than the other models. They suggest using more similar methods for improving the system performance.

In 2016, Isong et al. [11] introduced a change impact by analyzing the system for software fault detection. They focused on the changes that happen in software when failure happens. The system detects the class changes and at the same time maintenance is also initiated to overcome the major failure. They built the prediction model with a dataset available in public domains like NASA dataset. They are not testing their system with real-time data; their model is trained and tested with dataset from the public domain. This can be considered as one of the main limitations of their system. Also, their system is not automated and does not achieve sufficient accuracy.

Sandhu et al. [12] developed an early failure prediction model using decision tree and K-means clustering in a combined form. They mostly focused on the improvement of software quality. They checked the availability of metrics in the late and early life cycle of the software, and fault-prone modules are identified by combining the metrics in two life cycles. They used a real-time fault dataset called CM1 provided by NASA. Their system has a high accuracy rate and cross-validation score of 100%.

Jin et al. [13] proposed a fault detection system using support vector machine (SVM) and artificial neural network (ANN) algorithms application. They used four faulty datasets for the training model. The performance of the proposed SVM model is compared with four models: logistic regression (LR), K-nearest neighbor (KNN), decision tree (DT), and naive Bayes (NB). The dataset used in this project is downloaded from NASA public repository. The dataset includes four critical mission software which is written in a procedural programming language. They got satisfactory results for their approach compared with other algorithms.

Hamdan et al. [14] introduced a system for detecting fake news online. They implemented a systematic approach with machine learning for separating real and fake news. They were using convolutional neural network (CNN) and NB classifiers for decision-making in the proposed work. They mostly focused on the combined result rather than an individual. Their system output shows that they got more accuracy for the combined model.

We conclude that the use of different models in a combined form will give more accuracy for prediction. Table 36.1 shows all the methods mentioned in different works which we choose for review. The dataset considered by them and the performance evaluation result percentage are also compared in the table. We are motivated by some methods and added those techniques in our proposed system for enhancing accuracy.

36.3 Proposed System

The implementation of machine learning in the field of software fault detection helps to improve the quality of software development. We designed an ensemble learners-based system for detecting a failure in the newly developed software. In this section, a detailed explanation of the entire system is deliberated. Figure 36.1 shows the overall working of the system with internal blocks. We can start the explanation from the data acquisition block. For every machine learning project, the dataset

Table 36.1 Methods involved and their details

Authors	Algorithm	Dataset	Performance
Sandeep et al. [9]	Neural network (NN) and genetic programming (GP)	PROMISE data repository (PROP V4, PROP V40, PROP V85, PROP V121, PROP V157, PROP V185, Xerces 1.2, Xerces 1.3, Camel 1.2, Camel 1.4)	Average error—0.35 Average recall value GP—25 to 45% NN—6 to 30% Completeness—100%
Rodrigo et al. [10]	Particle swarm optimization algorithm (PSO)	NASA IV&V metrics data program (MDP)	PSO mean accuracy PSO-E: 79.34%PSO-M: 82.21%
Bassey et al. [11]	Software change impact analysis (SCIA), logistic regression (LR)	KC1 NASA dataset	Accuracy—M1: 97%, M2: 90%, M3: 99%
Parvinder [12]	Decision tree, K-means clustering	NASA metrics data program (MDP), CM1	Class precision—100% Class recall—100%
Jin et al. [13]	Artificial neural network (ANN), support vector machine (SVM), decision tree (DT), K-nearest neighbor (KNN), logistic regression (LR), and naive Bayes (NB)	NASA IV&V facility metrics data program, (PC1, CM1, KC1 and KC3)	ANN—98% SVM—93.2% DT—94.14% KNN—91.9% LR—91.6% NB—88%
Yasir et al. [14]	Convolutional neural network (CNN) and NB	News articles from the web scrapping	Accuracy—96.8% Sensitivity—85.7% Precision—81.1%

is the most important content. It requires previous data with or without labels for training (learning) models for predicting outputs. The camel-1.6 dataset from the PROMISE repository is used in this system for machine learning training and testing [15]. Explanation of the dataset is discussed in the section below. The dataset is downloaded and saved in the local folder as a .csv file. This dataset is loaded by the system and the system reads the data for further processing. Before passing a dataset to training and testing it needs to be free of errors (noises, corrections, etc.). For this system, the data is calibrated and saved for ML model creation. The preprocessed data is now ready for splitting. The second important part of machine learning is dataset splitting. The entire dataset is split into two sections. One for the training model and the other for the testing model. The dataset is sampled with oversampling and undersampling for adjusting data classes. From the sampled dataset the features for machine learning training are extracted. The features relevant for failure prediction are selected from the extracted features and it is fed to model training.

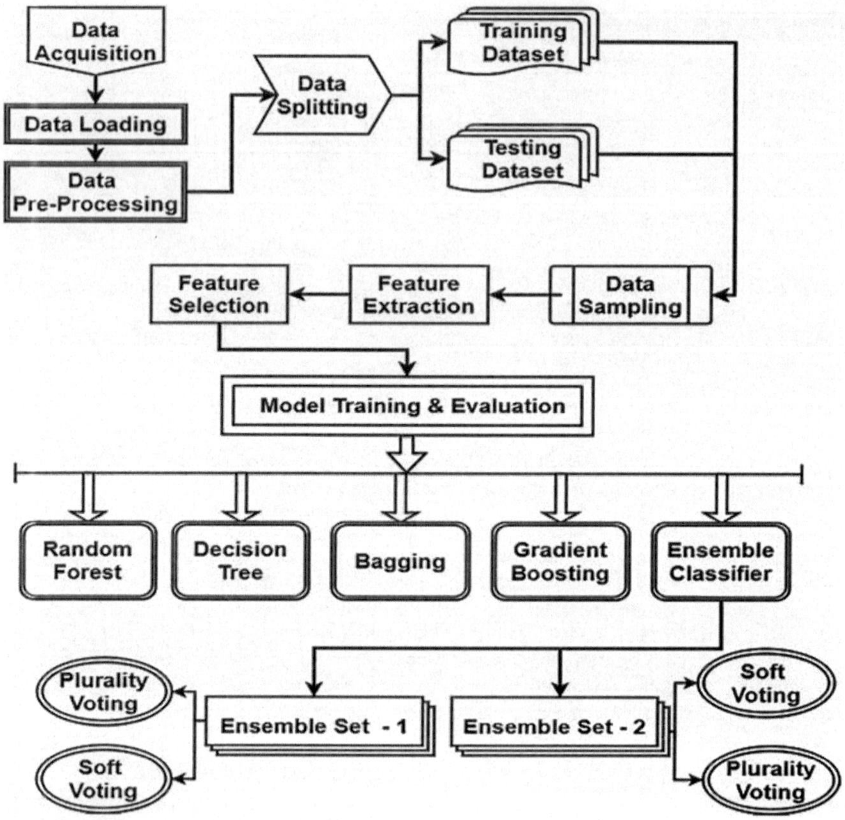

Fig. 36.1 System architecture

According to the proposed design we are using ensemble classifiers for prediction and we have also used other ML algorithms individually for comparing with the proposed method to find out the difference in performance. For the individual training, each algorithm is trained, tested, and evaluated separately. The first type of classifier used is random forest. The model is trained with the training dataset and tested with the testing dataset. The original output is compared with predicted output and evaluation results are plotted. These training and testing steps are also done with the other three types of algorithms one by one. Decision tree classifier, bagging classifier, and gradient boosting classifier algorithms are trained, tested, and plotted. As we already discussed, the ensemble method is a combination of several algorithms. In this proposed system we are using two sets of ensemble learners (set one and set two). Set one contains algorithms such as linear regression, decision tree, random forest, bagging, neural network, AdaBoost as its base learners. Set two contains algorithms XGB classifier, gradient boosting as base learners, and combined set one base learners. The prediction technique of ensemble classifiers is the voting method.

The two sets of ensemble learners are fed into soft voting and plurality voting for final prediction according to the learner with more votes.

The user needs to give input values for predicting manually and also the predicted output needs to be displayed to the user in a simple and effective way. A GUI is developed using the Tkinter module in python. The GUI has the text input field for entering parameters (WMC, DIT, NOC, CBO, RFC, LCOM) manually. Also, a dashboard is developed using Plotly for showing the output in a different way. Both GUI and Dashboard display the detected fault details accordingly.

A. *Dataset*

Dataset was obtained *from a public* repository called PROMISE. The dataset contains 188 fault modu es in total of 965 modules. The *fault was recorded* after delivering the project and it includes 20 *object-oriented* metrics [9]. The dataset also *provides a number* of faults but that parameter is not considered *as a feature* in the proposed system. The metrics that we have considered from our dataset for our study are WMC, DIT, NOC, CBO, RFC, LCOM. Considering their properties of identification of classes, the semantics of classes, and the relationship between classes these metrics are considered.

B. *Proposed Algorithms*

The first stage of algorithm implementation is the individual training and testing of ML algorithms.

Decision tree classifier, bagging classifier, and gradient boosting classifier are the algorithms selected for this stage. Using a training dataset these ML algorithms are trained and the model is saved. Using a test dataset the saved model is evaluated and the accuracy (performance) of each model is noted. The second stage of implementation is the building of ensemble learners by combining different ML algorithms. The ensembling method will increase the performance of algorithms that have less accuracy [10]. One of the crucial tasks is to select appropriate learners for the ensemble. In this system, we considered two sets of ensembles for prediction and compared them to find which one yielded better performance.

The first set of ensembles is the combination of logistic regression (LR), decision tree (DT), random forest (RF), bagging classifier (BC), neural network (NN), and AdaBoost (AD) learners. Figure 36.2 represents the first set of ensemble learners and its output. The second set combines all the learners of inset one and also with XGB classifier and gradient boosting learners. We are using the result of the first ensemble set in the second to experiment with the improvement in performance. Figure 36.3 represents the second set of learners and output. The next important task is the combination of learners in each set, for achieved by the approach called voting. Two types of voting are utilized in this system. They are plurality voting which selects the class with more votes and soft voting which selects the class with more probability. The learner's selection process finds the best algorithm for model creation. These algorithms give some promising results for individual prediction but each algorithm's

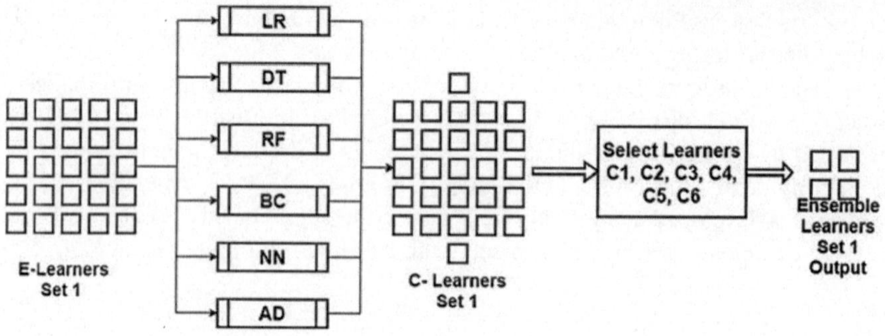

Fig. 36.2 Ensemble learner set 1

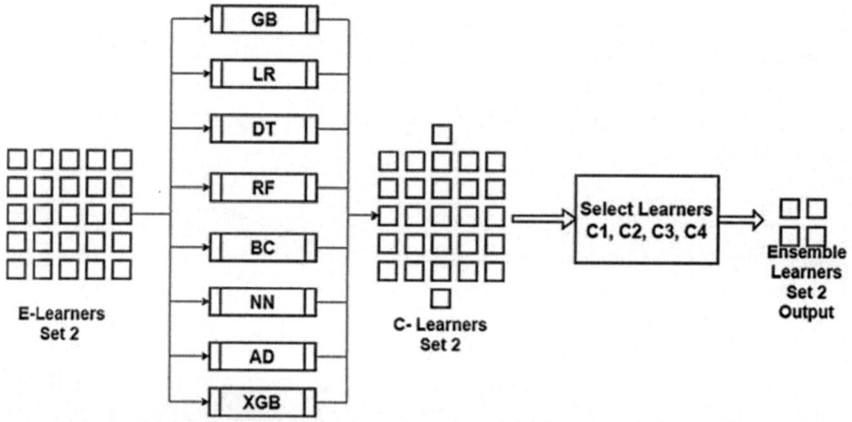

Fig. 36.3 Ensemble learner set 2

performance is not related. At the time of prediction, the most voted class is selected by plurality voting. This leads to increased performance and is predicted correctly. The base learner confidence and prediction are combined using probability again to increase the better performance of the system. Same as the plurality voting, the class with the highest probability is checked at the prediction time for soft voting.

36.4 Result and Discussion

As we discussed in the above section, the ML model is created with individual algorithms and also with ensemble classifiers. Each trained model is tested with a test dataset for evaluation. Using that test dataset fault predicted by models and the accuracy of each algorithm are calculated. The performance of each ensemble set is shown in Figs. 36.4, 36.5, 36.6 and 36.7. Figure 36.4 represents the evaluation result of ensemble set 1 with soft voting. Figure 36.5 represents the evaluation result of ensemble set 1 with plurality voting. Figure 36.6 represents the evaluation result of ensemble set 2 with soft voting. Figure 36.7 represents the evaluation result of ensemble set 2 with plurality voting. The graphs shown in each figure deliver the accuracy, precision, recall, F1-score, specificity, and sensitivity values in percentage

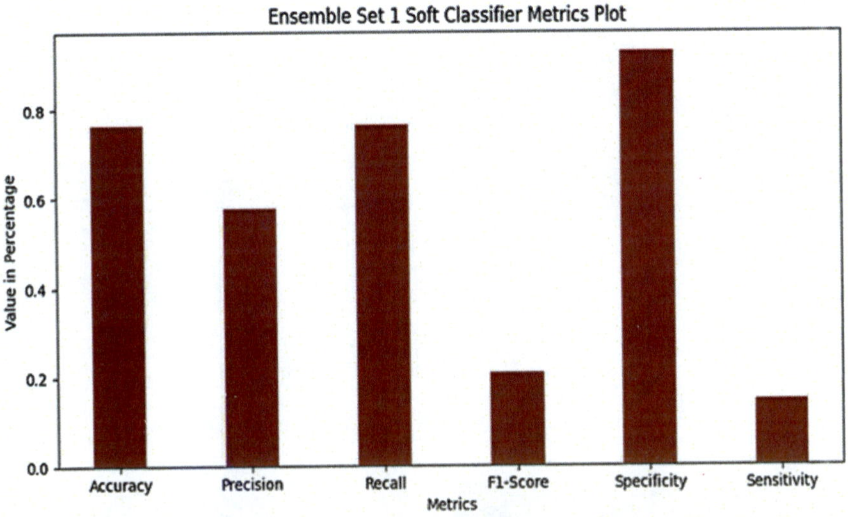

Fig. 36.4 Ensemble learner set 1 soft voting evaluation graph

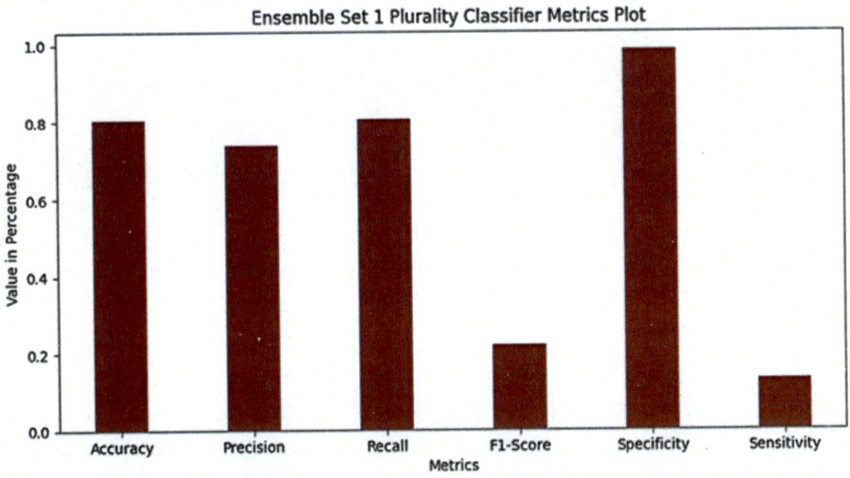

Fig. 36.5 Ensemble learner set 1 plurality voting evaluation graph

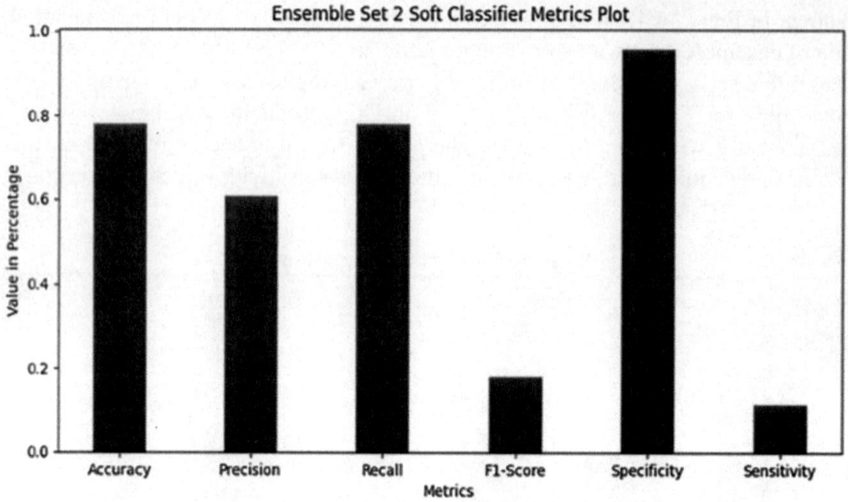

Fig. 36.6 Ensemble learner set 2 soft voting evaluation graph

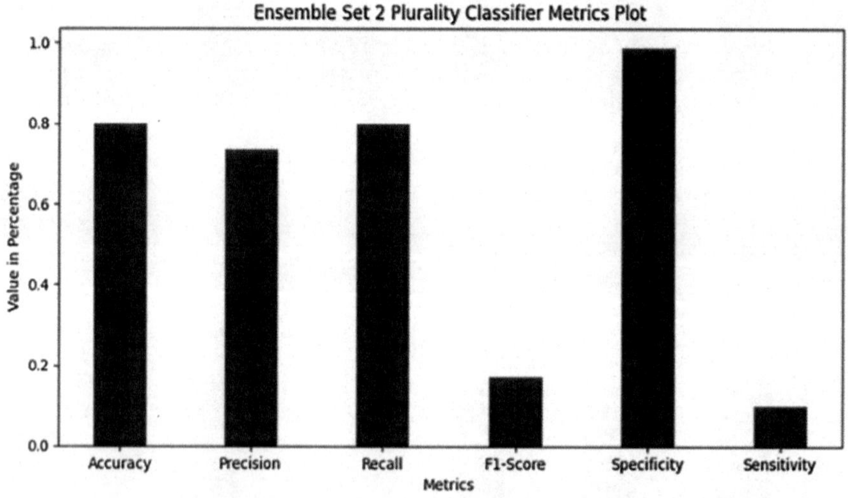

Fig. 36.7 Ensemble learner set 2 plurality voting evaluation graph

of each learner. The evaluation data is also represented in the numerical form, as shown in Fig. 36.8.

The GUI developed for the system is shown in Fig. 36.9. The input for prediction is given through the GUI input field. After inserting particular values the predict button is pressed. The system will automatically take the data inserted in GUI and process

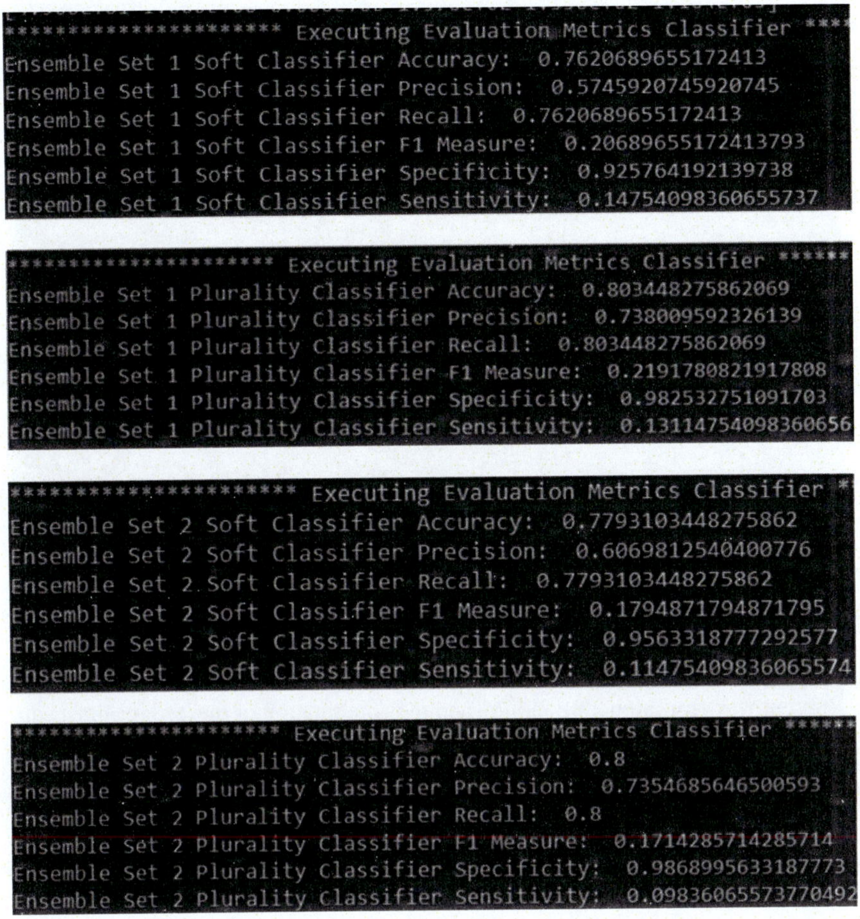

Fig. 36.8 Ensemble evaluation

it in the background for predicting the output. The predicted output is displayed on the dashboard shown in Fig. 36.10. The fault prediction output is displayed as zero which means that there is no fault in the system. If it is 1 then there is a fault in the system.

The measurements of different machine learning algorithms which are used in the proposed system are tabulated for comparison. Table 36.2 shows the accuracy, sensitivity, precision, recall, F1-score, and specificity after evaluation.

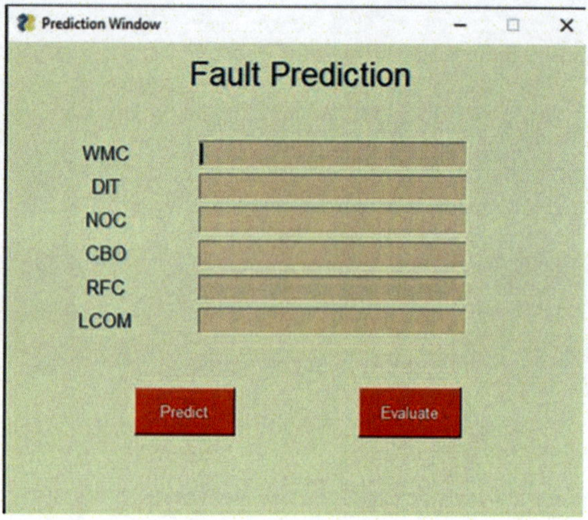

Fig. 36.9 GUI

36.5 Conclusion

The implementation of ML in predicting failure is now proven as the best choice. It predicts software failure accurately without any delay. Different algorithms perform in different ways for predicting failure, and the combination of these algorithms can create a better system for failure prediction and it can guarantee the best quality software. The proposed system is a pool of eight different algorithms and techniques which have the capacity to predict problems in different parts from different perspectives. We observed that our system performance is higher compared to other individual techniques. Ensemble set one achieved 76.5% accuracy for soft voting and 80.37% accuracy for plurality voting. Ensemble set two achieved 77.93% accuracy for soft voting and 80% accuracy for plurality voting. The second ensemble is the best ensemble for prediction after comparison. The system performed well in early detection of failure and it displayed over GUI and dashboard, respectively, without any delay. Moreover, we observed that the combination of learners contributes to each other to amplify the correct output.

In the future, more datasets can be included for training ML models which will lead to improved accuracy. For different software the fault is also different. Our system needs more learners and datasets for detecting those failures. We are planning to include that to make our system more reliable.

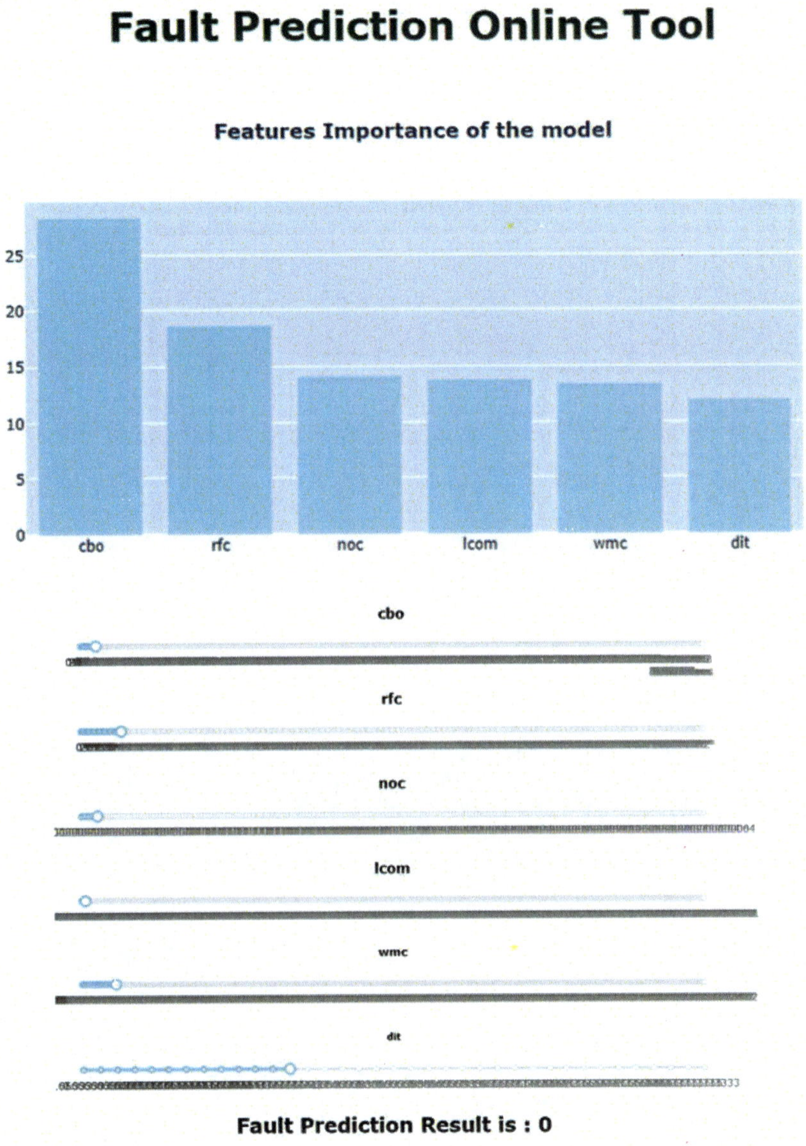

Fig. 36.10 Dashboard for predicted output

Table 36.2 Comparison of machine learning algorithms measurements

Method	Accuracy in %	Precision in %	Recall in %	F1 measure in %	Specificity in %	Sensitivity in %
EnsembleSet1 soft voting	76.20	57.45	76.20	20.68	92.57	14.75
EnsembleSet1 plurality voting	80.30	73.80	80.34	21.91	98.25	13.14
EnsembleSet2 soft voting	77.93	60.69	77.93	17.94	95.63	11.47
EnsembleSet2 plurality voting	80	73.54	80	17.14	98.68	9.83

References

1. Liu, A.G., Musial, E., Chen, M.H.: Progressive reliability forecasting of service-oriented software. In: ICWS'11: Proceedings of the International Conference on Web Services (2011)
2. Bug prediction at google. http://google-engtools.blogspot.com/. [Online]
3. Menzies, T., Milton, Z., Turhan, B., Cukic, B., Jiang, Y., Bener, A.: Defect prediction from static code features: current results, limitations, new approaches. Autom. Softw. Eng. **17**, 375–407 (2010)
4. Targett, C.B.R. (ed.): Amazon outage: estimated $99 million lost. https://www.cbronline.com/news/amazon-outage-lost-sales. Last Accessed 10 Oct 2019. [Online]
6. Baraniuk, B.C.: GPS error caused "12 Hours of Problems" for companies. https://www.bbc.com/news/technology-35491962. Last Accessed 10 Oct 2019. [Online]
6. Boehm, B.W., Papaccio, P.N.: Understanding and controlling software costs. IEEE Trans. Softw. Eng. **14**, 1462–1477 (1988)
7. Jyotsna, C., Amudha, J., Rao, R., Nayar, R.: Intelligent gaze tracking approach for trail making test. J. Intell. Fuzzy Syst. (2019)
8. Manjeri, A.S., Kaushik, R., Ajay, M.N.V., Nair, P.C.: A machine learning approach for detecting malicious websites using URL features. In: 2019 3rd International conference on Electronics, Communication and Aerospace Technology (ICECA), Coimbatore, India. IEEE, pp. 555–561 (2019). https://doi.org/10.1109/ICECA.2019.8821879
9. Rathore, S.S., Kuamr, S.: Comparative analysis of neural network and genetic programming for number of software faults prediction. In: National Conference on Recent Advances in Electronics & Computer Engineering (RAECE) (2016)
10. Coelho, R.A., dos RN Guimarães, F., Esmin, A.A.: Applying swarm ensemble clustering technique for fault prediction using software metrics. In: 13th International Conference on Machine Learning and Applications (2015)
11. Isong, B., Ifeoma, O. and Mbodila, M.: Supplementing object-oriented software change impact analysis with fault-proneness prediction. In: IEEE/ACIS 15th International Conference on Computer and Information Science (ICIS) (2016)
12. Sandhu, P.S., Goel, R., Brar, A.S., Kaur, J., Anand, S.: A model for early prediction of faults in software systems. In: The 2nd International Conference on Computer and Automation Engineering (ICCAE) (2010)
13. Salunkhe, P. et al.: Recognition of multilingual text from signage boards. In: 2017 International Conference on Advances in Computing, Communications and Informatics (ICACCI). IEEE (2017)

14. Porter, A., Selby, R.: Empirically guided software development using metric-based classification trees. IEEE Softw. **7**, 46–54 (1990)
15. Salfner, F., Lenk, M., Malek, M.: A survey of online failure prediction methods. ACM Comput. Surv **42**, 1–42 (2010)

Chapter 37
Hybrid Feature-Based Invasive Ductal Carcinoma Classification in Breast Histopathology Images

Vukka Snigdha and Lekha S. Nair

Abstract Breast cancer appears to be the most dangerous disease infecting women. Invasive ductal carcinoma is the most prevalent type of cancer that affects the breast. One in eight ladies suffer from carcinoma and even it leads to the removal of their breast. This study describes a hybrid feature-based approach for detecting invasive ductal carcinoma in WSI images. An optimal set of hybrid features is extracted from the images using deep learning models and handcrafted feature extraction techniques. In this work, we have done a comparison of feature sets considering the combination of handcrafted features like grey level co-occurrence matrix, Gabor filter and linear binary pattern with different deep learning models. A set consists of 9 handcrafted and 1000 deep learning features that are combined and used to classify the image patches as IDC or non-IDC by using a k-NN classifier. The experiment was performed on a Kaggle dataset, having breast histopathology images with 277,524 patches. Experimental results have shown that the handcrafted features with the MobileNet model features gave better classification accuracy, F1-score, precision and recall (91%, 87%, 88% and 85%, respectively).

37.1 Introduction

Breast cancer is one of the cancer types that affect women worldwide. Invasive ductal carcinoma is the leading cause of death in women with breast cancer. Invasive ductal carcinoma is an obtrusive disease in which abnormal malignancy cells that began interpreting within the milk pipes have spread past the channels into various parts of the breast tissue. One of the most important factors in determining treatment stages for women with malignant tumours is early detection of carcinoma. The diagnosis of

V. Snigdha (✉) · L. S. Nair
Department of Computer Science and Engineering, Amrita Vishwa Vidyapeetham, Amritapuri, India

L. S. Nair
e-mail: lekhas@am.amrita.edu

J. I.-Z. Chen et al. (eds.), *Machine Learning and Autonomous Systems*, Smart Innovation, Systems and Technologies 269, https://doi.org/10.1007/978-981-16-7996-4_37

carcinoma is typically accomplished through manual examination or regular examination via diagnostic procedure and ultrasound [1]. A diagnostic test, for instance, could be the gold standard for examining breast tissue. Pathologists can examine the microscopic structures of the breast tissue by using breast tissue biopsy [2]. However, the analysis of histopathological images is time-consuming and difficult because a pathologist scans massive images of tissues, and thus the outcomes might also be influenced by the pathologist's expertise.

Deep learning has acquired a positive response in the recent few years in a variety of AI fields, including computer vision, data mining, speech recognition, reinforcement learning and other related disciplines. Deep learning is a machine learning strategy in which artificial neural networks, or algorithms stimulated by the human brain, analyse a large amount of data [3]. Deep learning algorithms extract relevant high-level features from data, and deep learning models attempt to make correct estimates on their own [4].

The motive of this work is to combine various handcrafted features and deep learning models and categorize breast cancer subtypes as IDC or non-IDC and automate methods to save time and decrease error.

37.2 Literature Review

Many attempts are created to automatize the method of recognizing IDC in histopathological imaging of breast carcinoma. Methods using machine learning, and significantly deep learning, have thus far been productive in advancing the task of cancer detection in histopathological imaging. Using wavelet components from Log-Gabor complex, Niwas et al. [5] analysed the colour texture features of the segmental nucleus in biopsy images. This method produces several Log-Gabor filter convolution maps for various scales and orientations, from which first-order and second-order statistics can be derived. Agner et al. [6] used 15 statistics grey-level features, graph-based features, 64 Gabor filter features, 24 nuclear features and 16 second-order math features in each of three HSI colour channels, each with a different window size. Conventional classification methods are based on the feature extraction approach, which is based on domain knowledge for specific problems. In order to solve many problems related to feature-based methods, deep learning methods have become an important alternative. Classification of breast biopsy images stained with haematoxylin and fluorescein. It has been projected to use a convolutional neural network [7]. Radia Touahri et al. [8] investigated the power of the local binary pattern (LBP) texture and also the CNN architecture to form the automatic classification of breast tumour images as an efficient component, wherever the extraction of significant info from the input image does not need feature extractors.

Reference [9] developed a pre-trained ResNet50 and DenseNet-161 architecture which is primarily based on deep learning. The DenseNet-161 model was trained on the final layers of pre-trained deep networks, yielding an F1-score of 92.38% and a balanced accuracy value of 91.57%, and ResNet-50 architecture yielded an F1-score of 94.11% and a balanced accuracy value of 90.96%. The projected model achieved accuracy to diagnose the IDC in breast tissue lesion two-dimensional image [10]. The pre-trained GoogleNet model was used to predict the ductal carcinoma in situ and the polynomial kernel SVM was used as the classifier with 0.70 AUC. For the categorization of cancer images, the CNN, long short-term memory and a mix of convolutional neural network and LSTM are implemented. After the features were extracted with the new projected deep neural network models, the Softmax and SVM layers were employed for the decision segment.

Angel Cruz-Roa [11] devised a method for automatically detecting and segmenting glands and cell nuclei in tissue samples of BCa. To achieve framing potency, the CNN architecture is compared with various handcrafted features [fuzzy colour bar chart, grey bar chart, HSV colour histogram, JPEG constant histogram, local binary pattern, MPEG7 edge histogram, RGB histogram, graph-based and Haralick features] for recording varied visual properties in microscopic histology images. Mohammad Abdolahi [12] evaluated the convolutional neural network (CNN) model and transfer learning both through extraction of features and fine-tuning. VGG-16 was used as feature extraction and fine-tuning. In [13], various CNN architectures were trained on WSI for IDC cancer diagnosis, including Inception V3, MobileNetV2, ResNet50, VGG16, Xception and VGG19. Every model was validated on a similar dataset for overall performance evaluation, and nonheritable validation accuracy revealed that the projected CNN model outperformed with reduced validation loss with an F1-score (90%).

The current research proposes a breast cancer detection system that combines handcraft and deep learning techniques. For the identification of invasive ductal carcinoma, the proposed method analyses multiple DL architectures. The methodology first extracts the handcrafted features and then applies them to four different DL architectures, all of which were guided by a large dataset of approximately 275,000, 50 × 50-pixel RGB image patches. Training and validation tests will be used to evaluate the quantitative findings. The rest of this paper is structured as follows: The third section describes materials and methods, the fourth section discusses results and discussion and the final section discusses conclusion.

37.3 Materials and Methods

37.3.1 Dataset

In this work, we used the image dataset publicly released on the Kaggle coding website (https://www.kaggle.com/) [14]. The dataset is composed of 2040 * 1536 pixel images of 162 females who have been diagnosed with invasive ductal carcinoma. All slides were digitized with a constant scanner at 0.25 μm/picture element resolution (40 × magnification). Images are scanned and stored as 50 × 50-pixel patches. Images are labelled as "1" or "0" by manual delineation by a pathologist [15–19]. Figure 37.1a shows the IDC (−) patches and Fig. 37.1b shows the IDC (+) patches (Table 37.1).

The dataset was extremely imbalanced. From Fig. 37.2a, the negative patches were 2.5 times higher than the positive patches. To train a model, an imbalanced dataset is an issue. To solve this issue, two data augmentation techniques were used. Flip the images horizontally and vertically to increase the number of positive patches. From Fig. 37.2b, it was clear that the dataset distribution looks much better. The amount of positive and negative patches was very close.

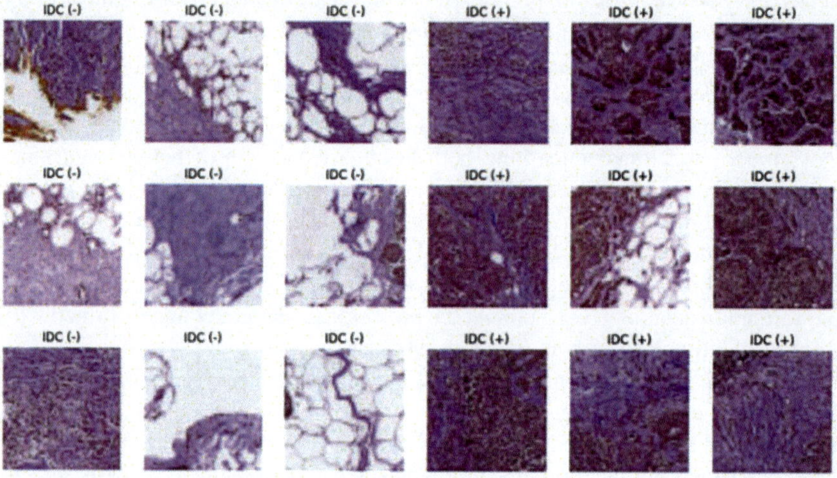

Fig. 37.1 **a** IDC (−) patches, **b** IDC (+) patches

Table 37.1 Ratios of a dataset

Total number of image patches	277,524
Total number of negative patches	198,738
Total number of positive patches	78,786

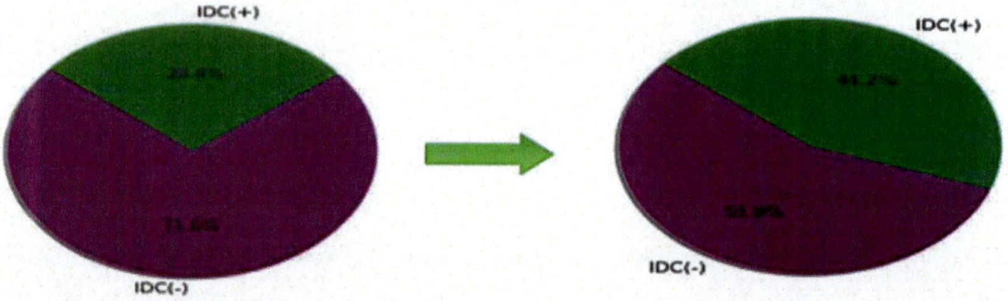

Fig. 37.2 **a** Before augmentation, **b** After augmentation

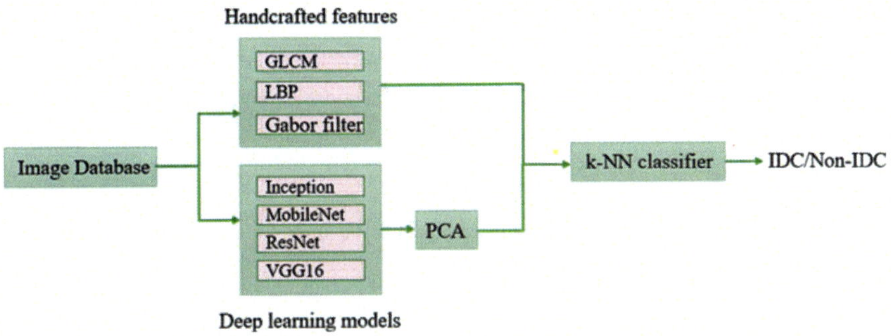

Fig. 37.3 Schematic diagram

37.3.2 Methodology

Our work involves a combination of an optimal set of handcrafted features and deep learning features from breast histopathological images. The handcrafted features are extracted using linear binary pattern, Gabor filter and grey level co-occurrence matrix. Inception, MobileNet, ResNet and VGG16 are the deep learning models used here for deep feature extraction. As the number of the output features varies for different deep learning models, we have applied PCA on the deep learned features to extract 1000 relevant features. A set consists of 9 handcrafted and 1000 deep learned features that are combined and used to classify the image patches as IDC or non-IDC by using k-NN classifier. Figure 37.3 demonstrates a system architecture of a methodology.

37.3.3 Feature Extraction

Feature extraction may be a selected type of dimensionality modification. The purpose is to obtain the necessary data attributes and interpret them in a space with extremely low dimensionality. A learning-based merging of handcrafted and deep learning features within the feature vector is projected during this framework. The following are the extracted handcrafted features:

Gabor filter: A Gabor filter is a sinusoidal plane altered by a Gaussian envelope with a particular direction and frequency. Gabor energy and Gabor entropy are two different sets of features we extracted.

Linear Binary Pattern: LBP is a term used to define the texture characteristics of a surface. The probability of texture patterns can be summarized into a histogram using LBP. Entropy and energy are the two features extracted from each image.

Gray Level Co-occurrence Matrix: GLCM is a method of analysing texture that considers spatial relationship among pixels. Dissimilarity, contrast, entropy, energy and homogeneity are the five features extracted from the co-occurrence matrices.

$$Energy = \sum_{i,j} (P(i, j))^2$$

$$\text{Entropy} = \sum_{i,j} P(i, j) \log_2 P((i, j))$$

$$\text{Correlation} = \sum_{i,j} \frac{(i - \mu_i)(i - \mu_j) P(i, j)}{\sigma_i \sigma_j}$$

$$\text{Contrast} = \sum_{i,j} |i - j|^2 P(i, j)$$

$$\text{Homogenity} = \sum_{i,j} \frac{P(i, j)}{1 + |\mathrm{i} - \mathrm{j}|}$$

where P (i, j) would be the probability density function of GLCM, (i,j) is a value of GLCM when i and j were referred to rows and a column, respectively. $\mu_i and \mu_j$ were the mean values. $\sigma_i and \sigma_j$ are the standard deviation values for i and j, respectively.

Convolution layers, pooling layers and fully connected layers are the basic components of deep learning techniques. Here we used the features extracted from the convolution and pooling layers before passing them to the fully connected layers. We have done a comparison of Inception, MobileNet, ResNet and VGG16 architectures.

Inception: It achieves an accuracy of top-five on ImageNet. It reduces the computational value to a tremendous extent without compromising the speed and accuracy. The input layer contains 128 filters with a kernel size of 3 × 3, which tends to result in an input size of (100 × 100 × 3), where three refers to RGB colour channels. In a convolution operation, every input channel is hooked up to every output channel. Therefore, here 1 × 1 convolutions are applied after actual convolution and the depth is increased to match the size of the output image with the input image. These 1 × 1 convolutions also reduce computation.

MobileNet: MobileNet makes use of depth-wise separable convolution which plays convolution on every coloured layer in preference to combining all three and flattening it. In a single step, a typical convolution filter merges the inputs into a group of substitute outputs. Convolution separates this into two stages, one for filtering and the other for combining. As a result of this factorization, the model's computation and size are significantly decreased.

ResNet50: ResNet50 comprises 50 layers of neural network. The main idea of residual network (ResNet) skips connections. By stacking skip connections, the resultant deep network should give the same performance as its shallow counterpart.

VGG16: The VGG16 employs a max-pooling layer with a 3×3 filter size and a stride of 2. VGG takes a $224 \times 224 \times 3$ RGB image as an input, passing through the first and second convolutional layers with 64 feature filters of the same size and pooling.

The feature set obtained from the various deep learning models is of varying sizes, 1208 features from Inception, 1040 features from VGG16, 1096 features from ResNet and 1126 features from MobileNet. As PCA is a linear dimension reduction technique that can select a smaller set of uncorrelated variables called the principal components while capturing as much variation from the original data as possible, we use PCA for the ideal collection of 1000 features from deep learning models.

37.3.4 Classification

k-NN classifiers assign a value to an object based on the majority of votes cast by its immediate neighbours. As a consequence, the object is assigned to the class that appears most frequently among its k-NN, which is usually a slight positive integer. Hence, we used the k-NN classifier to train the IDC and non-IDC patches using 1009 hybrid features obtained in the feature extraction.

37.4 Results and Discussion

We have used 193,412 patches for training and 71,552 patches for validation. The images have been selected randomly from both the positive and negative IDC patches. Each model was trained with 80 epochs. A total of 1,000 features were obtained from deep learning models and 9 features were extracted from handcrafted features. Both handcrafted and deep learning models yielded a total of 1009 features. The results of the ResNet50, Inception, VGG16 and MobileNet models are shown in Table 37.2. Recall, F1-score and precision are used to calculate the overall performance of a particular network.

$$\text{Recall} = \frac{\text{True positive}}{\text{True Positive} + \text{False Negative}}$$

Table 37.2 Precision, recall and F1-score on deep learning models

Model	Precision	Recall	F1 score
Gabor filter + GLCM + LBP + ResNet	0.84	0.89	0.87
Gabor filter + GLCM + LBP + Inception	0.87	0.86	0.86
Gabor filter + GLCM + LBP + VGG16	0.83	0.82	0.83
Gabor filter + GLCM + LBP + Mobile Net	0.88	0.85	0.87

$$\text{Precision} = \frac{\text{True positive}}{\text{True Positive} + \text{False Negative}}$$

$$\text{F1 Score} = \frac{2{*}(\text{Recall} * \text{Precision})}{\text{Recall} + \text{Precision}}$$

Figures 37.4, 34.5, 34.6 and 37.7 provide training accuracies, training loss, validation accuracies and validation loss obtained by the four deep learning models, respectively. From Table 37.2, MobileNet with LBP, GLCM and Gabor filter yielded best in terms of F1-score, precision and recall (0.87, 0.88 and 0.85, respectively).

The performance of ResNet is tested on train and validation dataset with 80 epochs. For both Fig. 37.4a, b, the X-axis is taken as the number of epochs and the Y-axis

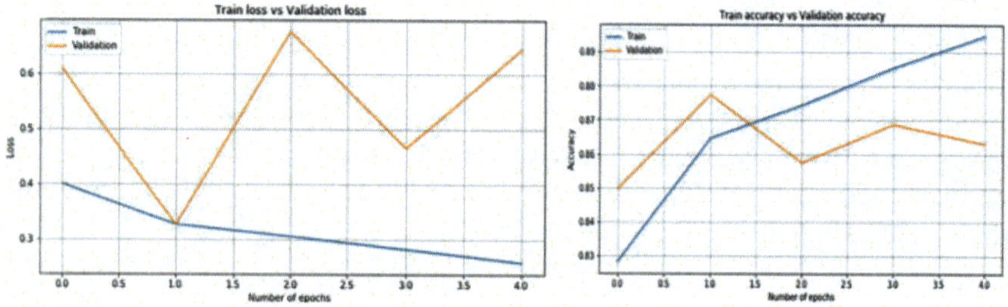

Fig. 37.4 **a** ResNet train versus validation loss, **b** ResNet train versus validation accuracy

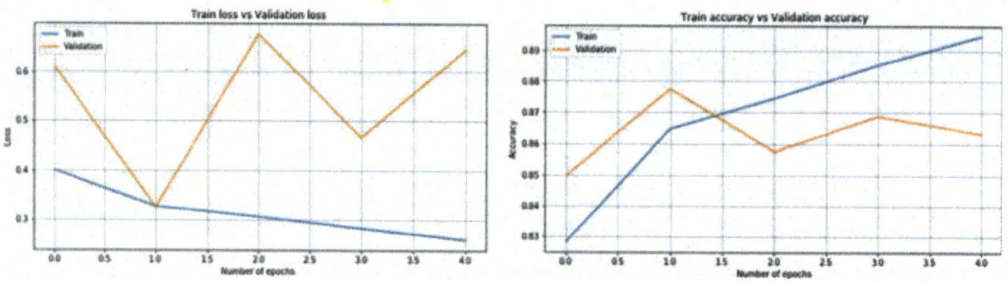

Fig. 37.5 **a** Inception train versus validation loss, **b** Inception train versus validation accuracy

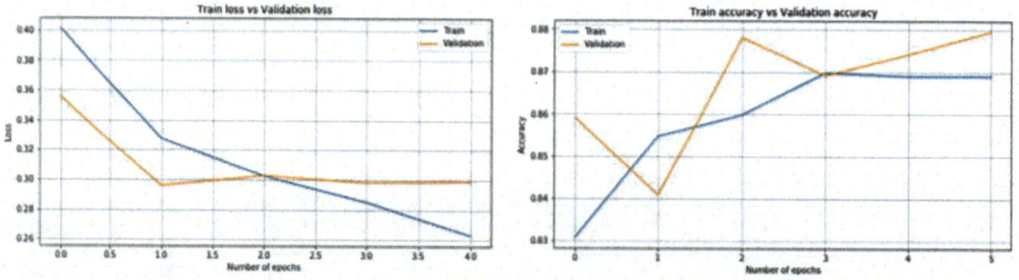

Fig. 37.6 **a** VGG16 train versus validation loss, **b** VGG16 train versus validation accuracy

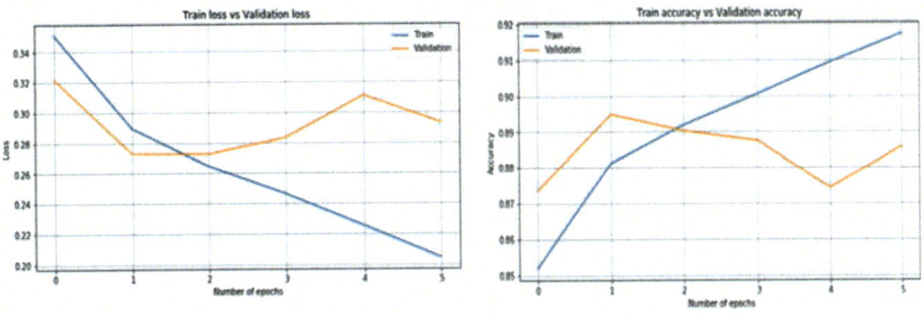

Fig. 37.7 **a** MobileNet train versus validation loss, **b** MobileNet train versus validation accuracy

is taken as loss in Fig. 37.4a and accuracy in Fig. 37.4b. By increasing the number of epochs, the validation loss also gets increased whereas the train loss got slightly decreased. In Fig. 37.4b, the train accuracy gradually increased by increasing the number of epochs.

The performance of Inception is tested on train and validation dataset with 80 epochs. For both Fig. 37.5a, b, the X-axis is taken as the number of epochs and the Y-axis is taken as loss in Fig. 37.5a and accuracy in Fig. 37.5b. By increasing the number of epochs, the validation loss also gets increased, whereas the train loss got slightly decreased. In Fig. 37.5b, the train accuracy gradually increased by increasing the number of epochs.

The performance of VGG16 is tested on train and validation dataset with 80 epochs. For both Fig. 37.6a, b, the X-axis is taken as the number of epochs and the Y-axis is taken as loss in Fig. 37.6a and accuracy in Fig. 37.6b. By increasing the number of epochs, the validation loss became constant at a point, whereas the train loss got slightly decreased. In Fig. 37.6b, the train accuracy gradually increased and remains constant by increasing the number of epochs.

The performance of MobileNet is tested on train and validation dataset with 80 epochs. For both Fig. 37.7a, b, the X-axis is taken as the number of epochs and the Y-axis is taken as loss in Fig. 37.7a and accuracy in Fig. 37.7b. By increasing the number of epochs, the validation loss also gets increased and decreased, whereas the train loss got slightly decreased. In Fig. 37.7b, the train accuracy gradually increased by increasing the number of epochs.

37.5 Conclusion

The most potentially harmful type of breast cancer is invasive ductal carcinoma. The breast histopathology images dataset from Kaggle was used. According to that dataset, the data was highly imbalanced. So, data augmentation techniques were used to balance the dataset. The main aim of this work was to classify IDC and non-IDC cells from breast histopathological ımages by extracting various handcrafted features and deep learning models. Nine handcrafted features extracted from histopathological images are two from LBP (energy, entropy), two from Gabor (energy, entropy) and five from GLCM (energy, entropy, contrast, correlation and homogeneity). ResNet, MobileNet, Inception and VGG16 deep learning models are extracted from histopathological images. Deep learning models were fed into PCA to extract a set of optimal features. A k-NN classifier was given a set of 1009 features. In this study, we noticed that MobileNet than any other handcraft with deep learning models.

References

1. Unger, K.: Challenges to the early diagnosis and treatment of breast cancer in developing countries. World J. Clin. Oncol. **5**(3), 465 (2014)
2. Basavanhally, A., Shih, N., Feldman, M., Mies, C., Madabhushi, A., Ganesan, S., Tomaszewski, J.: Multi-field-of-view framework for distinguishing tumor grade in ER+ breast cancer from entire histopathology slides. IEEE Trans. Biomed. Eng. **60**, 2089–2099 (2013)
3. Yang, S.D., Wu, G., Suk, H.I.: Deep learning in medical image analysis. Annu. Rev. Biomed. Eng. 19:221–248 (2017)
4. Arau jo, T., Castro1, E., Aguiar, P., Aresta, G., Rouco, J., Eloy, C., Polo nia, A., Campilho, A.: Classification of breast cancer histology images using Convolutional Neural Networks (2017)
5. Issac Niwas, S., Palanisamy, P., Isa, N.A.M., Zhang, W.J., Chibbar, R.: Log-gabor wavelets-based breast carcinoma classification using least square support vector machine. In: 2011 IEEE International Conference on Imaging Systems and Techniques], pp. 219–223, May, 2011
6. Naik, S., Agner, S., Doyle, S., Feldman, M., Madabhushi, A., Tomaszewski, J.: Automated gland and nuclei segmentation for grading of prostate and breast cancer histopathology. In: 2008 5th IEEE International Symposium on Biomedical Imaging: From Nano to Macro, pp. 284–287, May 2008
7. Ghaznavi, F., Madabhushi, A., Evans, A., Feldman, M.: Digital imaging in pathology: Whole-slide imaging and beyond. Annu. Rev. Pathol. **8**(1), 331–359 (2013). PMID:23157334
8. Touahri, R., Azizi, N., Aldwairi, M., Benaida, F., Hammami, N.E.: Automated breast tumor diagnosis using local binary patterns (LBP) based on deep learning classification. In: 2019 International Conference on Computer and Information Sciences (ICCIS)], pp. 309–313, April 2019
9. Celik, Y., Talo, M., yıldırım, Ö., Karabatak, M., Rajendra Acharya, U.: Automated ınvasive ductal carcinoma detection based using deep transfer learning with whole-slide ımages. Pattern Recogn. Lett. **133**, 232–239
10. Siddiqui, F., Gupta, S., Dubey, S., Murtuza, S., Jain, A.: Classification and diagnosis of ınvasive ductal carcinoma using deep learning. In: 2020 10th International Conference on Cloud Computing, Data Science & Engineering (Confluence), Noida, India, pp. 242–247 (2020)
11. Cruz-Roa, A., Basavanhally, A., Gonz´alez, F., Gilmore, H., Feldman, M., Ganesane, S., Shih, N., Tomaszewski, J., Madabhushi, A.: Automatic detection of invasive ductal carcinoma in whole slide images with convolutional neural networks. Medical Imaging (2014)
12. Abdolahi, M., Salehi, M., Shokatian, I., Reiazi, R.: Artificial intelligence in automatic classification of invasive ductal carcinoma breast cancer in digital pathology images. Med. J. Islamic Republic of Iran (2020)
13. Kumar, K., Rai, A., Islam, N., Saeed, U.: IDC Breast Cancer Detection Using Deep Learning Schemes, Advances in Data Science and Adaptive Analysis (2020)
14. https://www.kaggle.com/paultimothymooney/breast-histopathology-images

15. Unni, A., Vinod, S., Eg, N., Nair, L.S.: Tumour detection in double threshold segmented mammograms using optimized GLCM Features fed SVM. In: 2018 International Conference on Advances in Computing, Communications and Informatics (ICACCI), Bangalore, India (2018)
16. Ancy, C.A., Nair, L.S. : Tumour classification in graph-cut segmented mammograms using GLCM features-Fed SVM. İn: Intelligent Engineering Informatics, Advances in Intelligent Systems and Computing, Singapore (2018)
17. Ali Mehrabi, A.-A.N.M., Kong, Y.: Histopathological breast cancer ımage classification by deep neural network techniques guided by local clustering. BioMed. Res. Int. (2018)
18. Sathya, S., Joshi, S., Padmavathi, S.: Classification of breast cancer dataset by different classification algorithms. In: 2017 4th International Conference on Advanced Computing and Communication Systems, ICACCS (2017)
19. Shastri, S.S., Nair, P.C., Gupta, D., Nayar, R.C., Rao, R., Ram, A.: Breast cancer diagnosis and prognosis using machine learning techniques. International Symposium on Intelligent Systems Technologies and Applications (2017)

Chapter 38
Hybrid Model Using K-Means Clustering for Volumetric Quantification of Lung Tumor: A Case Study

Ranjitha U. N. and M. A. Gowtham

Abstract Early detection of cancerous lung tumors has a greater influence in increasing the life expectancy and even in the prognostic aspects. The challenges associated with the diagnosis of the tumor is all the more complicated when the tumor is attached to blood vessels or lung wall. Identification of the size of the tumor also has its role in staging and prognosis of lung cancer. Medical image processing is certainly a crucial approach in improving the accuracy of the diagnosis. Considering the importance of aiming at accurate image analysis and immense complex data volume in medical image analysis, the benefits of the cooperation of different complementary approaches are appreciable. In this paper a hybrid model of segmentation using K-means clustering with morphological operation is proposed. A comparison based on the sequence of the process implemented is presented which extends its scope in volumetric quantification of lung tumors.

38.1 Introduction

Among the cancer-related deaths around the world, death due to lung cancer is one of the major concerns. Lung cancer which is categorized as small cell lung cancer and non-small cell lung cancer accounts for a share of 87%. In spite of sophisticated imaging technologies, the chances of increasing the life expectancy are very bleak, scaling to around 15% [1]. This situation extends the scope for perseverant improvement in methodologies to detect lung tumors at an early stage. CT scan is considered to be the modality of choice over MRI for lung image analysis as the soft tissue scan and the movement of lungs poses the challenges [2]. In the context of improving the accuracy of detection of malignant lung nodules, the knowledge of nodule volume has a perceptible impact. According to some physicians, determining the size of the NSCLC tumor influences further treatment approaches including surgery decisions.

Ranjitha U. N. (✉)
REVA University, Bangalore, India

M. A. Gowtham
Adichunchanagiri Institute of Technology, Chickkamagalur, India

J. I.-Z. Chen et al. (eds.), *Machine Learning and Autonomous Systems*, Smart Innovation, Systems and Technologies 269, https://doi.org/10.1007/978-981-16-7996-4_38

American Joint Committee on Cancer (AJCC) proposes a TNM system, where T focuses on the size and extent of the main tumor. Quoting Makoto Nonaka et al. in their work put forward that the tumors with a maximum dimension of 3 cm are categorized as T1, whereas those greater than 3 cm are T2. Determining the size of the tumor not only influences a better prognostic approach but also emphasizes the non-invasive evaluation strategies of a malignant lung nodule. Considering the importance of aiming at accurate image analysis and immense complex data volume in medical image analysis, the benefits of the cooperation of different complementary approaches yield better results. Attributing to a huge volume of data especially in medical diagnosis, the computer-aided diagnosis supports diagnostic radiology and medical imaging [3]. While variation with biological and pathological aspects introduces uncertainties to medical image processing, it still serves as a crucial approach in improving the accuracy of the diagnosis. The work in this paper focuses on a hybrid method of segmentation and morphological operation to analyze the size of the cancerous lung tumor through the volumetric analysis of lung nodule.

The process of segmentation is an important step in image processing which results in non-intersecting partitions exhibiting homogeneity and the partitions are heterogeneous by themselves. According to Pal and Pal [4], there is no single method that can be considered as a universal approach for dealing with segmentation for all kinds of images. In this paper, a strategy of segmentation on lung cancer images is presented which fuses K-means clustering and morphological operations prompting to serve as a methodology for volumetric analysis of lung tumors. This approach yields a substantial PSNR value.

The choice of K-means clustering for segmentation in this paper is justified, as K-means requires less computation time when compared to other algorithms. However, the selection in the value of K and initial centroids do pose a challenge in the usage of the K-means algorithm. Many researches have targeted this flaw in the K-means algorithm and have proposed the hybrid model or enhanced version of K-means [5, 6].

The structural properties of the lung lesions are enhanced by using mathematical morphological operation which treats images as sets. The principle of morphological operations is related to the sliding of the defined kernel over the image of consideration [7]. In our work we focus on the volumetric analysis of the contouring action of the kernel used [8]. Emphasizing the importance of validations and metrics in measuring the results of medical image processing is crucial [9], the fact that a method must not deteriorate the interpretation capability is upheld. In our work we use PSNR to compare the results of the case study done.

38.2 Related Work

Computed tomography serves as the standard for pulmonary structures and surrounding anatomy for most of the diagnoses. The predecessor for the success of diagnoses associated with CT scans lies in effective segmentation. Segmentation of lung CT images extends the scope to quantitative computed tomography in densitometry [10], lesion detection for diagnosis [11–14], and staging of lesion [15, 16] for prognostic work. The prognostic efficiency is increased with the input from the method called staging, where the location and spread of the cancer tumor are described.

Computer-assisted segmentation in general can be classified based on many criteria. According to Matthew S. Brown [17], pixel-based segmentation involves various facts of contouring, thresholding, etc. It can be better guided by knowledge-based methods which use implicit knowledge with an intention of increasing robustness and automation. Considering pixel-based segmentation, clustering is one of them. Segmentation aims at creating partitions exhibiting homogeneity and one of the approaches in finding the homogeneity or similarity of characteristics is clustering [18]. Clustering helps in finding the relationship between data points so they can be segmented. Most of the algorithms in clustering use the concept of distance, to measure similarity [19]. While high scalability and efficiency are the reasons for the usage of partitional clustering in many pattern recognition applications, K-means clustering has a setback in handling categorical data [20]. However, K-means clustering is used as an effective approach in segmentation [21, 22]. The efficiency of the K-means algorithm in segmentation partially relates to the choice of number of clusters and the initialization of the centroid [23]. Madhu yedla et al. in their work [24] have proposed an enhanced K-means approach in which better initial centroids are found by transforming the negative value attributes to the positive space and claims to execute the assignment of data points to relevant cluster in an efficient way. With K-means having a high value of K, the time complexity is high as many clusters have resulted and in vice versa case, the segmentation may not be a complete one. In the case of images, the choice of the number of clusters is based on the histogram [25], where the relation between the image gray level distribution and histograms is considered and Otsu's thresholding is considered for addressing the randomness of initial cluster centers. K-means technique is reformulated [26] to make the method less sensitive to initialization of the centers by adopting ranked set sampling. The ranked set sampling uses the combination of simple random sampling under probabilistic sampling design and K-means. The role of segmentation has its implication in the prediction of tumor contour, position, and volume for diagnostic and prognostic purposes.

K-means clustering has been used in combination with other techniques for contour detection, which in turn aids in volumetric quantification. The diagnosis and recovery of lung tumor are influenced by accurate volumetric analysis. Kavita et al. [27] in their work proposed a framework for accurate segmentation and classification of lung tumors. Segmentation is done using K-means and classification is considered based on TNM. The nodules attached to the walls of the lungs are challenging from the diagnostic perspective as most of the time they get overlooked by the CAD system. Multilevel processing has to be implemented to capture the nodule required for volumetric analysis. M Javid et al. [28] have chosen K-means clustering and morphological operation in candidate nodule selection and segmentation. Many articles confirm the fact that precise measurements of lung cancer tumors can be used in better prognostic trials.

38.3 Proposed Work

In this paper, we use the K-means clustering algorithm for image segmentation and morphological operations, erosion, and dilation in smoothening of the segmented image. With this approach, a case study is carried out in comparing the PSNR value obtained. The comparative study of this process is performed considering two cases: (i) Using K-means segmentation prior to morphological operation and (ii) carrying out K-means segmentation after the morphological operation. On comparing the results of both the cases, the morphological operation being performed post segmentation has yielded a better PSNR value. The approach serves as the base for using this strategy for effective volumetric quantification of lung tumors. The choice of K-means clustering for segmentation in this paper is justified, as K-means requires less computation time when compared to other algorithms like Fuzzy-c means. Considering its speed and simplicity, 100 iterations are performed by parameterizing the accuracy required. Defining the number of clusters is a prime thing in K-means clustering. Here the segmentation results in three clusters which are defined in advance.

38.3.1 K-means Clustering

Among feature space clustering algorithms, the K-means clustering algorithm is widely used. The algorithm creates partitions or clusters by grouping the objects based on similarity. The similarity calculation is based on the distance calculation between the objects. The similarity and the distance show an inverse relation between each other.

The steps of the algorithm are listed below:

Step1: From the overall N categories, K clusters are arbitrarily selected as cluster centers.

Step2: Distance to each cluster center for each other category is calculated and the category is classified as the nearest cluster center.

Step3: Recompute the cluster centers of every class that has been segregated.

Step4: Repeat steps (2) and (3) till the new center is less than or equal to the specified threshold, and then the algorithm terminates.

Considering that there are M data points in total, they have to be divided into k clusters. The K-means clustering aims at minimizing E

$$E = \sum_{m=1}^{M} \sum_{k=1}^{K} \|x_m - b_k\|^2_{\alpha_{mk}} \tag{38.1}$$

Here according to the equation, the parameter E is minimized if the data points are grouped to the closest center, i.e., $\|x_m - b_k\|^2$ should possess minimum value as it measures the distance.

When data point m is classified into cluster (k), then

$$\alpha_{mk} = \begin{cases} 1 \text{ if} \|x_m - b_k\| = \min_j \|x_m - b_k\| \\ 0 \text{ otherwise} \end{cases} \tag{38.2}$$

Firstly, we need to settle with the value of b_k and select optimal α_{mk}. Later we need to fix α_{mk} and find the best b_k. The derivative of E w.r.t. b_k tends the derivative to 0. With the smallest value of E, α_{mk} is as follows:

$$b_k = \frac{\sum_m \alpha_{mk} x_m}{\sum_m \alpha_{mk}} \tag{38.3}$$

where b_k represents the mean of all data points in the cluster. The K-means is bound to converge during every iteration by reducing the value of E.

38.3.2 Morphological Operations

Morphological operations are a few of the simplest operations which impact the shape of the object, particularly closed shape. One of the promising applications of morphological application is contour detection. Its usage is endorsed in medical imaging with the scope of extending to the volumetric analysis of ROI. In our proposed case study after extracting the ROI from the pre-processed image through k-means clustering, the image is subjected to morphological operation of erosion and dilation. In the process, the kernel which is also referred to as the structure element (SE) is made to slide over the image.

In the process of sliding the defined SE over the image, the pixel value of the image within the SE is equated to 1, else it is 0. Setting of the size of the SE depends on the requirement. In our proposed method, the size of eroding as well as dilating kernel is set to 5 as a result of trial and error. Mathematically, erosion, for a gray-scale image (x, y), with kernel $k(s, d)$, which shrinks the foreground image, is defined as

$$(F\Theta K)(x, y) = min\{F(x + s, y + d) - k(s, d)\} \tag{38.4}$$

To increase the size of the foreground object, dilation is used. Here any pixel value inside the SE is "1" otherwise it's a zero. Depending on the value of pixel, the decision is taken in the original image, the equation is given by

Fig. 38.1 Input image

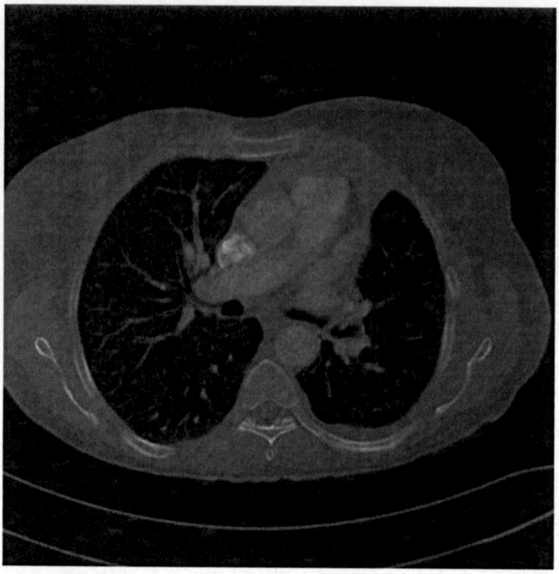

$$(F\Theta K)(x, y) = max\{F(x - s, y - d) + k(s, d)\} \quad (38.5)$$

38.4 Experimental Result

In this section, a comparison of the two cases, viz., (i) morphological operation after K-means clustering and (ii) morphological operation performed before clustering is done using the PSNR metric. The images for the experimentation are taken from the LIDC, CT scan dataset repository In the result obtained, the procedure of implementing erosion and dilation after the K-means clustering (i.e. case1) has given an improved PSNR value, which proves to be an effective approach in the volumetric analysis of lung tumors (Figs. 38.1, 38.2, 38.3 and 38.4).

The results of both case (i) and case (ii) are tabulated in Tables 38.1 and 38.2, respectively.

38.5 Conclusion

In this paper, K-means clustering is used in the segmentation of CT lung images with ROI being the tumors. The morphological operations were cascaded to K-means clustering as a pre- and post-processing operation. The case was analyzed comparing it with PSNR, which showed that carrying out morphological operations yield better

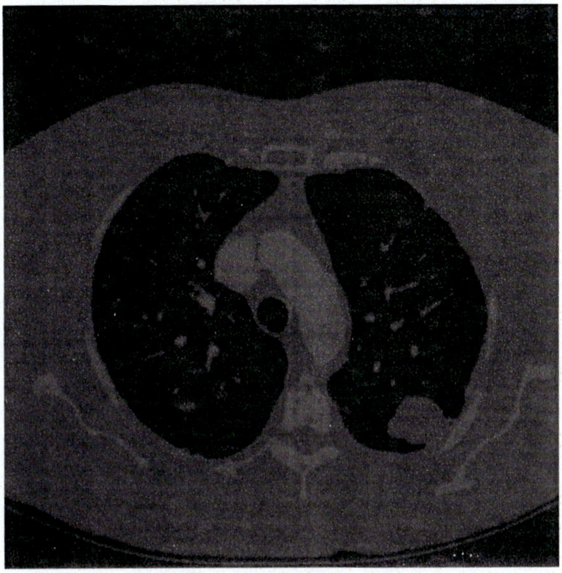

Fig. 38.2 Segmented image

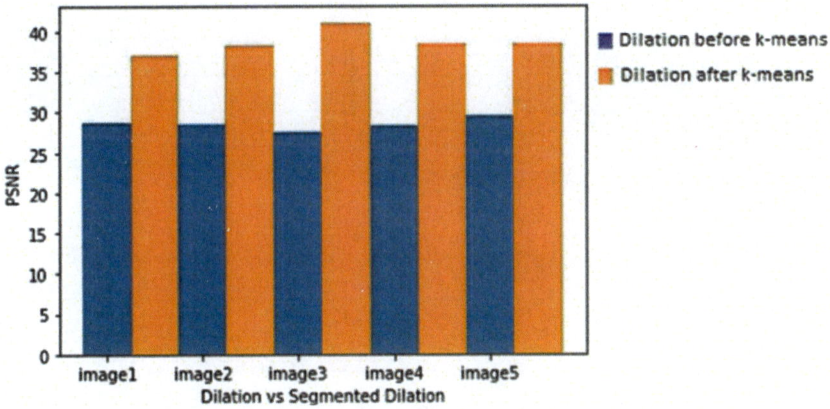

Fig. 38.3 Plot of dilation on case (i) and case (ii) v/s PSNR

PSNR. The case serves as the base in adopting the methodology in contour detection for volumetric quantification of lung tumors. However, improvisation in K-means w.r.t selection of the number of k value and initial k value can be considered. In the case of morphological operation improved kernel selection can yield promising results in the volumetric analysis of lung tumors.

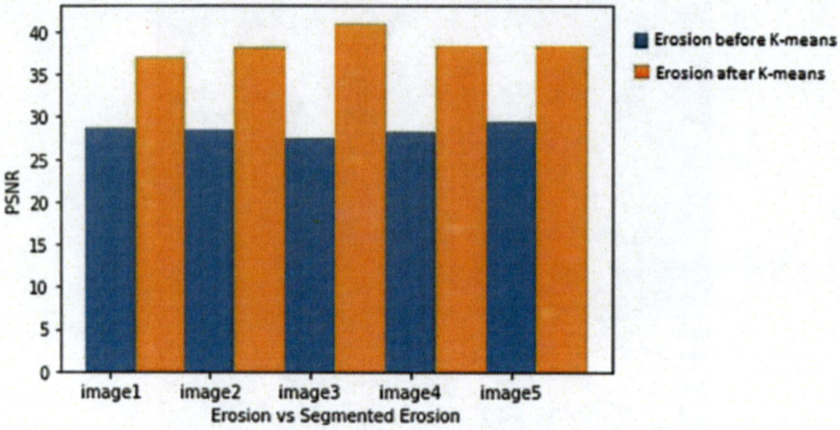

Fig. 38.4 Plot of erosion on case (i) and case (ii) v/s PSNR

Table 38.1 PSNR values for dilation and erosion before segmentation

Input image	PSNR(dB) value for dilation before segmentation	PSNR(dB) value for erosion before segmentation
1	28	32
2	28	33
3	27	36
4	28	31
5	29	31

Table 38.2 PSNR values for dilation and erosion after segmentation

Input image	PSNR(dB) value for dilation before segmentation	PSNR(dB) value for erosion before segmentation
1	34	37
2	37	38
3	40	41
4	37	38
5	37	38

Affiliated institutions are to be listed directly below the names of the authors. Multiple affiliations should be marked with superscript Arabic numbers, and they should each start on a new line as shown in this document. In addition to the name.

References

1. Purandare, N.C., Rangarajan, V.: Imaging of lung cancer: implications on staging and management. Indian J. Radiol. Imaging **25**(2), 109 (2015)
2. Laurent, F., Montaudon, M., Corneloup, O.: CT and MRI of lung cancer. Respiration **73**(2), 133–142 (2006)
3. Doi, K.: Computer-aided diagnosis in medical imaging: historical review, current status and future potential. Comput. Med. Imaging Graph. **31**(4–5), 198–211 (2007)
4. Pal, N.R., Pal, S.K.: A review on image segmentation techniques. Pattern Recogn. **26**(9), 1277–1294 (1993)
5. Ng, H.P. et al.: Medical image segmentation using k-means clustering and improved watershed algorithm. 2006 IEEE Southwest Symposium on İmage Analysis and İnterpretation. IEEE (2006)
6. Sulaiman, S.N., Mat Isa, N.A.: Adaptive fuzzy-K-means clustering algorithm for image segmentation. IEEE Trans. Consumer Electron. **56**(4), 2661–2668 (2010)
7. Firoz, R. et al.: Medical image enhancement using morphological transformation. J .Data Anal. Inform. Proc. **4**(1), 1–12 (2016)
8. Natarajan, P. et al.: Tumor detection using threshold operation in MRI brain images. 2012 IEEE International Conference on Computational Intelligence and Computing Research. IEEE (2012)
9. Jannin, P., Krupinski, E., Warfield, S.K.: Validation in medical image processing. IEEE Trans. Med. Imaging **25**(11), 1405–1409 (2006)
10. Kemerink, G.J. et al.: On segmentation of lung parenchyma in quantitative computed tomography of the lung. Med. Phys. **25**(12), 2432–2439 (1998)
11. Senthil Kumar, K., Venkatalakshmi, K., Karthikeyan, K.: Lung cancer detection using image segmentation by means of various evolutionary algorithms. Comput. Math. Methods Med. **2019** (2019)
12. Skourt, B.A., El Hassani, A., Majda, A.: Lung CT image segmentation using deep neural networks. Procedia Comput. Sci. **127**, 109–113 (2018)
13. Memon, N.A., Mirza, A.M., Gilani, S.A.M.: Segmentation of lungs from CT scan images for early diagnosis of lung cancer. Proceedings of World Academy of Science, Engineering and Technology, vol. 14 (2006)
14. Helen, R. et al.: Segmentation of pulmonary parenchyma in CT lung images based on 2D Otsu optimized by PSO. 2011 International Conference on Emerging Trends in Electrical and Computer Technology. IEEE (2011)
15. Bhakta, R., Aowlad Hossain, A.B.M.: Lung tumor segmentation and staging from CT ımages using fast and robust fuzzy C-means clustering. Int. J. Image Graph. Signal Process. **12**(1), 38 (2020)
16. Lavanya, M.: Lung cancer segmentation and diagnosis of lung cancer staging using MEM (modified expectation maximization) algorithm and artificial neural network fuzzy inference system (ANFIS) (2018)
17. Brown, M.S. et al.: Knowledge-based segmentation of thoracic computed tomography images for assessment of split lung function. Med. Phys. **27**(3), 592–598 (2000)
18. Abonyi, J. et al.: Fuzzy clustering based segmentation of time-series. International Symposium on Intelligent Data Analysis. Springer, Berlin, Heidelberg (2003)
19. Masulli, F., Schenone, A.: A fuzzy clustering based segmentation system as support to diagnosis in medical imaging. Artif. Intell. Med. **16**(2), 129–147 (1999)
20. Salem, S.B., Naouali, S., Chtourou, Z.: A fast and effective partitional clustering algorithm for large categorical datasets using a k-means based approach. Comput. Electr. Eng. **68**, 463–483 (2018)
21. Gonzalez, R., Woods, R.: Digital imaging processing. Prentice Hall, Englewood Cliffs, NJ, USA (2002)
22. Gose, E., Johnsonbaugh, R.: Pattern recognition and image analysis. Prentice Hall, Upper Saddle River, NJ, USA (1996)

23. Dhanachandra, N., Manglem, K., Chanu, Y.J.: Image segmentation using K-means clustering algorithm and subtractive clustering algorithm. Procedia Comput. Sci. **54**, 764–771 (2015)
24. Yedla, M., Pathakota, S.R., Srinivasa, T.M.: Enhancing K-means clustering algorithm with improved initial center. Int. J. Comput. Sci. İnform. Technol. **1**(2), 121–125 (2010)
25. Yao, H. et al.: An improved K-means clustering algorithm for fish image segmentation. Math. Comput. Modell. **58**(3–4), 790–798 (2013)
26. Ayech, M.W., Ziou, D.: Segmentation of Terahertz imaging using k-means clustering based on ranked set sampling. Expert Syst. Appl. **42**(6), 2959–2974 (2015)
27. Kavitha, M.S., Shanthini, J., Karthikeyan, N.: Volumetric analysis framework for accurate segmentation and classification (VAF-ASC) of lung tumor from CT images. Soft. Comput. **24**, 18489–18497 (2020)
28. Javaid, M. et al.: A novel approach to CAD system for the detection of lung nodules in CT images. Comput. Methods Programs Biomed. **135**, 125–139 (2016)

Chapter 39
Human Classification in Aerial Images Using Convolutional Neural Networks

K. R. Akshatha, A. K. Karunakar, and B. Satish Shenoy

Abstract Automatic detection of people in aerial images has potential applications in traffic monitoring, surveillance, human behavior analysis, etc. However, developing an algorithm for detection of human locations in aerial images is challenging because of the small target size, cluttered background, and varying appearance of humans. Deep learning-based object detections frameworks internally use the standard convolutional neural network (CNN) based classifiers for feature extraction and classification. Though these pre-trained classifiers perform image classification tasks with very good accuracy, they are computationally complex and hence require huge computation time. In this work, we custom-designed CNN-based classifiers to perform the human classification in aerial images and compared the performance with the standard VGG-16 based human classifier. Custom-designed classifier with fewer number of layers achieved a reduced computation time while maintaining good accuracy.

39.1 Introduction

Automatic detection of people present in the images is a key problem of computer vision because of its broader applications in surveillance, search and rescue missions, intelligent vehicles, robotics, etc. [1]. Detecting humans in aerial frames is

K. R. Akshatha (✉)
Department of Electronics and Communication Engineering, Centre of Avionics Manipal Institute of Technology, Manipal Academy of Higher Education, Manipal, Karnataka 576104, India
e-mail: akshatha.kr@manipal.edu

A. K. Karunakar
Department of Data Science and Computer Applications, Manipal Academy of Higher Education, Manipal, Karnataka 576104, India
e-mail: karunakar.ak@manipal.edu

B. Satish Shenoy
Department of Aeronautical and Automobile Engineering Manipal Institute of Technology, Manipal Academy of Higher Education, Manipal, Karnataka 576104, India
e-mail: satish.shenoy@manipal.edu

J. I.-Z. Chen et al. (eds.), *Machine Learning and Autonomous Systems*, Smart Innovation, Systems and Technologies 269, https://doi.org/10.1007/978-981-16-7996-4_39

challenging because of small target size, cluttered background, varying angle and the orientation of the camera, shape, size, etc. [2]. Object detection, in general, comprises mainly three steps, i.e., candidate region selection, feature extraction, and classification. The traditional approaches of object detection use features like HoG [3], DPM [4] with classifiers like support vector machines(SVM), artificial neural network(ANN), k-nearest neighbor (kNN), etc. The use of traditional approaches of manually identifying the features is a tedious job due to the variations in pose, size, and appearance of humans. Identifying and extracting the features for human classification can be made automatic using deep learning techniques.

Recently due to the advancements in deep learning the performance of various computer vision tasks like classification, detection, and segmentation, etc., has been improved significantly [5, 6]. Most of the efficient deep learning-based object detection frameworks internally use CNN-based standard classifiers like VGG net [7], Resnet [8], etc., to find the presence of objects in a specific region of interest. Classification is part of the detection problem, in which several region proposals generated from the image are classified using classifiers to check the presence of the object of interest. This work focuses on developing a convolutional neural network-based classifier with an aim of reducing the computational complexity of the object detection frameworks. Hence we designed two classifiers and compared their performance with the standard CNN based VGG-16 classifier for human classification in aerial images. This step is essential to build an efficient human detection system.

Image classification takes the input image and outputs the probability of object belonging to a particular category. A CNN-based classifier has three layers, i.e., convolutional, pooling, and fully connected layers as illustrated in Fig. 39.1. The purpose of the convolution layers is to extract the various features of the images. This layer is the first layer in CNN, in which a 2-D convolution is carried out between the input image and a filter of a particular size M*M. These filters are slid over the entire image to compute the feature maps. Activation functions are applied on these feature maps to add non-linearity to the network and to learn more complex features. ReLU, leaky ReLU, tanh, etc., are commonly used activation functions and each has a specific usage. These feature maps are fed to the other CNN layers to learn various

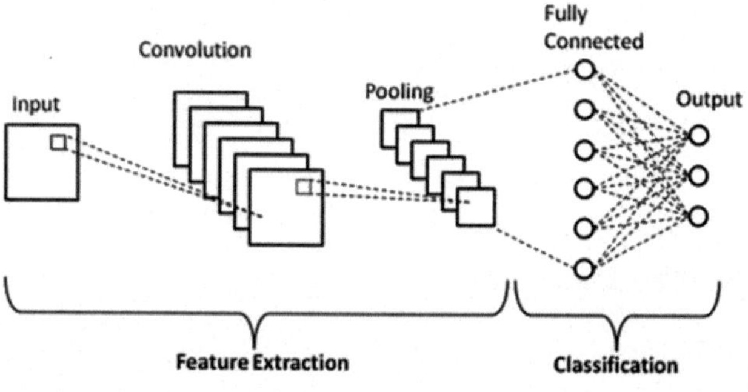

Fig. 39.1 Layers used in CNN

features from the input image. Pooling layers are placed following the convolution layers to reduce the feature dimension and hence to decrease the computational costs. Max Pooling or average pooling is generally used. The output from pooling layers is used by fully connected (FC) layers to predict the category of the objects. The FC layer has neurons with weights and biases, usually placed at the end of the CNN architecture. The feature maps from the earlier layers need to be flattened before connecting them to the FC layer. Sigmoid activation is used for binary classification and softmax is used for multiclass classification. The design of image classifiers can be done either from scratch using the appropriate convolutional, pooling, and FC layers in sequence or by using the standard pre-trained networks like Alexnet, VGGnet, or Resnet, etc., in a transfer learning approach and the process is explained in Sect. 39.3.

39.2 Related Work

This section explains the various approaches discussed in the literature to perform the human classification task. Munder et al. [9] performed a detailed experimental study on pedestrian classification. Various types of features such as Haar wavelets, local receptive fields(LRF), PCA coefficients with various classifiers like feed-forward neural networks, support vector machines, k-nearest neighbor classifiers were analyzed and their ROC performance and efficiency were compared. The correlation between training sample size and classification performance is tested in their experiment. They have shown that SVMs with LRF features performed best for human classification. Enzweiler et al. [10] performed pedestrian classification using the multilevel Mixture-of-Experts approach, which combines the details of various features and cues for improving the classification result. On pose-level, shape cues that were based on the Chamfer shape matching were considered. On the modality level, images were represented using image intensity, depth, and flow as features. At the feature level, local binary patterns (LBP) and histograms of oriented gradients(HOG) were used as features. Linear support vector machines and multilayer perceptrons were used for classification. The improved classification results were obtained using this approach. Tayara et al. [11], used single stage object detection framework with dense feature pyramid networks for VHR aerial images, that resulted multi-scale feature maps. They conducted the experiments using VGG-16, Resnet-50, and Resnet-101 backbones. AlDahoul et al. [12] developed a deep learning-based model for the detection of humans in aerial videos captured using UAVs. Optical flow with 3 deep models, i.e., pre-trained CNN feature extractor, supervised convolutional neural network (S-CNN), and hierarchical extreme learning machine are used to detect humans in UAV captured aerial videos with varying altitudes. UCF ARG aerial datasets are used for training and testing of the model. Digging, waving, throwing, walking, and running are the different actions considered for performance evaluation. Zhu et al. [13] proposed a deep learning approach for pedestrian attribute classification using a multi-label convolutional neural network(MLCNN). The classification was performed based on age, gender, and clothing, etc., which were used as soft biometric traits to recognize people. The training of a multi-label convolutional neural network was done using raw pixels instead of using hand-crafted features. The trained model was able to predict multiple attributes simultaneously. Melotti et al. [14] performed deep learning based study on pedestrian classification problems for monocular camera and a 3D LIDAR sensor data in separate and fusion approaches. CNN was used as a classifier for this work in distinct situations, with the input from single sensor data and the other was by fusing the data from two sensors. Range (distance) and intensity (reflectance) data obtained from LIDAR were treated as separate channels and were used to generate dense maps using sensor coordinate transformation and spatial filtering. This data was fed to CNN, which allowed the implementation of both sensor's data. The outputs from individual CNNs were combined using learning and non-learning approaches for the fusion-based approach. Evaluation of pedestrian classification was performed on the binary classification dataset which was created from the KITTI Vision Benchmark Suite.

Due to the recent advancements in AI and deep learning, state-of-the-art techniques for human classification and detection utilizes deep learning architectures. This has given very good results when compared to traditional feature-based approaches. From the literature, it is found that deep learning methods have proven to be efficient in classifying humans, but they are less explored for aerial images. And the existing deep learning-based object detection approaches make use of the networks like AlexNet [15], VGGNet [7], ResNet [8], etc., as classifiers, that are computationally heavy. This work focuses on developing a deep CNN-based human classifier from scratch and compared the performance with the pre-trained VGG-16 model in a transfer learning approach to classify humans and non-humans in aerial images.

39.3 Methodology

The following two approaches of deep learning are used in the present work to perform human classification in aerial images.

1. Using transfer learning approach of CNN to perform classification
2. Developing a CNN based classifier from scratch

39.3.1 Transfer Learning Approach

Transfer learning is the most common approach used to solve various deep learning tasks. Design of deeper neural network from scratch requires a huge amount of data and processing to avoid overfitting and to obtain better results. To overcome this large data requirement, one can directly start with a pre-trained network having learned

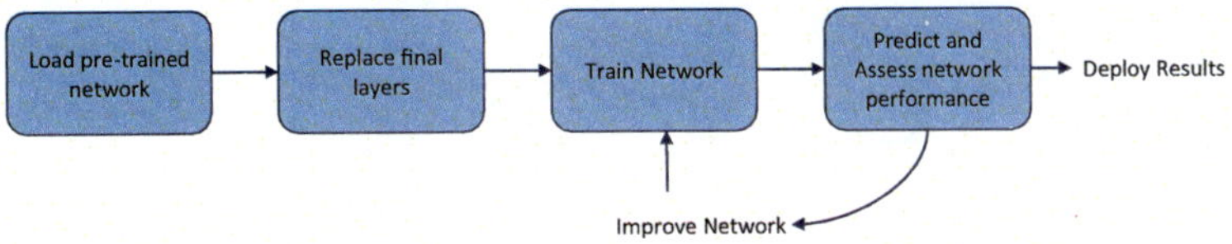

Fig. 39.2 Block diagram representation of transfer learning

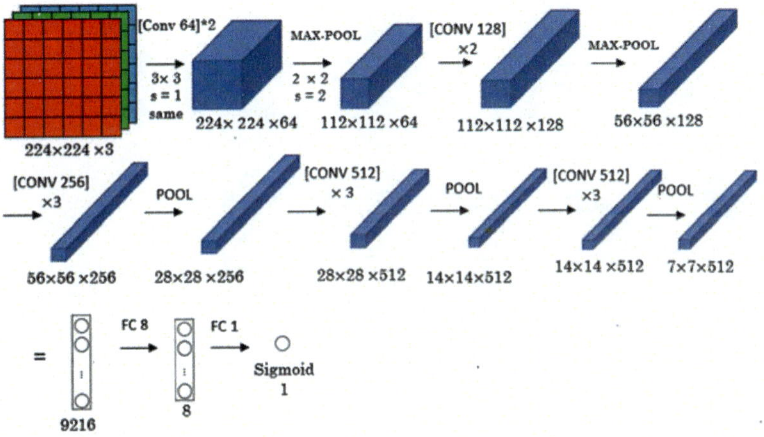

Fig. 39.3 Modified VGG-16 architecture for human classification

weights of some other task to learn a new task. In this approach of transfer learning fine-tuning is faster compared to training the network from scratch with random weights. As learned features of one task are transferred to a new task, with small number of training samples it is called transfer learning. Figure 39.2 shows the block diagram of the transfer learning approach used in some classification problems.

This work uses VGG-16, a COCO pre-trained deep model with modified final 2 layers to perform human classification task which is shown in Fig. 39.3. VGG-16 is a simple architecture compared to Alexnet [16], Lenet [17], etc., which uses 3∗3 filter with a stride of 1 and the same convolution, for max pool layer, it uses 2*2 with a stride of 2. It is a simplified neural architecture with 16 layers with 138 million parameters.

39.3.2 *Building of CNN Model from Scratch*

Developing a neural network model for a specific application is an iterative process. Designing a network starts by using a single layer, which consists of a convolution layer followed by the pooling layer and fully connected layer. Parameter tuning and evaluation of the result are necessary before adding further layers. In this work we aim

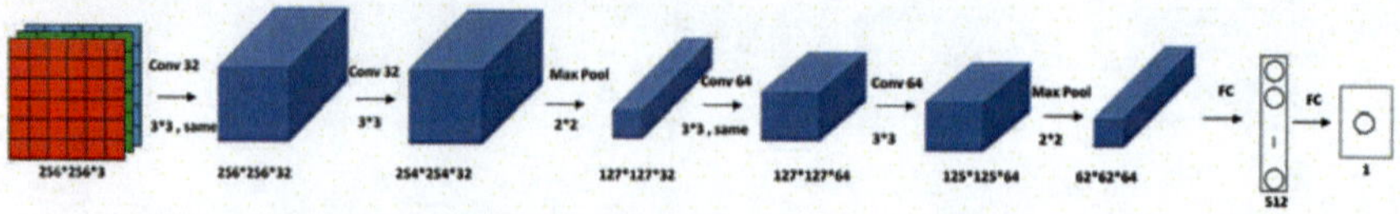

Fig. 39.4 Model-1 developed for human classification

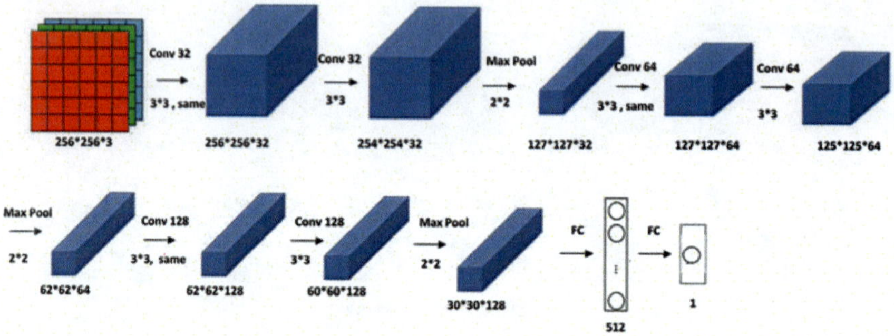

Fig. 39.5 Model 2 architecture

to develop a human classification model from scratch, using a single convolutional layer, pooling layer, and fully connected layer to see if it is learning the task of classifying humans. The iterative process is repeated to choose the best model to perform a classification task.

Model 1: Here the classification model is built using the following 6 layers in sequential order, which is shown in Fig. 39.4.

- Input layer of size (256, 256, 3)
- Convolution layer with 32 filters of size (3, 3) with same padding and Relu activation
- Convolution layer with 32 filters of size (3, 3) with Relu activation
- Max-pooling layer of size (2, 2)
- Convolution layer with 64 filters of size (3, 3) with same padding and Relu activation
- Convolution layer with 64 filters of size (3, 3) with Relu activation
- Max-pooling layer of size (2, 2)
- Fully connected layer with 512 neurons
- FC layer with 1 output and sigmoid function

Model 2: Another model is formed by adding 2 more CNN layers to the existing layers to make it as large network to perform human classification. This model is named as model 2 and is shown in Fig. 39.5. All these models are trained on a training dataset of aerial images and the performance of the trained model is evaluated on the validation dataset. The results are discussed in Sect. 39.4.

39.4 Experiments and Results

The experiment has been conducted using HP Z230 Tower Workstation having 16GB RAM. It has an NVIDIA Quadro K2000 Graphic Processing Unit(GPU) with 2GB RAM. The algorithm is implemented in python 2.7 using Keras and tensor-flow framework.

39.4.1 Dataset Details

In this experimental study, images from UCF-ARG dataset [18] and Stanford drone dataset [19] are used which has aerial video clips captured using aerial camera and drone camera having varying poses and sizes of humans. The sample images of the dataset are shown in Fig. 39.6.

39.4.2 Pre-processing

The frames are extracted from the videos and then from these frames, the portion containing humans is extracted. The frames are selected from the videos comprising of all human actions. Hence the new dataset contains images of humans with varying poses and appearances. Also from the aerial view, the human size and shape looks highly non-rigid and non-uniform. The new dataset consists of 500 images for each class (humans and non-humans/ positive and negative) for experimenting. 70% of the data from each class is used for training and the remaining 30% is used for testing/validation. The sample images from the dataset belonging to each class are shown in Fig. 39.7. All the images are normalized before performing training.

Fig. 39.6 Sample images of UCF-ARG dataset [18]

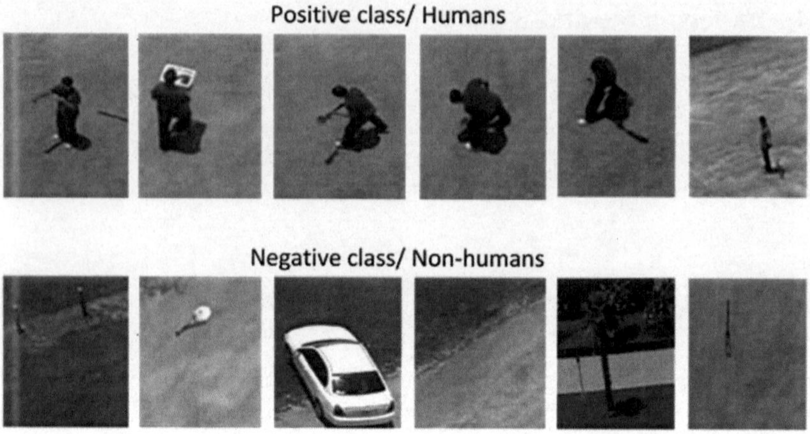

Fig. 39.7 Sample images of humans and non-humans from aerial view

39.4.3 Classification Results

This section shows various experiments conducted on aerial images and their results.

Transfer Learning Using VGG-Net 16: The VGG-16 model with pre-trained weights of Imagenet is used in a transfer learning approach for human classification. It has 16 layers and the last FC layer is changed into two neurons with sigmoid activation to perform the binary classification task. The modified set of layers of VGG-Net 16 for human classification is shown in Fig. 39.3. The layers and their parameters are shown in Fig. 39.8.

The progress in the training of this model is shown in Fig. 39.9. The classification result obtained for transfer learning using VGG-16 model is shown in Tables 39.1 and 39.2. The average accuracy of 98% is obtained for this approach.

Building of CNN Model for Classification The 6 layer classification model as shown in Fig. 39.4 is trained from scratch and the different layers and the parameters associated with the model are shown in Fig. 39.10.

The trained model-1 is tested for human classification with parameter settings as listed below.

- Batch normalization is done for each layer
- The dropout factor is kept at 0.25
- Binary cross-entropy loss function
- Adam optimization algorithm
- Number of epochs = 10
- Batch size = 32

During training model 1, it is observed that the loss and accuracy is not improved with training iterations. The average accuracy of 75.6% is obtained using model 1

```
input_1 (InputLayer)         (None, 224, 224, 3)       0
block1_conv1 (Conv2D)        (None, 224, 224, 64)      1792
block1_conv2 (Conv2D)        (None, 224, 224, 64)      36928
block1_pool (MaxPooling2D)   (None, 112, 112, 64)      0
block2_conv1 (Conv2D)        (None, 112, 112, 128)     73856
block2_conv2 (Conv2D)        (None, 112, 112, 128)     147584
block2_pool (MaxPooling2D)   (None, 56, 56, 128)       0
block3_conv1 (Conv2D)        (None, 56, 56, 256)       295168
block3_conv2 (Conv2D)        (None, 56, 56, 256)       590080
block3_conv3 (Conv2D)        (None, 56, 56, 256)       590080
block3_pool (MaxPooling2D)   (None, 28, 28, 256)       0
block4_conv1 (Conv2D)        (None, 28, 28, 512)       1180160
block4_conv2 (Conv2D)        (None, 28, 28, 512)       2359808
block4_conv3 (Conv2D)        (None, 28, 28, 512)       2359808
block4_pool (MaxPooling2D)   (None, 14, 14, 512)       0
block5_conv1 (Conv2D)        (None, 14, 14, 512)       2359808
block5_conv2 (Conv2D)        (None, 14, 14, 512)       2359808
block5_conv3 (Conv2D)        (None, 14, 14, 512)       2359808
block5_pool (MaxPooling2D)   (None, 7, 7, 512)         0
sequential_6 (Sequential)    (None, 1)                 200721
=================================================================
Total params: 14,915,409
Trainable params: 14,915,409
Non-trainable params: 0
```

Fig. 39.8 Layers and parameters of modified VGG-16 architecture

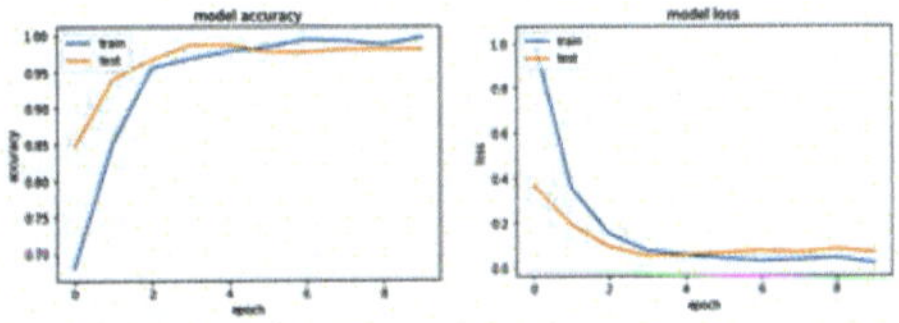

Fig. 39.9 Loss and accuracy graph for VGG-16 model

Table 39.1 Confusion matrix for VGG-16 model

	Non-human	Human
Non-human	**148**	2
Human	4	**146**

Table 39.2 Classification performance measures for VGG-16

	Precision	Recall	f1-score	support
Non-human	0.97	0.99	0.98	150
Human	0.99	0.97	0.98	150
Avg/total	0.98	0.98	0.98	300

Fig. 39.10 Layers and parameters of Model-1

Layer (type)	Output Shape	Param #
conv2d_23 (Conv2D)	(None, 256, 256, 32)	896
conv2d_24 (Conv2D)	(None, 254, 254, 32)	9248
max_pooling2d_12 (MaxPooling	(None, 127, 127, 32)	0
batch_normalization_12 (Batc	(None, 127, 127, 32)	128
dropout_16 (Dropout)	(None, 127, 127, 32)	0
conv2d_25 (Conv2D)	(None, 127, 127, 64)	18496
conv2d_26 (Conv2D)	(None, 125, 125, 64)	36928
max_pooling2d_13 (MaxPooling	(None, 62, 62, 64)	0
batch_normalization_13 (Batc	(None, 62, 62, 64)	256
dropout_17 (Dropout)	(None, 62, 62, 64)	0
flatten_5 (Flatten)	(None, 246016)	0
dense_9 (Dense)	(None, 512)	125960704
dropout_18 (Dropout)	(None, 512)	0
dense_10 (Dense)	(None, 1)	513

Total params: 126,027,169
Trainable params: 126,026,977
Non-trainable params: 192

Fig. 39.11 Layers and parameters of Model-2

Layer (type)	Output Shape	Param #
conv2d_19 (Conv2D)	(None, 256, 256, 32)	896
conv2d_20 (Conv2D)	(None, 254, 254, 32)	9248
max_pooling2d_10 (MaxPooling	(None, 127, 127, 32)	0
batch_normalization_10 (Batc	(None, 127, 127, 32)	128
dropout_13 (Dropout)	(None, 127, 127, 32)	0
conv2d_21 (Conv2D)	(None, 127, 127, 64)	18496
conv2d_22 (Conv2D)	(None, 125, 125, 64)	36928
max_pooling2d_11 (MaxPooling	(None, 62, 62, 64)	0
batch_normalization_11 (Batc	(None, 62, 62, 64)	256
dropout_14 (Dropout)	(None, 62, 62, 64)	0
conv2d_23 (Conv2D)	(None, 62, 62, 128)	73856
conv2d_24 (Conv2D)	(None, 60, 60, 128)	147584
max_pooling2d_12 (MaxPooling	(None, 30, 30, 128)	0
batch_normalization_12 (Batc	(None, 30, 30, 128)	512
dropout_15 (Dropout)	(None, 30, 30, 128)	0
flatten_4 (Flatten)	(None, 115200)	0
dense_7 (Dense)	(None, 512)	58982912
dropout_16 (Dropout)	(None, 512)	0
dense_8 (Dense)	(None, 1)	513

Total params: 59,271,329
Trainable params: 59,270,881
Non-trainable params: 448

and the performance has not been improved even after increasing the iterations or by varying parameters. Hence we added two more layers to model 1 to improve the results.

Model 2 Results: Model 2 is created by adding 2 more CNN layers to the existing layers of model 1 to make it as large network to perform human classification, which is shown in Fig. 39.5. The different layers and parameters of model-2 are shown in Fig. 39.11. The progress in the training for 100 epochs is shown in Fig. 39.12. The

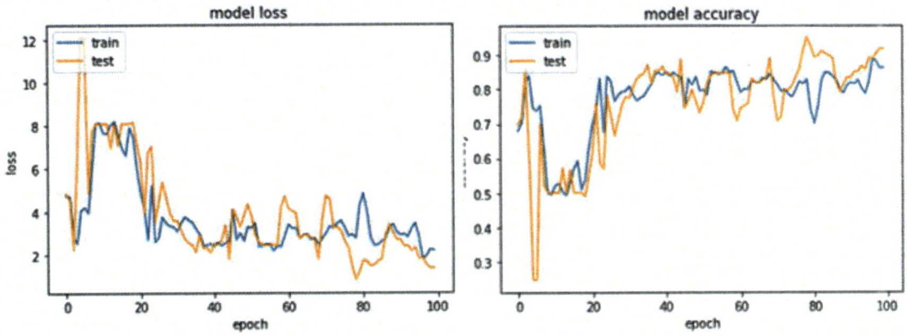

Fig. 39.12 Graph showing loss and accuracy versus epochs for model-2

Table 39.3 Confusion matrix for model-2

	Non-human	Human
Non-human	**108**	42
Human	5	**145**

Table 39.4 Performance measures for model-2

	Precision	Recall	f1-score	Support
Non-human	0.96	0.72	0.82	150
Human	0.78	0.97	0.86	150
avg/total	0.87	0.84	0.84	300

Table 39.5 Comparison of performance between various models

Model	Layers	Time/epoch	Accuracy
Model-1	6	187s	75.66%
Model-2	8	373s	84.3%
VGG-16	16	1149 s	98%

classification results are shown in Tables 39.3 and 39.4. The average accuracy of 84% is obtained for this model.

The comparison between two custom-designed CNN models and the VGG-16 network for human classification is shown in Table 39.5.

39.4.4 Discussion

From the experiment it is found that both deep learning models, i.e., model 2 and VGG 16 model performed satisfactorily for the human classification task. The designed models can classify humans even if they appear with different appearances, shapes, etc. The VGG-16 has resulted in the highest accuracy with a fewer number of training epochs due to the use of pre-training weights. The custom-designed model caused comparable results to that of VGG 16, using reduced computation time. Hence it is found that even the smaller networks can perform better for some applications, as compared to larger networks, which need higher computation requirements. Thus we can reduce the model complexity and the computation time.

39.5 Conclusion

In this work, we have demonstrated two approaches of designing a CNN-based human classifier, i.e., custom design of a classifier from scratch and a VGG-16 based classifier in a transfer learning approach. Both the networks can classify humans with reasonably good accuracy. With the transfer learning approach, learning takes place with fewer iterations and high accuracy is obtained due to the use of pre-trained weights. The experiment has demonstrated the reduction in computation time by designing a simple custom classifier for a human classification task. Satisfactory results have been obtained for the aerial dataset having variation in human pose, size, and shape, etc. The research work can further be extended to utilize such custom-designed classifiers with standard object detection frameworks so that computations can be reduced by a large extent.

References

1. Gerónimo, D., López, A.M., Sappa, A.D., Graf, T.: Survey of pedestrian detection for advanced driver assistance systems. IEEE Trans. Pattern Anal. Mach. Intell. **32**(7), 1239–1258 (2010)
2. Liu, Y., Sun, P., Wergeles, N., Shang, Y.: A survey and performance evaluation of deep learning methods for small object detection. Expert Syst. Appl. 114602 (2021)
3. Dalal, N., Triggs, B.: Histograms of oriented gradients for human detection. In: 2005 IEEE Computer Society Conference on Computer Vision and Pattern Recognition (CVPR'05), vol. 1, pp. 886–893. IEEE (2005)
4. Felzenszwalb, P.F., Girshick, R.B., McAllester, D., Ramanan, D.: Object detection with discriminatively trained part-based models. IEEE Trans. Pattern Anal. Mach. Intell. **32**(9), 1627–1645 (2009)
5. Liu, L., Ouyang, W., Wang, X., Fieguth, P., Chen, J., Liu, X., Pietikäinen, M.: Deep learning for generic object detection: a survey. Int. J. Comput. Vision **128**(2), 261–318 (2020)
6. Voulodimos, A., Doulamis, N., Doulamis, A., Protopapadakis, E.: Deep learning for computer vision: a brief review. Computat. Intell. Neurosci. **2018** (2018)
7. Simonyan, K., Zisserman, A.: Very deep convolutional networks for large-scale image recognition. arXiv preprint arXiv:1409.1556 (2014)
8. He, K., Zhang, X., Ren, S., Sun, J.: Deep residual learning for image recognition. In: Proceedings of the IEEE Conference on Computer Vision and Pattern Recognition, pp. 770–778 (2016)

9. Munder, S., Gavrila, D.M.: An experimental study on pedestrian classification. IEEE Trans. Pattern Anal. Mach. Intell. **28**(11), 1863–1868 (2006). https://doi.org/10.1109/TPAMI.2006.217
10. Enzweiler, M., Gavrila, D.M.: A multilevel mixture-of-experts framework for pedestrian classification. IEEE Trans. Image Process. **20**(10), 2967–2979 (2011). https://doi.org/10.1109/TIP.2011.2142006
11. Tayara, H., Chong, K.T.: Object detection in very high-resolution aerial images using one-stage densely connected feature pyramid network. Sensors (Switzerland) **18**(10) (2018). https://doi.org/10.3390/s18103341
12. AlDahoul, N., Md Sabri, A.Q., Mansoor, A.M.: Real-time human detection for aerial captured video sequences via deep models. Comput. Intell. Neurosci. **2018**, 1–14 (2018)
13. Zhu, J., Liao, S., Lei, Z., Li, S.Z.: Multi-label convolutional neural network based pedestrian attribute classification **58**, 224–229 (2017) https://doi.org/10.1016/j.imavis.2016.07.004, http://dx.doi.org/10.1016/j.imavis.2016.07.004
14. Melotti, G., Premebida, C., Goncalves, N.M.d.S., Nunes, U.J., Faria, D.R.: Multimodal CNN pedestrian classification: a study on combining lidar and camera data. In: 2018 21st International Conference on Intelligent Transportation Systems (ITSC), pp. 3138–3143. IEEE (2018)
15. Krizhevsky, A., Sutskever, I., Hinton, G.E.: Imagenet classification with deep convolutional neural networks. In: Advances in Neural Information Processing Systems, pp. 1097–1105 (2012)
16. Krizhevsky, A., Sutskever, I., Hinton, G.E.: Imagenet classification with deep convolutional neural networks. Adv. Neural. Inf. Process. Syst. **25**, 1097–1105 (2012)
17. Al-Jawfi, R.: Handwriting arabic character recognition lenet using neural network. Int. Arab J. Inf. Technol. **6**(3), 304–309 (2009)
18. UCF-ARG Dataset. https://www.crcv.ucf.edu/data/UCF-ARG.php. Accessed 19 Nov. 2018]
19. Robicquet, A., Sadeghian, A., Alahi, A., Savarese, S.: Stanford drone dataset http://cvgl.stanford.edu/projects/uav_data/

Chapter 40
Multi-sensor Fusion-Based Object Detection Implemented on ROS

Pranay Mathur, Ravish Kumar, and Rahul Jain

Abstract 3D Perception of the environment in real-time is a critical aspect for object detection, obstacle avoidance, and classification in autonomous vehicles. This paper proposes a novel 3D object classifier that can exploit data from a LIDAR, a RADAR, and a monocular camera image after orthogonal projection. To achieve this, a learnable architecture is designed end-to-end, which fuses the detection results from multiple sensor modalities initially and exploits continuous convolution subsequently to achieve the desired levels of accuracy. An adaptive algorithm for prediction in real-time is used to automatically increase weightage to prediction results from a particular sensor modality which aids in keeping accuracy invariant to scene changes. To prevent the bias, we are using a training strategy which provides attention to the specific type of sensor. This strategy is inspired by dropout. The entire algorithm has been implemented on the Robot Operating System to make it easier to deploy and transfer. We have experimentally evaluated our method on the NuScenes dataset.

40.1 Introduction

3D perception in real-time is an essential capability of mobile robots and autonomous vehicles, enabling robust localization, mapping, obstacle avoidance, and path planning in challenging real-world environments. Reliable detection of other road users in the environment and other obstacles poses a challenge for self-driving cars. Fusion of data has multiple benefits in the field of object detection for autonomous driving

Supported by KPIT.

P. Mathur
BITS Pilani, K.K Birla Goa Campus, Sancoale, India
e-mail: f20170487@goa.bits-pilani.ac.in

R. Kumar (✉) · R. Jain
KPIT Technologies, Pune, India
e-mail: ravishk2@kpit.com

R. Jain
e-mail: rahul.jain2@kpit.com

J. I.-Z. Chen et al. (eds.), *Machine Learning and Autonomous Systems*, Smart Innovation, Systems and Technologies 269, https://doi.org/10.1007/978-981-16-7996-4_40

[1–3]. In the present scenario, autonomous vehicles are often equipped with different sensors to perceive the environment. A multi-sensor fusion considers the output from each sensor and displays more robust and reliable information than an individual sensor would be capable of. Considering that humans perceive their surrounding while driving mainly via their visual sense, it is a natural assumption that an artificial perception system for autonomous vehicles should rely on similar information. But in conditions like, heavy fog, heavy snow or rain, the visibility is reduced, and driving safely becomes more challenging. Additionally, based on season changes, camera sensors easily get affected by noise, especially while driving in the tunnel or driving in the night. The camera can be rendered unusable through weather-induced occlusions, such as droplets of water in front of the camera lens restrict the view. Unlike camera sensors, RADAR sensors seem to be more robust especially in the conditions where camera sensors convincingly fail [4].

For the case of object detection, range scans offer an advantage over image data streamed from a monocular camera as they provide spatial coordinates of the object stored in a set of points, known as a point cloud. This makes detection of pose and shape of the object easier to detect, which is a crucial aspect for subsequent stages like planning and control. The RADAR receives information related to distance and is particularly adept at acquiring the radial velocity of objects. As a disadvantage, a RADAR only locates objects in a 2D ground parallel plane and also it does not capture information related to height.

LIDARs provide highly accurate results on obstacle positions and less accurate results on their velocities. Simultaneously, RADARs provide more accurate and higher precision results on obstacle speeds and lesser precision results on obstacle distances. An advantage of LIDAR sensors is they provide a significantly higher resolution when compared to automotive RADAR sensors. In addition to data elucidating position, LIDAR also captures surface reflectivity. Lidars are widely famous about its precision for estimation the positions of objects. By fusing the multi-sensor data for performing sensor fusion, the goal is to achieve more reliable perceptions for a wider range of tasks especially in the areas of environment perception in autonomous vehicles overcoming any shortcomings offered by individual sensors. The contributions of this paper can be listed as follows:

- An end-to-end to deep learning architecture that uses a combination of loosely coupled fusion of bounding-box estimates using visual data and bounding-box estimates from tightly coupled fusion of LIDAR and RADAR data, which ensures proper weightage is given to each sensor modality depending on the scene.
- Implementation of our framework on ROS and using the Black-In technique of training to ensure no bias exists in weightage given to each modality.

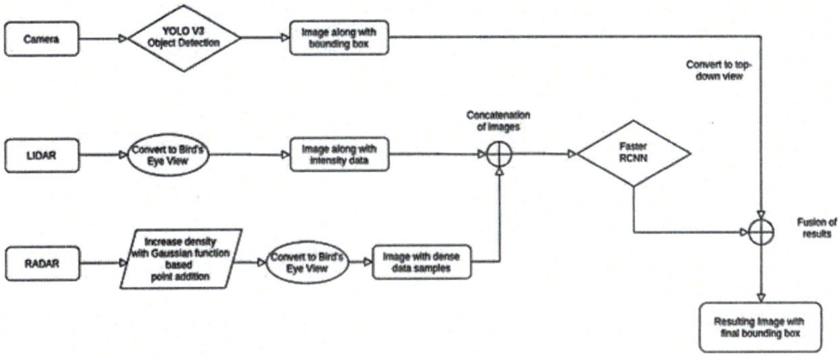

Fig. 40.1 Block-diagram of our proposed approach

40.2 Related Works

There have been numerous attempts in the past at improving object detection performance through sensor fusion. While certain newer methods use a variety of sensor-modalities such as thermal cameras, monocular cameras and RADARs [5] we limit our comparison to methods that use either a single sensor modality or a combination of LIDAR, monocular camera, and RADAR. We first examine methods using LIDAR pointclouds. A majority of methods convert the 3D pointcloud data into a 2D representation by projection onto an image plane along the vertical axis and resulting in the horizontal plane. This is known as a Bird's-eye view(BEV) representation. It exploits existing 2D object classification architectures used with images along with some other information such as height, density and reflectivity encoded in channels [6, 7]. Alternatively some methods project the LIDAR data along the horizontal axis on the vertical plane. The second method is shown in [8] where the pointcloud is projected onto a perpendicular image plane. This is then clubbed with the image representation of camera to perform the object detection task. A similar approach is used in [9] while working with a bird's-eye view projection for the two sensors. The second method is shown in [10] where LIDAR data is projected onto the two-dimensional ground plane as a bird's-eye view and then the camera data is clubbed for object detection using three-dimensional overview. Caltagirone et al. [11] mixes camera and LIDAR data using a neural network and performs segmentation to estimate the boundaries of the road. In this research work, proposal of a network structure is hypothesized. The branches responsible for fusion and concatenation of these two streams are trainable as well to ensure the network possesses the ability to memorize the patterns of an optimized depth level in the network.

Object detection using 2D images is one of the most widely researched areas. All methodologies related to object detection can be categorized into two main types. First is region proposals. Faster R-CNN [12] is one such approach. A separate region proposal network (RPN) uses CNN features to suggest possible regions where the objects can be present. The rest work is to classify the proposed regions. These processes are computationally complex and significantly decrease the processing speed. Second approach is to use one network for both label classification and predicting potential regions. An example of such an architecture is you only look once (YOLO) [13, 14]. Here, the input image is first divided into coarse grids. YOLO is way faster than Faster R-CNN, but fails for smaller objects.

Certain methods use RADAR data as well. Ji and Prokhorov [15] uses detections observed in RADAR to identify important regions in image streams and classify objects present in it using NN. Other works which uses RADAR information to facilitate the task of object detection are, [16–19]. Jha et al. [20] proposes a novel idea of fusing the independent detections of the camera and RADAR. Kim et al. [21] proposes to club independently tracked detections of every sensor and extract the final estimation and incorporate the readings of both the sensors. Blanc et al. [22] uses conventional Kalman-filter algorithm to fuse LIDAR-RADAR readings. It helps in sensing the environment better and provides a marginal reduction in false alarms. Similarly, in [23] LIDAR-RADAR fusion was shown with cars running on roads. Evidence theory-based LIDAR, RADAR, and Camera fusion approach appear in [24] which is used to track objects. Rangesh and Trivedi [25] focused on the fusion of more than one camera and LIDAR and demonstrated results on real-road data. Via this they actually showed the reliability of their methodology. Blanc et al. [26] proposes to use Cramer-Rao lower bound.

In our paper, we propose to use the BEV(Birds Eye View) obtained from a LIDAR and RADAR as well as the monocular camera by fusion of detection results obtained from purely visual means by carrying out an orthographic feature transform. We use a convolutional neural network for computing a high-resolution depth map given a single RGB image with the help of transfer learning based on the work of [27] following a standard encoder-decoder architecture, initializing their encoder along with augmentation and training strategies that lead to more accurate results. We also use the inverse projection function and the depth estimate to transform data from the monocular image plane to the plane parallel to the BEV frame as obtained using the LIDAR.

40.3 Methodology

This paper uses three separate input branches for LIDAR, monocular camera, and RADAR data. We opt for a bird's-eye view representation of LIDAR and RADAR data as shown in Figs. 40.2 and 40.3 respectively. The pointcloud of the LIDAR is projected onto the ground plane. The advantage of transferring the data from multiple sensor modalities is that they are agnostic to the frame of the sensor itself and provide only top-down semantic information. This raw data undergoes pre-processing and is published on ROS nodes. The monocular camera image undergoes orthogonal projection so that detected results are projected in a plane parallel to that of the LIDAR and RADAR. The data is then trained to overcome any bias toward camera images and such that LIDAR and RADAR data are given optimal weightage. We

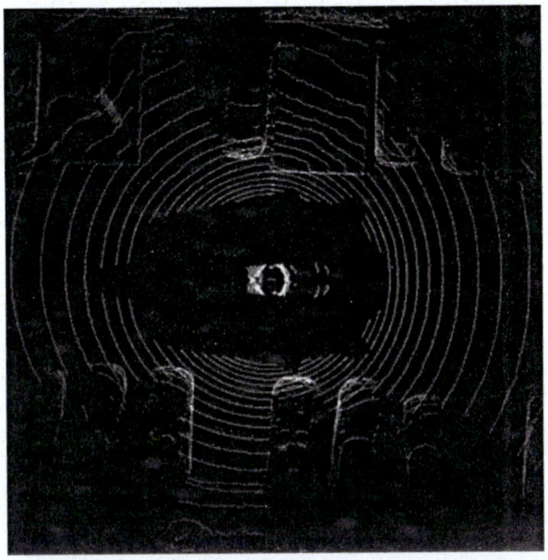

Fig. 40.2 BEV representation of LIDAR data

Fig. 40.3 Sparse raw RADAR data visualized

also introduce an adaptive real-time algorithm for dimly lit scenes that gives higher weightage to LIDAR and RADAR data predictions to ensure high accuracy levels invariant to scene changes. Our proposed approach is shown in Fig. 40.1.

We show our results using the NuScenes dataset. The NuScenes dataset consists thousands of 20 s long videos captured in cityscapes like Singapore and Boston which makes it more diverse. For annotation, frequency is set at a rate 2 Hz, which essentially implies per second there will be 40 frames annotated. Six cameras are used to capture this dataset, spread around the boundary of the car, full LIDAR spin point clouds, and 3-Dimensional annotated bounding boxes along with 23 object classes. The details are shown in Table 40.1.

Table 40.1 Sensor data

Sensor	Qty.	Update Freq. (Hz)	Range (m)	Resolution
Camera	6	12	–	1600*900
RADAR	5	13	250	77GHz
LIDAR	1	20	100 ± 0.02	360°HFOV +10° –30°VFOV

40.3.1 *Preprocessing*

40.3.1.1 LIDAR

The LIDAR data is projected onto the ground plane along the vertical axis and a bird's eye view of LIDAR data is obtained as seen in Fig. 40.2. Let the set of points in a pointcloud at a time instant t, be denoted by $P_k^w = (x, y, z)^T$ in the world frame. Top-down grid features are used to represent our input LIDAR. Generation of these features is done by projecting lidar points onto the top-down grid reference frame and encoding data such as intensity, reflectivity and density in three channels. For each input frame, we have total dimensions, $\mathbb{R}^{W*H*3}$, where W and H denote the width and height of the frame such that intensity is given by a function proportional to z. We scale the pointcloud and covert it to a 2D image using the following functions where u and v denote the pixel coordinates.

$$u = W * \frac{x + |x_{min}|}{x_{max} + |x_{min}|} \tag{40.1}$$

$$v = H * \frac{y + |y_{min}|}{y_{max} + |y_{min}|} \tag{40.2}$$

$$intensity = 255 * \frac{(z + z_{min})}{z_{max}} \tag{40.3}$$

40.3.1.2 RADAR

We have used RADAR data in our approach and we will discuss the projection of this data in this section. Projection requires spatial calibrations and is described here. We also describe an approach to alter the sparsity of the radar data. The point cloud coming from the radar sensor, characterized by extreme sparsity and limited to two dimensional plane with associated radar characteristics. A bird's eye view of RADAR data is obtained. Since the data is sparse as seen in Fig. 40.3, sparsity of RADAR data is decreased and density increased, and the final result is as shown in Fig. 40.4. Pixel values mapped in the range of [0, 255] are used to store the properties of the

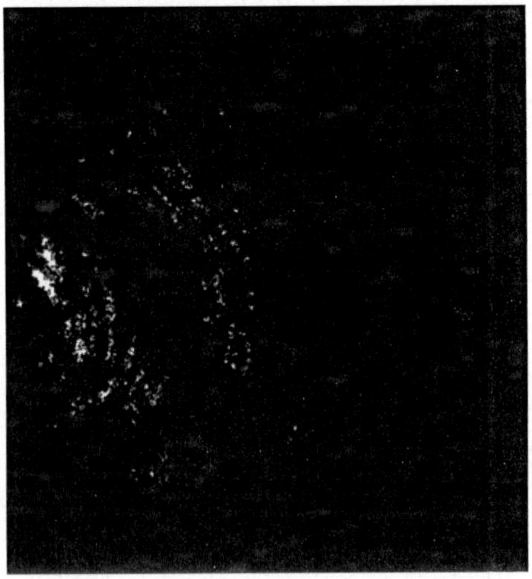

Fig. 40.4 RADAR data after pre-processing

RADAR return in the augmented image. The pixel values in image, (u, v) where there is absence of RADAR returns, radar channel values are set to be 0. We increase the density of clustered points by checking if $density$ of points in a region is above a threshold and add points using Gaussian distribution to increase pointcloud density. The Gaussian is characterized by $\mathcal{N}(\mu = (u, v), \sigma = 0.01)$

$$u = W * \frac{x + |x_{min}|}{x_{max} + |x_{min}|} \tag{40.4}$$

$$v = H * \frac{y + |y_{min}|}{y_{max} + |y_{min}|} \tag{40.5}$$

40.3.1.3 Monocular Vision

Similar to that obtained from the other two sensors, output representation is a top-down view which is the desired format. Vision inputs are images in perspective view. We would be converting the pixel space to top-down space, using the method described in [27] along with the inverse projection function. This method allows us to learn features in pixel space which can then be decoded into a top-down output. For a pixel $(u, v)^T$ we use the estimated depth y_{est} to construct a 3D point $\mathbf{p}$. We use the inverse projection function π such that

$$\mathbf{p} = \pi^{-1}(x, y, \mathbf{Y}(x, y)) \tag{40.6}$$

$$\mathbf{p} = \mathbf{Y}(x, y)\left(\frac{u + c_x}{f_x}, \frac{v + c_y}{f_y}, 1\right)^T \tag{40.7}$$

where **Y**(x,y) is the depth of the pixel, and f_x,f_y and $c_x c_y$ denote the focal length and optical center of the pinhole camera model, respectively. Only the regions containing object detection results above a parametric threshold undergo orthographic transformation.

40.3.2 Training

The dataset is split in a ratio of 60:20:20, for training, validation, and testing, respectively. Dropout technique is an inspiration for the proposed strategy [28]. The method is similar to the one used in [29] which is referred to as BlackIn. Blank images with the pixel value = 255 are used to deactivate all input neurons for random training steps. Rate is kept at 0.2 for all the training samples for this. In case of no camera input data, algorithm forces the network to rely more on the rest sensors. We target being teaching the network the value of the information possessed by the high-resolution LIDAR data and sparse radar data independent of camera representation which is comparatively highly dense (Fig. 40.5).

40.3.2.1 Model and Implementation

Our neural network is RCNN Inception v2 architecture which is a region-based convolutional neural network. R-CNN consists of three main steps: image scanning and generating region proposals using selective search algorithm for possible objects. On each of these proposals, a CNN is subsequently run. Then the output is fed to Support Vector Machine based classifier for classifying the regions and then a linear regressor model for tightening of bounding boxes. This approach is very slow and hence a variant of RCNN is the Faster R-CNN [32] algorithm. Instead of extracting features after finding the regions, it extracts features over the image before proposing regions. Thus over the entire image just one CNN is required instead of multiple CNN's over multiple overlapping regions. It does not use SVM to classify, instead uses softmax layer for predictions. A more faster algorithm named *Faster R-CNN* is also there. It uses a fast neural network RPN (Region proposal network) instead of slow selective search algorithm. We subscribe to our LIDAR data and RADAR data for bounding box prediction using the above model. Simultaneously we run the YOLOv3 [33] architecture on monocular camera data for visual object detection. YOLO(You only look once) is an architecture that uses only convolutional layers, and is therefore referred to as a fully convolutional network (FCN). The authors of YOLO v3 presented Darknet-53 architecture, a deeper architecture to extract the features. It has fifty-three convolution layers and followed by a BN layer and activation function as *Leaky ReLU*.

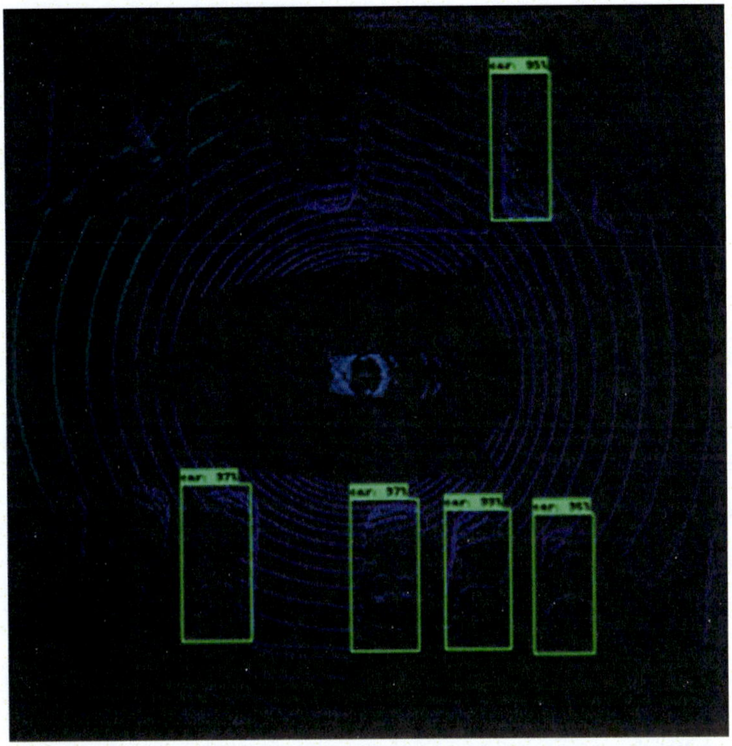

Fig. 40.5 Object detection in BEV representation after encoding depth and intensity

Architecture of YOLO-V3 and Faster-RCNN is represented in Figs. 40.6 and 40.7, respectively. Any form of pooling is avoided and in addition a CNN layer with stride = 2 is introduced for the purpose of feature maps downsampling. Hence it prevents the loss of low-level features associated with pooling. We then project the detected regions into a plane parallel to the LIDAR and RADAR BEV represented plane and use an aggregation function that combines these results. The entire detection network has been implemented using ROS that subscribes to incoming data streams from different sensor modalities and pre-processes them before republishing them for prediction. We also use an adaptive weightage mechanism during classification where LIDAR data classification is given higher preference if the lighting conditions are adverse for object detection based on monocular camera data.

Our aggregation function resizes all the results to 600 by 600 images and then uses the confidence of each prediction to make the final prediction represented in the BEV view.

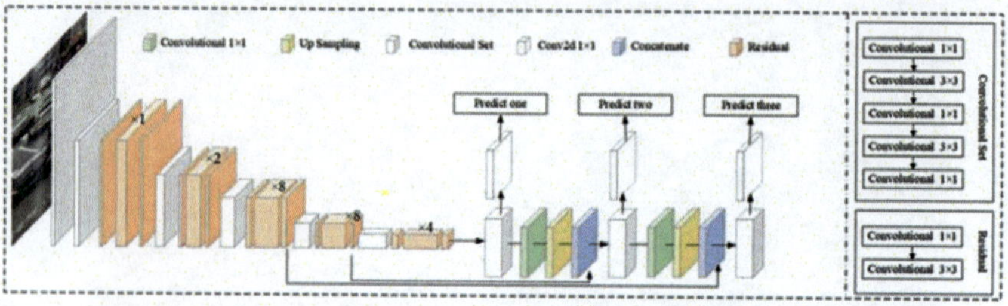

Fig. 40.6 YOLOv3 architecture [30]

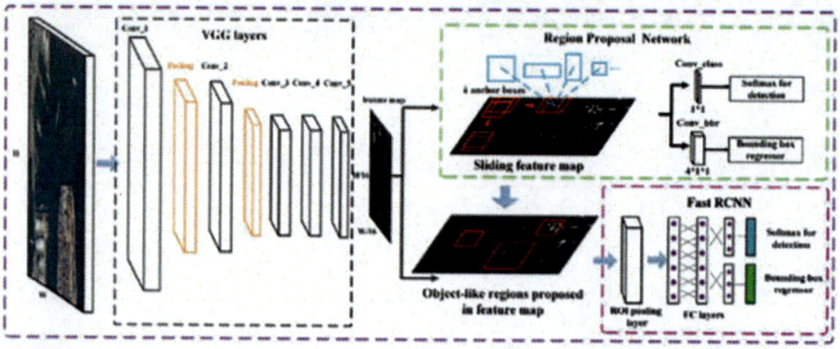

Fig. 40.7 Faster RCNN architecture [31]

Table 40.2 Accuracy

NuScenes Set	Loss	Time (s)	Accuracy%
Day	0.115	0.422	89.23
Day (Occlusion)	0.367	0.184	81.15
Night	0.579	0.559	78.27
Night (Rain)	0.267	0.591	77.32
Night (Peds)	0.767	0.414	68.11

40.4 Results

We test our results on an Intel Core i3–4010U dual-core processor clocked at 1.7 GHz. The results are shown in Table 40.2. We verify our results on the NuScenes [34] dataset and the accuracy is shown in Table 40.2. Apart from standard evaluations, we highlight the results for corner cases such as partial occlusion, in conditions of diminished illumination, both with and without rain. While diminished illumination causes a reduction of 11.04% in accuracy as opposed to during daylight, the decline in performance is minimized as opposed to methods where a single sensor modality is used. The fusion of multiple sensor modalities provides a high degree of accuracy. The plots for training loss and accuracy are shown in Fig. 40.8a and b respectively.

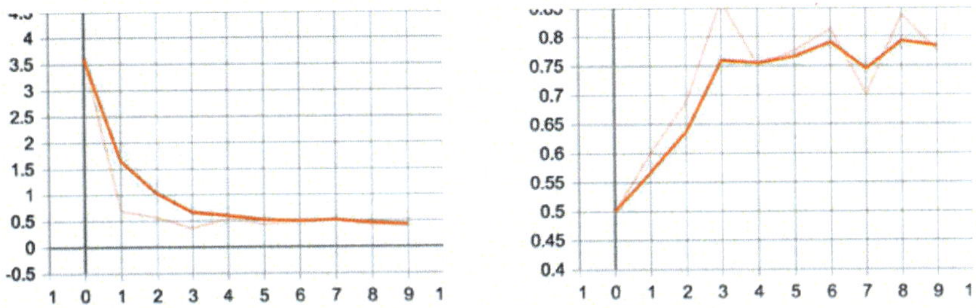

(a) Loss while training (b) Accuracy change over epochs

Fig. 40.8 Training results

40.5 Discussion and Conclusions

The goal of this research paper was to demonstrate a robust system for object detection using multi-sensor fusion. We presented a model architecture that uses orthogonal projections of monocular camera object detection and an aggregation function that combines the results obtained from all three sensor modalities. The implementation ensures an improved real-time performance and scene invariant accuracy. The proposed hypothesis is validated over the NuScenes dataset and achieved decent accuracy when tested for different daytime or weather time. While our method predicts bounding boxes that give the location of an object. Bounding box prediction should also specify the pose of the object detected so that its orientation can be known. Future work can also include modifying the monocular camera sensor modality and using a stereo camera or depth camera to ensure additional depth information can be incorporated in our model architecture. Along with using this additional information to improve accuracy research should also focus on improving real-time performance and decreasing power requirements to improve deployment.

Acknowledgements The authors would like to acknowledge the Chief Technology Officer Department of KPIT Technologies Ltd., Pune (India), as the work was done during the internship period

References

1. Rangesh, A., Trivedi, M.M.: No blind spots: Full-surround multi-object tracking for autonomous vehicles using cameras and LIDARs (2018). https://arxiv.org/pdf/1802.08755.pdf
2. Mitchell, H.B.: Multi-Sensor Data Fusion: An Introduction, 1st edn. Springer Publishing Company, Incorporated, Berlin (2007)
3. Blanc, C., Checchin, P., Gidel, S., Trassoudaine, L.: Data fusion performance evaluation for range measurements combine with cartesian ones for road obstacle tracking. In: 11th IEEE International Conference on Intelligent Transportation Systems (2007)

4. Gohring, D., Wang, M., Schnurmacher, M., Ganjineh, T.: RADAR/LIDAR sensor fusion for car-following on highways. IEEE Int. Conf. Autom. Robot. Appl. 407–412 (2011)
5. Hall, D.L., McMullen, S.A.H.: Mathematical Techniques in Multisensor Data Fusion. Artech House (2004)
6. Yu, D., Xiong, H., Xu, Q., Wang, J., Li, K.: Multi-stage residual fusion network for lidar-camera road detection. IEEE Intelligent Vehicles Symposium (2019)
7. Deng, Z., Sun, H., Zhou, S., Zhao, J., Lei, L., Zou, H.: Multi-scale object detection in remote sensing imagery with convolutional neural networks. ISPRS J. Photogram. Remote Sens. **145** (2018). https://doi.org/10.1016/j.isprsjprs.2018.04.003
8. de Ponte Müller, F.: Survey on ranging sensors and cooperative techniques for relative positioning of vehicles. Sensors (Basel, Switzerland) **17**(2) (2017)
9. Nobis, F., Geisslinger, M., Weber, M., Betz, J., Lienkamp, M.: A deep learning-based radar and camera sensor fusion architecture for object detection. In: Sensor Data Fusion: Trends, Solutions, Applications (SDF)vol. 2019, pp. 1–7. Bonn, Germany (2019). https://doi.org/10.1109/SDF.2019.8916629
10. Mitchell, H.B.: Multi-Sensor Data Fusion: An Introduction, 1st edn. Springer Publishing Company, Incorporated (2007)
11. Caltagirone, L., Bellone, M., Svensson, L., Wahde, M.: Lidar-camera fusion for road detection using fully convolutional neural networks. Robot. Autonom. Syst. **111**, 125–131 (2018)
12. Ren, S., He, K., Girshick, R., Sun, J.: Faster R-CNN: towards real-time object detection with region proposal networks. IEEE Trans. Pattern Anal. Mach. Intell. **39**, 1137–1149 (2017)
13. Alhashim, I., Wonka, P.: High quality monocular depth estimation via transfer learning (2019). https://arxiv.org/pdf/1812.11941.pdf
14. Beltran, J., Guindel, C., Moreno, F.M., Cruzado, D., Garcia, F., De La Escalera, A.: Birdnet: a 3d object detection framework from lidar information. In: 2018 21st International Conference on Intelligent Transportation Systems (IISC), pp. 3517–3523 (2018)
15. Kocic, J., Jovicic, N., Drndarevic, V.: Sensors and sensor fusion in autonomous vehicles. In: TELFOR 2018, pp. 420–425. Telecommunications Society and Academic Mind, Belgrade (2018)
16. Ku, J., Mozifian, M., Lee, J., Harakeh, A., Waslander, S.: Joint 3d proposal generation and object detection from view aggregation. In: IROS (2018)
17. Caesar, H., Bankiti, V., Lang, A.H., Vora, S., Liong, V.E., Xu, Q., Krishnan, A., Pan, Y., Baldan, G., Beijbom, O.: nuScenes: a multimodal dataset for autonomous driving. arXiv:1903.11027
18. Kim, K.-E., Lee, C.-L., Pac, D.-S., Lim, M.-T.: Sensor fusion for vehicle tracking with camera and radar sensor. In: 2017 17th International Conference on Control, Automation and Systems: Proceedings (2017)
19. Caltagirone, L., Bellone, M., Svensson, L., Wahde, M.: Lidar-camera fusion for road detection using fully convolutional neural networks. Robot. Autonom. Syst. **111**, 125–131 (2018)
20. Srivastava, N., Hinton, G., Krizhevsky, A., Sutskever, I., Salakhutdinov, R.: Dropout: a simple way to prevent neural networks from overfitting. J. Mach. Learn. Res. **15**, 1929–1958 (2014). http://jmlr.org/papers/v15/srivastava14a.html
21. Redmon, J., Divvala, S., Girshick, R., Farhadi, A.: You only look once: unified, real-time object detection. In: Proceedings of the IEEE Conference on Computer Vision and Pattern Recognition, vol. 27–30, pp. 779–788. Las Vegas, NV, USA (2016)
22. Redmon, J., Farhadi, A.: YOLO9000: better, faster, stronger. In: Proceedings of the IEEE Conference on Computer Vision and Pattern Recognition, vol. 21–26, pp. 6517–6525. Honolulu, HI, USA (2017)
23. Redmon, J., Farhadi, A.: YOLOv3: An Incremental Improvement (2018)
24. Chavez-Garcia, R.O., Aycard, O.: Multiple sensor fusion and classification for moving object detection and tracking. IEEE Trans. Intell. Transp. Syst. **17**(2), 252–534 (2015)
25. Ren, S., He, K., Girshick, R., Sun, J.: Faster R-CNN: towards real-time object detection with region proposal networks. IEEE Trans. Pattern Anal. Mach. Intell. **39**, 1137–1149 (2017)
26. Han, S., Wang, X., Xu, L., Sun, H., Zheng, N.: Frontal object perception for intelligent vehicles based on radar and camera fusion. In: Chen, J., Zhao, Q., (eds.), Proceedings of the 35th Chinese Control Conference, pp. 4003–4008. IEEE, Piscataway (2016)

27. Sun, H.-M., Liu, Y.-B., Jia, R.-S.: Mini-YOLOv3: Real-Time Object Detector for Embedded Applications. IEEE Access, p. 1 (2019). https://doi.org/10.1109/ACCESS.2019.2941547
28. Ren, S., He, K., Girshick, R., Sun, J.: Faster R-CNN: towards real-time object detection with region proposal networks. IEEE Trans. Pattern Anal. Mach. Intell. **39**(6), 1137–1149 (2017). https://doi.org/10.1109/TPAMI.2016.2577031
29. Yu, S., Westfechtel, T., Hamada, R., Ohno, K., Tadokoro, S.: Vehicle detection and localization on bird's eye view elevation images using convolutional neural network. In: 2017 IEEE International Symposium on Safety, Security and Rescue Robotics (SSRR), Oct 2017, pp. 102–109 (2017)
30. Zeng, S., Zhang, W., Litkouhi, B.B.: Fusion of obstacle detection using radar and camera, Patent US 9.429,650 B2 (2016)
31. Chen, X., Ma, H., Wan, J., Li, B., Xia, T.: Multi-view 3d object detection network for autonomous driving
32. Ren, S., He, K., Girshick, R., Sun, J.: Faster R-CNN: towards real-time object detection with region proposal networks. IEEE Trans. Pattern Anal. Mach. Intell. **39**(6), 1137–1149 (2017). https://doi.org/10.1109/TPAMI.2016.2577031
33. Bar-Shalom, Y., Fortmann, T.E.: Tracking and Data Association, Series Mathematics in Science and Engineering, vol. 179. Academic (1988)
34. Ji, Z., Prokhorov, D.V.: Radar-vision fusion for object classification. In: 11th International Conference on Information Fusion (2008)

Chapter 41
Fault Diagnosis Using VMD and Deep Neural Network

A. R. Aswani and R. Shanmughasundaram

Abstract A Self-diagnosis automotive system is capable of diagnosing its fault and corrects it by itself. The drive train part of automotive system consists of numerous rotating components which helps to control the power flow. Hence the regular monitoring of rotating components would help to prevent the severe damage. The condition monitoring of rotatory systems especially toward bearing fault is a trending topic nowadays. In this project, the bearing fault diagnosis is done using vibration signal analysis. Outer race fault and Roller faulted experimental data are taken for analysis. Signal Decomposition is the foundation of any signal processing techniques. An Variational Mode Decomposition method (VMD) transforms the vibrational signal into several Intrinsic Mode Functions (IMF) and the features corresponding to these IMFs give better fault information. A Deep Neural Network is constructed using unsupervised Auto Encoder for dimensionality reduction and stacked the auto encoder with softmax classifier for bearing fault diagnosis.

41.1 Introduction

Maintenance allows industry manufacturers to extend equipment life, lower costs, reduce downtime and increase production quality by solving problems before they cause equipment failures. Predictive maintenance technique is used to determine any machinery system's health condition. It continuously monitors the daily activities and thereby we can diagnose the system behavior. It is used to predict any future failure so that it is easy to take any remedial actions such as replace the component and maintenance which could be predictive or preventive. It increases the reliability of the components, decreases the down time, and increases the efficiency. In preventive

A. R. Aswani (✉)
Department of Electrical and Electronics Engineering, Amrita School of Engineering, Coimbatore, India

R. . Shanmughasundaram
Amrita Vishwa Vidyapeetham, Vengal, India
e-mail: r_shanmughasundaram@cb.amrita.edu

J. I.-Z. Chen et al. (eds.), *Machine Learning and Autonomous Systems*, Smart Innovation, Systems and Technologies 269, https://doi.org/10.1007/978-981-16-7996-4_41

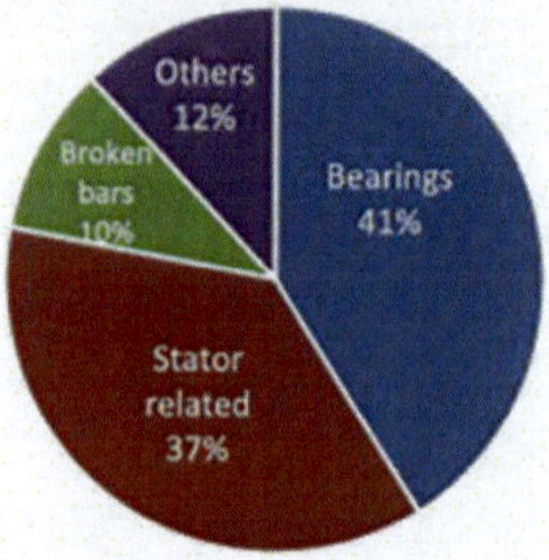

Fig. 41.1 Fault classification

maintenance, the maintenance service gets applied once it gets damaged or after the fault occurred. It leads to the need of labor service, increased downtime, and loss of productivity. But predictive maintenance will diagnose the fault priorly and takes necessary remedial action to cure the fault in real time basis.

Motors are important for locomotive systems, which are used to generate the power, amplify the power, and transmit properly. The analysis of fault in a motor could be difficult process as it includes different types of moving components and observing each rolling elements will be a laborious process. Hence it is necessary to diagnose the fault in motor with less processing time. The analysis of rotating component can be done using various signal processing and machine learning approaches. Faults that occur inside Induction Motor are namely [1] Motor fault, Fault in stator winding, Eccentricity fault, and bearing fault. Stator winding connection failure, Open or short circuit winding, and winding insulation failure are the majot faults that occur in the stator windings. Thermal analysis and Motor current analysis method are used to analyze these faults. Bearing faults include Outer race fault, Train effect, and Inner race fault. Current analysis, Vibration analysis, and Acoustic signal analysis are the various methods used to analyze the bearing fault. Eccentricity related faults occur due to unequal air-gap between stator with rotor and motor current analysis is used to solve eccentricity related fault. Among all the mechanical faults 40–45 % of faults that occur in a motor is due to bearing fault. Figure 41.1 shows the classification of different faults occurrence in an motor.

If the power from the motor is not properly transmitted to wheels, then the vehicle cannot be controlled by driver. In order to find bearing fault three methods which are widely used are vibration analysis, current analysis, and acoustic signal analysis. Vibration, current, and Acoustic give more optimum result than the other methods.

There are different types of fault in bearing, namely Inner race, Roller element, and Outer race. Motor current analysis is easy to implement but sensitive to broken rotor bar faults and its sidebands are self-reliant of load and speed. Acoustic signal analysis [2] influences to bearing fault but susceptible to external noise sources easily. Vibration signal analysis [3] gives variable sideband amplitudes with load and speed. Also it has strong harmonics and interferences but necessary permanent installation.

Vibration data contains rich information about the characteristics of fault location [4]. The time and frequency domain analysis of vibration data is used to take the characteristic features. Most time domain features are mean value, root mean square, and standard deviation. Frequency-domain features include the standard deviation of frequency, root variance frequency, and frequency center. Frequency characteristics can be calculated using Fast Fourier Transform, Hilbert transform, etc. The combination of time and frequency domain features will give better information regarding the fault characteristics. Hence Time–Frequency domain analysis is now most widely used feature extraction technique. Time–Frequency methods are effective in extracting the features which include Wavelet transform, local characteristic-scale decomposition (LCD), local mean decomposition (LMD), empirical mode decomposition (EMD) [5], variational mode decomposition (VMD) [6], and empirical wavelet transform (EWT) [7].

VMD [8], EMD and wavelet transform are commonly used Time–Frequency analysis. EMD is a self-adaptive method. It is more applicable when the vibration signals are nonlinear and non-stationary. EMD automatically estimates the number of modes. The EMD overcomes the intrinsic limitations of wavelet approach. EMD shows problems such as mode mixing that is EMD overestimates the number of modes and seperates some information which is originally part of the same component. EWT and VMD overcome that issue by decomposing the input signal into discrete number of intrisic mode functions. To select the sensitive features from the large amount of features both traditional machine learning as well as deep learning algorithms are used [9, 10]. The Traditional machine learning algorithms are less complex compared to deep learning which require large amount of dataset to train the algorithm. But when we take consideration about accuracy deep learning gives better accuracy compared to the traditional methods. The machine learning has less feature learning ability and it requires effective feature extraction methods. Commonly used deep learning algorithms are Restricted Boltzmann machines (RBMs), Auto encoder (AE), and Convolutional neural networks (CNNs). K-nearest neighbor (K-NN) classifier, feed-forward neural network (FNN) classifier, support vector machine (SVM), and Softmax are widely used machine learning algorithms. Softmax is the most widely used one for better classification result in the case of condition monitoring and also it takes less time.

41.2 Background

41.2.1 Variational Mode Decomposition

Variational Mode Decomposition is a noval latest decomposition technique which was developed by Dragomiretskiy and Zosso [11]. EMD was the earliest decomposition method before VMD method adopted. EMD was widely used for rotatory fault diagnosis but it still has a problem of mode mixing and end effect. VMD overcomes this issue with discrete number of modes. VMD is a non recursive method which decomposes the input signal into discrete number of Variational mode functions [12] with separate bands corresponding to it [13]. The frequency band and the corresponding modes are determined adaptively which helps to reduce the errors between the modes.

Weiner filtering and Hilbert transform are the two important concepts where weiner filtering is equivalent to central frequency of the IMF and Hilbert transform is used to calculate the envelop signal of each IMF components. Each mode is accessed using Guassian smoothness of the shifted signal and each mode is calculated using quadratic penality and Lagrange multiplier. It consists of balancing parameter of the data fiedility constraint. This equation is solved using alternative direction method of multipliers.

The u_k represents the intrinsic modes with each having a central frequency w_k. The modes are determined by taking a consideration of sum of all the modes should be equal to input signal and the bandwidth corresponding to each mode should be as minimum as possible. Bandwidth is continuously adjusted during the whole iteration until it satisfies the above criteria. Ensemple modes and the balancing parameter are the two important parameters in the decomposition process.

$$min(\Sigma_k \|\partial[(\partial(t) + j/\pi t) * u_k(t)]e^{-}jwt\|) \tag{41.1}$$

where u_k represents variational mode functions and w_k represents central frequency. $\partial(t)$ is Dirichlet distribution function. The Lagrange multiplier gives the strictness of the constrains. The extended Lagrange multiplier is as follows,

$$\begin{aligned} L(u_k)(w_k\chi) &= \alpha \Sigma_k \|\partial_t[(\partial(t) + j/\pi t) * u_k(t)]e^{-}jwt\| \\ &+ |f(t) - \Sigma_k(t)\| + (\chi(t), f(t) - \Sigma_k u_k(t)) \end{aligned} \tag{41.2}$$

where α is penalty factor and χ is Lagrangian multiplier. The continuous updation of $u_k(n+1)$, $w_k{}^(n+1)$ and χ leads to saddle point of the extended Lagrangian multiplier. The$u_k and w_k$are as followes,

$$U_k^{n+1}(w) = (\overline{f(w)} - \Sigma_{i \neq k} \widehat{u_i}(w) + \widehat{\chi(w)/2} * 1/(1 + 2\alpha(w - w_k)^2 \tag{41.3}$$

$$w_k^{n+1} = \frac{\int_0^\infty w|\widehat{u_k(w)}|^2 dw}{\int_0^\infty |\widehat{u_k(w)}|^2 dw} \tag{41.4}$$

41.2.2 Auto Encoder

Auto encoder is used to reduce the dimension of the feature set or to find out the most sensitive features from the high dimensional feature set. It is an unsupervised feature

Fig. 41.2 Autoencoder structure

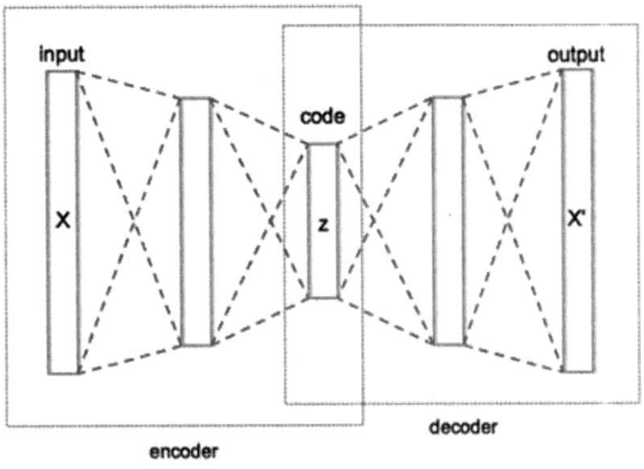

learning algorithm. Auto encoder consists of encoding and decoding section with a bottleneck region. Bottleneck or the code portion is the one which connects encoding and decoding section. Encoding and decoding layers consist of equal number of nodes which is same as the number of features in the high dimensional feature set. Decoding layer is used for reconstructing the original input image. Auto encoder architecture is shown in Fig. 41.2.

The encoder layer have an input of size n f =(f1, f2, f3,.....fn) and the inner layer with nodes will be equal to the number of feature dimension which we want to reduce. Let it be l =(l1, l2, l3,......lm) where m>n. Input layer and hidden layers are connected through feed forward path as per in the mapping.

$$l(j) = sigmoid(wf(i) + b) \tag{41.5}$$

where l is the hidden layer outputs and f is the input layer features. w is the weight matrix. b is the bias vector. Back mapping is done through the decoder layer f with a feature representation of the input features.

$$F(i) = sigmoid(WL(j) + B) \tag{41.6}$$

where W represents weights and bias vector is represented using B. The optimized auto encoder is obtained by constructing algorithm with suitable parameters such as w, W, B, b.

The feature set extracted from the hidden layers of the auto encoder contains reduced number of features which have most fault information of the input data. The output from the hidden layer is given as an input to the another layer stacked to this auto encoder which is a classifier called softmax classifier.

41.2.3 *Softmax Classifier*

Constructing the proper architecture of the classifier is essential for the design of optimized deep neural network. The feature set obtained from the last encoded layer of the auto encoder neural network can be give as an input to the softmax classifier which is the labeled feature set. It uses cross entropy-based loss function and when compared to other loss functions, cross entropy performs well for multiclass classification. It is also a type of sigmoid activation function but most commonly used for classification problems. It gives the probabilities to define each class of the input. Softmax function maps the output to the value 0 to 1.

41.3 Experimental Setup

The system methodology is shown in the Fig. 41.3 which consists of four steps namely data collection, feature extraction, feature reduction and classification. Data acquisition includes collecting data of vibration signal using a data acquisition system. This requires some prepossessing stages before moving onto fault diagnosis which includes feature extraction and formation of feature set. VMD algorithm gives a discrete amount of IMF and applying features into reach IMF will give a feature set which is capable of representing the information contained in the signal.

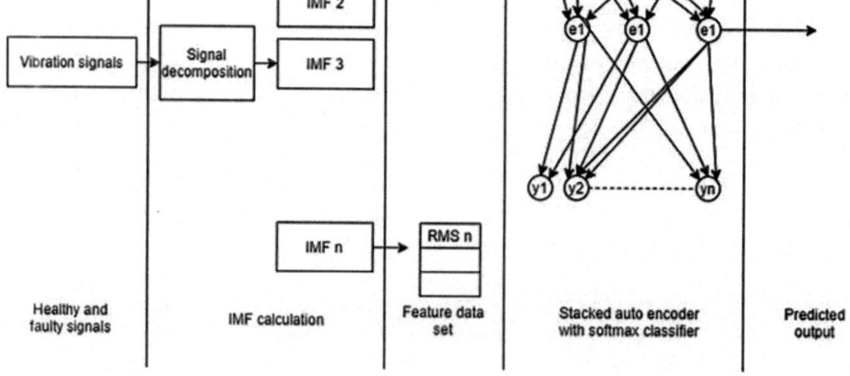

Fig. 41.3 Methodology

Feature set can be formed by grouping features representing each IMF and Auto encoder is used for finding most suitable features among whole feature set. Auto encoder reduces the dimenationality and appending a softmax classifier layer along with auto encoder will classify the data into normal or healthy and faulted data.

41.4 Data Acquisition

The first phase of implementation is to acquire data and find the corresponding features from it. The data collected contain vibration data of three phase induction motor with four load conditions. Figure 41.4 represents experimental setup and Dytran triaxile accelerometer is used to measure vibration data. The experiment setup consists of a 2 HP 3 phase AC Induction motor with an accelerometer. An accelerometer is placed at the magnetic base of the motor and data is collected for both healthy as well as fault condition.

Normal and roller element fault bearing are placed at 3 o'clock position and faulty data were grouped accordingly. In addition to that motor with different load condition is considered such as no load, 25% load condition, 50%load condition, 75% load

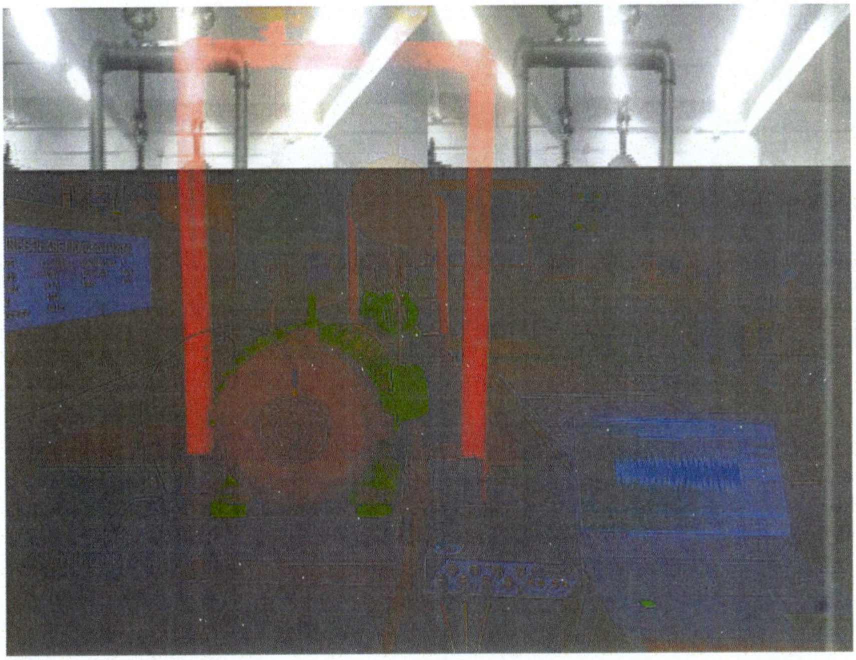

Fig. 41.4 Experiment setup

Table 41.1 Collected data

Algorithm	No	EMD		EWT		VMD	
		Precision	Recall	Precision	Recall	Precision	Recall
0	1	0.99	0.96	0.75	0.82	0.03	0.01
1	2	1.00	1.00	0.95	0.91	0.03	0.01
2	3	0.96	0.99	0.72	0.69	0.04	0.01

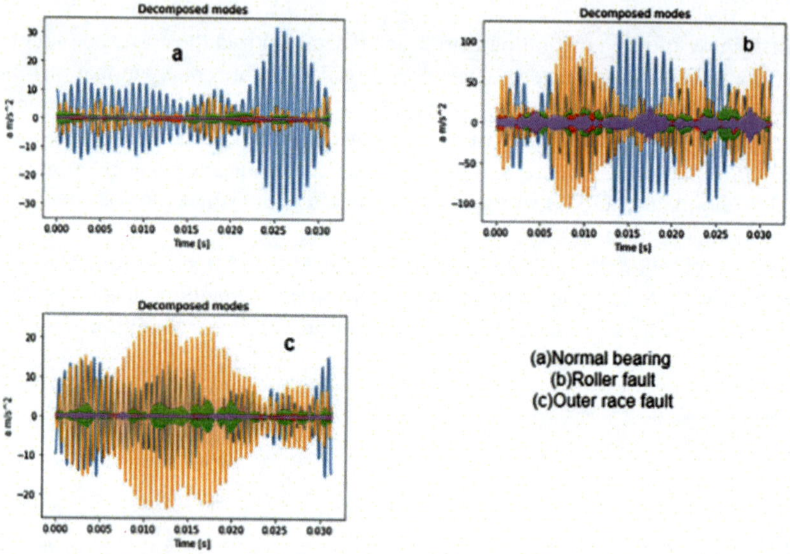

Fig. 41.5 VMD analysis

condition and full load. The sampling frequency of the platform 32768 Hz. The data can be arranged in a way that each sample contain 1000 sample points (Table 41.1).

41.4.1 Feature Extraction

VMD is performed to decompose the signal into several IMF. Vibration data with 1000 samples are decomposed into 5 IMFs. Figure 41.5 represents the VMD results of normal vibration data, Roller fault, and Outer race fautlt. The entropy of IMF will reduce for higher order IMF. Hence, the lower order IMF will contain more information than higher order IMFs.

The feature set required for training the deep neural architecture is constructed using five time domain parameters and two frequency domain features. The time domain parameters are Skewness (Sk), Kurtosis (Kr), Root mean square value (RMS),

Impulse factor (Im), and Crest factor (Cr). Frequency variance and Standard deviation of frequency (Stf) represent the frequency domain parameters. A high dimensional feature set is formed by applying time frequency features to each sample. Time frequency domain features are listed below.

$$RMS = \sqrt{1/N}(\sum_{i=1}^{N} W_i^2) \tag{41.7}$$

$$Skewness = (1/(N-1)St^3)\sum_{i=1}^{N}(W_i - \bar{W})^3 \tag{41.8}$$

$$Kurtossis = (1/(N-1)St^4)\sum_{i=1}^{N}(W_i - \bar{W})^4 \tag{41.9}$$

$$Impulsefactor = max(W)/\bar{W} \tag{41.10}$$

$$Crestfactor = max(W)/\sqrt{1/N}(\sum_{i=1}^{N} W_i^2) \tag{41.11}$$

$$Standarddeviationoffrequency = 1/N(\sum_{i=1}^{N}(f_1 - \bar{f})^2) \tag{41.12}$$

$$Frequencyvariance = (\sum f_i^2/\sum f) - (\sum f_i/\sum f)^2 \tag{41.13}$$

Figure 41.6 represents the data distributions of healthy and faulty data for each features. Red colored data which represents the Roller element fault, is easily separated from the other two informations. Hence the separation of Healthy and Outer race fault is the challenging for the classifier algorithm. The data which contain higher order IMF shows less valuable information compared to that of lower order IMF. Hence first five features are taken for creating the feature set.

These 7 features are applied to the first 5 IMF and obtained a total of 35 features. The whole 35 features may not represent the information content in the signal. The features which carry useful information can be understood by analyzing the feature plot shown in the Fig. 41.6. Hence a dimensionality reduction technique is used to find out the most sensitive features from the 35 feature set.

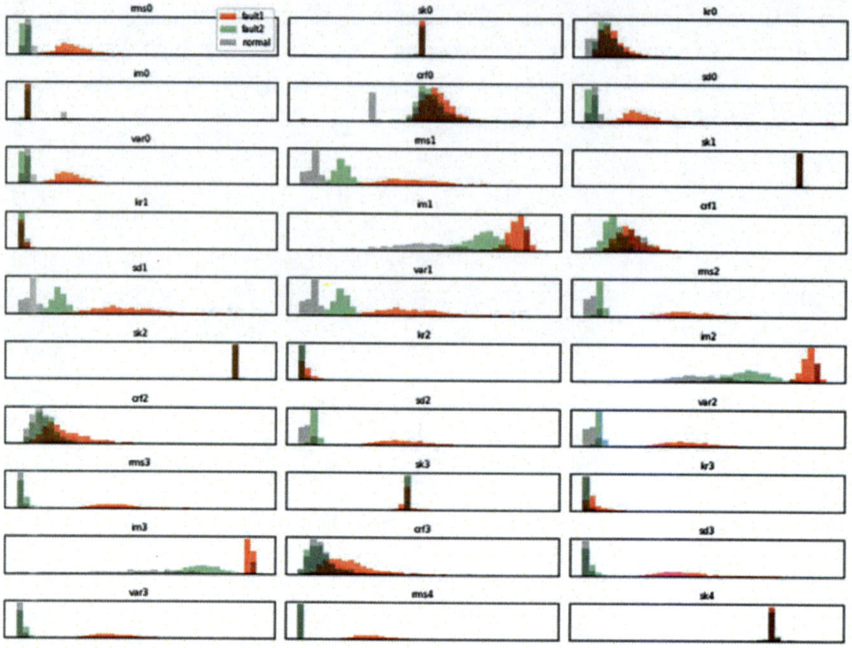

Fig. 41.6 Data distribution on features

41.4.2 *Feature Reduction and Classification*

Auto encoder is used to reduce the dimensionality of the feature set. A lower dimensional feature set obtained from the hidden layer of auto encoder is given as an input to the softmax classifier. The Extracted feature set is the input of Auto encoder which reduces the dimension of features. It contains several neural layers which reduces the feature size and the number of features obtained from the inner most layer represents the sensitive features, i.e., from the hidden layer or last encoded layer. The reduced feature set can be given to the softmax classifier as the input to detect the fault. The implementation of auto encoder can be divided into 3 stages namely Pretraining, Fine tuning, and Classification. The innermost layer has output shape equal to reduced number of features.

The fine tuning stage consists of appending the decoding layers above the encoding layers. Decoding layers reconstruct the reduced feature set into original feature set. This stacked auto encoder is trained until the reconstruction error reduces to minimal value. The model is trained for 100 epochs and by plotting the history it is clear that training and validation loss are decreasing with synchronization. It means that model is not over fitting and also both the training and validation loss almost merged together. Figure 41.7 shows the training loss and validation loss of training auto encoder by considering one fault condition that is roller element fault.

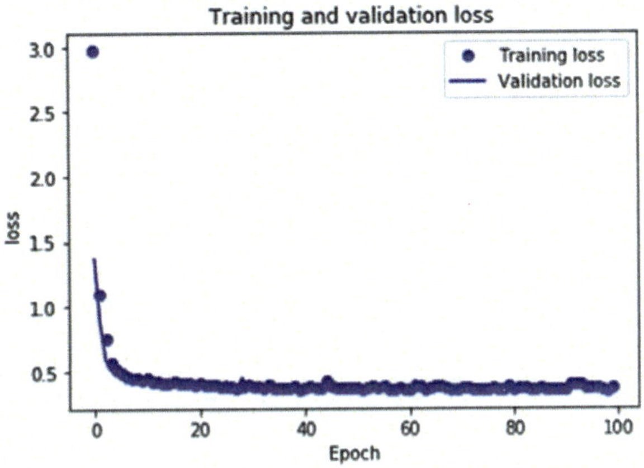

Fig. 41.7 Training and Validation loss

After training the encoding-decoding layers of auto encoder the output from the innermost encoding layer is given as input to the softmax classifier. During classification stage the softmax layer is appended on the top of hidden most layer of encoding layers. Softmax layer gives output as a range of values corresponding to the probability of occurrence of fault. The deep stacked layer of auto encoder is trained until the error is reduced to zero. Figure 41.8 represents the train and validation accuracy of binary classification. The classification accuracy obtained is 98.8%. The quantitative analysis of the result will be obtained after the calculation of the confusion matrix. Out of 3 classes which includes Normal, Outer race, and Roller element fault only 7 data of healthy data are misclassified to Outer race fault. One data of Outer race fault is missclassified to healthy data and Roller element fault data. None of the data are missclassified from Roller element fault. While comparing Roller element fault and Outer race fault, Outer race fault is not contributing much bearing fault information. It is not making much disturbances while rotating the bearing. Hence it is very difficult to distinguish between Healthy as well as Outer race faulty data. VMD is a strong signal decomposing technique which decomposed the vibrational data effectively into several IMF. The analysis of each IMF seperatly will give the information about the fault signal.

41.5 Conclusion

A bearing fault diagnosis system using VMD and DNN architecture is proposed in this work. The experimental vibration data is collected from a 3 phase AC Induction motor with two fault conditions. VMD decomposes the vibration signal effectively and applied both statistical time and frequency domain parameters. The obtained whole feature set is given as an input to the deep neural network which acts as an unsupervised learning structure. Auto encoder is used as the dimensionality reduction

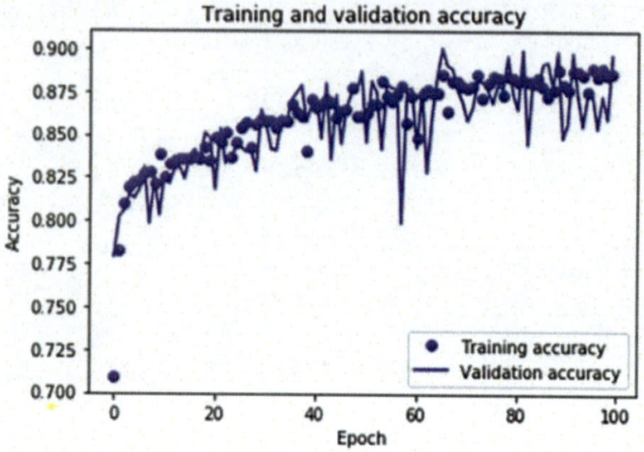

Fig. 41.8 Training and validation accuracy

algorithm to reduce the feature size. The Softmax classifier together with auto encoder acts as a supervised learning model. The testing accuracy is 98.9%. While comparing the classification performance with the other reduction techniques and classifiers, it is observed that the Auto encoder with softmax performs better than the other algorithms. Applying an improved version of VMD with auto encoder would give better performance for classifying several bearing fault cases.

References

1. Lanigan, P.E., Kaulya, S., Narasimhan, P.: Diagnosis in automotive systems: a survey': research by general motor company CMU-PDL-11-110 (2010)
2. Kang, M., Kim, J., Wills, L.M., Kim, J.-M.: Time-varying and multiresolution envelope analysis and discriminative feature analysis for bearing fault diagnosis. IEEE Trans. Indust. Electron. **62**(12) (2015)
3. Zhang, X., Liang, Y., Zhou, J.: A novel bearing fault diagnosis model integrated permutation entropy, ensemble empirical mode decomposition and optimized SVM'. Measurement **69**, 164–179 (2015)
4. Vinothraj, C., Praveen Kumar, N., Isha, T.B.: Bearing fault analysis in induction motor drives using finite element method. Int. J. Eng. Technol. **7**(3,6), 30–34 (2018)
5. Nguyen, V.H.: An architecture of deep learning network based on ensemble empirical mode decomposition in precise identification of bearing vibration signal. J. Mech. Sci. Technol. **33**(1), 41–50 (2019)
6. Liu, H., Li, D., Yuan, Y., Zhang, S., Zhao, H., Deng, W.: Fault diagnosis for a bearing rolling element using improved VMD and HT. Appl. Sci. **9**(7), 1439 (2019)
7. Wardana, A. N.: A comparative study of EMD, EWT and VMD for detecting the oscillation in control loop. In: International Seminar on Application for Technology of Information and Communication (ISemantic), pp. 58–63. IEEE (2016)
8. Han, H.: Fault diagnosis using improved complete ensemble empirical mode decomposition with adaptive noise and power-based intrinsic mode function selection algorithm. Electronics **7**(2), 16 (2018)
9. Yi, H.: A study on deep neural networks framework. In: IEEE Advanced Information Management, Communicates, Electronic and Automation Control Conference. IEEE (2016)

10. Zhao, R.: Deep learning and its applications to machine health monitoring. Mech. Syst. Signal Process. **115**, 213–237 (2019)
11. Dragomiretskiy, K., Zosso, D.: Variational mode decomposition. IEEE Trans. Signal Process. **62**, 531–544 (2014)
12. Krishna, K.R., Ramachandran, K.I.: Machinery Bearing Fault Diagnosis Using Variational Mode Decomposition and Support Vector Machine as a Classifier. IOP Conference Series: Materials Science and Engineering, vol. 310. No. 1. IOP Publishing (2018)
13. Mohan, N., Soman K.P.: Power system frequency and amplitude estimation using variational mode decomposition and chebfun approximation system. In: Twenty Fourth National Conference on Communications (NCC), Hyderabad, India (2018)

Chapter 42
A Smart and Precision Agriculture System Using DHT11 Plus FPGA

R. Jenila, C. Kanmani Pappa, and C. Supraja

Abstract Precision agriculture is a type of innovative agriculture, based on new technologies, which aim to streamline the agricultural process. Parameters such as temperature, humidity, soil characteristics and nutrients all play a role in plant development. Numerous sensors are available to measure the environmental factors and depending on datum, the devices are developed to monitor and control the parameter changes for each crop. Some existing technologies are more complex and very complicated for farmers to implement. The main goal of this project is to produce a precision agriculture-based grid with high performance and low energy consumption. This is achieved by using a DHT11 sensor for the measurement of temperature and humidity in the surrounding environment for plant growth. With the help of Basys3 Artix-7 FPGA Board, it is easy to monitor and an alert signal makes you aware of controlling those environmental parameters.

42.1 Introduction

In the past few years, all applications related to fields in modern technology have been developed like automation and the Internet of Things. In India, most of the population depends on agriculture (production and farming). Due to traditional farming, the crop was damaged causing loss to the farmer and thus precision farming was introduced based on the concept of automation [1]. The use of precision agricultural crop production will increase and reduce crop spoilage. In precision cultivation, all parameters responsible for plant growth can be monitored and vary according to the plant's growth needs.

R. Jenila (✉) · C. K. Pappa · C. Supraja
ECE Department, VelTech Rangarajan Dr. Sagunthala R&D Institute of Science and Technology, Avadi, India

J. I.-Z. Chen et al. (eds.), *Machine Learning and Autonomous Systems*, Smart Innovation, Systems and Technologies 269, https://doi.org/10.1007/978-981-16-7996-4_42

Fig. 42.1 Tradition versus precision agriculture

It is important to measure and monitor many interacting physical variables in greenhouse crops in order to improve their quality and productivity. Only control systems with built-in applications are capable of completing these tasks. It is costly to build a greenhouse. The high cost of greenhouse technology prevents many farmers from adopting it. The greenhouse approach to environmental control is the focus of our project. Via input sensors and procedures, with the greenhouse controller, changes in temperature (dry temperature, wet temperature) can be sensed and the necessary actions can be taken. Real-time monitoring offers accurate and timely crop and soil status information, which is crucial in making crop production improvement decisions [2]. Figure 42.1 shows the comparison of traditional and precision agriculture.

In traditional agriculture, many problems arise like excess water, pesticides, temperature etc. Due to this inorganic farming occurs. However, in precision agriculture, various sensors like P-mod HYGRO, DHT11, TMP3 and so on are used [3, 4]. By using the sensor data and input to the FPGA, one would be able to control the various problems in the agriculture field. Kavianand et al. proposed a smart drip irrigation system for sustainable agriculture. ARM9 processor is used for controlling and monitoring the irrigation system [5]. Similarly, Sabo and Qaisar [6] use IoT with an intelligent wireless sensors network in order to monitor the health of plants and population of larva in a remote crop field. The sensor data is collected by the front-end sensing node, developed with an ARM processor.

Many researchers proposed various automated precision agriculture system for increasing the production of crop yield and plant growth. Table 42.1 shows the comparison of different smart agriculture methods based on the Internet of Things and embedded processors.

The aim of the proposed system is to develop an automated precision green monitoring system that will crucially minimize human efforts and also improve the yield and product quality. For the crops, this system will provide a greenhouse environment [13, 14, 15]. The sensor data can be used to ensure that the crops get exactly what they need in terms of water content, temperature and so on. This is implemented with Basys3 Artix-7 FPGA Board with the software Vivado 2016.4 and Xilinx SDK. The

Table 42.1 Comparison of various precision agriculture systems using IoT and ARM-based processors

Research	Methodology	Specifications
Kavianand et al. [5]	Automated drip irrigation system using ARM processor	pH value and nitrogen content of the soil—monitor and control by GSM module and ARM9 processor AT91SAM9G45
Sabo and Qaisar [6]	IoT with ARM-based wireless sensor network	Measures ACI, temperature, humidity and moisture with STM32F407VG ARM processor and ZigBee Interface
Dholu and Ghodinde [7]	Application of cloud-based IoT in agriculture	Temperature, humidity, moisture, light intensity—monitor using MCU with Wi-Fi module through android application
Patokar et al. [8]	Multiparameter monitoring system in precision agriculture with IoT	Monitor and control—low power Intel's Galileo Gen-2 platform Quark processor
Sudharson et al. [9]	Smart agriculture using IoT technology with solar energy	Prevention of crops from damage due to rain and recycling of rain water—Atmega328 & GSM, Wi-Fi module
Adetunji et al. [10]	Development of a cloud-based monitoring platform to monitor agriculture resource	Sensor assembly logger development to sense temperature, soil moisture—4D system 4duino
Lakshmisudha et al. [11]	Precision agriculture system with Raspberry Pi	Soil moisture and humidity—monitor using XBee interface with Raspberry Pi
Lai et al. [12]	Fog computing with IoT	Cloud computing interconnected with Internet of Things

goal of this proposed system is to produce a precision agriculture-based grid with high performance and low energy consumption. Several researches were performed for smart and automated agricultural developments. Table 42.2 gives the comparison of various smart agriculture systems using field-programmable gate array programmed with HDL code.

This section gives the introduction and the related works. The remaining paper is organized as follows. Section 42.2 details the proposed methodology and the design flow, while Sect. 42.3 gives the results and discussions of the methodology and Sect. 42.4 provides conclusions and some ideas for future works.

Table 42.2 Comparison of precision agriculture systems using FPGA

Research	Methodology	Specifications
Patidar et al. [1]	Smart and precision agriculture system to develop ASIC	Water supply, pH value, nutrients ratio, temperature, humidity, moisture, light intensity—Xilinx ISE 14.7, FPGA
Sarwade et al. [2]	An automatic monitoring system using wireless sensor networks (WSN)	Greenhouse controller—ADC and FPGA with GSM module
Ranade et al. [16]	Design of a smart wireless sensor network (WSN) for an agricultural field	Field moisture conditions—field sensor node interface with FPGA + ZigBee
Proposed method	Monitor the agricultural lands using FPGA	Temperature, moisture, humidity—Basys 3 Artix-7 FPGA Board

42.2 Smart Precision Agriculture

To achieve low power consumption, two methods have been designed using the digital temperature and humidity sensors (P-mod HYGRO and DHT11). With the help of Xilinx Vivado, an IP Integrator using Micro blaze and P-mod HYGRO port has been designed. By using these methods inbuilt .c code is generated in inbuilt drivers. This .c code is dumped into the Basys3 Artix-7 FPGA Board by using the SDK tool. Using the terminal emulator, either Tera term or serial output terminal or some hardware like LCD display, seven-segment display, the output is verified.

Also, using the DHT11 sensor, Verilog flow is designed and commands are given based on the threshold values of the particular crop. The code is dumped into Basys3 Artix-7 FPGA Board with a DHT11 sensor. Based on the sensor data the action needed is displayed in the console window. The updates are notified by the farmer and the necessary actions are to be taken further for the growth of plants. Figure 42.2 shows the block diagram of the proposed precision agriculture system.

The data sets are to be collected from *"The new seed starters"* handbook by the University of California, on agriculture and natural resources. Many environmental and room factors influence seed germination, the condition of the seed, presence of water, sufficient air, temperature, light, soil conditions and organic matter. A great success will be achieved when optimum conditions are provided for seed germination. The optimum soil temperatures and the finest humidity range are given in Table 42.3 for seed germination.

The optimal temperature and humidity range are to be maintained for the growth of plants. A software and hardware tool is required to monitor and control the optimal temperature and humidity values. With the help of Basys3 Artix-7 FPGA Board, a system is developed by which the environmental parameters can be easily monitored and controlled by the farmers. For the seeds to germinate, the day and night optimum temperature is controlled by a heater and cooler. The moisture value is

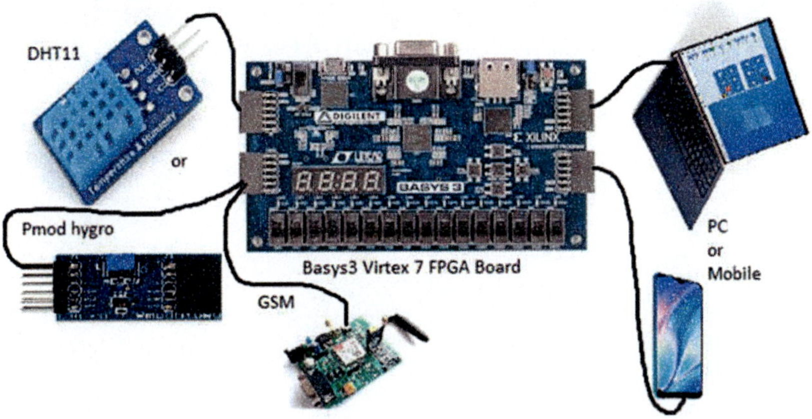

Fig. 42.2 Proposed block diagram

Table 42.3 Temperature and humidity range for seed germination

S. no	Name	Temperature (°F)	Humidity
1	Green leaves (Spinach)	45–75	50–60%
2	Asparagus	60–85	50–60%
3	Cauliflower	45–85	45–65%
4	Corn	60–95	60–70%
5	Tomatoes	60–85	50–60%

controlled by giving a proper water supply with the help of a motor pump. The optimal temperature and humidity in plant are sensed by the digital temperature and humidity sensor (DHT11 or P-mod HYGRO). If the ranges are not in appropriate limit, further controlling actions to be taken by the gadgets (heater, cooler, and motor pump) by the farmer with the commands given by the software on PC through Basys3 Artix-7 FPGA Board.

42.2.1 Finite State Machine Design Using DHT11 Sensor

In this, the sensor DHT11 senses the data of temperature (° C) and humidity. These sensed data is sent to Basys3 Artix-7 FPGA Board along with the GSM module to send the message through a mobile application. The following steps are involved in the module design:

1. If temp < 25

   Then "Action needed: Turn ON heater"

   Else if temp>30

   Then "Action needed: Turn ON cooler"

   Else

   "No action needed"

2. If humd < 50

   Then "Action needed: Turn ON motor" Else if humd > 60

   Then "Action needed: Turn OFF motor" Else

   "No action needed"

This is simulated and synthesized using Vivado 2016.4 and the process when repeated again and again gives continuous monitoring to the farmland. Based on the sensor data, the farmer controls the environmental parameters with the necessary hardware tools. As there is no need for ADC conversion by using DHT11 and P-mod HYGRO sensors, the time delay is also low when compared with existing methods.

42.2.2 Design Flow Using P-mod HYGRO

The following steps describe the design flow of the IP integrator using P-mod HYGRO with the software Vivado 2016.4 and the process works with Basys3 Artix-7 FPGA Board.

STEP 1: Create a new block project in Vivado 2016.4 and a new block design is selected where a micro blaze processor and sys clock are added.

STEP 2: From project settings, add Vivado library and add system clock  select P-mod-connector JA (P-mod HYGRO) and also select UART.

STEP 3: Do build automation and connection automation. Save the design, regenerate the layout and validate the design.

STEP-4: Select bin file and generate bit stream. Check the power consumption.

STEP 5: Export the project and from files select launch SDK. Open new application project.

STEP 6: Create an application and select an empty application.

STEP 7: Copy the P-mod HYGRO.c from drivers and paste it into the created application. Save the application and select program FPGA.

STEP 8: Run the application, simulate and verify the output.

42.2.3 GSM Module

With the GSM module, one can establish a connection with a remote network. P-mod Wi-Fi provides Wi-Fi access to the microchip MRF24WG0MA Wi-Fi radio transceiver module. Researchers can interface with the IEEE 802.11 g compatible chip over SPI and obtain data rates up to 54 Mbps (Fig. 42.3).

42.3 Results and Discussions

The proposed methodology is simulated and synthesized using Xilinx Vivado 2016.4 software tool in Verilog HDL. After synthesis, implementation is performed with Basys3 Artix-7 FPGA Board, and the output results (power consumption and bonded IOBs) of the two sensors are compared. By using P-mod sensors, the performance of the proposed system is greatly increased when compared with the earlier techniques.

Fig. 42.3 P-mod Wi-Fi

42.3.1 Simulation Results Using P-mod HYGRO

Figure 42.4 is the IP integrator window using a micro blaze processor (micro blaze processor supports Basys3 Board) where the power consumption is to be calculated and .c file generated for P-mod HYGRO which is taken from inbuilt drivers using SDK tool. This file is dumped into Basys3 Artix-7 FPGA Board and the output is verified using the Tera term terminal emulator. Figure 42.5 gives the temperature and humidity data in the terminal emulator (Tera term).

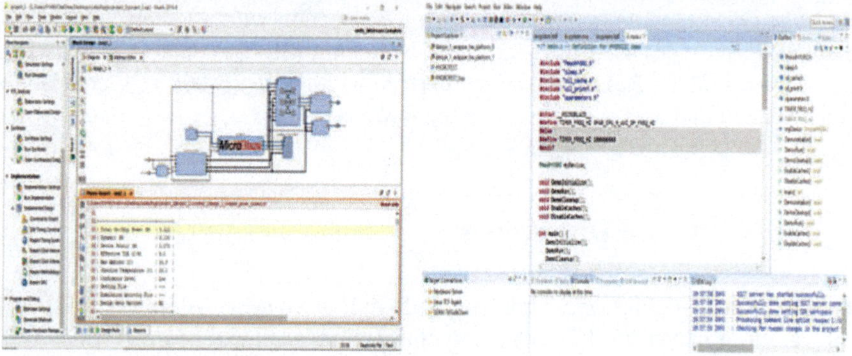

Fig. 42.4 IP integrator and P-mod HYGRO .c file

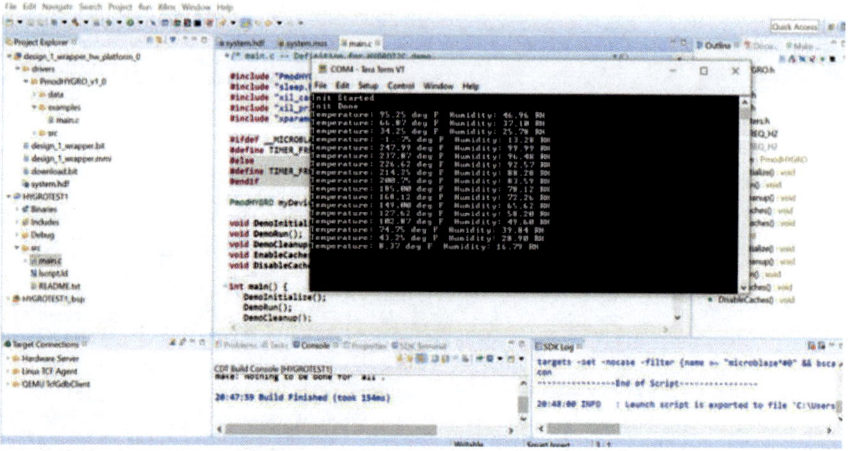

Fig. 42.5 P-mod HYGRO output

42.3.2 *Simulation Results Using DHT11 Sensor*

The simulation and synthesis are performed using the RTL design in the HDL software tool. Figure 42.6 shows the waveform of the RTL design. The necessary actions needed for the control of temperature and humidity are displayed in the Tc console window as shown in Fig. 42.6. After synthesis and implementation, the power consumption is calculated. Figure 7a gives the implementation result where 67 nets and 48 leaf cells are used.

The gate-level schematic diagram of the RTL design is shown in Fig. 42.7b, where 17 cells, 27 I/O ports, 43 nets and 19 leaf cells are used. This gives the elaborated design of the proposed system and the verification is done using automation processes by the software tool.

In the utilization report, one can be able to find the slice LUTs, Logic Slice Registers, Bonded IOB's and so on. Figure 42.8 gives the device utilization summary of the proposed precision agriculture system.

The approximate power consumption and bonded IOBs for the smart system by P-mod sensor and DHT11 are given in Table 42.4.

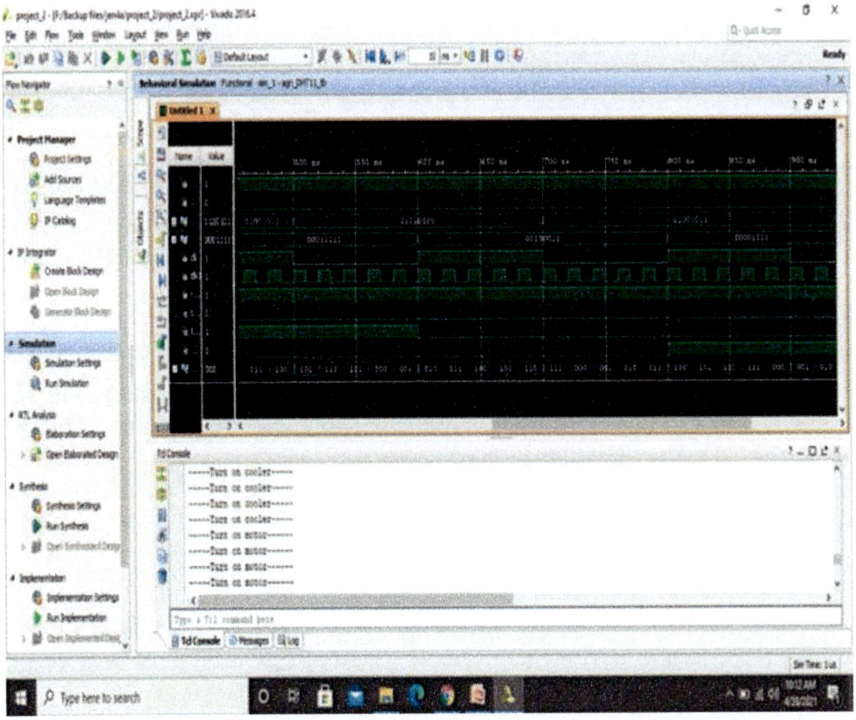

Fig. 42.6 Waveform of the RTL design

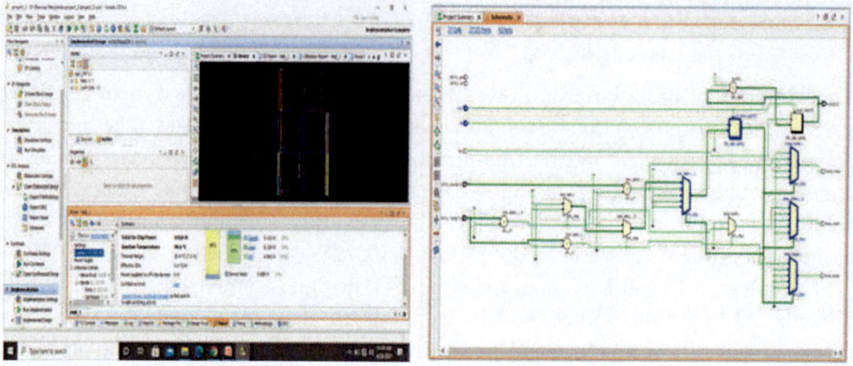

Fig. 42.7 **a.** Implementation results. **b.** Schematic diagram

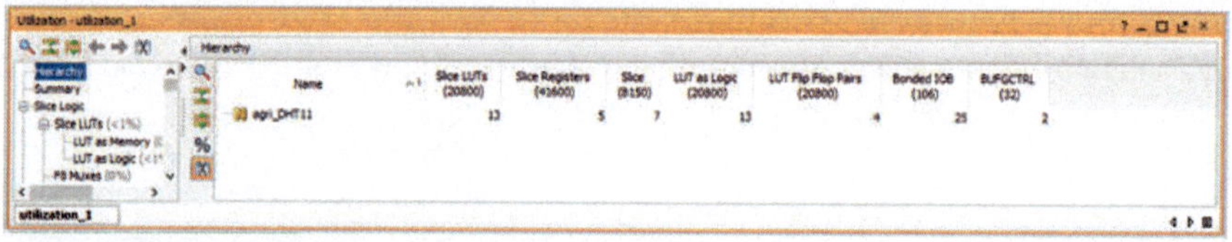

Fig. 42.8 Device utilization summary

Table 42.4 Utilization results

Sensors	Power consumption	Bonded IOBs
P-mod HYGRO	0 W–1 W	12–15
DHT11	4 W–6 W	20–30

42.4 Conclusion

In this work, power utilization by the Basys3 Board using the P-mod sensor and DHT11 sensor is compared. The power utilized by the P-mod sensor is low compared to the DHT11 sensor. By using P-mod sensors, the performance of the precision system is also highly increased. As there is a need for ADC conversion, the time delay is also less compared to the existing methods. In the future, this project could be implemented by considering other parameters like moisture and soil characteristics, and the complexity of the precision agriculture system can be reduced.

References

1. Patidar, J., Khatri, R., Gurjar, R.C.: Precision agriculture system using verilog hardware description language to design an ASIC. In: International Conference on Electronics, Materials Engineering and Nano-Technology, IEEE Proceedings (2019)
2. Sarwade, P., Shinde, N., Tingre, S.: Fpga based real time monitoring system for agricultural field. Int. J. Adv. Res. Innov. Ideas Educ. **3**, 2766–2774 (2017)
3. Review on IoT based smart grid architecture implementations: S. Mugunthan., T. Vijayakumar. J. Electr. Eng. Autom. **1**, 12–20 (2019)
4. Zen, J.I,Z., Chang, J-T.: Route Choice behaviour modeling using IoT integrated artificial intelligence. J. Artif. Intell. Caps. Netw **2**(4), 232–237 (2020)
5. Kavianand, G., Nivas, V.M., Kiruthika, R., Lalitha, S.: Smart Drip Irrigation system for sustainable agriculture. In: International Conference on Technological Innovations in ICT for Agriculture and Rural Development, pp. 19–22. IEEE Proceedings (2016)
6. Sabo, A.S., Qaisar, M.: An ARM based wireless sensors network for monitoring of plants health: In: IEEE Proceedings, pp. 80–85 (2018)
7. Dholu, M. Ghodinde, K.A.: Internet of Things (IOT) for precision agriculture application. In: IEEE Proceedings on the 2nd International Conference on Trends in Electronics and Informatics, pp. 339–342 (2018)
8. Patokar, A.M., Gohokar, V.: Precision Agriculture System Design Using Wireless Sensor Network. Springer Nature Singapore Private Ltd., Singapore (2018)
9. Smart agriculture monitoring and protection system using IOT: Sudarshan K G., Rakshith Ramesh Hegde., Sudarshan K., Siddesh J, Shilpa Patil. Proceedings of national conference on emerging trends in VLSI. Embed. Network. **2**, 308–310 (2019)
10. Adetunji, K.E., Joseph, M.K.: Development of a cloud based monitoring system using 4duino: applications in agriculture. In: IEEE Proceedings (2018)
11. Lakshmisudha, K., Hegde, S., Kale, N., Iyer, S.: Smart precision based agriculture using sensors. Int. J. Comput. Appl. **146**(11), 36–38 (2016)
12. Lai, K.L., Chen, J.I.Z.: Development of smart cities with fog computing and internet of things. J. Ubiquit. Comput. Commun. Technol. **3**(1), 52–60 (2021)
13. Abbasi, A.Z., Islam, N., Shaikh, Z.A.: A review of wireless sensors and networks' applications in agriculture. Comput. Stand. Interf. **36**(2), 263–270 (2014)
14. James, K.: Remote sensing and control of an irrigation system using a distributed wireless sensor network. IEEE Trans. Instrum. Measur. **57**(7), 1379–1387 (2008)
15. Dursun, M., Ozden, S.: A wireless application of drip irrigation automation supported by soil moisture sensors. Sci. Res. Essays **6**(7), 1573–1582 (2011)
16. Ranade, P., Takele, S.B.: Smart irrigation system using FPGA based wireless sensor networks. Int. Res. J. Eng. Technol. **3**, 2229–2232

Chapter 43
Cloud-Based CVD Identification for Periodontal Disease

K. G. Rani Roopha Devi, R. Murugesan, and R. Mahendra Chozhan

Abstract People are surrounded by numerous life-threatening disorders in this incredibly quick world, and one of them is cardiovascular disease (CVD). CVD has been linked to plenty of other problems, including periodontal disease. Gum disease, also known as periodontal disease, increases the risk of a heart attack, stroke, or other severe cardiovascular event by two to three times. However, there might not be a clear connection. The modified ANFIS and deep learning modified neural network were used to test cloud-based periodontal and CVD in this study. The cloud has been used as the centralised storage for all the information such as periodontal patient information, periodontal blood samples, recordings, and other relevant things in the data management model. The novelty of the work incorporated with the cloud based patient monitoring with the help of cognitive environment and machine learning. The cloud data management and intelligent CVD have been designed with the Cloud in this work. Quality measurements were used to assess the overall performance of accuracy, precision, recall, sensitivity, specificity, NPV, FPR, FNR, MCC, and FRR for intelligent identification of periodontal and CVD disease identification. The response time, efficiency, and other related performance have been carried out for cloud-based identification.

43.1 Introduction

Patients in conventional hospitals (and hospice situations) go to the hospital or doctor's office when they are sick or, in the worst-case scenario, when they need immediate treatment and are taken straight to Emergency. Patients can self-diagnose conditions using automated wearable devices such as smartwatches, digital blood

K. G. Rani Roopha Devi (✉)
Madurai Kamaraj University, Madurai, Tamilnadu, India

R. Murugesan
Department of Computer Science, C.P.A College, Bodinayakanur, Tamilnadu, India

R. Mahendra Chozhan
Chozhan Dental Clinic, Periyakulam, Tamilnadu, India

J. I.-Z. Chen et al. (eds.), *Machine Learning and Autonomous Systems*, Smart Innovation, Systems and Technologies 269, https://doi.org/10.1007/978-981-16-7996-4_43

pressure monitors, and weight scales long before they become acute cases requiring emergency treatment. Self-diagnosis, on the other hand, can be risky, so it's often better to have doctors look at remote sensor data and make a diagnosis before going to the clinic or hospital [1]. The problem is making such information accessible to doctors in real time. Patients will usually be doing their regular activities outside of the hospital (at home, at work, or in a senior facility, for example). Experts in the field of cardiovascular treatment predict the following for the future of cardiovascular care: Remote monitoring for cardiovascular activities for periodontal patients using patient facial video These remote physiologic monitors, which are enabled by IoT and Cloud Technology and used in combination with best practices at the clinic, have the ability to minimise patient hospitalisations and improve treatment [2].

Cloud computing and fog are tightly related. In nature, fog is closer to the ground than clouds; in the technical world, fog is closer to end-users, bringing cloud capabilities closer to the earth. According to the definition, fog is a cloud computing extension that comprises of many edge nodes that are directly connected to physical devices. Such nodes, unlike centralised data centres, are physically closer to devices, allowing for instant connections [3]. Edge nodes have a lot of processing power, so they can handle a lot of data without having to send it to distant servers. Cloudlets, which are small but efficient data centres positioned at the network's edge, may also be included in fog. Their goal is to assist resource-intensive IoT programmes that require minimal latency. Cloud computing is a centralised system, whereas fog computing is a distributed decentralised infrastructure.

In this graphic, a cloud health plant is defined and analysed. As illustrated in Fig. 43.1, health IOT has been created and deployed for data collection, administra-

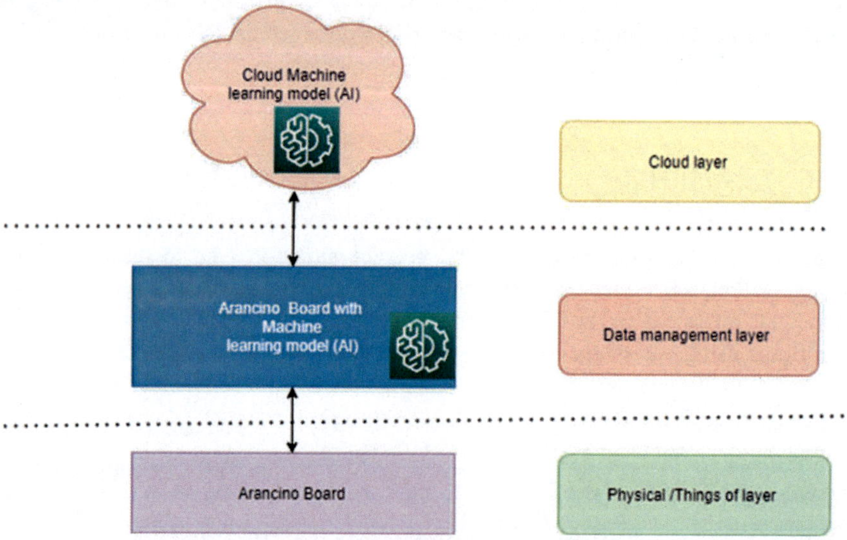

Fig. 43.1 Layer architecture of IoT with machine learning model

tion, and analysis. From a structural aspect, the infrastructure is separated into three layers: the physical layer, the data storage layer, and the cloud layer. In terms of the physical layer, it is the one that is closest to the plant; here, the microcontrollers on numerous boards act as transducers, transforming the electric current. Figs. 43.2 and 43.3 describe the layer architecture of the IOT model with machine learning model.

Sensors' signals and sending the digital measurements to the top stage [4]. These layers are in charge of data storage and analysis using an AI module that predicts cardiovascular disease activities by Patient face video submitted by the things of

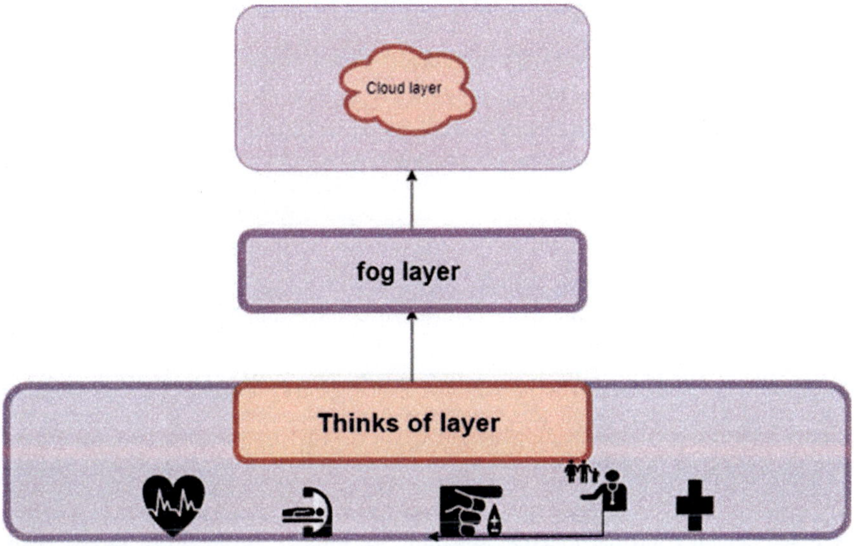

Fig. 43.2 Architecture of the IOT with Fog and Cloud

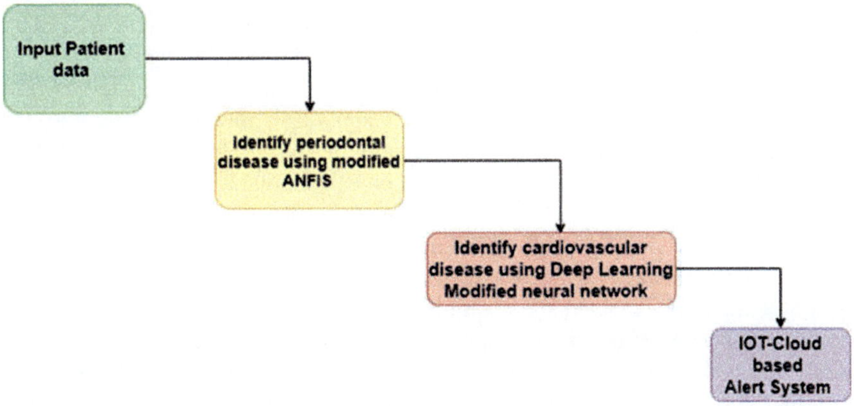

Fig. 43.3 Architecture of the proposed work

layer [5]. with patient info. In addition, the cloud serves as a data repository, keeping patient history data such as periodontist details alongside other test reports [6].

Fog computing [7, 8] is a type of computing that serves as a link between devices and remote servers. It determines which data is forwarded to the server and which is handled locally. Fog is an intelligent gateway that offloads cloud tasks, making data storage, processing, and analysis more efficient [9]. Fog networking isn't a new architecture, and it doesn't replace cloud computing; rather, it complements it by getting as near to the source of data as feasible. Another technique to data processing that is similar to fog computing is edge computing. In essence, rather than being transferred to other nodes or data centres, data is cached on computers [10]. Edge computing is very beneficial for IoT projects since it minimises bandwidth usage while also improving data security [23, 24].

The individual development of the periodontal disease identification and CVD identification over the machine learning has been presented in the Refs. [23] and [24]. The main aim of this paper is to incorporate the model into cognitive cloud-based development.

43.2 Related Work

Balakrishnan et al. [11] suggested a pulse estimate approach based on BCG and face footage in 2013. The goal of this technique was to separate the video's automatic head movement. The next speeding feature was focused on the face area and extracted the heart rate from the facial footage. To follow the microscopic development of the skull, the pace of the component points was divided. The head's feature purposes were segregated from the makers' specified location of between est (ROI). Shan et al. [12] proposed a [11]-based BCG extraction technique. The developers chose to use a single ROI rather than a mixture of ROIs in this case. The brow district was only employed to eliminate feature points since the area underneath the eye line was much more prone to movement curios, which were caused by non-unbending developments of the facial muscle as a result of talking or expressing emotions.

By applying deformable face fitting computation, Haque et al. [13] developed a BCG-based pulse evaluation methodology by applying a melding corner with the aims of good highlights and 49 specified facial milestones. The developers used an element combination-based approach to overcome the limitations of [11, 12] in terms of the loss of highlights due to facial development or obstruction. The KLT following calculation was used to follow the element focuses, and the vertical segments of the directions were deleted.

Poh et al. [15] extended their approach in 2011 to refine the PPGsignal by performing a series of transient separations. The filtering included a section where the signals were de-moved and smoothed using moving normal channels. Using ICA [14], the cycle yielded a much-refined set of indicators for dazzling source detachment (BSS). The best looking independent source sign to a PPG resembling the sign was chosen as the PPG signal, similar to previous work [14]. The chosen PPG signal was further improved by bandpass filtering at an operating recurrence of 0.7 to 4 Hz, which corresponded to a heart rate of 40 to 240 beats per minute. By extracting the PSD of the chosen PPG signal, the pulse was calculated. Manju and Sneha examined the Wiener and Kalman filters in the year 2020 to assess how effective they were at denoising ECG signals. According to the reproduction results, the Wiener filter is a better sifting approach than the Kalman filter in terms of SNR, PSD, MSE, and PRD. Due to its limited application to non-straight structures, the Kalman channel's presentation is inefficient.

In 2019, [29]. experimented with directed channel multi-lead model-based ECG signal denoising for MA and white Gaussian noise removal. When compared to the GA-EEMD, New MP-EKF, Butterworth guided channel, the proposed technique's presentation has been validated at multi-level commotion and offers unparalleled results. Furthermore, in a range of complex scenarios, the proposed method has a good denoising effect.

Altered ECG signal that was framed using EMD analysis and then given to the characterisation task in the year 2020 [30]. The modified ECG is created by combining the first three IMF signals obtained from EMD inspection, and it is demonstrated that the modified ECG has better denoising properties. Then, for characterisation purposes, a one-dimensional CNN network is created to familiarise itself with the signal's highlights. Aside from that, the proposed strategy is compared to a variety of preparation methods using various IMF signal combinations. The proposed method has been proven to outperform all other methodologies. The technique is robust and capable of treating a larger spectrum of cardiovascular disorders because it is based on three publically available data sources.

Reference [31] propose an ECG denoising technique based on a specified range investigation (SSA) in 2020, which addresses the bulk of concerns prior to the defects stated. A direction lattice of measurement K L is formed using the initial one-dimensional ECG sign of length N. The boundary L, also known as the window-length, is related with the sign's investigating recurrence and plays a crucial part in SSA activity. The morphological component of the ECG signal has a substantial effect on the assessment of L in this investigation. The framework is then degraded using a unique value decay process, and the first ECG signal's critical segments (PC) are determined. The recreated segments (RC) are then determined using the PCs, and all of the RCs are sorted using Butterworth bandpass and indent channels. The worthiness of isolated RCs determines the quantity that should be preserved. Finally, the denoised ECG signals are obtained by summing the held RCs. The evaluation results show that the proposed method surpasses best-in-class ECG denoising processes; in particular, the denoised signal's mean assessment score falls into the 'excellent' category according to the emotional measure of highest quality level [32].

De Haan et al. [16] developed a PPG evaluation technique based on chrominance highlights from the advanced camera's RGB spectrum. A mechanism for measuring the proportion of two symmetrical chrominance signals was incorporated in the suggested technique. The method works by subtracting two symmetrical chrominance signals from the RGB range's suggestions. Using a tentatively eliminated fixed coefficient, RGB colours were converted to chrominance hues. The proportion between the symmetrical chrominance colours was used to measure the PPG signal.

43.3 Architecture of Cloud-Based Cardiovascular Diseases Identification for Periodontal Disease Patients

43.3.1 Data Collection

The data was collected from dental clinics in Chozhan, Theni, Periyakulam, and Kodaikanal. Photos with different noises and haziness that are partially available or severely blurred are listed here. Teeth with four roots can require root canal therapy and surgery with root resection in cases of serious caries, as well as a full restorative crown made from anatomical structures. Fig. 43.3 describes the over all architecture of the work.

Table 43.1 describes the data set features of periodontal disease identification and Table 43.2 describes the CVD data set features.

The x-ray images of periodontal patient as well as the facial x-ray can also be uploaded and processed with this model Fig. 43.4 represents the sample image of periodontal disease.

The facial video has been converted into frames and the sample frames have been represented in Fig. 43.5

Table 43.1 Features of periodontal dataset

Parameters
pID
GI-values
MB-values
PPD-values
AL-values
ABL-values
pSex
Smoking-values
Drug-values
Genetic
Hormonol
PI
pAge
BVMI

Table 43.2 Features of CVD Identification

Parameters
PAge
PSex
Cp
Tresbps
Chol
Fbs
Restecg
Exang
Oldpeak
Slope
Ca
thal

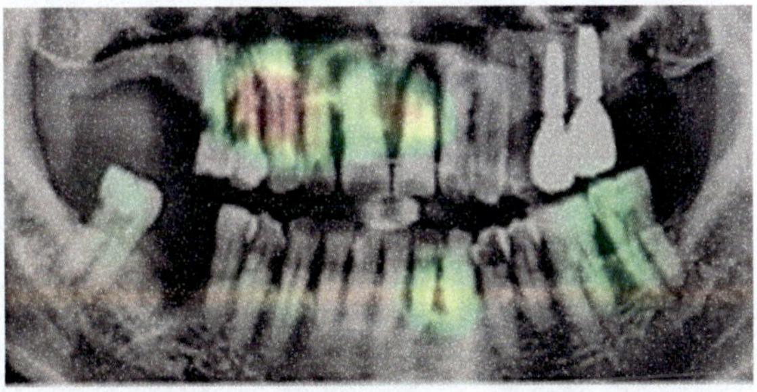

Fig. 43.4 Periodontal bone loss x-ray

The suggested model has a three-tier architecture, which is achieved by merging sensor nodes, building fog networks, and executing the suggested network's machine learning algorithms.

Tier 1: Tier I includes communication technologies including WiFi, Ethernet, and RS485. Tier-I architecture collects patient data and communicates with fog gateways.

Tier 2 FAN gateways are constructed using nodes and communication topologies in this scenario, for example, MQTT is utilised in a FAN for a quick reaction time and less computational analysis.

Tier 3 Cognitive fogs are created when machine learning algorithms are used to detect periodontal disease patients with cardiovascular disease (CVD).

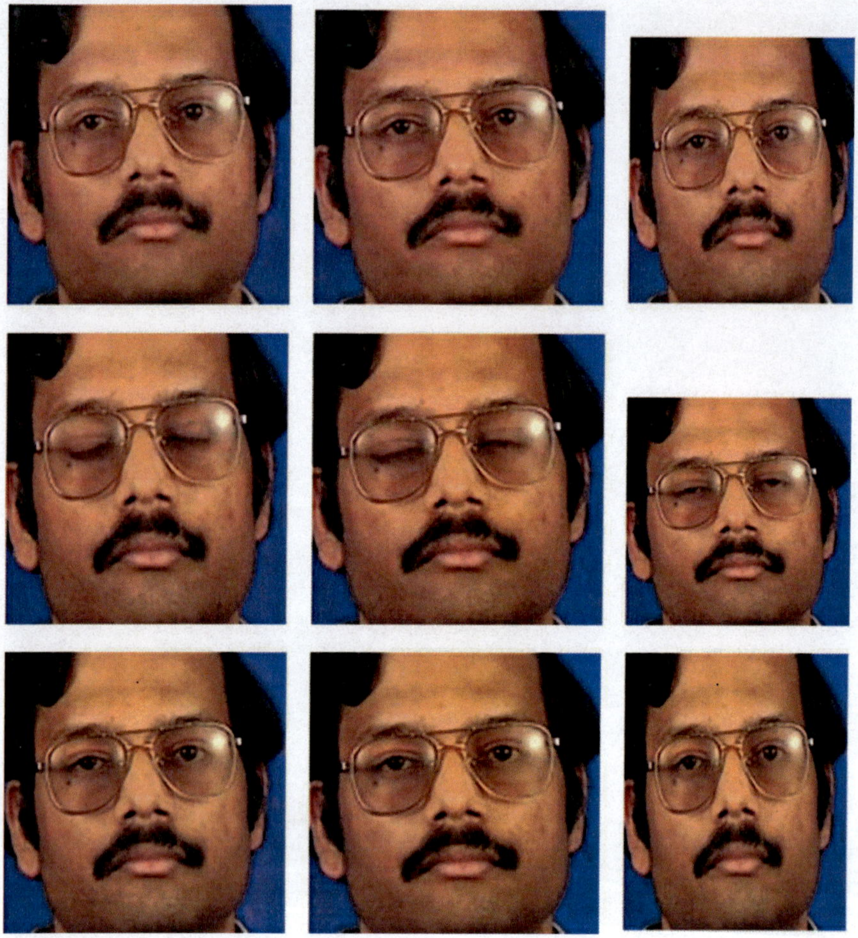

Fig. 43.5 Facial video frames sample

43.3.2 System Components

The components of the proposed system for detecting periodontal disease patients with CVD. The above-mentioned various media are used to create the communication media. Patient data is sent to fog networks through WIFI/ Ethernet.

43.3.2.1 Cognitive Fogs

Figure 3: The suggested architecture's whole framework Design and deployment of CfoTS networks in hospitals… In the advising tradition, buy-in/distribution is used

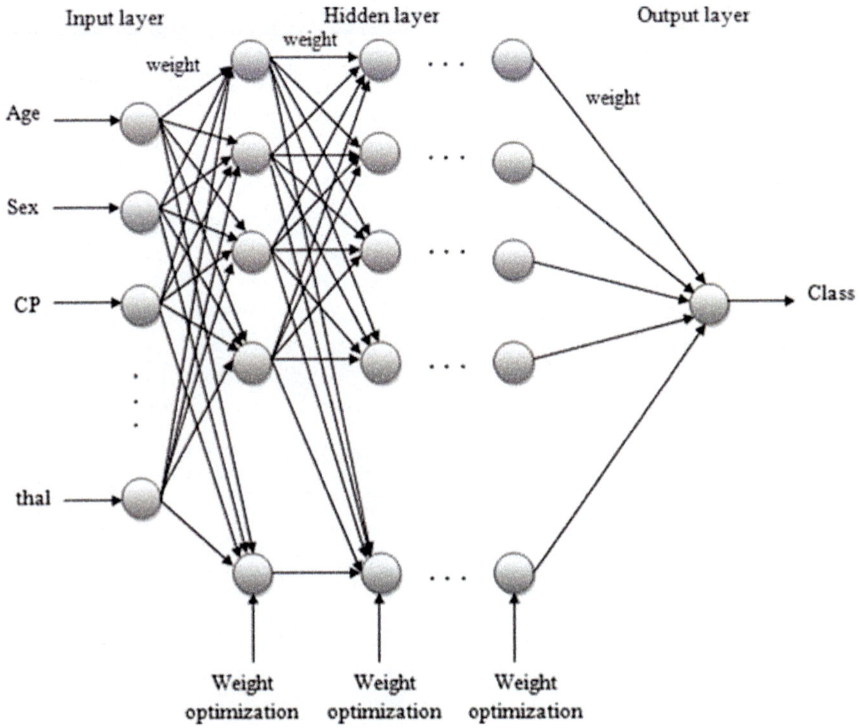

Fig. 43.6 Deep learning modified neural network

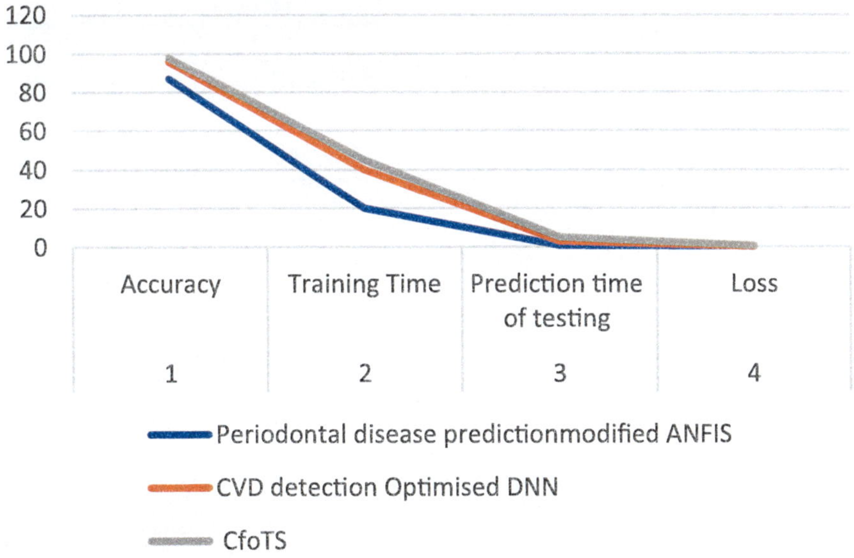

Fig. 43.7 Performance analysis of proposed CfoTs system

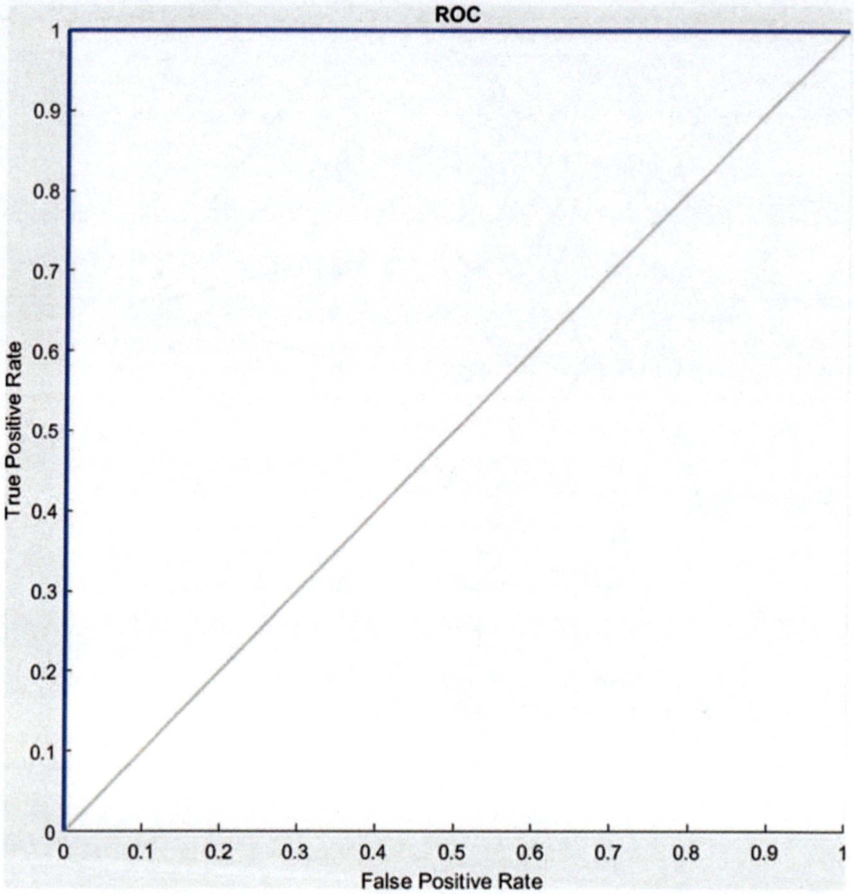

Fig. 43.8 False positive rate versus true positive rate

Fig. 43.9 Performance analysis of proposed cognitive learning

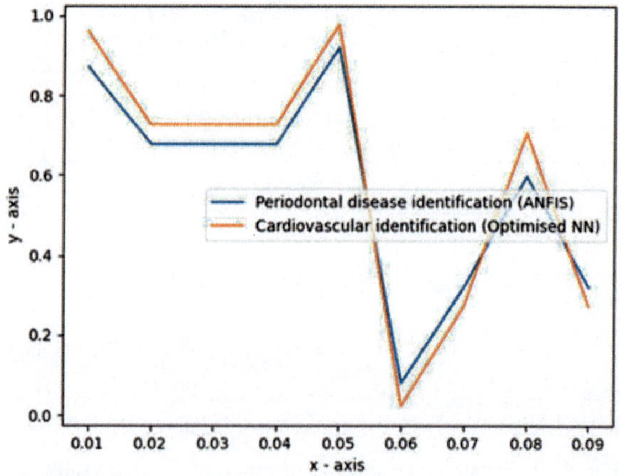

for minor code impressions. The educating server transfers messages from the physical node contraptions to the screen devices in the central and described model of a message prepared middleware [7]. Fogs in cognition Machine learning algorithms are commonly used to identify, forecast, and forecast periodontal/cardiovascular disease. These algorithms are often divided into two categories: supervised and unsupervised learning algorithms. Clustering mechanisms go into the first group, whereas neural networks go into the second. Collaborative filtering is a method that combines supervised and unsupervised elements.

43.3.2.2 Identify Periodontal Using ANFIS Classifier Model

Modelled as an improved version of traditional NN combined with fuzzy logic, the Adaptive Neuro-Fuzzy Inference System [16, 22]. IF–THEN rules are used to build a non-linear map between input and output. There are two inputs and one output on this unit. ANFIS is made up of five layers, each with its own feature [17]. The first layer receives data and evaluates each node's output using membership functions as shown in Eq. (43.2):

$$O_{li} = \mu A_i(x) \tag{43.1}$$

Where, i = 1, 2, $O_{li} = \mu_{Bi\text{-}2,\, I = 3,4}$

$$\mu\ (x) = \hat{e} - \left(\frac{x - \rho_i}{\alpha_i}\right)^2 \tag{43.2}$$

The membership values Ai and Bi are used to compute the output of nodes in the second layer.

$$O_{2i} = \mu_{A_i}(x) * \mu_{B_{i-2}}(y) \tag{43.3}$$

A node in the third layer evaluates output using the normalising strength of the layer shown in Eq. (43.4):

$$O_{3i} = \overline{w_i} = \frac{\omega_i}{\sum_{(i=1)}^{2} \omega_i} \tag{43.4}$$

An adaptive node over the fourth layer evaluates output based on O 3 1 as in Eq. (43.5):

$$O_{4,i} = \overline{\omega_i} f_i = \overline{\omega_i}(p_i x + q_i y + r_i) \tag{43.5}$$

The parameters of ri, qi and pi are parameters over ith node. Finally, ANFIS output computed with Eq. (43.6):

$$Output = \sum_i \overline{w_i} f_i \tag{43.6}$$

k-medoid ANFIS

The k-cluster is represented by an arbitrary selection of k-data items in this algorithm. The other elements are a cluster that contain the medoid that is closest to it. Better clustering is expected as a result of this. Clusters with the nearest medoid are given the entire dataset [18]. The medoid can adjust over all iterations. It sets a limit on the number of differences between data items and corresponding medoids [19, 20] This repeats until no variation is found, signalling the end of the operation. Additionally, final cluster findings with medoids have been achieved. Medoids are used to construct 'k' clusters, and full data members are distributed around a suitable cluster base [21].

43.3.2.3 Identifying Cardiovascular Disease Using Modified Deep Neural Network

ABC-PSO

Optimisation problems play an essential role in a variety of fields, including logical and design domains, for example. With the purpose of addressing a difficult enhancement problem, recent research has concentrated on the development of multiple knowledge computations, such as the ant colony optimisation (ACO) algorithm, the particle swarm optimisation (PSO) method, and the artificial bee colony (ABC) algorithm. [7] Calculations based on swarm-intelligent algorithms, according to analysts, offer a lot of promise and have gotten a lot of attention because of their simplicity.

The ABC computation has been used to shed light on a wide range of mathematical and nonnumerical optimisation problems in the years since its inception. Our comparison and contrast of the ABC calculation with a variety of other intelligent computations, such as GA and differential advancement calculation, revealed a number of fascinating parallels and differences (DE). ABC computing is superior than, or at least comparable to, other techniques [8, which are addressed more below]. Since then, tens of thousands of versions of ABC calculations have been created to improve the presentation of ABC calculations on a computer screen. Aside from that, a small number of mixed ABC computations based on PSO calculations have been done due to PSO's high exploitation capability. In the current situation, ABC's solution search condition is deficient, which is acceptable during the inquiry stage but insufficient during the exploitation stage [9, 10]. To overcome this problem, a new counterfeit honey bee state calculation is proposed that is based on the molecular swarm search component of the method. Rather than employing irregular determination immediately away, this algorithm uses valid statement set hypothesis to construct the underlying populace. The calculation's combination speed is boosted by using valid statement set hypothesis. Honey bees, observers, and scouts are used as an additional measure to increase exploitation capacity, and the PSO component, which is included in the PSO package, is used to find new rival installations.

GFT-SDM

Specifically, the proposed technique for recognising facial highlights points use a "Good Features Track" in conjunction with a "Supervised Descent Method" to arrive at its conclusions. Using these strategies, it is possible to overcome the limitations of current frameworks for human resource estimate frameworks that make use of face video, which are now in use. Current frameworks, such as improper constraints on object development and the aiding issue, are described above in order to distinguish between cyclic and non-cyclic trajectories of the observed points and delete non-cyclic components. By using the face quality appraisal approach, the likelihood of obtaining inaccurate findings is reduced significantly. It is required to follow the element focuses of the face and produce direction by recording the head movement in the video that is caused by the heartbeat and then putting this information together. When used in conjunction llwith sign preparation, the goal of direction extraction is to identify the cyclic and non-cycling directions of the followed focuses and to delete the non-cycling components. When it comes to finding facial highlights spots, the technique that has been suggested is a combination of the "Good Features Track" and the "Supervised Descent Method." When employing the aforementioned technique, it is feasible to overcome the constraint imposed by the present frameworks used for human resource estimation frameworks that make use of face video. Further problems with current frameworks have been identified, including unreasonable constraints on object development and the issue of help.

$$\text{F(GFT)} = \sum\nolimits_{x=px}^{px+wx} \sum\nolimits_{y=py}^{py+wy} 1(x) - g(a + s) \tag{43.7}$$

GFT feature has been generated using Eq. 43.7. And the SDM has been calculated using Eq. 43.8.

$$\text{F(sdm)}(*0 + \wedge \text{x}) = ||\text{f}(\text{x}0 + \wedge \text{x})) - 0^{*}| \tag{43.8}$$

The above formula was used to find SDM feature. Where x_0 is the initial configuration of the landmarks in a facial image, f(x) indexes the landmarks configuration (x) in the image, g is a non-linear feature extractor, o*f(x), and x_* is the configuration of the true landmarks By participating in the face quality evaluation procedure, the chances of receiving erroneous results are greatly reduced. By shooting the head movement caused by the heartbeat and ensuring that the video follows these aspects and directions, the video's success is dependent on its ability to follow the element focuses of the face and creative direction. Direction extraction aims to determine the cyclic and non-cyclic directions of the followed focuses and to eliminate the noncycling segments in the same way as sign preparation aims to detect the cyclic and non-cyclic directions of the followed focuses and to eliminate the noncycling segments.

PCA-DCT

There are several causes of head movement, but only those generated by blood flow through the aorta have been linked to the rate of heat production. The filtered trajectories (FT) were subjected to the Principal Component Analysis (PCA) algorithm in order to identify the causes of head movement. By calculating the orthogonal components PI and adding a load matrix LM to the original FT, PCA converts FT into another organising framework. The one that shares a common position with the heartbeat is the most intermittent of these isolated segments [27].

After that, the Discrete Cosine Transform (DCT) is used for all of the (PI) segments to find the most period one and choose the major consonant to achieve the HR. As a result, the proposed strategy [28], which combined SDM and GFT-based approaches, became more practical.

DLMNN

In this study, it was decided to employ a deep learning modified neural network classifier, which comprises three layers: a convolution layer, a pooling layer, and a completely connected layer. Weights and biases are included in the neural network learning structure; in this case, the weights and biases were applied using a variant of the PSO method.

There has been a lot of research into using facial video to measure heart rate. The Ref. [19] examined numerous study work based on heart rate estimation using video in 2017, with the majority of the work employing Photoplethysmography and Ballistocardiography, which are methods for determining heart rate using human facial video [20].

43.4 Result and Discussion

The accuracy and other performances are compared with those of cloud loading data systems. Table 43.3 and Fig. 43.9 show the accuracy comparison for the two scenarios that were explained in Sect. 3.3 Besides the accuracy, other parameters such as precision, recall, F-measure,

NPV, FPR, FNR, MCC, and FRR have been calculated and subsequently tabulated in Table 43.3. Table 43.4 clearly shows that the proposed architecture, The CfoTS network prediction time with training time and loss has been given in Table 43.4.

The various performance metrics of accuracy Precision, recall, F-measure, NPV, FPR, FNR, MCC FRR for identification of periodontal disease as well as CVD identification has also been evaluated, both models outperformed with conventional approaches.

Table 43.3 Performance metrics analysis of proposed work

S. no	Parameters	Periodontal disease prediction modified ANFIS	CVD detection Optimised DNN	CfoTS
1	Accuracy	87.1	96	98
2	Training Time	20	40	45
3	Prediction time of testing	0.75	3	5
4	Loss	0.34	0.12	0.51

Table 43.4 Performance analysis of proposed cognitive learning

Performance metrics	Periodontal disease identification (ANFIS)	Cardiovascular identification (Optimised NN)
Accuracy	0.8714	0.96
Precision	0.6785	0.72727
Recall	0.6785	0.72727
F-measure	0.6785	0.72727
NPV	0.9196	0.97842
FPR	0.0803	0.021853
FNR	0.3214	0.27273
MCC	0.5982	0.70569
FRR	0.3214	0.27273

43.5 Conclusion

The proposed Cloud-Based Cardiovascular Diseases Identification For Periodontal Disease Patients was effectively executed in this study using the cognitive fog system architecture. Based on the findings, the suggested design can share the load of the cloud architecture, which deals with millions of data points. The recommended design will aid automation, the food industry, and the beverage industry. Future IoT devices will benefit from the CofTS design's low power consumption, secure and reliable data transport, and, most importantly, quick response.

References

1. Civerchia, F., Bocchino, S., Salvadori, C., Rossi, E., Maggiani, L., Petracca, M.: Industrial Internet of Things monitoring solution for advanced predictive maintenance applications. J. Ind. Inf. Integr. **7**, 4–12 (2017)
2. Verma, P., Sood, S.K.: Fog assisted-IoT enabled patient health monitoring in smart homes. IEEE Internet Things J. **5**(3), 1789–1796 (2018)

3. Wang, Y., Yan, J., Sun, Q., Jiang, Q., Zhou, Y.: Bearing intelligent fault diagnosis in the industrial internet of things context: a lightweight convolutional neural network. IEEE Access **8**, 87329–87340 (2020)
4. Debauche, O., Mahmoudi, S., Manneback, P., Assila, A.: Fog IoT for health: a new architecture for patients and elderly monitoring. Proc. Comput. Sci. **160**, 289–297 (2019)
5. Moradbeikie, A., Jamshidi, K., Bohlooli, A., Garcia, J., Masip-Bruin, X.: An IIoT based ICS to improve safety through fast and accurate hazard detection and differentiation. IEEE access **8**, 206942–206957 (2020)
6. Vilela, P.H., Rodrigues, J.J.P.C., Solic, P., Saleem, K., Furtado, V.: Performance evaluation of a fog-assisted IoT solution for e-Health applications. Fut. Gener. Comput. Syst. **97**, 379–386 (2019)
7. Huang, H., Ding, S., Zhao, L., Huang, H., Chen, L., Gao, H., Ahmed, S.H.: Real-time fault detection for iiot facilities using gbrbm-based dnn." IEEE Internet Things J **7**(7), 5713–5722 (2019)
8. Monteiro, K., Rocha, E., Silva, E., Santos, G.L., Santos, W., Endo, P.T.: Developing an e-health system based on IoT, fog and cloud computing. In: 2018 IEEE/ACM International Conference on Utility and Cloud Computing Companion (UCC Companion), pp. 17–18. IEEE (2018)
9. Gia, T.N., Jiang, M., Sarker, V.K., Rahmani, A.M., Westerlund, T., Liljeberg, P., Tenhunen. H.: Low-cost fog-assisted health-care IoT system with energy-efficient sensor nodes. In: 2017 13th International Wireless Communications and Mobile Computing Conference (IWCMC), pp. 1765–1770. IEEE (2017)
10. Paul, A., Pinjari, H., Hong, W-H., Seo, H.C., Rho, S.: Fog computing-based IoT for health monitoring system. J. Sens. **2018** (2018)
11. Balakrishnan, G., Durand, F., Guttag, J.: Detecting pulse from head motions in video." In Proceedings of the IEEE Conference on Computer Vision and Pattern Recognition, pp. 3430–3437 (2013)
12. Shan, L., Yu, M.: Video-based heart rate measurement using head motion tracking and ICA. In: 2013 6th International Congress on Image and Signal Processing (CISP), vol. 1, pp. 160–164. IEEE (2013)
13. Haque, M.A., Irani, R., Nasrollahi, K., Moeslund, T.B.: Heartbeat rate measurement from facial video. IEEE Intell. Syst. **31**(3), 40–48 (2016)
14. Poh, M.-Z., McDuff, D.J., Picard, R.W.: Non-contact automated cardiac pulse measurements using video imaging and blind source separation. Opt. Express **18**(10), 10762–10774 (2010)
15. Comon, P.: Independent component analysis, a new concept? Signal Process. **36**(3), 287–314 (1994). Poh, M.-Z., McDuff, D.J., Picard, R.W.: Advancements in noncontact, multiparameter physiological measurements using a webcam, IEEE Trans. Biomed. Eng. **58**(1), 7–11 (2011)
16. de Haan, G., Jeanne, V.: Robust pulse rate from chrominance-based RPPG. IEEE Trans. Biomed. Eng. **60**(10), 2878–2886 (2013)
17. Ronneberger, O., Fischer, P., Brox, T.: U-net: convolutional networks for biomedical image segmentation. In: International Conference on Medical Image Computing and Computer-Assisted Intervention, pp. 234–241. Springer (2015)
18. Lee, J.-H., Hyung Kim, D., Jeong, S.-N., Choi, S.-H.: An overview of deep learning in the field of dentistry. In: Diagnosis and Prediction of Periodontally Compromised Teeth Using a Deep Learning-Based Convolutional Neural Network Algorithm, vol. 48, pp. 114–123 (2018)
19. Krois, J., et al.: Deep learning for periodontal bone loss radiographic detection. Sci. Rep. **9**, 8495 (2019)
20. Lin, T.-Y., Goyal, P., Girshick, R., He, K., Dollár, P.: Focal loss for dense object detection. In: Proceedings of the IEEE International Conference on Computer Vision, pp. 2980–2988 (2017)
21. Lee, J.H., Kim, D.H., Jeong, S.N., Choi, S.H.: Diagnosis and prediction of periodontally compromised teeth using deep learning-based convolutional neural network algorithm. J. Periodontal Implant Sci. **48**, 114–123 (2008)
22. Lee, J.H., Kim, D.H., Jeong, S.N., Choi, S.H.: Detection and diagnosis of dental caries using a deep learning-based convolutional neural network algorithm. J. Dent. **77**, 106–111 (2018)

23. Rani Roopha Devi, K.G., Mahendra Chozhan, R., Karthika, M.: Prediction of periodontitis in patients using big data analytics. In: Advances in Natural And Applied Science, pp. 194–199 (2016). ISSN 1995-0772
24. Rani Roopha Devi, K.G., Mahendra Chozhanand, R., Murugesan, R.: Analysis of various machine learning algorithms to predict periodontitis disease, p. 279, ICCIASH2020/850 (2020). ISSN 978-93-88096-42-3
25. Sklan, J.E., Plassard, A.J., Fabbri, D., Landman, B.A.: Toward content-based image retrieval with deep convolutional neural networks. Proc. SPIE Int. Soc. Opt. Eng. (2015)
26. Rajpurkar, P., Hannun, A.Y., Haghpanahi, M., Bourn, C., Ng, A.Y.: Cardiologist-level arrhythmia detection with convolutional neural networks. arXiv e-print 2017 :arXiv:1707.01836
27. Litjens, G., Kooi, T., Bejnordi, B.E., Setio, A.A., Ciompi, F., Ghafoorian, M., et al.: A survey on deep learning in medical image analysis. Med Image Anal (2017)
28. Kim, T.S., Obst, C., Zehaczek, S., Geenen, C.: Detection of bone loss with different X-ray techniques in periodontal patients. J. Periodontol (2008)
29. Manju, B.R., Sneha, M.R.: ECG denoising using wiener filter and Kalman filter. Proc. Comput. Sci. **171**, 273–281 (2020)
30. Hao, H., Liu, M., Xiong, P., Du, H., Zhang, H., Lin, F., Hou, Z., Liu, X.: Multi-lead model-based ECG signal denoising by guided filter. Eng. Appl. Artif. Intell. **79**, 34–44 (2019)
31. Hasan, N.I., Bhattacharjee, A.: Deep learning approach to cardiovascular disease classification employing modified ECG signal from empirical mode decomposition. Biomed. Signal Process. Control **52**, 128–140 (2019)
32. Mukhopadhyay, SK., Krishnan, S.: A singular spectrum analysis-based model-free electrocardiogram denoising technique. Comput. Methods Progr. Biomed. **188**, 105304 (2020).

Chapter 44
PalmNet: A CNN Transfer Learning Approach for Recognition of Young Children Using Contactless Palmprints

Kanchana Rajaram, Arti Devi, and S. Selvakumar

Abstract Recognition of children especially infants has received more attention nowadays. These are used in many applications such as law enforcement, health medical services, and vaccination tracking. Palmprint recognition has many unique advantages compared to other biometrics like richness of features, high user-friendliness, suitability for private security, etc. Acquiring the palmprints in a contactless way is very important in the context of pandemic situation even otherwise to avoid infection. In this paper, we created a dataset of palmprints of children using a mobile phone and proposed an approach for palmprint recognition using CNN transfer learning. Our approach is tested with IITD dataset having adult palmprints as well as compared with a state-of-the-art approach that uses infant palmprints. It is verified that our approach exhibit a good performance with a matching accuracy of 96% and Equal Error Rate (EER) of 0.02%.

44.1 Introduction

Traditional methods of human recognition include passwords, security pins, and tokens (e.g. ID cards, passports, etc.) and are based on any unique trait that helps in personal identification. Traditional methods have several drawbacks like more than one identity for a person, lost/stolen/forgotten passwords/tokens. Hence, traditional methods are not effective in achieving the security requirements. In the recent years, biometric gained more attention over the classical methods in identification and verification. Biometric features can be physiological such as face [1], iris [2], fingerprint [3], and palmprint [4] or can be behavioral which can be recognizable by a person's interaction such as signatures [5] and gaits [6]. Biometric systems are used in various fields such as banking, law enforcement, mobile access and authentication, home

K. Rajaram (✉)
Department of Computer Science and Engineering, Sri Sivasubramaniya Nadar College of Engineering, Chennai, Tamil Nadu, India
e-mail: rkanch@ssn.edu.in

A. Devi · S. Selvakumar
IIIT Una, Una, Himachal Pradesh, India

J. I.-Z. Chen et al. (eds.), *Machine Learning and Autonomous Systems*, Smart Innovation, Systems and Technologies 269, https://doi.org/10.1007/978-981-16-7996-4_44

access, schools, companies, and for many other surveillance purposes. Biometric-based recognition provides high security and assurance and is a convenient, fast method saving a lot of time. Face and fingerprint are commonly used methods in applications like unlocking/locking mobile phones, attendance, etc. Face includes several unique features such as Gabor Wavelet [7], SIFT [8], HoG [9] in facial recognition. A model based on VGG network [10] was fine-tuned with loss function to achieve a 98.95% accuracy in facial recognition. Karras et al. proposed a model [11] using both the generator and discriminator of GAN (Generative Adversarial Network). Face recognition has a major drawback that changes in facial expressions due to aging factors make it quite a challenging task.

Fingerprint Recognition consists of many traits like ridges and valleys that form unique patterns. In fingerprints, a feature called minutiae is useful for recognition. There are a few works that use deep learning models for fingerprint minutiae extraction such as MENet [12] for contact-based fingerprints and a method based on CNN for contactless 3D fingerprint recognition [13].

Iris and retina are two other important biometrics used for identification. It includes the texture of the iris. Iris recognition gained popularity in 1990s. Early works in Iris recognition used handcrafted methods for feature extraction. William et al. [14] used Hamming distance method for iris recognition. Another work was based on deep scattering convolutional features on IIT Delhi dataset and achieved a high accuracy rate [15]. There is also a limitation of using iris recognition system that there can be two persons of similar eye texture which make the identification process difficult.

Voice or speaker recognition is another process of human identification using the characteristics of voice. It is used in several security applications like Google Assistant. Speaker recognition can be done using vector approach [16] or a DNN-based acoustic model [17]. A method for speech recognition using encoder discriminator architecture like GAN is proposed in [18].

Palmprint biometric can be used for recognition and authentication purposes. Palmprints include several unique features like valley points, principal lines, and wrinkles. The advantages of acquiring palmprints in a contactless way are multi-fold: (i) no need to clean the hand with alcohol or harmful chemicals (ii) hygienic method and the infection is not carried over (iii) no residue of latent prints. Hence, a CNN-based approach namely **PalmNet** is proposed for palmprint recognition of young children. The approach uses two VGG16 networks and transfer learning method for extracting the features of palmprints.

The rest of the paper is organized as follows: Sect. 2 surveys the existing works on palmprint recognition. Section 3 describes the proposed approach, and the experimental results are provided in Sect. 4. Section 5 concludes the work and suggests future enhancements.

44.2 Literature Review

Nowadays, palmprint has received an attention in biometric identification since palmprint based recognition is more reliable in real-world applications due to the following advantages: (i) unique features; (ii) wide area of Region Of Interest (ROI); (iii) features are not affected by age; and (v) does not require high resolution images. Existing work for palmprint recognition is categorized into four types:

(i) Palmprint recognition of adults using images acquired with contact-based devices
(ii) Palmprint recognition of children using images acquired with contact-based devices
(iii) Palmprint recognition of adults using images acquired with contactless devices
(iv) Palmprint recognition of children using images acquired with contactless devices

Palmprint recognition of adults using images acquired with contact-based devices

A CCD camera-based palmprint acquisition device [19] which acquires images by touching the scanner was designed for adults. The device was less user friendly and the elder ones and those people having some disease had difficulty placing the hand on the device. A method for palmprint recognition [20] built a deep belief net using unsupervised training from top-to-down and then labeled testing samples in the end using deep learning models. This work is compared with traditional methods, such as PCA and LBP. Zhong et al.'s method used two VGG16 networks [21] to extract two input palmprint images with an EER of 0.2819%. When applied the same method on self-captured contactless dataset, it achieved an EER of 4.559%. It was observed that Zhang et al.'s method works better with the benchmark contact-based dataset. Shao et al. [22] proposed a method using Graph Neural Networks (GNNs) and achieved an accuracy of 99.90%. They proposed another method for cross-domain palmprint recognition [23] using transfer autoencoder. Experiments showed improved accuracy of cross-domain identification by 23.26% and more than 99% of recognition accuracy.

Minaee et al. proposed deep scattering network for palmprint recognition [24] and used Principal Component Analysis (PCA) to reduce the dimensionality of the extracted features. This method achieved 99.9% accuracy rate on SVM. Zhao et al.'s [25] Joint Deep Convolutional Feature Representation (JDCFR) method with CRC classifier for hyperspectral palmprint recognition achieved an accuracy of 99.62% with EER of 0.1%. However, since images were captured using hyper-spectral imaging device in near infra-red (NIR) light, it leads to health issues. Michele et al. [26] used MobileNet CNN and SVM classifier for palmprint recognition and tested with PolyU dataset. A combination of biometrics such as iris, palmprint, face, fingerprint was used in [27] and achieved an accuracy of 94% with an EER of 0.008%.

CNN was used for feature extraction with feature level fusion and score fusion. All the above works that use contact-based palmprints have many drawbacks related to health.

Palmprint recognition of children using images acquired with contact-based devices

Weingaertner et al. [28] designed a contact-based palmprint device that can acquire the palmprint images of resolution 140 dpi and used texture-based methods. Images were acquired from 250 newborn babies and achieved reasonable recognition accuracy. Contact-based palmprint systems have several drawbacks such as the need of sensors, need to press babies' palms on the sensors that results many hygiene issues like infections.

Palmprint recognition of adults using images acquired with contactless devices

In Izadpanahkakhk et al.'s approach [29] for palmprint verification used a pretrained CNN with transfer learning on PolyU dataset. ROI and features were extracted using pretrained Alexnet model and classification was performed using three different classifiers such as SVM, KNN, RF. This work achieved an ERR of 0.0125% on SVM classifier. Samai et al. [30] proposed a method to build biometric recognition system by combining 2D and 3D palmprints and extracted the features using Discrete Cosine Transform Net (DCT Net). Experiments were performed on PolyU 2D and 3D datasets containing 8000 samples. The performance of this method was tested with unimodal and multimodal biometric. This method achieved EER of 0.0806% on Euclidean distance and 0.0889% on cosine similarity.

Another method based on the approaches of hash coding and knowledge distillation [31] achieved EER of 0.607% and accuracy of 97.49% by experimenting with a dataset of size 30,000 acquired using 5 smartphones. This work adapted a deep hashing network and knowledge distillation-based DHN using VGG-16 and achieved identification accuracy of 97.49% and EER of 0.607%. Xie et al. [32] proposed a VGGNet method that uses palmprints for identifying the gender. It achieved 80.09% recognition accuracy. For investigating its performance, multispectral dataset was used and observed that the blue spectrum is most suitable for gender classification. A CNN-based method was employed in [33] and tested with PolyU hyperspectral palmprint database to achieve an accuracy of 99.9%.

Palmprint recognition of children using images acquired with contactless devices

Raghavendra et al. [34] have alone worked on children palmprints. Raghavendra et al. have proposed a method that performs fine-tuning of pretrained deep CNN that achieved EER of 0.3%. Experiments were performed on a small self-acquired dataset from only 50 infants. ROIs were extracted manually in this method and feature extraction was performed using pretrained Alexnet. The performance of this method was compared using seven state of art methods.

Our work aims to recognize children based on their palmprints acquired in contactless way avoiding infection. The challenge in recognition of children palmprints includes difficulty in image acquisition and covering of infant's palm with a white substance.

44.3 Proposed System

The proposed system for palmprint recognition of children, PalmNet uses a deep CNN architecture inspired by VGG16 [35] model. The architecture of the CNN is shown in Fig. 44.1. The VGG16 model is pretrained on ImageNet data set. ImageNet is a large data set having high-resolution images of over 15 million belonging to around 22,000 categories. The VGG16 has 16 layers containing three different parts as shown in Fig. 44.2. It has two sets of two convolution layers each followed by pooling, three sets of three convolution layers each followed by pooling, and a set of two fully connected layers and a softmax layer. In total, it has 13 convolution layers, 2 fully connected layers, and 1 softmax layer. In the first set, 2 convolution layers have 64 kernel filters, In the second set, two convolution layers have 128 kernel filters each. The three convolution layers in the third set have 256 kernel filters and the six convolution layers in the fourth and fifth set has 512 kernel filters. All the kernel filters are of same size 3X3. In each set, a max pool layer of stride 2 follows the convolution layers. The last set has two fully connected dense layers of 4096 units and the last layer is a soft max layer of 1000 units.

The workflow of the proposed PalmNet approach is shown in Fig. 2. The PalmNet involves the following steps: image acquisition, image preprocessing and finding ROI, feature extraction, and matching. The input images are preprocessed, and the ROI is extracted from them using palmprint localization. Since CNN-based approaches have shown best performance in object localization, it is proposed to use VGG16 network for palmprint localization as well as feature extraction. Lastly, the feature vectors are input to the matching module that uses SVM classifier and score fusion method. The steps involved in palmprint recognition are elaborated in the subsequent Sub Sections.

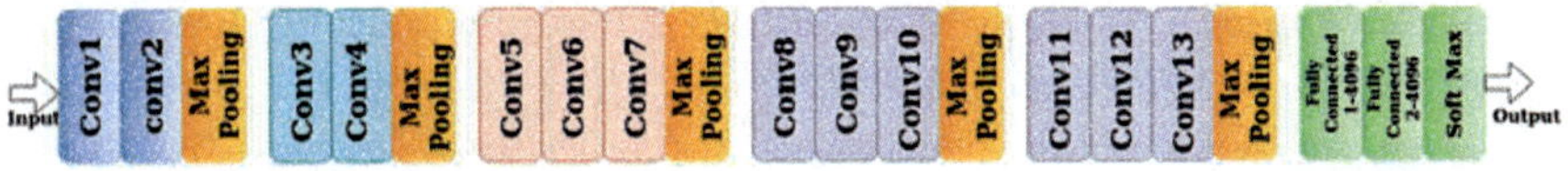

Fig. 44.1 Architecture of VGG16

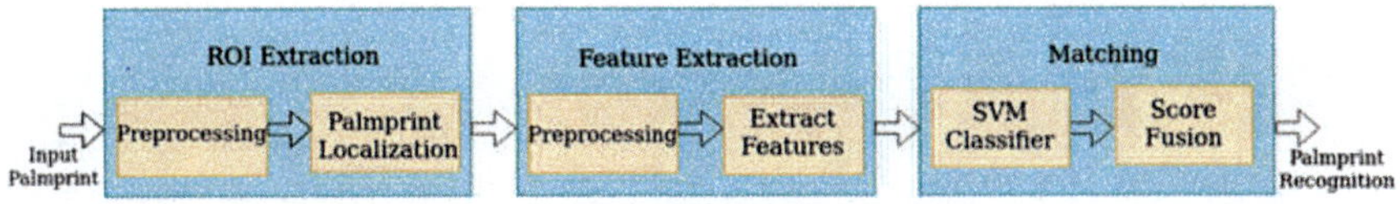

Fig. 44.2 PalmNet—our proposed model for palmprint recognition

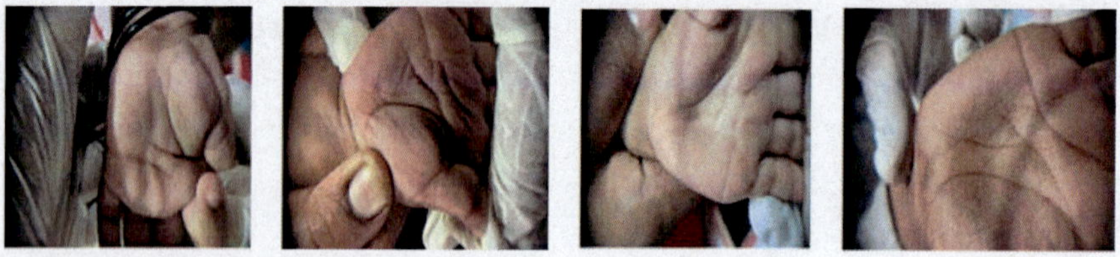

Fig. 44.3 Palm print acquisition of babies captured using a mobile phone

44.3.1 Image Acquisition

The palmprint images have been acquired using the mobile phone camera, creating our own dataset, namely CL-PalmNet. Palm images are collected from 100 children including newborn babies in a hospital. For data collection, iPhone SE is used in regular illumination under normal room conditions. Around 4–5 images were collected from each of the 100 children. The data collection has several challenges like the shaking of hands, crying of infants, blurred images, non-cooperation of the toddlers, etc. However, the parents play a vital role in assisting to capture fine quality samples. Figure 44.3 depicts data collection from a few children.

44.3.2 ROI Extraction

The extraction of ROI refers to the extraction of the palm area without any background and is shown in Fig. 44.4. The input palmprint images are preprocessed by resizing them to the pixel size of $224 * 224 * 3$ as required by the CNN that extracts the ROI. The resizing is carried out using nearest neighbor interpolation method. Nearest neighbor interpolation calculates an average value by generating some intermediate values. This method determines the nearest neighbor pixel and assumes its intensity value.

Palmprint localization is the process of drawing a rectangle around the ROI in the palmprint images. A VGG16 model-based CNN is proposed for ROI extraction.

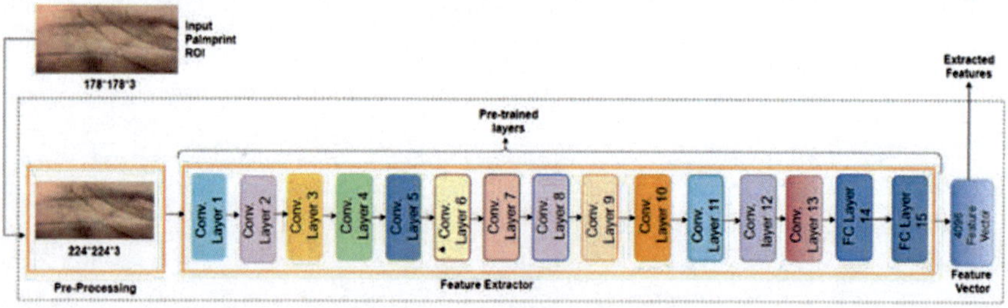

Fig. 44.4 ROI extraction

Transfer learning is useful when it is not possible to perform complete training of CNN, due to inadequate training data. With transfer learning [36], all the layers of CNN except the last one are trained with a larger dataset and the knowledge is transferred to the last layer which is trained on a smaller dataset. Transfer learning approach is used in PalmNet, since the size of the CL-PalmNet dataset is small having around 800 images of 100 children. The first fifteen layers of VGG16 network are pretrained on ImageNet and the last layer is replaced with a four neuron-layer unit which is trained with CL-PalmNet database resulting in a bounding box. A constant learning rate is set and a loss function is used to evaluate the performance. The loss function is computed using Mean Squared Error (MSE) which is measured as the average of squared differences between predictions and actual observations. The MSE is computed using Eq. 44.1 where, *xi* are the actual values and *yi* are the predicted values.

$$\text{MSE} = \sum_{i=1}^{n} (xi - yi)^2/n \tag{44.1}$$

44.3.3 Feature Extraction

In the proposed PalmNet, the distinctive features of palmprint ROIs are extracted in the form of feature vectors and labels using VGG16 model as shown in Fig. 44.5. Since the network requires palmprint ROI images of size $224 * 224 * 3$, the output images of size $178*178*3$ from ROI extraction step are pre-processed to adjust the size. VGG16 is a pretrained model and it is already trained on ImageNet. In VGG16, the first 15 layers are frozen, and the last layer of fixed size ($224 * 224 * 3$) is initialized with the weights learnt from the ImageNet to implement the transfer of parameters. Extracted ROI images are passed to the last layer in batches of 32 to get divided into features (X) and labels (Y). A label encoder is used to encode all the categorical variables to ensure that the learning model interprets the features correctly. Each

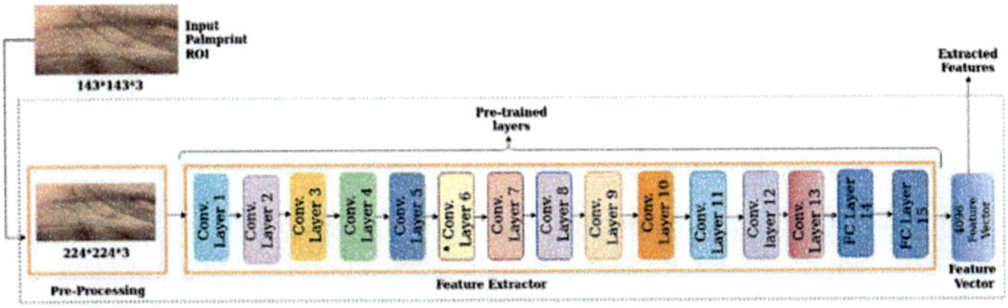

Fig. 44.5 Feature extraction module

batch of output feature vectors has class labels in the first entry of each row. The same process is repeated for all the batches until all features are extracted.

44.3.4 Matching of Palmprints

PalmNet uses SVM classifier for the matching process. A support vector machine (SVM) is a supervised machine learning model used for classification problems. Given a set of labeled training data for each class, the SVM builds a linear model that categorizes test samples to one of the classes.

The proposed PalmNet approach is applied on left and right palmprints separately to build the learning models during the training phase. The left and right test palmprints are matched against respective set of trained images. The output score is fused using a sum rule to recognize the child. The palmprint images in our self-captured dataset are labeled with a unique ID and serve as ground truth images. The results of matching are verified for correctness using these labels.

44.4 Experimentation and Results

In this work, experiments were performed with two datasets. The IITD palmprint image database [37] consists of palmprint images from 230 adults in the age group of 12–57 years. For each person, there are five to six images of each palm. Second dataset is the self-captured CL-PalmNet dataset which consists of palmprint images of 100 children of age 3 months to 8 years. Around 4 sample images each of left and right palm from each child are available with a total of around 800 images. All images have been captured in normal lighting environment without using any external flashlight. The size of the acquired images is normalized to $224 * 224 * 3$ to make them compatible with the CNN model.

The CNN was implemented on the Tensorflow framework using Python language. All the parameters of the VGG16 network are set to their default values. In our work, we used parameters like Adam optimizer with 0.0001 learning rate and MSE as a loss function. We removed one sample each of left and right palmprint for testing. Thus, 600 images (75%) are used for training and 200 (25%) images are used for testing.

44.4.1 Performance Evaluation of ROI Extraction

For ROI extraction with CL-PalmNet database, the learning rate is assumed to be a constant value $\mathrm{Lr} = 10^{-4}$. The training phase was completed after 50 epochs. Training with more than 50 epochs resulted in the overfitting of the model. Figure 44.6 shows

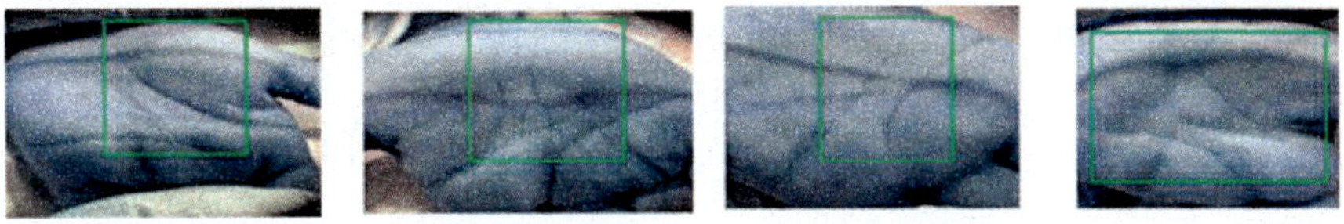

(a) 2 Epochs **(b) 10 Epochs** **(c) 30 Epochs** **(d) 50 Epochs**

Fig. 44.6 Examples of extracted ROIs for different epochs

the predicted bounding boxes with different epochs in the training procedure and it is evident that the bounding box fits the intended ROI. Loss function is used to evaluate the ROI extraction process and it is calculated using MSE (mean squared error). Figure 44.7 shows the graph where the learning parameters of each epoch are plotted against loss function. It is realized that the cost of palmprint localization decreases during initial steps but becomes steady after 20 epochs.

In the testing phase, 200 test images were used and the results of ROI phase as shown in Fig. 44.8, verify that bounding boxes capture the correct ROI.

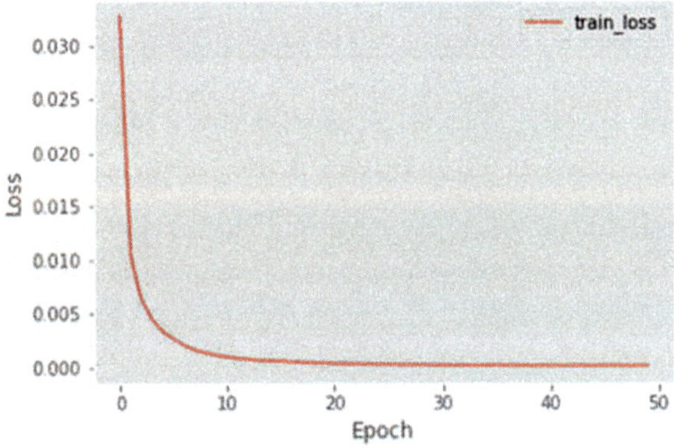

Fig. 44.7 Palmprint localization loss rate versus epochs

Fig. 44.8 Extracted ROIs in the testing phase

44.4.2 Performance Evaluation of PalmNet

The proposed approach of PalmNet is trained and tested with IITD database containing contactless adult palmprints and then with CL-PalmNet database with contactless children palmprints. The left and right palmprints of each child in the test set are matched with the trained images left and right palmprints using SVM classifier. The performance of the matching process in the proposed PalmNet approach is compared with two different datasets using the metrics such as precision, accuracy, recall, F1 score value and EER based on the following four concepts:

- TP (True Positives—the number of images correctly identified with a match),
- TN (True Negatives—the number of images correctly identified as having no match),
- FP (False positives—the number of images incorrectly identified as having a match)
- FN (False Negatives—the number of images incorrectly identified as not having a match).

Accuracy is an important metrics used for evaluating classification models and it is calculated as the ratio of number of correct predictions to the total number of input samples as given by the following formula:

$$\text{Accuracy} = \frac{TP + TN}{TP + TN + FP + FN} \tag{44.2}$$

The performance metric precision represents what proportion of positive identifications are actually correct. High precision indicates low false positive rate. Precision is calculated as follows:

$$\text{Precision} = \frac{TP}{TP + FP} \tag{44.3}$$

The metric recall represents what proportion of actual positives are identified correctly. Mathematical representation of recall is as follows:

$$\text{Recall} = \frac{TP}{TP + FN} \tag{44.4}$$

F1 score is the harmonic mean of precision and recall. Mathematical formula for computing F1 score is as follows:

$$\text{F1 score} = 2 * \frac{Recall * Precision}{Recall + Precision} \tag{44.5}$$

Experimental results in terms of the above four metrics are tabulated in Table 44.1. Figure 44.9 shows the confusion matrix between actual values and the predicted values. The correct and incorrect predictions are shown. It is observed that the

Table 44.1 Performance of the PalmNet approach using different datasets

Dataset	Recall	Precision	F1 score	Matching accuracy (%)
IIT Delhi (Contact less palmprint)	0.954	0.939	0.945	95
CL-PalmNet (Self-acquired)	0.913	0.883	0.892	**96**

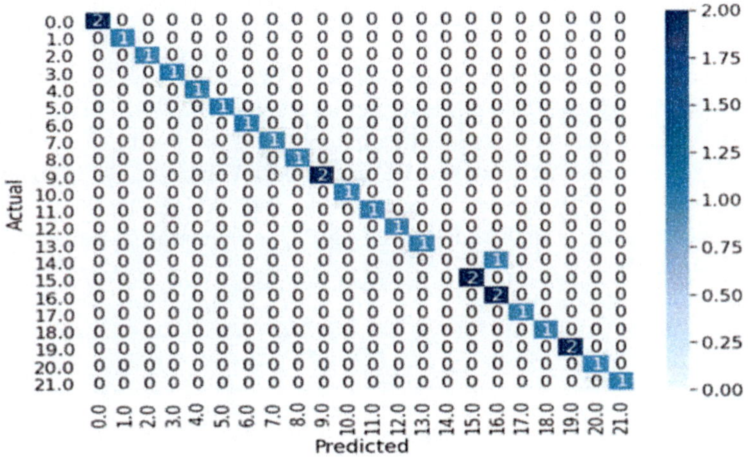

Fig. 44.9 Palmprint matching—predictions versus actuals

proposed PalmNet approach has achieved 96% accuracy with the contactless children palmprint database (CL-PalmNet dataset) against 95% with contactless adult palmprint dataset (IITD dataset). However, our approach exhibits less precision, recall, and F1 score when compared to IITD database. Data acquisition from children was quite challenging especially with infants who keep their palm closed and their palms are covered with a white substance. The toddlers keep their hands moving. Hence, difficulty in acquiring images decreases the quality of the images which caused this variation. However, the co-operation from parents is highly acknowledged in collecting good quality palmprint images.

EER is an essential metric to check the performance of any work related to biometrics. The EER is the point where False Acceptance Rate (FAR—percentage of unauthorized users incorrectly accepted) and False Rejection Rate (FRR—percentage of authorized users incorrectly rejected) meet and represent the best threshold to choose. FAR, FRR, and EER are calculated as follows:

$$\mathrm{FAR} = \frac{FP}{FP + TN} \tag{44.6}$$

$$\mathrm{FRR} = \frac{FN}{TP + FN} \tag{44.7}$$

Table 44.2 The performance comparison of PalmNet and the state of art method

Approach	EER (%)
Raghavendra et al.'s method [33]	0.31
PalmNet	**0.0236**

$$\mathrm{EER} = \frac{FAR + FRR}{2} \tag{44.8}$$

The performance of our proposed method PalmNet using children palmprints database is compared with Raghavendra et al.'s approach [33] in Table 44.2 that uses self-acquired dataset of children palmprints. It is observed that our PalmNet approach achieves a lesser EER of 0.02% against 0.31% achieved by [33]. Thus, it is proved that our PalmNet approach has achieved a reasonable palmprint recognition performance, since it is applied on the left and right palmprints separately and the results are fused.

44.5 Conclusion

We have recognized the children based on their palmprints by using transfer learning CNN model. The acquired palmprint images of children are preprocessed, ROIs and features are extracted using pre trained CNNs. Our method is applied on left and right palmprint sets. For the test images, SVM classifier is used to match the features of left and right palmprints and resulting scores are fused. The proposed PalmNet approach is applied on self-acquired contactless children palmprint database as well as IITD dataset which contains contactless adult palmprints. It is found that PalmNet shows better accuracy of 96%. Further, our PalmNet approach is compared with a state-of-art method that uses self-acquired contactless palmprints of infants and found that our method has a better performance with EER of 0.023%. The salient features of our works comprising of using VGG16 network for ROI and feature extraction and matching of palmprints using fusion of scores have helped achieving better performance.

The ongoing work involves strengthening our dataset with more palmprints of newborn babies and acquiring palmprints of a given child at different ages. Further, the impact of growth of textural pattern on children palmprints will be investigated in future.

Acknowledgements This work was supported by Grand Challenges India (GCI) for Immunization Data: Innovating for Action (IDIA) funded by BIRAC and jointly funded by Department of Biotechnology, Government of India, and Bill & Melinda Gates foundation.

References

1. Chu, Y., Ahmad, T., Bebis, G., Zhao, L.: Low-resolution face recognition with single sample per person. Signal Process. **141**, 144–157 (2017)
2. Daugman, J.: How iris recognition works. In: The Essential Guide to Image Processing, pp. 715–739. Academic (2009)
3. Rivaldería, N., Gutiérrez-Redomero, E., Alonso-Rodríguez, C., Dipierri, J. E., Martín, L.M.: Study of fingerprints in Argentina population for application in personal identification. Sci. Just. **57**(3), 199–208 (2017)
4. Jia, W., Zhang, B., Lu, J., Zhu, Y., Zhao, Y., Zuo, W., Ling, H.: Palmprint recognition based on complete direction representation. IEEE Trans. Image Process. **26**(9), 4483–4498 (2017)
5. Elhoseny, M., Nabil, A., Hassanien, A. E., Oliva, D.: Hybrid rough neural network model for signature recognition. In: Advances in Soft Computing and Machine Learning in Image Processing, pp. 295–318. Springer, Cham (2018)
6. Wang, J., She, M., Nahavandi, S., Kouzani, A.: A review of vision-based gait recognition methods for human identification. In: 2010 International Conference on Digital Image Computing: Techniques and Applications, pp. 320–327. IEEE (2010)
7. Kong, W.K., Zhang, D., Li, W.: Palmprint feature extraction using 2-D Gabor filters. Pattern Recogn. **36**(10), 2339–2347 (2003)
8. Lowe, D.G.: Distinctive image features from scale-invariant keypoints. Int. J. Comput. Vision **60**(2), 91–110 (2004)
9. Dalal, N., Triggs, B.: Histograms of oriented gradients for human detection. In: 2005 IEEE Computer Society Conference on Computer Vision and Pattern Recognition (CVPR'05), vol. 1, pp. 886–893. IEEE (2005)
10. Parkhi, O.M., Vedaldi, A., Zisserman, A.: Deep Face Recognition (2015)
11. Karras, T., Aila, T., Laine, S., Lehtinen, J.: Progressive growing of gans for improved quality, stability, and variation (2017). arXiv:1710.10196
12. Sero, D., Garachon, I., Hermens, E., Liere, R.V., Batenburg, K.J.: The study of three-dimensional fingerprint recognition in cultural heritage: trends and challenges. J. Comput. Cult. Herit. (JOCCH) **14**(4), 1–20 (2012)
13. Lin, C., Kumar, A.: Contactless and partial 3D fingerprint recognition using multi-view deep representation. Pattern Recogn. **83**, 314–327 (2018)
14. Williams, G.O.: Iris recognition technology. In: 1996 30th Annual International Carnahan Conference on Security Technology, pp. 46–59. IEEE (1996)
15. Minaee, S., Abdolrashidi, A., Wang, Y.: Iris recognition using scattering transform and textural features. In: 2015 IEEE Signal Processing and Signal Processing Education Workshop, SP/SPE 2015, pp. 37–42 (2015)
16. Dehak, N., Kenny, P.J., Dehak, R., Dumouchel, P., Ouellet, P.: Front-end factor analysis for speaker verification. IEEE Trans. Audio Speech Lang. Process. **19**(4), 788–798 (2010)
17. Lei, Y., Scheffer, N., Ferrer, L., McLaren, M.: A novel scheme for speaker recognition using a phonetically-aware deep neural network. In: 2014 IEEE International Conference on Acoustics, Speech and Signal Processing (ICASSP), pp. 1695–1699. IEEE (2014)
18. Ravanelli, M., Bengio, Y.: Learning speaker representations with mutual information (2018). arXiv:1812.00271
19. Zhang, D., Lu, G.: 3D Biometrics. Springer, New York (2013)
20. Xin, Z.D.P., Xin, P., Xiaoling, L., Xiaojing, G.: Palmprint recognition based on deep learning (2015)
21. Zhong, D., Yang, Y., Du, X.: Palmprint recognition using siamese network. In: Chinese Conference on Biometric rRcognition, pp. 48–55. Springer, Cham (2018)
22. Shao, H., Zhong, D.: Few-shot palmprint recognition via graph neural networks. Electron. Lett. **55**(16), 890–892 (2019)
23. Shao, H., Zhong, D., Du, X.: Cross-domain palmprint recognition based on transfer convolutional autoencoder. In: 2019 IEEE International Conference on Image Processing (ICIP), pp. 1153–1157. IEEE (2019)

24. Minaee, S., Wang, Y.: Palmprint recognition using deep scattering network. In: 2017 IEEE International Symposium on Circuits and Systems (ISCAS), pp. 1–4. IEEE (2017)
25. Zhao, S., Zhang, B., Chen, C.P.: Joint deep convolutional feature representation for hyperspectral palmprint recognition. Inf. Sci. **489**, 167–181 (2019)
26. Michele, A., Colin, V., Santika, D.D.: Mobilenet convolutional neural networks and support vector machines for palmprint recognition. Proc. Comput Sci. **157**, 110–117 (2019)
27. Vijayakumar, T.: Synthesis of palm print in feature fusion techniques for multimodal biometric recognition system online signature. J. Innov. Image Process. (JIIP) **3**(02), 131–143 (2021)
28. Weingaertner, D., Bellon, O.R.P., Silva, L., Cat, M.N.: Newborn's biometric identification: can it be done? In: VISAPP, vol. 1, pp. 200–205 (2008)
29. Izadpanahkakhk, M., Razavi, S.M., Taghipour-Gorjikolaie, M., Zahiri, S.H., Uncini, A.: Deep region of interest and feature extraction models for palmprint verification using convolutional neural networks transfer learning. Appl. Sci. **8**(7), 1210 (2018)
30. Samai, D., Bensid, K., Meraoumia, A., Taleb-Ahmed, A., Bedda, M.: 2d and 3d palmprint recognition using deep learning method. In: 2018 3rd International Conference on Pattern Analysis and Intelligent Systems (PAIS), pp. 1–6. IEEE (2018)
31. Shao, H., Zhong, D., Du, X.: Efficient deep palmprint recognition via distilled hashing coding. In: Proceedings of the IEEE/CVF Conference on Computer Vision and Pattern Recognition Workshops (2019)
32. Xie, Z., Guo, Z., Qian, C.: Palmprint gender classification by convolutional neural network. IET Comput. Vis. **12**(4), 476–483 (2018)
33. Jalali, A., Mallipeddi, R., Lee, M.: Deformation invariant and contactless palmprint recognition using convolutional neural network. In: Proceedings of the 3rd International Conference on Human-agent Interaction, pp. 209–212 (2015)
34. Raghavendra R., Raja, K.B., Venkatesh, S., Hegde, S., Dandappanavar, S.D., Busch, C.: Verifying the newborns without infection risks using contactless palmprints. In: 2018 International Conference on Biometrics (ICB), pp. 209–216. IEEE (2018)
35. Tammina, S.: Transfer learning using vgg-16 with deep convolutional neural network for classifying images. Int. J. Sci. Res. Publ. (IJSRP) **9**(10), 143–150 (2019)
36. Pan, S.J., Yang, Q.: A survey on transfer learning. IEEE Trans. Knowl. Data Eng. **2010**(22), 1345–1359 (2010)
37. IITD contactless adult palm print data set. https://www4.comp.polyu.edu.hk/~csajaykr/IITD/Database_Palm.htm

Chapter 45
Multi-Model Convolution: An Innovative Machine Learning Approach for Sign Language Recognition

Swaraj Rathi and Vedant Mehta

Abstract Hearing loss impairs DHH (Deaf and Hard of Hearing) people's language abilities and speech development. It is always recommended to perform early detection and begin training in early childhood. This study investigates a real-time sign language recognition system by developing a profitable machine learning technique. The proposed system makes use of the DCNN (Deep Convolutional Neural Network) model. The trained model is further transformed into an useful real-time application by using computer vision to accept real-time input for DCNN. The optimal CNN architecture for the proposed DCNN is selected from a multitude of alternatives, and it is then modified and tuned to make the model as modest as possible. It extracted spatial as well as temporal features from data. The innovative concept of multi-model convolution is built for accurate recognition of real-time signs and showcased in the implemented potential improvements section of the paper. With this novel technique, the model's accuracy is increased to 83%, ensuring that every ASL character is accurately detected.

45.1 Introduction

Hearing-impaired people face various problems on multiple frontiers but the most major one arises in communication, especially between signers and non-signers. This impairment also negatively affects their academic progress and achievements along with social and emotional interactions. The rate of getting unenrolled from schools is very high due to the incapability of coping up with the regular academic curriculum. Hence unemployment remains an unresolved problem due to the lack of qualifications, which in turn affects financial and social stability. Hence to curb such adverse effects, a helpful product or an application has to be instigated. Although

S. Rathi (✉) · V. Mehta
Shri Ramdeobaba College of Engineering and Management, Nagpur, India
e-mail: rathism_1@rknec.edu

V. Mehta
e-mail: mehtava@rknec.edu

J. I.-Z. Chen et al. (eds.), *Machine Learning and Autonomous Systems*, Smart Innovation, Systems and Technologies 269, https://doi.org/10.1007/978-981-16-7996-4_45

recently sign language is pervasive, the challenge of communication between signers and non-signers retains itself. With the latest advances in the domain of computer vision and deep machine learning, the progress has been bright and encouraging in the fields of pattern, gesture, motion detection, and recognition using techniques based on computer vision, machine learning, and deep learning.

In order to overcome this barrier and encourage effectual communication, a robust real-life application has been presented for sign language recognition, which utilizes Deep Convolutional Neural Network [DCNN] for the recognition of signs. While Neural Networks [NNs] have been used for any character recognition from ASL in the past with accuracies that are unfailingly over 90%, several of them necessitate a 3D capture component with Microsoft Kinect or motion-tracking gloves, and only one of the both delivers real-time classification [1]. The impediments inflicted by the additional necessities shrink the feasibility and scalability of the solution. The grounds for using Deep Learning are as follows: Convolutional neural networks are exemplary for image recognition and classification problems [2]. Deep learning requires the capability to automatically absorb features from the data, which is commonly useful when lots of training data are present for problems with inputs like images [3]. Training a ConvNet from abrasion on a diminutive dataset will result in not-so-great results [4].

For structural optimization and establishing progressive performance, rudimentary research on CNNs was performed by Bengio et al. [5]. LeCun et al. in [6] emphasized that deep CNN is a revolutionary headway in the image, video, audio, speech processing, and classification. Till the date, no substantial research has been executed which inspects deep CNN for real-time sign language recognition in any real-life application, that too using the concept of multi-model convolution. The focus of this paper and research is to explore the DCNN performance in recognizing real-time sign language gestures with the help of computer vision.

There are two main phases in constructing an automatic recognition system for human gestures in Spatio-temporal data [7]. Extraction of features from the frame sequences is the very first step. It results in a representation comprising of single or multiple feature vectors, which are also termed descriptors. This representation will support the computer to differentiate among the possible classes of signs. Classification of the signs is the next step. These representations will be used by a classifier to distinguish various signs or actions. In the proposed DCNN, the extraction of features is labor-saving with the help of convolutional neural networks and the classification is performed by an artificial neural network.

The research work and its evaluation is carried out in 3 phases. In phase-1, the evaluation of the outputs for the initially developed DCNN is performed. Then after a few modifications and improvisations, the DCNN is again evaluated in phase-2. Final evaluation is exercised after implementing the multi-model convolution to check the output and worth of the innovated approach.

45.2 Related Work

The literature survey for the problem statement indicates that there have been multiple ways to resolve the issue of gesture recognition. The challenge of sign language recognition is usually assigned to the field of computer vision. Research in [8–10] shows computer vision as a tool to make the solutions working for run-time implementations. It has added a new horizon to real-life problems, which are ought to be solved with the help of deep learning but with additional achievements, there always comes a glitch. Models did not perform well when there was a color variation. Factors like noise in ROI, brightness level, skin color, background color always interrupt the accuracy. Maybe reducing such background noises by isolating hand gestures from the images in a well-confined ROI, which would help accuracy. But for the concern of this research, consistent color for all components was used. Using multicolored backgrounds can be an issue here, which is tried to be resolved in the proposed work.

Most of the research works done and proposed in this field belongs to the domain of deep learning [1, 8, 11–15]. Here, the concept of Convolutional Neural Network [CNN] is considered as a boon. In most of the above-mentioned research works, the authors propose 2D convolutional models that limit the processing and classification of the images to grayscale. Here, multiple CNN architectures like Unet, Inception, etc. have been tried and tested. In most situations, this technique gives excellent accuracy; but, in many cases, hyperparameter tuning and fine-tuning are required to get respectable or at least similar accuracy to the state-of-the-art model. At times, CNNs are also accompanied by Recurrent Neural Networks (RNNs) depending upon the logic and approach of the author, dataset for the model and architecture of the model. In such cases, the spatial features are recognized by CNN and RNN is used to train the temporal features. Here, in many cases though the respectable accuracy is achieved, the proposed research work gives the accurate prediction of each ASL character.

In [16], the authors propose CNN as an Extreme Learning Machine (ELM). It offers very rapid learning with consistency in the performance since it has only one tuning parameter. Primary experimental results showcased that Extreme Learning Machine [ELM] is capable of producing palatable generic performance as a classifier. The dimensionality of data if increased, better separation has been achieved. Research, results, and use cases of 3D CNNs are presented in [17, 18]. Here, the processing and classification of colored images are dealt with. Which lately is very useful and is a cliché in recent times. It suffers in real-time implementation even though background noise with computer vision input is a significant challenge in real-time implementation, and the computational complexity and computational power required are high. However, 3D CNNs achieve a reasonable level of accuracy.

45.3 Methodology

45.3.1 Dataset

For initial purposes, the pre-built dataset named as "American Sign Language Dataset" [19] is used in this work. This dataset only consisted binary grayscale images and it was discreetly considered since the work proposes 2D convolution. From it, selectively 100 binary grayscale images were used in training dataset for each alphabet in ASL, i.e., a total of 2600 images and in the testing dataset selectively 10 binary grayscale images were included for each alphabet in ASL, i.e., a total of 260 images. In the subsequent phase of the work, the above said dataset was modified. The addition of a large number of real-time training and testing images was executed to enhance the performance of the model. The added images included augmented images, where images with a palatable level of noise and different backgrounds were captured manually. Finally, the training dataset was increased to 13,000, i.e., 500 binary grayscale images per alphabet, and the testing dataset was increased to 3250, i.e., 125 binary grayscale images per alphabet.

45.3.2 Convolutional Neural Network (CNN)

CNN is a form of neural network, which is most commonly used to analyze and visualize the imagery. The basic architecture of convolutional neural network comprises of 3 layers, i.e., convolutional layer, pooling layer, and fully connected layer.

Fundamental layer of a CNN is the Convolutional layer. It requires some input including the number of neurons, input image shape, activation function, filter size, and strides. These hyperparameters can also be tuned according to the need of the model to improve the accuracy. Convolutional layer when passes the data ahead, downsampling for it is exercised by the Pooling Layer. A filter with no assigned weights moves across the input and provides the output. Max and Average are the 2 main forms of pooling. In Max pooling, the filter contains an aggregate function that selects the maximum value from the input pixel and transfers it to the output array whereas in average pooling the function calculates the average value, and this value is sent to the output array. In the pooling layer, a lot of information loss is acknowledged which helps in complexity reduction, fall in the risk of overfitting and increase in the accuracy. Lastly, the fully connected layer, i.e., classifier which performs the classification based on the features extracted from the previous layers using the softmax function.

When multiple layers are used for convolution and pooling in a CNN, then a shallow neural network becomes Deep. Such deep network for convolution is called a Deep Convolutional Neural Network (DCNN). Fundamental working of a DCNN and a CNN is exactly the same. As an extrapolation, DCNN can be called as a special case of the CNN.

45.3.3 Proposed DCNN Architecture

The main aim of the CNN is to efficiently and effectively classify and predict the alphabets of ASL (American Sign Language). A fully connected layer in last produces output, the Pooling layer in the middle and the Convolutional layer in the beginning are used for obtaining more precise and explicit results. The CNN architecture that is basically employed and referred to is LeNet-5 but for improvement in the accuracy, it was customized as per requirements. The pooling layers are varied for the sake of accuracy. Hence the architecture of the proposed work is Customized LeNet.

The input image size of the binary grayscale image is 64 × 64 to the convolutional layer. The convolutional layers contain 16 filters each with a 2 × 2 kernel with a Relu activation function. The output dimension is reduced by passing it through the pooling layer of pool size 2 × 2. The second convolutional layer with 32 filters and a 3 × 3 kernel is followed by another pooling layer of pool size 3 × 3. Finally, the second convolutional layer is with 64 filters and the 5 × 5 kernel is followed by a third pooling layer of pool size 5 × 5. *The features are extracted with each Convolution and Maxpooling layer as mentioned above, these extracted features are then assembled to form a matrix.*

The obtained matrix then is flattened and converted into a vector adding a fully connected layer with the softmax classifier since there are 26 output labels. This ended by compiling the model with "Adam" optimizer and loss function for multiclass classification as "categorical cross entropy". *This is how the updating of filters and sample training is executed in the DCNN* (Fig. 45.1).

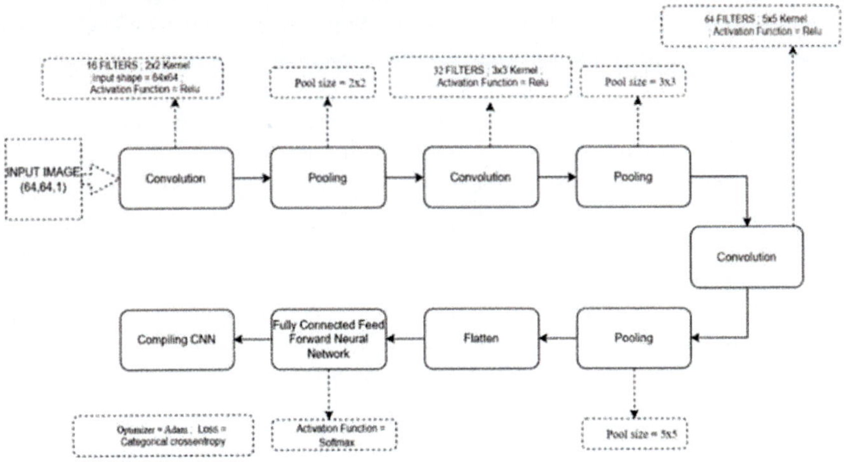

Fig. 45.1 Customized LeNet—proposed DCNN architecture

Fig. 45.2 Region of Interest (ROI)

45.3.4 Interfacing Computer Vision with Proposed Architecture

The input provided to the machine learning model is through the real-time video input (cv2.VideoCapture). The captured frame is flipped, simulating the mirror image. Region of Interest, i.e., ROI is plotted with blue color to show the region to the user to place hands as shown in Fig. 45.2. *ROI is Region of Interest, i.e., a frame on the screen, and input will be only captured through the ROI. The features will be extracted from the gestures inside the ROI frame and no other component will be considered for input, if outside the ROI.* The increment and decrement of 1 in the frame are done to compensate for the bounding box. Lastly, performing ROI re-coordinating for feeding input to the model for prediction.

Input frame image is captured, and it is resized to 64 × 64. Also, cv2.cvcolor is used to convert the real-time RGB image to a Binary grayscale image. ROI is being displayed in the mask with black pixels representing the input image area and the background and white representing the unwanted areas. Amongst all the output labels sorting is done based on the top prediction and the prediction is displayed on the mainframe. Figure 45.2 shows the ROI designed and implemented in the proposed work. Figures 45.7 and 45.8 also demonstrate the real-time usage of ROI.

45.4 Experimental Evaluation

During evaluation of the model some of the challenges came into existence. The problems periodically faced during the experimental testing are mentioned below.

Table 45.1 Letter recognition accuracy (Phase-1)

Letter	Prediction status (0- Incorrect prediction) (1- Correct prediction)	Accuracy (%)
A	0	12.07
B	1	70.78
C	1	94.91
D	0	9.23
E	0	23.67
F	1	71.11
G	1	83.22
H	1	77.81
I	0	38.93
J	1	76.19
K	0	17.85
L	0	49.08
M	1	84.22
N	1	87.89
O	1	98.27
P	1	71.33
Q	0	16.53
R	1	70.19
S	0	38.27
T	1	71.43
U	1	86.75
V	0	48.61
W	0	61.29
X	1	70.91
Y	0	13.37
Z	0	24.94

1. Initially the training dataset was of size 2600 binary grayscale images, i.e., 100 images per alphabet and the test data set contained 260 binary grayscale images, i.e., 10 images per alphabet. In phase-1 of implementation when the images from the training dataset were passed to the model, it could not classify and sort them based on the top prediction thus providing unsatisfactory results as shown in Table 45.1.
2. In Second Phase, the Accuracy Was Improved but the Model Was Still Getting Confused Between a Few Alphabets and Was Predicting Those Alphabets Error Fully as Shown in Table 45.2.

Table 45.2 Letter recognition accuracy (Phase-2)

Letter	Prediction status (0- Incorrect prediction) (1- Correct prediction)	Accuracy (%)
A	0	36.94
B	1	70.11
C	1	92.79
D	1	75.88
E	0	58.63
F	1	74.47
G	1	81.24
H	1	88.31
I	0	49.81
J	1	77.82
K	0	44.29
L	0	59.35
M	1	84.68
N	1	88.37
O	1	96.84
P	1	78.29
Q	1	85.64
R	1	73.59
S	1	79.11
T	1	74.66
U	1	89.01
V	0	63.97
W	0	64.48
X	1	90.04
Y	0	59.95
Z	1	76.33

3. During the second phase, the model could classify, recognize and predict all the ASL alphabets accurately except for 8 letters namely—[A, E, I, Y] and [K, L, V, W]. Model was displaying wrong output labels for these letters.
4. Also the background noise in the ROI was interrupting with the correct prediction of the letter and so with the accuracy.

These problems stood as stumbling challenges in dragging the model out from the state of confusion to the state of confidence.

45.5 Implemented Potential Improvements

For overcoming the challenges faced while achieving the accuracy of the designed model various potential improvements were deciphered from the experimental evaluation. After successful execution of the potential improvements, the accuracy of the model was acceptably increased. The implemented potential improvements were as follows:

1. Initially the result of experimental evaluation showcased the inadequacy of the model. This issue arose due to the tiny and repetitious nature of the dataset. Also, the background noise was not considered. These were the reasons for not so favorable outcomes. A solution to such problems was simply to transform the dataset by taking (1) The size of the dataset (2) Noise and Image augmentation (3) The background into account. Dataset modification and image augmentation was performed. The number of images was appreciably increased. The images were captured in real time with various backgrounds and multiple augmentations and then were converted to binary grayscale images.
2. For overcoming the problem of background noise, color, and light fluorescence: a method called Simple Thresholding was employed and also the ROI was re-coordinated.
3. Hyperparameter tuning was performed with all possible hyperparameters such as number of neurons, number of hidden layers, size of filters, activation functions.
4. Lastly to achieve the state of high confidence and low confusion for the model, an innovative approach was thought of and was tried and tested in the proposed system. The proposed work suggests a name for that approach as Multi-Model Convolution. This is the simple approach to reduce the confusion in the output. Here a separate DCNN model is developed which will be trained and tested only on the data of the alphabets among which its recognition and prediction fails.

Primarily, sets of all the alphabets have to be formed among which the model fails to be accurate. Then for each set of alphabets a separate DCNN model has to be trained. The dataset of the DCNN for a separate set of alphabets will only have the images of the alphabets in that respective set. For this proposed method, 2 separate DCNN models with exact same architectures were programmed and developed, out of which one was trained for A, E, I,Y, and the other was trained on K, L, V,W dataset providing the desired outputs. Algorithm for the implemented Multi-Model Convolution is showcased by a flowchart from Fig. 45.3.

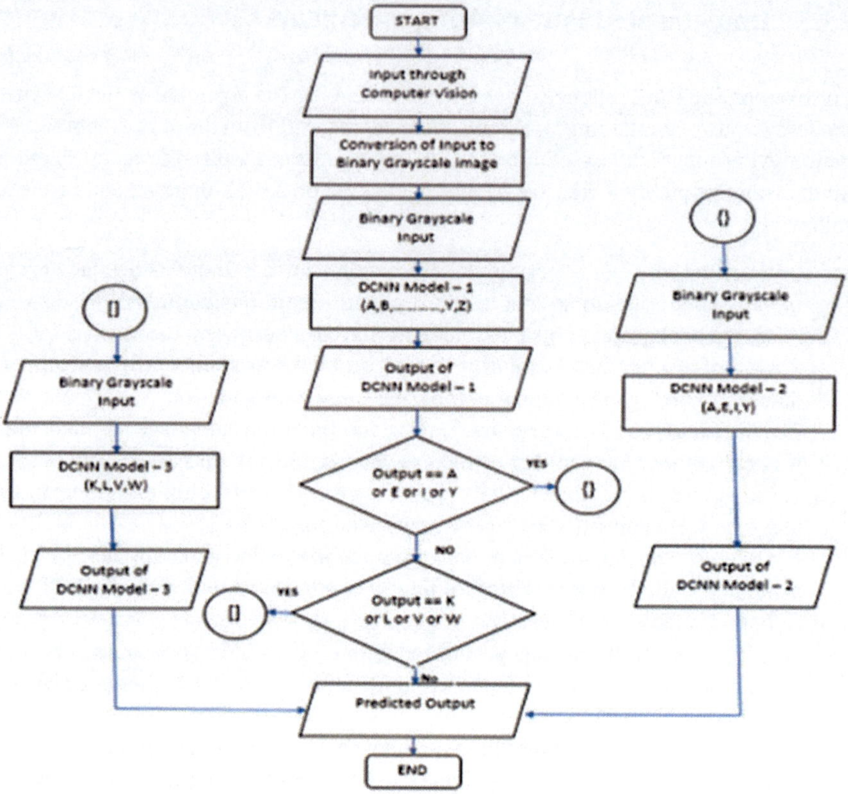

Fig. 45.3 Working flowchart of multi-model convolution algorithm

45.6 Results and Analysis

After implementation of all the potential improvements, the model was found to be modest with increased accuracy. Still it was getting confused between 2 sets of alphabets, i.e., A, E, I, Y and K, L, V, W. Hence to achieve the accurate prediction, the concept of Multi-Model Convolution was instigated. It made the proposed architecture a robust model. Table 45.3 shows the final predictions made by the DCNN model with the accuracy.

The most appreciable achievement for the proposed work was that *it could classify, recognize and predict each and every alphabet of ASL correctly* with the averaged accuracy of 83.28653846%, i.e., 83% rounded-off accuracy. The model was dragged out of the state of confusion to the state of confidence. The confidence motion of the proposed model was varied and was finalized as 70%.

The final recognition can also be traced from Fig. 45.4 for the developed classification matrix from the outputs of the model when it is tested for 15 run-time iterations,

Table 45.3 Final letter recognition accuracy (Phase-3)

Letter	Prediction status (0- Incorrect prediction) (1- Correct prediction)	Accuracy (%)
A	1	72.07
B	1	71.86
C	1	92.49
D	1	75.82
E	1	83.76
F	1	74.43
G	1	81.32
H	1	88.17
I	1	93.38
J	1	77.91
K	1	84.57
L	1	73.44
M	1	84.42
N	1	89.98
O	1	97.78
P	1	79.81
Q	1	84.13
R	1	71.99
S	1	80.95
T	1	74.61
U	1	88.75
V	1	89.64
W	1	96.38
X	1	89.72
Y	1	91.45
Z	1	76.62

PREDICTED LABELS

INPUT LABELS

	A	B	C	D	E	F	G	H	I	J	K	L	M	N	O	P	Q	R	S	T	U	V	W	X	Y	Z
A	15	00	00	00	00	00	00	00	00	00	00	00	00	00	00	00	00	00	00	00	00	00	00	00	00	00
E	00	00	00	00	15	00	00	00	00	00	00	00	00	00	00	00	00	00	00	00	00	00	00	00	00	00
I	00	00	00	00	00	00	00	00	15	00	00	00	00	00	00	00	00	00	00	00	00	00	00	00	00	00
K	00	00	00	00	00	00	00	00	00	00	15	00	00	00	00	00	00	00	00	00	00	00	00	00	00	00
L	00	00	00	00	00	00	00	00	00	00	00	15	00	00	00	00	00	00	00	00	00	00	00	00	00	00
V	00	00	00	00	00	00	00	00	00	00	00	00	00	00	00	00	00	00	00	00	00	15	00	00	00	00
W	00	00	00	00	00	00	00	00	00	00	00	00	00	00	00	00	00	00	00	00	00	00	15	00	00	00
Y	00	00	00	00	00	00	00	00	00	00	00	00	00	00	00	00	00	00	00	00	00	00	00	00	15	00

Fig. 45.4 Multiple input-output classification mtrix

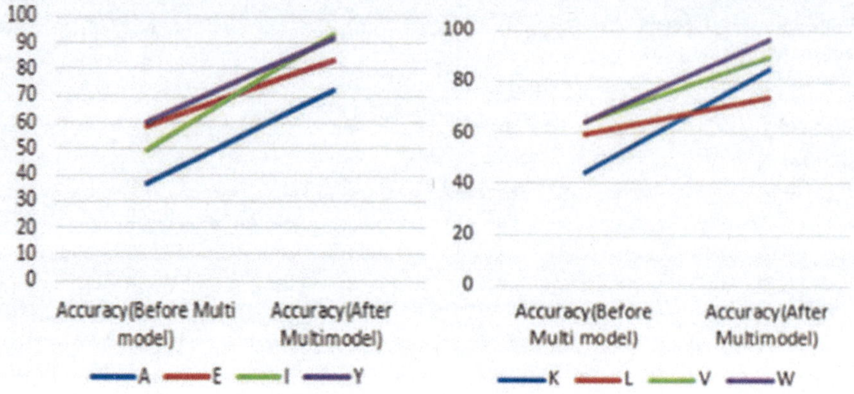

Fig. 45.5 Variation in accuracy before and after multi-model convolution

specifically for the letters A, E, I, Y and K, L, V, W in which the prediction of the model was classically improved.

Figure 45.5 shows improvement in accuracy with respect to the implementation of Multi-Model Convolution. The graphs are divided into two phases, i.e., accuracy before implementation of the multi-model convolution and the accuracy after employing the multi-model convolution. It can also be seen from the graphs in Fig. 45.5 that the accuracy for these 8 letters has risen substantially, resulting in flawless identification, categorization, and prediction of the signs for the alphabets.

Statistics for the prediction accuracy of all the letters from A to Z can be observed in Fig. 45.6. From this, it can be inferred that the accuracy of each letter is palatably increased in the traversal from Phase-1 to Phase-3. The accuracy of the confusing letters, namely, A, E, I, Y, K, L, V, W is drastically improvised in Phase-3, i.e., after

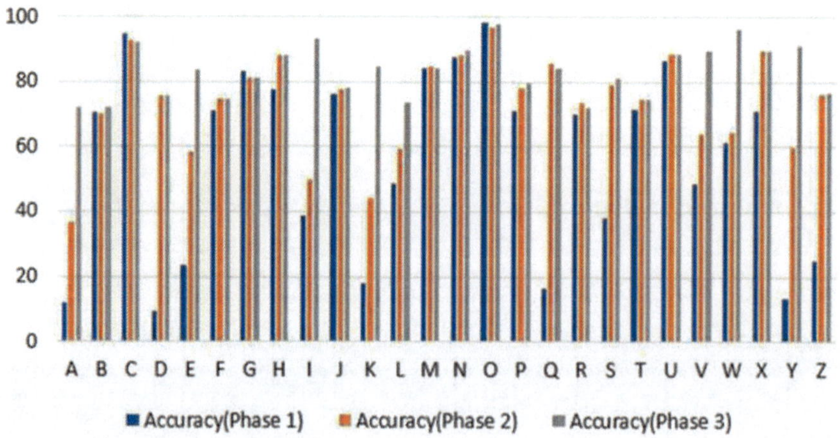

Fig. 45.6 Prediction statistics for all letters

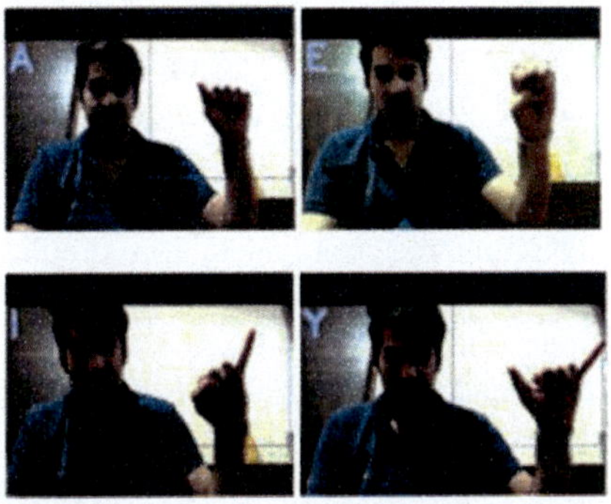

Fig. 45.7 Final outputs for A, E, I, Y

Fig. 45.8 Final outputs for K, L, V, W

Multi-Model Convolution. Also the accuracy for other letters is good and similar for phase-2 and phase-3 as the Multi-Model Convolution is not utilized for this letters since these letters are already righteously predicted by the DCNN model even before applying Multi-Model Convolution.

As shown in Figs. 45.7 and 45.8, the letters in which the accurate real-time prediction was missing are achieved with the help of Multi-Model Convolution. Hence our proposed work is hereby supported with this very proof of the images showing real-time accurate prediction of those letters in Figs. 45.7 and 45.8.

45.7 Conclusion

An ASL recognizer was trained and built with a Deep CNN and implemented in a GUI at backend. CNN is a powerful algorithm for Imagery, pattern recognition, and classification. In this paper, a CNN architecture for predicting and classifying the American Sign Language gestures for all the alphabets is proposed. The name for the experimented DCNN architecture is proposed as Customized LeNet. Different phases and batches of training are performed to explore the sturdiness of massive modes of training required for DCNN. For the proposed DCNN architecture, the amount of validation and training loss is observed to be low. Initial results show that DCNN learning as an appropriate learning algorithm while performing classification, recognition, and prediction on sign language characters, and making it a modest model. For achieving the modesty of the model, processes like data incrementation, image augmentation, simple thresholding, variation in confidence motion of the model, ROI re-coordination and hyperparameter tuning are performed. Still the model was getting confused between A, E, I, Y and K, L, V, W. Hence the approach of Multi-Model Convolution was innovated. For developing a robust model from modest model, deeper optimizations are required. Hence with the help of Multi-Model Convolution the authors are able to generate a robust model for the whole ASL which righteously classifies, recognizes, and predicts all the alphabets with the accuracy of 83%. This approach is unexplored in the reviewed literature and stands as a novelty for the proposed research work.

Ackmowledgement The authors are thankful to the Electronics Department, Shri Ramdeobaba College of Engineering and Management for the support in research work.

References

1. Garcia, B., Viesca, S.A.: Real-time American sign language recognition with convolutional neural networks. Convolut. Neural. Netw. Visual Recogn. **2**, 225–232 (2016)
2. Krizhevsky, A., Sutskever, I., Hinton, G.E.: Imagenet classification with deep convolutional neural networks. Adv. Neural. Inf. Process. Syst. **25**, 1097–1105 (2012)
3. Yosinski, J. et al.: How transferable are features in deep neural networks?. arXiv:1411.1792 (2014)
4. Pan, S.J, Yang, Q.: A survey on transfer learning. IEEE Trans. Knowl. Data Eng. 22.10: 1345–1359 (2009)
5. Bengio, Y.: Learning deep architectures for AI. Now Publishers Inc (2009)
6. LeCun, Y., Bengio, Y., Hinton, G.: Deep learning. Nature **521**(7553), 436–444 (2015)
7. Poppe, R.: A survey on vision-based human action recognition. Image Vis. Comput. **28**(6), 976–990 (2010)
8. Thakur, A., Budhathoki, P., Upreti, S., Shrestha, S., Shakya, S.: Real time sign language recognition and speech generation. J. Innovat. Image Process. **2**(2), 65–76 (2020)
9. Bantupalli, K., Xie, Y.: American sign language recognition using deep learning and computer vision. In: 2018 IEEE International Conference on Big Data (Big Data). IEEE (2018)
10. Shinde, A., Kagalkar, R.: Advanced Marathi sign language recognition using computer vision. Int. J. Comput. Appl. **975**, 8887 (2015)
11. Pigou, L., et al.: Sign language recognition using convolutional neural networks. In: European Conference on Computer Vision. Springer, Cham (2014)
12. Rao, G.A. et al.: Deep convolutional neural networks for sign language recognition. In: 2018 Conference on Signal Processing And Communication Engineering Systems (SPACES). IEEE (2018)
13. Hossen, M.A., et al.: Bengali sign language recognition using deep convolutional neural network. In: 2018 Joint 7th İnternational Conference on İnformatics, Electronics & Vision (iciev) and 2018 2nd İnternational Conference on İmaging, Vision & Pattern Recognition (icIVPR). IEEE (2018)

14. Tolentino, L.K.S., et al.: Static sign language recognition using deep learning. Int. J. Mach. Learn. Comput. **9**(6), 821–827 (2019)
15. Elakkiya, R.: Machine learning based sign language recognition: a review and its research frontier. J. Ambient Intell. Humaniz. Comput. 1–20 (2020)
16. Sole, M.M., Tsoeu, M.S.: Sign language recognition using the extreme learning machine. IEEE Africon'11. IEEE (2011)
17. Liang, Z., Liao, S., Hu., B.: 3D convolutional neural networks for dynamic sign language recognition. Comput. J. **61**(11), 1724–1736 (2018)
18. Huang, J., et al.: Sign language recognition using 3d convolutional neural networks. In: 2015 IEEE İnternational Conference on Multimedia and Expo (ICME). IEEE (2015)
19. American Sign Language Dataset – Github.: https://github.com/shadabsk/Sign-Language-Recognition-Using-Hand-Gestures-Keras-PyQT5-OpenCV/tree/master/Source%20Code/Dataset

Chapter 46
Ant Lion Optimization for Solving Combined Economic and Emission Dispatch Problems

H. Vennila and R. Rajesh

Abstract This work shows how to apply the ant lion optimization (ALO) method, which was recently created and inspired by the natural hunting mechanism of ant lions, to solve the combined economic and emission dispatch problem in power systems with the valve point impact. The suggested method's major goal is to simultaneously reduce generation costs and pollutant emissions while meeting load demand and operational restrictions while accounting for the weighting factor. The suggested algorithm's performance was evaluated on a variety of test systems, and the results were compared to those described in recent literature. The results demonstrate that the proposed method can provide superior outcomes.

46.1 Introduction

The Economic Dispatch (ED) optimization issue is one of the most pressing concerns in a power device. This contains a detailed strategy for the output generated by each linked generator in an energy device so that the overall value remains optimal, meeting all patron load requirements, and providing the desired quality. Electrical power systems are mostly operated on the basis of convention, with the goal of lowering operational costs while maintaining machine restrictions [1]. Fuel, exertions, resources, and preservation expenses are all included in the total generator operating fee. We consider the cost of gasoline to be the sole variable fee for the sake of simplicity, because the costs of labour, materials, and upkeep are usually constant chances of the fuel price. The operating charge isn't always that significant in the case of thermal and nuclear plant life; nevertheless, nuclear plants are operated at regular output ranges, and the operating fee isn't always that significant in the case of hydro stations where the strength cost is ostensibly loose. While valve point effects

H. Vennila (✉)
Department of EEE, Noorul Islam Centre for Higher Education, Kanyakumari 629180, Tamil Nadu, India

R. Rajesh
Rohini College of Engineering and Technology, Kanyakumari 629180, Tamil Nadu, India

J. I.-Z. Chen et al. (eds.), *Machine Learning and Autonomous Systems*, Smart Innovation, Systems and Technologies 269, https://doi.org/10.1007/978-981-16-7996-4_46

are taken into account in the real world, the objective feature of an ED problem has non-differentiable components. As a result, the goal function is composed of a group of difficult cost functions, and the impact appears to be numerous stages. As a result of two significant cases, non-smooth value capabilities are accounted for. The first is the valve-factor loading problem, in which the goal function is treated as a superposition of quadratic and sinusoidal characteristics [2]. In the second situation, the objective feature can be stated as bitwise or piecewise quadratic pricing features, and the problem is treated as multiple -gasoline annoyance. There are a handful of minima in each of the answers in the examples above. As a result, there is no clear-cut answer, and the job of finding a global solution must be completed effectively. The generator with multi-valve steam turbines will have a different enter-output curve than the simple cost function because to the impact of the valves. The ripples that occur as steam valves are turned on have the most significant effect. Many sinusoidal capabilities are supplied to the earlier quadratic value functions to account for the valve point effects on generators. These days, fossil fuels are used to create mechanical strength for the rotor shaft of manufacturing machines. Because of the widespread emissions of carbon dioxide, sulphur dioxide, and nitrogen oxides, this pollutes the atmosphere. As a result of fossil-fired generating devices and the implementation of environmental standards, there was a lot of enthusiasm for pollution management. Furthermore, as public awareness of environmental pollution has grown, as has the passing of the Clean Air Act Amendments of 1990, utilities have been forced to change their layout or operational practises in order to reduce pollution and atmospheric emissions from thermal energy facilities. There are also tight guidelines for carbon credits. As a result of these circumstances, the cost of emissions or fines associated with emissions is equivalent to the cost of fuel. As a result, the financial dispatch problem must also consider environmental pollution issues and address them alongside gas cost optimization. Carbon dioxide emissions are attracting a lot of attention. As a result, the objective characteristic must be altered using weighting capabilities with an awesome and ultimate decision-maker, or a price penalty function must be included [3]. The monetary and emission dispatch problem transforms into a highly non-linear and combinatorial multi-objective non-convex problem. As a result, the most up-to-date Meta-heuristic techniques are employed to resolve the financial and emission dispatch issues. Fuzzy good judgement, artificial neural networks, mixed fuzzy and neural networks, Genetic algorithm (GA), Evolutionary Programming (EP), Genetic algorithm [6], Hybrid HB-Simulated Annealing (SA) [7], Tabu search, PSO algorithm, GA-PSO hybrid, Ant Colony algorithm, fireplace Fly(FF) algorithm, and synthetic Bee Colony (ABC) algorithm are some of the algorithms that can be widely used.

46.2 Problem Formulation

46.2.1 Economic Load Dispatch with Losses

Transmission losses are an important factor in delivering power to long-distance substations. As a result, it's critical to consider transmission loss as a major element while undertaking economic dispatch. However, this will exacerbate the economic dispatch problem [4]. Loss coefficients are referred to as B-Coefficients in this case. The supplied loss equation is shown below.

Minimize

$$F(P_i) = a_i P_i^2 + b_i P_i + C_i \qquad \text{Rs/h} \tag{46.1}$$

46.2.2 Equation of Energy Balance

$$\sum_{i=1}^{N} F_i(P_i) = P_D + P_L \tag{46.2}$$

Because it is impossible to have an electricity device without any losses, transmission losses must be taken into account while resolving any financial dispatch issue. The equation for the power system when considering transmission losses is as follows:

$$P_L = \sum_{I=1}^{N} \sum_{J=1}^{N} P_i B_{ii} + \sum_{i=1}^{n} B_{01} P_i + B_{00} \tag{46.3}$$

where

P_i—Load of ith generator in MW unit.

P_j—Load of jth generator in MW unit.

B_{ij}—Coefficients indicating transmission losses between ith and jth generating unit in MW.

P_L—loss in power represented in MW.

46.2.3 Capacity Constraints

$$P_{imin} \leq P_i \leq P_{imax} i = 1, 2 \ldots N \tag{46.4}$$

46.2.4 Economic Load Dispatch with Valve Point Loading

Adding the valve point impact the goal function will become

$$FC_T = \sum_{i=1}^{N} a_i + b_i P_i + c_i P_i^2 + |d_i sin(e_i (P_{imin} - P_i))| \tag{46.5}$$

The fuel cost coefficients of ith generator unit are a_i, b_i, c_i, d_i and e_i
P_i—is the load of ith generator in unit of MW.
N i—indicates total number of generators.
P_{imin}—is the minimum range of ith generator in MW unit.
FC_i—indicates the ith generator fuel cost function.
FC_T—indicates the total fuel cost,$ /hr.

46.2.5 Emission Dispatch Problem

The emission dispatch characteristic can be defined as minimizing the quantity of pollutants of the power gadget [5]. The whole emission function such as the coefficients of the device may be represented as

$$E_T = \sum_{i=1}^{N} E_i(P_i) \sum_{i=1}^{N} \alpha_i + \beta_i P_i + \gamma_i P_i^2 \tag{46.6}$$

The toxic gases generated out of the energy plant to get brought to the system. This gives the mixed emission feature as

$$E_T = \sum_{i=1}^{N} \alpha_i + \beta_i P_i + \gamma_i P_i^2 + \epsilon_i \exp(\vartheta_i P_i) \tag{46.7}$$

The coefficients and are treated as coefficients of ith generator unit.
P_i—indicates generated power of ith unit, represented MW.
N—indicates the number of generating units is the emission function of unit.
E_T—indicates the total emission of the system represented in tones /hr.

46.2.6 Combined Economic and Emission Dispatch

Blended financial and Emission dispatch can also be handled as a multi-goal trouble, which may be treated as sum of unmarried-objective features of financial and Emission Dispatch problems.

$$C_T = FC_T + Pf * E_T \quad (46.8)$$

46.3 Ant Lion Optimization Algorithm

46.3.1 Inspiration

The entire concept of the ALO set of rules stems from how an ant lion catches its prey by laying sand pit traps on the sand and capturing the ant. The size of the pit is related to the moon's mild. Because there is more light, the ant lion prey has more visibility and there are fewer chances of getting caught, allowing it to counteract the ant-increase lion's in the size of its pit.

46.3.2 Algorithm Use-Cases

The following procedures are taken into consideration when developing an ant lion optimization set of rules. i. Ants / Ant lion prey employ numerous random walks to move around inside the sand space (region below attention).

ii. These random walks are taken into account and applied to all of the locations on the beach.
iii. Ant lions create traps and those walks of ants get impacted by using the traps (sand pits) created with the aid of ant lions.
iv. Ant lions build pits as in keeping with its electricity level which is considered as its health function. That is, if ant lions have more fitness their traps may be bigger.
v. Ant lions with large sand pits have more chance / possibility to capture ants as their prey.
vi. Every ant / prey can be trapped by way of an ant lion in every of the iterations and it's dealt with that the fittest receives stuck.
vii. The random walks are decreased successfully to emulate sliding impact of ants towards ant lions.

    If the ant is fitter than the ant lion, it is considered to be stuck through the ant lion. In this example, the ant lion pulls the ant underneath the sand pits.
viii. After catching the ant, the ant lion transforms into a new function (or enhances the current sand pit) and makes significant changes to the current sand pit in order to capture every other ant.

46.3.3 Ant Lion Algorithm Mathematical Model

First step of the set of rules is to version random walks of ants.

X(t) = [0;cumsum(2r(t_1-1),cumsum(2r(t_2-1),.....Cumsum(2r(t_n-1).

'cumsum' suggests the cumulative sum of random walks.

'n' is the total range of iterations carried out.

't' shows the random stroll and to be unique step of an ant and r(t) is a first-rate characteristic described and r(t) is the principle feature for the same.

The location of ants with appreciates to ant lions is saved first. The following matrix suggests the equal.

$$M_{Ant} = \begin{vmatrix} A_{1,1} A_{1,2} \ldots\ldots A_{1,d} \\ A_{2,1} A_{2,2} \ldots\ldots A_{2,d} \\ \ldots\ldots\ldots\ldots\ldots\ldots\ldots\ldots \\ A_{n,1} A_{n,2} \ldots\ldots A_{n,d} \end{vmatrix}$$

Fitness of ants is considered because the opportunity of the same being stuck through ant lion. This health needs to be represented in matrix as properly.

$$M_{Ant} = \begin{vmatrix} \int A_{1,1} A_{1,2} \ldots\ldots A_{1,d} \\ \int A_{2,1} A_{2,2} \ldots\ldots A_{2,d} \\ \ldots\ldots\ldots\ldots\ldots\ldots\ldots\ldots \\ \int A_{n,1} A_{n,2} \ldots\ldots A_{n,d} \end{vmatrix}$$

n addition to the conduct of ants, its assumed that the ant lions also are hidden within the region. Their function turns into another essential criterion and the same is depicted as follows.

$$M_{Antlion} = \begin{vmatrix} AL_{1,1} AL_{1,2} \ldots\ldots AL_{1,d} \\ AL_{2,1} AL_{2,2} \ldots\ldots AL_{2,d} \\ \ldots\ldots\ldots\ldots\ldots\ldots\ldots\ldots \\ AL_{n,1} AL_{n,2} \ldots\ldots AL_{n,d} \end{vmatrix}$$

A fitness feature for ant lions is likewise to be considered for higher remedy of use case. MOAL is handled because the matrix indicating the fitness functions of each ant lion.

$$M_{Ant} = \begin{vmatrix} \int AL_{1,1} AL_{1,2} \ldots\ldots AL_{1,d} \\ \int AL_{2,1} AL_{2,2} \ldots\ldots AL_{2,d} \\ \ldots\ldots\ldots\ldots\ldots\ldots\ldots\ldots \\ \int AL_{n,1} AL_{n,2} \ldots\ldots AL_{n,d} \end{vmatrix}$$

Any other need is to keep the random walk inside the place chosen. To try this, a mix-max normalization method is applied

$$X_i^t = \frac{X_i^t - a \times (d_i - C_i^t)}{d_i^t - a_i} + C_i$$

where

a_i— is treated as the lowest of walks of variable i bi—indicates the highest of the walk with variable i C_i^t—indicates the lowest of ith variable at tth iteration.

d_i^t—indicates the highest of ith variable at tth iteration. This equation needs to be applied for each iteration.

$$C_i^t = Antlion_j^t + C^t$$

$$d_i^t = Antlion_j^t + d^t$$

Ant lions can build traps that are proportional to their fitness (instant). Ants are required to migrate in a random manner. While the ant lions are aware that the ants are being imprisoned, they toss sands out from the centre of the cone-shaped hole towards the floor, reducing the ants' chances of breaking free. The mathematical model of this behaviour can be completed by using gradually reducing the diameter of ants' walks search vicinity. The subsequent equations are introduced to compensation.

$$C^t = {C_t}/{i}$$

$$d^t = {d_t}/{i}$$

Ants completing the centre of the hole and an ant lion catching the same with its jaws signify the final step of the quest. Following this halt, the ant lions herd the ants closer to the pit's centre, where they are devoured. For the purposes of simulating the entire use case scenario, we'll assume that when an ant lion catches its prey, it becomes a lot more fit than an ant lion at that particular generation. After then, it was hoped that the ant lion would return to its current location while consuming ants in order to increase its chances of capturing additional ants. The subsequent equation is proposed for the same

$$Antlion_j^t = Ant_i^t if (Ant_i^t) > \int (Antlion_j^t)$$

This suggests gift iteration; Ant lion tj suggests the placement of decided on jth ant lion at tth new release, and Ant ti shows the location of ith at tth new release.

Elitism is a key phenomenon of many algorithms developed which makes the technique to continue its most beneficial result at any stage of the execution. The pleasant ant lion found beneath new release is saved in database and currently treated as the excellent answer. As the elite ant lion is dealt with as the excellent, it's going to affect the actions or walks of every ant in the course of iterations. This assumes that every and each ant walk in random was encircled with the selected ant lion.

$$Ant_i^t = \frac{R_A^t + R_E^t}{2}$$

where R_A^t indicates ant walk around the ant lion represented by the roulette wheel at the location of tth iteration,

R_E^t indicates the random walk around the elite at the location of tth iteration, Ant t_i indicates the position of ith ant at the position of tth iteration.

46.4 Test System and Results

Case study 19(a): Six Generator Test systems with ALO

The system with ALO turned into accomplished with the same old coefficients and general demand as 2.83 pu, to give you the results. The results including gasoline value and emission cost supplied proper optimization and it had shown splendid convergence traits too. Those are indicated within the table and parent given beneath (Table 46.1 and Fig. 46.1).

(b): Six Generator Test Systems with PSO and ALO.

The results consisting of fuel cost and emission cost are compared with the proven PSO algorithm (Table 46.2).

Table 46.1 Optimized cost for 6 generators system using ALO

Generators	Individual loads
P1	46.0512
P2	44.1675
P3	45.2571
P4	47.2645
P5	46.8234
P6	48.1302
Emission (t/hr)	**0.21067**
Total fuel cost ($/hr)	**619.565**

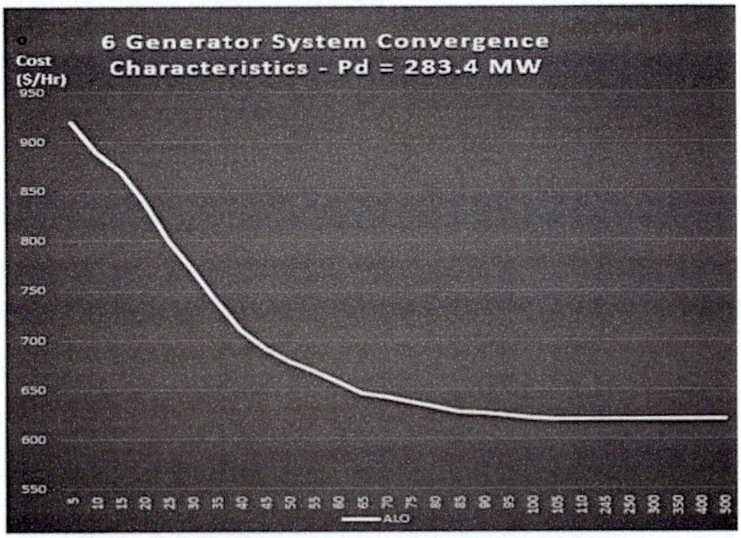

Fig. 46.1 Convergence characteristics of 6 generators system using ALO

Table 46.2 Optimized cost for 6 generators system using PSO and ALO

Generators	PSO	ALO
P1	45.7654	46.0512
P2	46.8007	44.1675
P3	48.4312	45.2571
P4	46.9782	47.2645
P5	45.4593	46.8234
P6	48.9532	48.1302
Emission(t/hr)	**0.21105**	**0.21067**
Total fuel cost($/hr)	**639.248**	**619.565**

Case study 2(a): Ten Generator Test systems with ALO.

In this case study, the base for generation is treated as 100 MVA and the total demand is considered as 10.36p.u (Table 46.3 and Fig. 46.2).

(b): Ten Generator Test Systems with PSO and ALO

All of the three selected algorithms furnished splendid outcomes in contrast with PSO and proven superb convergence traits making it suitable for the use with any combined financial Emission Dispatch issues (Table 46.4).

Table 46.3 Optimized cost for 10 generators system using ALO

Generators	Individual loads
P1	159
P2	144
P3	88.402
P4	71.044
P5	131.4517
P6	102.114
P7	129.115
P8	62.2375
P9	89
P10	52.2678
Emission (t/hr)	**0.39428**
Total fuel cost($/hr)	**61,490.73**

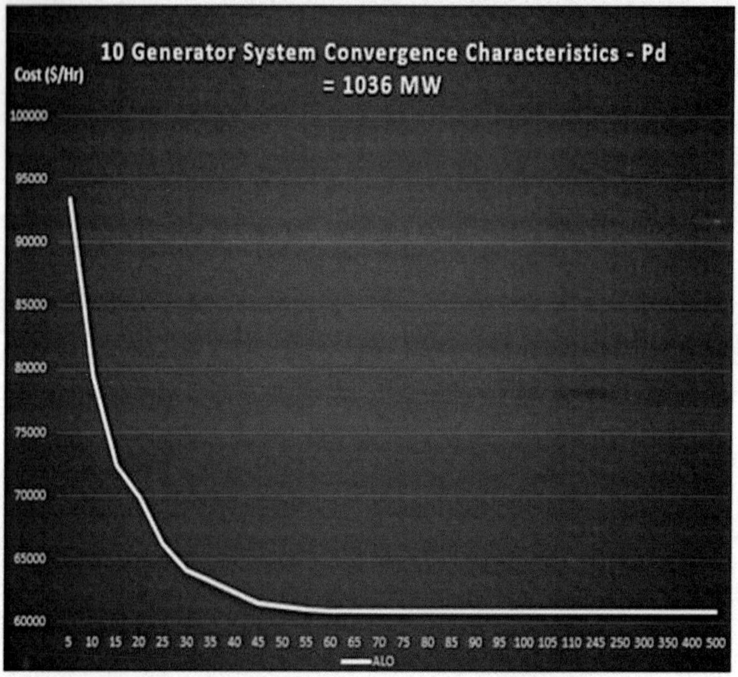

Fig. 46.2 Convergence characteristics of 10 generators system using ALO

Table 46.4 Optimized cost for 10 generators systems using PSO and ALO

Generators	PSO	ALO
P1	158	159
P2	143	144
P3	91	88.402
P4	72.505	71.044
P5	130.999	131.4517
P6	105.145	102.114
P7	132.009	129.115
P8	61.718	62.2375
P9	95.1355	89
P10	46.7054	52.2678
Emission(t/hr)	**0.387578**	**0.39428**
Total fuel cost ($/hr)	**61,845.72**	**61,490.73**

46.5 Conclusion

The ALO algorithm was used to address the Combined Economic and Emission Dispatch problem of producing units in this research, taking into account the valve point impact. The proposed technique produced a global solution in a six and ten generator systems that was superior to prior studies presented in the literature. The simplicity, reliability, and effectiveness of the suggested technique make it ideal for practical applications.

References

1. Al-Awami, A., Sortomme, E., El-Sharkawi, M.: Optimizing economic/environmental dispatch with wind and thermal units. In power & energy society general meeting, 1–6. IEEE (2009)
2. Basu, M.: Artificial immune system for dynamic economic dispatch. Int. J. Electr. Power Energy Syst. **33**(1), 131–136 (2011)
3. Brodsky, S.F., Hahn, R.W.: Assessing the influence of power pools on emission constrained economic dispatch. IEEE Trans. Power Syst. **1**(1), 57–62 (1986)
4. Vennila, H., Rajesh, R.: Combined static economic and emission dispatch by improved moth optimisation with valve point loading. Int. J. Enterp. Netw. Manag. 10 (2), 152–161
5. Mirjalili, S.: The ant lion optimizer. Adv. Eng. Softw. **83**, 80–98 (2015)
6. Vennila, H., Ruban Deva Prakash, T.: Economic emission dispatch of thermal generating units using genetic algorithm technique. Int. J. Enterpr. Netw. Manag. 4(4), 344–353 (2011)
7. Vennila, H., Ruban Deva Prakash, T.: A solution to the combined economic and *emission* dispatch using Hybrid HB-SA algorithm on large scale power system. Int. Rev. Model. Simul. **6**(4), 1256–1263 (2013)

Author Index

J. I.-Z. Chen et al. (eds.), *Machine Learning and Autonomous Systems*, Smart Innovation, Systems and Technologies 269, https://doi.org/10.1007/978-981-16-7996-4